This book is to be returned on or before
the last date stamped below.

International Federation of Automatic Control

AUTOMATION FOR
MINERAL RESOURCE DEVELOPMENT

IFAC Proceedings Series, 1986. Number 1

IFAC Proceedings Series

1985

No. 1   *PERKINS, MACFARLANE & LEININGER: Analysis and Synthesis of Control Systems

No. 2   *LJUNG & ASTROM: Identification, Adaptive and Stochastic Control

No. 3   *TITLI, SINGH, PACHTER, HAUTUS, KOKOTOVIC & RAUCH: Large-scale Systems, Decision-making, Mathematics of Control

No. 4   *NISENFELD, LEFKOWITZ, PAI & DY LIACCO: Process Industries, Power Systems

No. 5   *REMBOLD, KEMPF, TOWILL, JOHANNSEN, PAUL, HARRISON, EHRENBERGER, TABAK, LESKIEWICZ, FENTON & DEBRA: Manufacturing, Man–Machine Systems, Computers, Components, Traffic Control, Space Applications

No. 6   *VOSSIUS, WEED, RESWICK, COBELLI, HAIMES, TAKAMATSU, KAYA, NAJIM, MARTENSSON, CHESTNUT & LARSEN: Biomedical Applications, Water Resources, Environment, Energy Systems, Development, Social Effects, SWIIS, Education
(*Ninth Triennial World Congress)

No. 7   BARKER & YOUNG: Identification and System Parameter Estimation (1985)

1986

No. 1   NORRIE & TURNER: Automation for Mineral Resource Development

No. 2   CHRETIEN: Automatic Control in Space

No. 3   DA CUNHA: Planning and Operation of Electric Energy Systems

No. 4   VALADARES TAVARES & EVARISTO DA SILVA: Systems Analysis Applied to Water and Related Land Resources

No. 5   LARSEN & HANSEN: Computer Aided Design in Control and Engineering Systems

No. 6   PAUL: Digital Computer Applications to Process Control

No. 7   YANG JIACHI: Control Science & Technology for Development

No. 8   MANCINI, JOHANNSEN & MARTENSSON: Analysis, Design and Evaluation of Man–machine Systems

No. 9   GELLIE, FERRATE & BASANEZ: Robot Control "Syroco '85"

No. 10   JOHNSON: Modelling and Control of Biotechnological Processes

IFAC Related Titles

BROADBENT & MASUBUCHI: Multilingual Glossary of Automatic Control Technology
EYKHOFF: Trends and Progress in System Identification
ISERMANN: System Identification Tutorials (*Automatica Special Issue*)

AUTOMATION FOR MINERAL RESOURCE DEVELOPMENT

Proceedings of the First IFAC Symposium
Brisbane, Queensland, Australia, 9–11 July 1985

Edited by

A. W. NORRIE

Julius Kruttschnitt Mineral Research Centre
Indooroopilly, Queensland, Australia

and

D. R. TURNER

Victoria, Australia

Published for the

INTERNATIONAL FEDERATION OF AUTOMATIC CONTROL

by

PERGAMON PRESS

OXFORD · NEW YORK · TORONTO · SYDNEY · FRANKFURT

U.K.	Pergamon Press Ltd., Headington Hill Hall, Oxford OX3 0BW, England
U.S.A.	Pergamon Press Inc., Maxwell House, Fairview Park, Elmsford, New York 10523, U.S.A.
CANADA	Pergamon Press Canada Ltd., Suite 104, 150 Consumers Road, Willowdale, Ontario M2J 1P9, Canada
AUSTRALIA	Pergamon Press (Aust.) Pty. Ltd., P.O. Box 544, Potts Point, N.S.W. 2011, Australia
FEDERAL REPUBLIC OF GERMANY	Pergamon Press GmbH, Hammerweg 6, D-6242 Kronberg, Federal Republic of Germany
JAPAN	Pergamon Press Ltd., 8th Floor, Matsuoka Central Building, 1-7-1 Nishishinjuku, Shinjuku-ku, Tokyo 160, Japan
BRAZIL	Pergamon Editora Ltda., Rua Eça de Queiros, 346, CEP 04011, São Paulo, Brazil
PEOPLE'S REPUBLIC OF CHINA	Pergamon Press, Qianmen Hotel, Beijing, People's Republic of China

First edition 1986

Library of Congress Cataloging-in-Publication Data

Automation for mineral resource development
(IFAC proceedings series; 1986, no. 1)
Includes index.
1. Mines and mineral resources — Automation — Congresses. 2. Mines and
mineral resources — Mathematical models — Congresses. 3. Mines and mineral
resources — Data processing — Congresses. 4. Metallurgical plants — Automation —
Congresses. 5. Metallurgy — Mathematical models — Congresses. 6. Metallurgy —
Data processing — Congresses. I. Norrie, A. W. (Angus W.) II. Turner, D. R.
(David R.) III. International Federation of Automatic Control. Technical Committee
on Applications. IV. Institution of Engineers, Australia. V. International Federation of
Automatic Control. Committee on Systems Engineering. VI. Series.
TN275.A1A87   1986   622   85-21844

British Library Cataloguing in Publication Data

Automation for mineral resource development: proceedings of the lst IFAC
Symposium, Brisbane, Queensland, Australia, 9-11 July 1985.— (IFAC proceedings
series; 1986, no. 1)
1. Mines and mineral resources. 2. Mining engineering — Automation. 3. Metallurgical
plants — Automation. I. Norrie, A. W. II. Turner, D. R. (David R.) III. Series
622'.34'02853   TN153
ISBN 0-08-031663-8

Printed in Great Britain by A. Wheaton & Co. Ltd., Exeter

IFAC SYMPOSIUM ON AUTOMATION FOR MINERAL RESOURCE DEVELOPMENT

Organized by
The Australian Institute of Mining and Metallurgy

Sponsored by
IFAC Technical Committee on Applications
The Institution of Engineers, Australia

Co-sponsored by
IFAC Technical Committee on Systems Engineering

International Programme Committee

K. E. Mathews, Australia (Chairman)
R. Chaussard, France
S. Cierpisz, Poland
N. W. G. Cook, USA
R. J. Fraser, Australia
S. Granholm, Sweden
E. Hoek, Canada
P. C. Jain, India
M. Kassier, Australia

L. Kruszecki, Poland
D. W. McMahon, Australia
Y. Sawaragi, Japan
T. J. Sheer, S. Africa
C. A. Stapleton, Australia
T. Takematsu, Japan
B. Tamm, USSR
M. Thoma, FRG
J. C. Wilson, Canada

National Organizing Committee

A. J. Lynch (Chairman)
C. W. Bailey (Secretary)
D. Maconachie
K. E. Mathews
F. Nicolosi

A. W. Norrie
J. M. Rose
G. S. Stacey
C. A. Stapleton
D. R. Turner

Chief Executive Officer, The Aus. I.M.M.
Honorary Treasurer, The Aus. I.M.M.

FOREWORD

"Today, in most parts of the world, the human race has adopted a metal-dependent existence. With few exceptions, every sizeable group of people relies on metal artefacts to provide the food, shelter, energy, transport and industrial products that make life comfortable." (Out of the Fiery Furnace by Robert Raymond, McMillan, Australia, p. xi).

The ore deposits from which minerals are extracted and metals produced are becoming more difficult to mine and process. The reason is that deposits which are high grade, easy to process or readily accessible are being depleted and can no longer meet the market demands.

Consequently mines of today are frequently located in remote or hazardous areas, and techniques involving automation, remote control and automatic control are important in their development. For example, metalliferous nodules are now being experimentally mined on the ocean floor, minerals are mined on land at depths approaching 4km and in areas of extremely high temperatures, and coal is mined in areas prone to gas and rock outbursts, and roof falls. Mines are now being developed in even harsher environments ranging from hot, remote desert areas to the Arctic.

The purpose of this symposium is to review the role of advanced automation technology in the development of mineral resources. Mining and mineral processing are the major areas for discussion and the papers which will be presented provide an appropriate blend of case studies of existing systems, theoretical studies and predictions about the future.

On behalf of the International Federation of Automatic Control and the Australasian Institute of Mining and Metallurgy I would like to thank all associated with the organisation of the symposium and the preparation of the technical program. In particular, I would like to thank the editors, Angus Norrie and David Turner, and the Symposium secretary, Chris Bailey, for their untiring efforts.

A.J. LYNCH
Chairman — Organising Committee

CONTENTS

Contents

Contents

Trends in Instrumentation and Control

M. THOMA

President, International Federation of Automatic Control

It is amazing how Automatic Control in its widest sense did develop in the last 25 years. It has an increasing impact not only on technology but also on other fields like society, economy, developing countries and others. The present development of micro electronics is causing far-reaching changes which have in the past been characterized by the expression `Industrial Revolutions`. The start of industrialization saw engines replacing man´s muscle power and incorporated the first revolution. Feedback control indeed had already been used at that time - let me just mention the steam engines - but it took quite some time until Automatic Control was accepted as an independent special discipline. It was a laborious path before Automatic Control gained in the 40th year of this century recognition as a basic science and was granted appropriate consideration in the curricula of all technical universities. The 2nd industrial revolution whereby work was made easier and more efficient by the use of automation was characterized particularly by developments in control technology. Nobody at that time could ever predict the colossal progress which was achieved up to now where man´s intelligence is being supported by micro electronics and he is being released from tedious routine brain work. There is no need to point out that modern Automatic Control technology and theory is playing a major role in this 3rd industrial revolution, as it is sometimes called. We just need to think of modelling, of the use of modern algorithmic control methods, of output distributed computer control, and of automation of production processes with `intelligent robots`.

The classical control of single-input single-output servomechanism or fixed point control was for two reasons very effective. On one hand the classical methods like frequency response and root locus methods give for single loop control a very good insight in the real (physical) behaviour of the systems. In other words, the applied methods are very closely related to the systems dynamics. On the other hand the dynamic behaviour of single loop control systems is rather robust concerning parameter changes and/or small changes of the systems structure.

However, if one has to deal with complex systems like multivariable control systems or one likes to implement higher control strategies like optimal or adaptive control the classifical methods are no more sufficient. New methods for the analysis or synthesis of such systems were created which are usually characterized by the expression state space methods. However, they imply much more advanced mathematical methods whereby the intuition was more or less lost.

For economic reasons, however, one was forced to automize e.g. complete energy supply networks or production processes. But the use of single

control strategy did for complex systems lead to very obscure and inflexible structures so that the safety was reduced to a great extent. The results were large control rooms, where all control measurement and protection equipments were centralized. This trend was supported by the development of reliable and fast process computers.

Until a few years ago the process computer was a rather expensive equipment. Its use was therefore limited to processes with a very high turnover like hot strip rolling mills, large chemical plants, extensive electric energy supply systems and so on. For economic reasons one was forced to utilize fully the expensive process computer which did lead to a centralized concept.

But the explosion in micro computer technology did lead to a price decay. Nowadays the expenses for process computers are almost negligible. Sophisticated logic circuitry and minicomputers will soon be a fact in every day life. With the help of cheap computers more and more decentralized control concepts will be used in the future. Besides a better understanding of an entire plant high control strategies for a more economic and safe operation can easily be implemented.

Three examples are used to demonstrate how modern control strategies can be implemented with the help of process computers.

The first example is an energy and time optimal on-line control of a railway system: the train No. 1 is running with constant speed at the inner loop (long distance traffic). At any desired time the train No. 2 can be started and it should in the optimal way complete two full rounds (local traffic) in the opposite direction of train 1. Of course, the two trains should never meet during the three conflicting sections. With the help of an infra-red transmission a realtime feed forward control with constraints is realized. By the using of spline functions the optimal trajectory of the 2nd train is computed in 0.5 seconds. A feedback control algorithm is implemented in the process computer which forces the train No. 2 to follow the optimal trajectory. See Figures 1 and 2.

The second example is the control of a continuous stirred tank reactor. As an example the two component chemical reaction in liquid phase of

$$\text{ethyl acetate + sodium hydroxide} \rightarrow \text{sodium acetate + ethanol}$$

is used. Thereby the flow rate of ethyl acetate is the disturbance, the flow rate of the sodium hydroxide is the controller output, and the concentration of ethanol is the controlled variable. The necessary on-line measurement of the concentration is, however, rather difficult to obtain or at least very expensive. Instead a non-linear observer is used to observe the concentration by measurements of the pH value and the conductivity of the liquid in the reactor. See Figures 3 and 4.

The third example is the observer based control of a double pendulum, hinged to a cart on the rail. The aim is to stabilize the system in the neighbourhood of an equilibrium point. Only one

angle of the double pendulum and the position of the cart are measured. The crucial second angle is not measured, it is reconstructed with the help of an observer. Besides by means of a model reference control the position dependent friction on the rail can be neglected; the cart behaviour is forced to match the output. See Figure 5.

These examples demonstrate very clearly the power of modern control theory. However, without the use of computers these problems could not be solved economically.

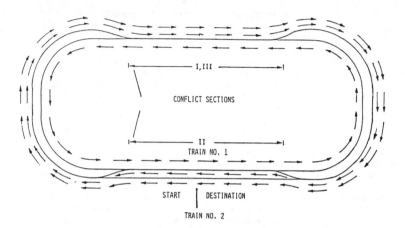

Fig. 1 Considered Railway System

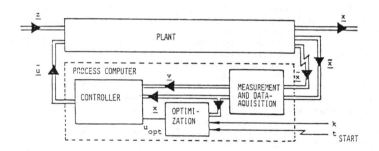

Fig. 2: Block Structure of the Controlled Process

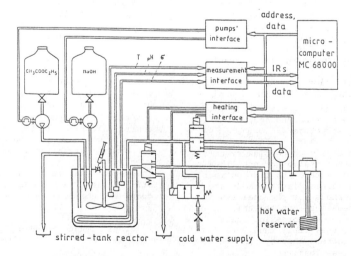

Fig. 3.: Schematic diagram of the plant and control equipment

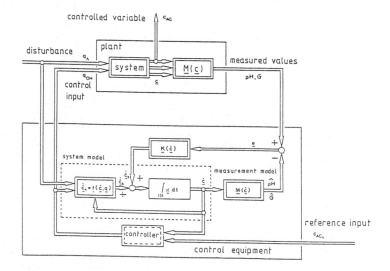

<u>Fig. 4.:</u> Block diagram of the controlled process

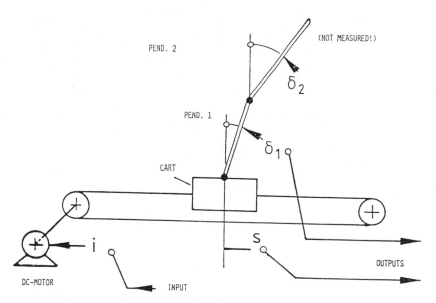

<u>Fig. 5:</u> Double Pendulum

Advanced Engineering and Mineral Development

R.S. DAVIE

President, Institution of Engineers, Australia

Engineering is essentially based on science, technology, and on what might be called engineering art, with the mixture of these three components varying from time to time and from industry to industry. Like many other aspects of engineering during recent times, engineering for the mining and minerals industry, by encompassing more of the science and technology, has moved away from an approach associated with the art of engineering and based on historical practices, to a more mathematically rational one. The dependence of mining and the minerals industry on advanced engineering has increased as much, if not more, than any other sector of industry.

Nowhere is this more evident than in the increasing utilisation of automatic control and automation in the development of mineral resources. This has been dependent both on team work and on intellectual rigour while, in general, the generation of suitable control strategies has been achieved by evolution through the interaction of the diverse skills of a number of professions. This symposium is, in itself, a testimony to the interaction of skills and to the team approach to the solution of problems. It is also a testimony to the way in which difficulties, peculiar to the development of mineral resources, have been and are being overcome.

The inimical environment frequently associated with mineral development requires robustness in any engineering work. This robustness should not be confused with a lack of sophistication, but rather seen as an additional barrier which must be overcome, and which is overcome, in the development of sophisticated control systems.

Other industries generally do not have to contend with non-deterministic control systems in which both the inputs and system-parameters are continually varying. Furthermore, in the mineral industry, the processes requiring control are often characterized by inadequate on-line, real-time sensing, for measuring both the composition of the material and the dynamic variations in systems parameters. In addition, it is often necessary to adopt sophisticated techniques for estimating these parameters. Such topics are covered by papers being presented to the conference and will undoubtedly lead to considerable discussion.

Because the design process, by comparison with other industries, is frequently intricate, great demands are placed on the designers of the associated control systems. The control engineer

---

The assistance of P.W. Higgins of the Department of Mechanical Engineering, Swinburne Institute of Technology, in the preparation of this address, is gratefully acknowledged.

has to call on the abilities of engineers from other branches of the profession, especially chemical and mechanical engineers, as well as seeking the substantial support of specialists from other professional disciplines; in particular, metallurgy and mathematics. In addition, there are often environmental constraints which are likely to involve the environmental scientist as well as the engineer with a specialist knowledge, such as fluid mechanics or other specialist fields, to tackle particular problems.

The move towards real-time control of processes has been both demanding and revealing. While rigour in objectivity and mathematical analysis can be onerous, the ability to be able to discern actual processes in real-time operation has permitted new insight into the world of dynamics. This has special relevance to the dynamics of chemical processes.

However, process control is merely one element in the overall requirements for operating a successful venture. The incorporation of economic factors into the optimization trajectories used for controlling the process is also important. While the majority of optimizing routines are based on straight technical criteria, such as feed rate and energy usage, the inclusion of external factors, such as costing data, is becoming more common. Most modern systems also feed operational data back to supervisory control centres, where the technical performance of plant may be monitored. These centres are generally ergonomically designed to ensure that vital details are communicated to the system operators in an easily comprehended manner.

As the degree of computerization of the industry increases these centres form but one link in a hierarchical chain. Computer-based management information systems are becoming a normal aid to management in a successful enterprise. These systems are even more effective when their operational data are continually updated by on-line information fed from the plant supervisory computer. To be able to set up these company-wide computer-based knowledge systems requires the assistance of wider professional advice than is required for the physical control of a plant. Some other professionals who may be required are the economists, the accountants, and the management consultants.

The ever increasing availability of digital controls at reasonable cost has made real-time control a reality. However, this has placed demands on the engineer in terms of increased rigour, both in objectivity and in the ability to model mathematically. Similar progress and problems may be observed in other industrial areas, but these are not areas which suffer the same environmental harshness as the mining industry. Installation of PLC's in automated warehousing, in a clean, dust free, environment is a very different matter from their installation on a conveyor system operating in an open and often inimical environment. By the same token, use of robotics and robots to carry out repetitive tasks where the required judgement can be minimised, such as happens in many industries, is not readily applicable in mining situations where intuitive human discrimination and adaptability is still a prime requirement.

The need for a diversity of disciplines converging on the issue of control is reflected in the organisational structure of IFAC itself, as is shown by consideration of a few of the technical committees. The Components and Instruments Committee is involved in microprocessor based intelligent components and instruments for decentralized control. The Mathematics of Control Committee has a strong interest in the interaction between modelling and solution techniques and the requirements of practical applications. The Economic and Management Systems Committee is interested in dynamic planning and management processes, considered from the point of view of economics, finance, organization, resource allocation and decisions. In particular it is interested in the interaction of these with various systems. Finally, and of increasing importance, is the Social Effects of Automation Committee which seeks to relate automation to its social environment.

This elaboration on the constituent structure of IFAC, emphasizes the fact that the future prosperity of the industry depends upon the utilization of well-trained professionals who are able to work together in multi-disciplinary teams with a high degree of intellectual rigour. In a small country with a limit to the number of skilled people, as is the case for Australia, vital disciplinary interaction, such as that provided by IFAC on a global basis, must be sought at a local level through bodies which enjoy broad professional support.

The Institution of Engineers, Australia, has recognized the specialist needs of engineering in mining. Because the Institution comprises all branches of engineering in its four Colleges of Chemical, Civil, Electrical and Mechanical Engineers, it has the background required to form, in association with the Australasian Institute of Mining and Metallurgy, the new National Committee on Engineers in Mining, a splendid co-operative effort, covering all areas of engineering associated with mining. The Institution believes strongly in acting in such a role. From its base of over 37,000 members, the Institution has proved able to work with other bodies, as it has for many years with the Australasian Institute of Mining and Metallurgy through the Australian Geomechanics Society, and as it intends to do in other fields, such as information technology.

The need for sharing is especially important where the interconnection of multifarious knowledge and experience extends beyond a single industry. This is obvious within the field of microprocessor based technologies. Through the use of microelectronics, general purpose equipment has replaced, with considerable effect, the older, specially designed hardware. Knowledge and skills developed in the application of process controllers in one field of industry are equally applicable in other fields. While the mineral and mining industry is able to borrow ideas from manufacturing, the reverse is equally applicable.

The same may be said of management information systems, machine condition monitoring, ergonomic design of control centres, and telecommunications.

A lead must be given and it must come, not only from eminent bodies orientated towards the industry, such as the Julius Kruttschnitt Mineral Research Centre, and from the learned societies such as The Institution of Engineers, Australia and the Australasian Institute of Mining and Metallurgy, but from industry itself.

While Australia is internationally competitive in pure scientific research, particularly that which is funded by the government, there is overwhelming need in Australia for much greater applied research and development. The funding of applied research by industry is abysmal. With our own substantial mining activity, we should by rights have a strong heavy engineering sector. Instead, while there has been huge investment in heavy engineering by the mining industry, the Australian industry has almost atrophied. To be able to win contracts the industry must develop competitive products through vigorous investment in the development of new techniques. It can only do this through the efforts of engineers and associated professionals, working together, on an energetic program of applied research, development and implementation.

Educational institutions should also play a strong role in the creation of an applied research base within Australian industry. There has been a call by government for the formation of strong co-operative research links between industry and the universities and institutes of technology. The concept of teaching companies has been well canvassed over recent years, and is now being implemented. The government has announced a measure, sought by the Institution of Engineers for many years, whereby industrial research and development will attract tax deductibility of 150%.

The multi-disciplinary outlook must also be extended to education. Courses must adopt an integrated approach. Nowhere is this better illustrated than in the teaching of automatic control where the interaction between control of the physical system needs to be linked, as part of the educational process, with computer aided design, computer aided manufacture, management information systems, machine monitoring and automatic data collection. If this synthesis is not achieved in the course, it will not be achieved in practice.

The combination of the Annual Conference of the Australasian Institute of Mining and Metallurgy and the International Federation of Automatic Control Symposium illustrates the importance with which a major Australian group, vitally linked to an export industry, views the nexus between competitiveness and dependence upon advanced engineering. In many mining operations in Australia today, this link is clearly visible. Its recognition by the two sponsoring bodies, and the way in which the conference and the symposium have been brought together, speaks highly of the abilities of the people involved in the conference organisation. On behalf of the Institution of Engineers, Australia, I am delighted that we have played a part in this valuable exercise.

Finally, I remind you that:

"The function of science is to know, of technology to know how to, of engineering to do."

It is vital that we do it now.

Prediction of Hazards in Underground Excavations

EVERT HOEK

Golder Associates, 224 W 8 Ave., Vancouver, Canada

Abstract. Exceptionally weak ground, usually associated with faulting, unanticipated groundwater pressures or flows and exceptionally high stress concentrations are some of the hazards which confront the underground excavation engineer. One or more of these hazards can seriously delay or even halt a tunnel or mining project, and methods for predicting these hazards are urgently required, particularly for projects involving high speed drill and blast or tunnelling machine excavations.

This paper explores the current availability and the potential for the development of tools and techniques for the prediction of these hazards. It is concluded that systematic and automated geological data collection and interpretation represent the most promising avenue for development. It is suggested that there is room for the use of expert systems in which the computer is used to assemble and organize practical underground excavation experience into an interactive decision making system.

Keywords. Geological hazards; underground excavations; rock mechanics.

INTRODUCTION

The cost per ton of ore recovered from the ground depends, to a significant degree, upon the ease with which this ore can be mined. Difficult mining conditions result in reduced rates of production and hence higher mining costs. As easy-to-find high grade deposits have been depleted and as the search for minerals has moved to more difficult terrain and/or to greater depths below surface, mining difficulties have tended to increase. These difficulties have been offset by technological advances which have resulted in more efficient mining equipment, blasting techniques and methods of rock support and hence mining costs have tended to remain reasonably stable.

In evaluating a new mine it is normal to consider a range of mining methods which are best suited to the type of orebody under consideration. The availability of suitable equipment and of skilled manpower need be considered together with the grade of the ore and the most suitable recovery processes . If, having considered all these factors, the mine is considered economically viable, detailed mine planning can commence and ultimately the mine will go into production.

Experience suggests that, however thorough the evaluation and planning of a mine, there will always be problems which were not or could not be anticipated and that these problems can have serious consequences upon the performance of the mine. Amongst the most serious of these problems are "bad ground" conditions caused by faulting, intense jointing, high groundwater pressures or flows, exceptionally high stress levels and other geologically related phenomena. These "bad ground" problems can be particularly serious if they are not anticipated and if a smoothly-running mining operation is brought to a halt by a large fall of ground, water inrush, rockburst or similar event.

GEOLOGICAL HAZARDS IN UNDERGROUND MINING

While in no way wishing to diminish the importance of hazards in surface mining, this paper will concentrate upon the geological hazards which can

have the most serious impact upon the safety and economic performance of underground mines. This choice has been made in order to limit the scope of the discussion and also because the prediction of these hazards is much more difficult in an underground environment than in the case of a surface mine.

One of the most obvious sources of "bad ground" problems in underground mining is the discontinuous nature of rock masses. In many cases the orebody is associated with major faulting and, in addition to concentrating the ore, these faults will also have caused a significant amount of fracturing in the surrounding rock mass. These fractures or joints reduce the strength of the rock mass by providing weak planes upon which movement is concentrated. Even in relatively massive rock, the presence of a few joints or bedding planes can reduce the strength of a pillar to about one quarter of that which is calculated on the basis of small scale laboratory tests on the intact rock material. In the case of heavily jointed rock masses, the rock mass strength may be one tenth or even one hundredth of the intact material strength.

The discontinuous nature of most rock masses is made worse by the presence of water. In near surface excavations, the downward migration of water carrying soil and clay will cause the deposition of weak in-filling material in the discontinuities and this will reduce their strength even further. At greater depths, water under high pressure can move considerable distances along faults and joints and can be very dangerous if a water bearing discontinuity is suddenly intersected by the excavation.

As the depth below surface increases, so does the magnitude of the in situ stress level in the rock. Creation of excavations in highly stressed rock further increases the local stress levels and a point is reached at which these local stresses exceed the strength of the rock mass. In the case of weak rocks this will result in closure or squeezing of the excavation and, while this may be inconvenient, it is not necessarily dangerous.

On the other hand, when a strong brittle rock such as a granite or quartzite fails, it may do so with considerable violence and the energy released by the failure of a large volume of strong rock can generate a rockburst. These rockbursts, which are extremely difficult to predict, can have a serious impact upon the operation of an underground mine, not only because of the physical danger associated with the event but also because of the uncertainty and fear generated by the phenomenon.

Other hazards in underground mining are associated with time-dependent deformation of evaporite rocks such as salt and potash; the presence of methane and other gases in rock associated with deposits of organic origin such as coal; high rock temperatures which occur in certain parts of the world with high thermal gradients and the danger of spontaneous combustion associated with oxidation of some sulphides .

PREDICTION OF GEOLOGICAL HAZARDS

Ideally, during the exploration of a mineral deposit, all potential hazards should be defined so that, in planning the mining operation, appropriate steps can be taken to avoid or to minimize the impact of these hazards. Imagine some form of radar device or seismic exploration tool which could penetrate the earth to a depth of several thousand metres and which could define faults and zones of intense fracturing, sub-surface groundwater reservoirs, zones of exceptionally high stress and, of course, the detailed three-dimensional shape of the orebody. The availability of such a tool or tools would be of enormous value to the mining industry but they exist only in science fiction and are likely to remain there for many decades to come.

Even on a local level, where penetration is limited to tens of metres, these remote sensing tools have not lived up to expectations and the author's personal experience with these devices suggests that a great deal of development remains to be done before practical tools become available for general use.

In the absence of remote sensing tools, what techniques are available for the geologist, the geotechnical engineer or the mining engineer to use for the prediction of the geological hazards discussed earlier in this paper ? In the opinion of this author, the most promising techniques are to be found in the systematic collection and interpretation of geological data during site exploration and during the mining of underground excavations. In order to understand the use and the potential of these techniques, it is necessary to examine what geological data are important, how these data are collected and how this information is used for the prediction of hazards.

Relevant geological data

The role of faults and joints in making the rock mass discontinuous has already been mentioned but it is not adequate to characterize a rock mass as faulted or jointed in order to understand its behaviour. The strength of the intact rock material, the nature of the infilling material in the faults or joints, the orientation and inclination of the discontinuities, the level of stress acting on the rock mass and the presence or absence of groundwater all have an influence upon the behaviour of the rock mass. Consqeuently, any attempt to predict geological hazards must take all of this information into account.
Hoek and Brown (1980) have discussed the question of interpretation of geological data for underground excavation design and have suggested

that the rock mass classification systems developed by Bieniawski (1974) and by Barton, Lien and Lunde (1974) are currently the most logical vehicles for this interpretation.

Bieniawski's Rock Mass Rating is made up as follows:

$$RMR = A + B + C + D + E - F$$

where
    A = compressive strength of intact rock
    B = Rock Quality Designation (RQD) – an index determined from diamond drill core (Deere (1964))
    C = Spacing of joints
    D = Condition of joints
    E = Groundwater conditions
    F = Adjustment for joint orientation

Numerical ratings are given to each of these parameters on the basis of matching the field observations with standard descriptions published in Bieniawski's paper. Although qualitative, these descriptions provide a very practical and easy to use guide which permits rapid and systematic classification of rock masses on the basis of diamond drill core logging and/or mapping of surface outcrops or exploration adits.

Barton, Lien and Lunde's Tunnelling Quality Index (Q) is calculated as follows:

$$Q = RQD/J_n \; x \; J_r/J_a \; x \; J_w/SRF$$

where
    RQD = Rock Quality Designation as above
    J_n = Joint set number
    J_r = Joint roughness number
    J_a = Joint alteration number
    J_w = Joint water reduction factor
    SRF = Stress reduction factor

Barton, Lien and Lunde published a detailed set of tables to allow the user to establish the value of each of these parameters in the field and, although not quite as easy to use as Bieniawski's classification, the Tunnelling Quality Index is a very useful practical tool.

Both of these classifications include information which describes the strength of the intact rock, the size of average blocks within the rock mass, the surface characteristics (and hence the strength) of the discontinuity surfaces and the influence of water. The Rock Mass Rating system described by Bieniawski includes an allowance for joint orientation which makes it suitable for the classification of relatively shallow rock masses – say to depths of 200 metres. Barton, Lien and Lunde's system allows for the influence of in situ stress and is more suited to the design of excavations at greater depth.

Once the geological data has been collected and assembled into one of these classification systems, the index obtained can be used to estimate rock mass strength or tunnelling conditions by means of empirical relationships derived from practical experience.

Systematic data collection

In order to make use of rock mass classification systems for the prediction of geological hazards which could be encountered in the mining of underground excavations, it is important that sufficient data be collected to ensure that the site is adequately covered. Since diamond drilling and the mining of exploration adits are expensive, it is essential that the exploration programme be planned very carefully and that all meaningful information be collected systematically. Practical limitations imposed by the weather, by unrealistically tight budgets and by uncooperative contractors can make it very difficult to collect this information but these limitations must be allowed for in planning the exploration programme.

Because of the large amount of information which must be collected for rock mass classification, most users of these systems will have thought about automating the data collection process. Unfortunately, the descriptive nature of the information and the current lack of international or even national standards make this virtually impossible. In this author's opinion, the best which can be hoped for at present and for several years to come is to have a trained observer such as an engineering geologist or geotechnical engineer collect the information and input into a central computer system by means of a portable computer or field terminal. The author is aware of one major civil engineering tunnelling contract where this is being done and where it is possible for the central computer to produce daily logs and drawings of the geological features mapped at the tunnel face.

It is relatively common when tunnelling into previously unmined ground to probe ahead of the tunnel face. Typically, a percussion drilled hole is extended several tunnel diameters ahead of the face to probe for high pressure water and to investigate the rock quality. During drilling, the rate of penetration , the thrust on the drill, the quality of the drill cuttings and of the cooling water return are monitored. When this information is compared with similar information obtained from previous probe holes, an indication can be obtained of possible changes in the rock ahead of the face. If major changes are indicated by the probe hole, the tunnel advance can be stopped while a horizontal hole is diamond drilled from the face. In addition to logging the core of such a hole, down-hole geophysics techniques can sometimes provide very useful supplementary information and, in extreme cases, borehole television cameras can be used to investigate potential trouble spots.

This probing ahead is particularly important when using a tunnelling machine or roadheader for excavation since these machines are far less flexible than drill and blast methods. Once a tunnelling machine has been set in motion and its gripping system and cutting mechanisms chosen, it is very difficult to change these at short notice. Consequently, as much advance warning of changing ground conditions as possible should be available to the machine operator.

Interpretation of geological data

Imagine an ideal operation in which a very carefully planned exploration programme has been executed, borehole core has been carefully logged, exploration adits and existing excavations have been mapped and all of the information has been assembled by computer into rock mass classification systems. It is planned, once excavation commences, to have a full-time engineering geologist on site to continue the data gathering process and to assist the owner and the contractor in the interpretation of this information.

How is this information to be interpreted in order to predict excavation stability conditions, to design support systems, to predict excavation advance rates and, most important, to predict potential hazards which could seriously disrupt the operation ? Unfortunately, at present, the interpretation of this information is based almost entirely upon judgement and experience. Fortunately, a great deal of this type of experience exists but very little of it has been formalized and presented in tables or charts which would permit

someone other than a tunnelling specialist to draw meaningful conclusions. Bieniawski (1974) and Barton, Lien and Lunde (1974) attempted to use their rock mass classifications as a basis for drawing together experience on underground excavation support and they presented tables which gave recommended support designs for different rock mass classifications. These papers represent a good start but even the authors would not claim that their recommendations are adequate.

In the author's opinion, the interpretation of systematically collected and classified geological information presents one of the most exciting challenges for the application of computer based expert systems or, as it is sometimes called, artificial intelligence. In developing these systems for use in underground excavation engineering, a group of experienced mining and/or tunnel engineers would be asked a series of very carefully structured questions related to their interpretation of a number of sets of inter-related observations. For example, "What conditions would you anticipate when tunnelling in massive gneiss when the rock mass at the face exhibits intense jointing (four or five joint sets with spacings of 10 to 100mm), staining, water seepage from several of the joints, continuous ravelling of small pieces of rock from the face ?" The answers to many such questions would be stored and sorted by computer and would be organized into an interactive system in which, in addition to providing answers, the user would be asked questions which would narrow down or amplify the original input, thereby permitting progressive refinement of the interpretation.

This process is basically a means of collecting and recording the interpretations which would be made by a very experienced engineer or geologist when examining a large body of relevant data in search of a particular answer. The diagnostic techniques which are used are peculiar to each individual and obviously depend upon the experience base from which that individual draws. Never-the-less, these interpretation techniques follow similar patterns and trends and it is these similarities which form the basis of the computer based expert systems.

Small expert systems have been successfully developed in some fields of geology (Campbell et al (1982)). No serious attempt has yet been made to develop systems for the prediction of the types of hazards discussed in this paper. Since the entire science of artificial intelligence or expert systems is still in its infancy, it is probable that it will be many years before effective systems for the prediction of geological hazards become available and that many more years will pass before these systems are accepted by the conservative industries for which they are intended.

CONTROL OF GEOLOGICAL HAZARDS

Suppose that a tunnel engineer or mine operator is lucky enough to have an experienced engineering geologist or geotechnical engineer on staff and that this individual can successfully predict most of the hazards which are likely to be encountered in underground excavations. What can be done, short of stopping the tunnelling or mining operation, to control these hazards ? The answer obviously depends upon the nature of the hazard and some of the more common problems which can be encountered underground are discussed below.

Rockfalls

Rockfalls occur when blocks of rock are released from the roof or sidewalls of an underground excavation by intersecting discontinuities in the rock. These discontinuities can be naturally occurring bedding planes or joints or they can be fractures induced by blast damage or high stress concentrations. These latter types of discontinuities will be dealt with later in the paper and the present discussion will concentrate upon naturally occurring geological features.

Techniques for analyzing structurally controlled failures in underground excavations have been discussed by Hoek and Brown (1980) and new and improved techniques are being developed at several universities and research organizations. The lack of adequate analytical tools is not a problem but there are problems with the adequacy of the input data and with techniques for controlling structural failures. The data input problem is the same as that which plagues the entire subject of geotechnical engineering – there is never enough reliable information on which to base accurate predictions and analyses. This problem has already been discussed in this paper and , while improved exploration tools and systematic data collection techniques will help, the interpretation will still have to rely upon judgement and experience for the foreseeable future.

The question of control is much more dependent upon developments in practical support techniques. Once a potential structural failure has been identified, the miner has two choices - let it fall or support it. In the case of small wedges and blocks, the former choice is usually the most practical although it does have the potential of creating a progressive ravelling problem if the block which falls is a key block which keeps the interlocking rock mass from falling apart. If the block is a key block or is of sufficient size that a fall could be dangerous, then support will be required.

The design of support to prevent structurally controlled failures is a relatively simple process since the driving force is the weight of the block and this driving force must be resisted by the installed support. The trick of good support for structurally controlled failures is to place it early. This is because, once a block or wedge is released, there is nothing to stop it falling and hence the principal aim of the support design is to prevent movement starting. There are many practical techniques for pre-supporting excavations in which structurally controlled failures are likely. One of the most effective is the pilot and slash method where a small pilot tunnel is driven ahead of the main tunnel to allow the installation of rockbolts before the full tunnel span is excavated. Others include spiling or forepoling where forward inclined bolts or grouted dowels are installed in such a way as to form an umbrella of pre-supported rock under which the tunnel face can be advanced. In the case of large fault controlled structural failures, the use of long grouted cables is usually required and the techniques for installing these are relatively well developed.

There is considerable potential for automating the equipment used to install the support in these cases and some equipment is already on the market. Ideally, a highly mobile rubber-tyred jumbo or similar unit, with an option for remote control, could drill holes in appropriate locations close to the face and then, without moving from a given location, install a mechanically anchored or resin grouted bolt or a friction anchored or expanded dowel. The advantages of such a unit in terms of safety and speed of operation are obvious and the need will become more urgent as mining conditions become more arduous and as it becomes more difficult to attract skilled miners to work under the crude conditions which are still common around the world.

Groundwater

While the presence of groundwater in underground excavations can aggravate stability problems, the principal threat posed by groundwater is that of flooding or of a sudden, unanticipated inrush. Systematic data collection and careful probing ahead of the face are the main predictive tools available while drainage and grouting are the main methods of control.

In the case of isolated pockets of groundwater, drainage is the obvious remedy to subsurface groundwater problems. Where a large water reservoir is likely to be tapped by the excavation process, grouting followed by local drainage is the preferred route. This grouting may be carried out from holes drilled ahead of the face to form an impermeable umbrella or it may be carried out from long holes aimed at specific water passages.

Currently, the practice of underground grouting is rather like shooting in the dark – there are very few techniques available to help the operator decide upon optimum location, grout viscosity, pressures and pumping rates during placing of the grout. The only successful test is that the water stops flowing but experience suggests that a great deal of grout is usually wasted before this test is passed.

The potential for the development of tools and automatic control systems for such grouting operations seems to be remote at present – mainly because it is very difficult to define the problem to be addressed. However, there is clearly a need for a better understanding of this process and this understanding will eventually lead to the development of suitable equipment for the control of underground grouting.

Stress induced fracturing

The creation of any underground excavation results in a redistribution of the stresses in the rock mass and this generally leads to a concentration of stresses in the rock immediately surrounding the opening. When these local stresses exceed the strength of the rock, failure of the rock occurs and, if this failure is severe enough, complete collapse of the excavation can occur. In extreme cases, when mining in very hard, strong and brittle rock, these failures can occur as rockbursts in which significant amounts of energy are released with explosive violence.

The ingredients in the stress induced fracture problem are the direction and magnitudes of the in situ stresses in the rock, the shape and size of the excavations created and the strength characteristics of the rock mass. A full discussion on these problems exceeds the scope of this paper and the interested reader is referred to the general review published by Hoek and Brown (1980) and to the proceedings a recent conference on rockbursts published by the Institution of Mining and Metallurgy in London (1984).

In the context of this paper, the question of control of stress induced fracturing and rockbursts comes back to a full understanding of the in situ stresses and rock mass characteristics, based upon the systematic data collection and interpretation discussed earlier, and the incorporation of this information into the excavation design process. In choosing the shape of the excavations to be created and the percentage extraction , the mine designer has some control over the stresses which will be induced around the excavations and in pillars between excavations. The sequence of mining is also important in that remnants and island pillars which tend to attract high stress should be avoided.

Numerical models which can assist mine planners in avoiding the creation of high stress concentrations are becoming available and are being used in an increasing number of mines. A good example of one such application has been presented by von Kimmelmann, Hyde and Madgwick (1984).

When conditions are such that it is not possible to avoid stress induced fracturing, a substantial amount of control is available by means of support using rockbolts, dowels, cablebolts, props, packs and backfill. In contrast to the support of structurally controlled failures, discussed earlier, some movement of the rock before the installation of support is desirable in order to allow a controlled amount of fracturing and the dissipation of some of the stored energy in the rock mass. In tunnelling, the time and deformation-dependent interaction between the rock mass and the installed support is reasonably well understood and goes under the names of the closure-confinement method, the rock-support interaction technique (Hoek and Brown (1980)) or the New Austrian Tunnelling Method (Rabcewicz (1964)). In the case of mining the problem is more complex because of the complex shape of most mine openings but practical techniques such as rockbolting and the use of rapid-yielding hydraulic props (Tyser and Wagner (1976)) have proved to be effective in controlling local problems of stress induced fracturing.

Rockbursts are a more serious problem and have many parallels to earthquakes. Not only are they extremely dangerous but their unpredictability gives rise to a fear of rockbusts which can be seriously disruptive in an operating mine or tunnel. Attempts to predict rockbursts have received a great deal of attention but have only been partially successful (Salamon (1984)). The monitoring of these events by means of microseismic techniques (Cook (1963)) has made an important contribution to our understanding of the location of rockbursts and refinements in microseismic monitoring techniques together with computer aided interpretation of the results offers some long term hope of further improvements in this understanding (Salamon (1984)).

CONCLUSIONS

The purpose of presenting this paper at a conference on Automation for Mineral Resource Development has been to give a brief overview of the current state of our understanding and ability to predict geological hazards in the creation of underground excavations. Water inrushes, falls of ground and rockbursts are some of these hazards and the author has attempted to show that, while our understanding of the origins of these problems is reasonable, our ability to predict them is severly limited.

There are no magic tools which enable us to look into the earth and to detect the presence of trapped groundwater, zones of intense fracturing or areas of exceptionally high stress - some of the factors which contribute to the hazards encountered in underground mining. The most useful of the currently available techniques is the systematic collection and interpretation of relevant geological data during exploration or in advance of the tunnel or mining face. The wide availability of computers will be of great assistance in this data collection and interpretation system. There appears to be considerable merit is the use of expert systems in which the knowledge of experienced miners, engineering geologists and geotechnical engineers is gathered into an ordered computer based information system and developed into an inter-active design tool.

The control of geological hazards in underground excavations is an equally difficult task but significant practical progress has been made by trial and error development of mining equipment and support systems . There is certainly room for some automation of such equipment as drills, mucking machines and rockbolting equipment in order to increase the productivity and to give the operators a safer and more attractive working environment. However, it is considered unlikely that automation will contribute significantly to the control of geological hazards which will remain as inescapable part of the process of underground mining.

REFERENCES

Barton, N., Lien, R. and Lunde, J. (1974) Engineering classification of rock masses for the design of tunnel support. *Rock Mechanics*, Vol 6, pp 189-236.

Bieniawski, Z.T. (1974) Geomechanics classification of rock masses and its application in tunnelling. In *Advances in Rock Mechanics, Vol.2, part A : proc. 3rd congress International Society for Rock Mechanics*, Denver, 1974, pp 27-32.

Campbell, A.N., Hollister, V.F., Duda, R.O. and Hart, P.E. (1982). Recognition of a hidden mineral deposit by an artificial intelligence program. *Science*, Vol. 217, pp 927-929.

Cook, N.G.W. (1963) The seismic location of rockbursts. *Proc. 5th Rock Mechanics Symposium.* Pergamon Press, Oxford, pp 493-516.

Deere, D.U. (1964) Technical description of rock cores for engineering purposes. *Rock Mechanics and Engineering Geology*, Vol. 1, No. 1, pp 17-22.

Hoek, E. and Brown, E.T.(1980) *Underground Excavations in Rock*, The Institution of Mining and Metallurgy, London.

Institution of Mining and Metallurgy, London (1984) *Rockbursts: prediction and control*, IMM, London

Rabcewicz, L.V. (1964) The new Austrian tunnelling method, *Water Power*, Vol. 16, 1964, pp 453-457 (part 1) and Vol. 17, 1965, pp 19-24 (part 2).

Salamon, M.D.G. (1984) Rockburst hazard and the fight for its alleviation in South African gold mines. *Proc. Symp. Rockbursts: prediction and control*, The Institution of Mining and Metallurgy, London, pp 11-36.

Tyser, J.A. and Wagner, H. (1976) Review of six years of operations with the extended use of rapid yielding props at the East Rand Proprietary Mines, Limited and experience gained throughout the industry. *Association of Mine Managers of South Africa*, Papers and Discussions, 1976-1977, pp 321-347.

von Kimmelmann, M.R., Hyde, B. and Madgwick, R.J (1984) The use of computer applications at BCL Limited in planning pillar extraction and the design of mining layouts. *Proc. ISRM Symp. Design and Performance of Underground Excavations*, Cambridge, UK, pp 53-63.

The Role of Automation in the Mining of Tabular Deposits at Depth

M.D.G. SALAMON

Research Organization, Chamber of Mines of South Africa, PO Box 91230, Auckland Park 2006, Johannesburg

Abstract. A brief introduction is given of the scale and of some special features of South African gold mining; it is noted that great depth of mining and narrowness of the reefs, and hardness and abrasiveness of the rock are the main sources of the problems of the industry. The most inhibiting problems are: Excessive rock pressure occcuring in conjunction with rockbursts, hot and humid working places due to high virgin rock temperature and auto-compression of the air, and low level of productivity. Conventional solutions applied ordinarily in mining do not promise to eliminate, or even alleviate, these difficulties.

However, research and development performed during recent years have produced promising solutions. To illustrate this progress some examples of advances are discussed. These include three-dimensional seismic networks used to alleviate the rockburst hazard, recirculation of air to improve the environment in the workings, automatic sorting underground to provide material for backfill and, at the same time, to upgrade the ore, and the development of the concept of 'hydro-power'. The application of these developments hinges on the employment of advanced technology, involving automation, remote control and so on.

Keywords. Deep level mining; rock pressure; rockburst; mine cooling; hydro-power; remote control; automated control; fail safe system.

INTRODUCTION

Most of the gold resources of South Africa are located at depths which exceed 1 000m. The weighted mean depth of mining is about 1 700m. Conventional mining technology is inadequate to sustain large scale operations at these depths. Safe and efficient deep level mining can only be envisaged if it is supported by the development of new technology.

The Research Organization of the Chamber of Mines of South Africa has assumed in recent years a leading role in the development of new technologies for mining at depth in hard rock. The Chamber of Mines is a voluntary private association of mining companies. The research and development performed by its Research Organization are guided and financed by the members of the Chamber. This intimate relationship between the research body and the industry ensures that the R & D carried out is relevant to the solution of the most pressing problems of the mines.

During the last quarter of a century, while the Research Organization has concentrated most of its resources on the problems of deep level mining, it has become obvious that the simple extrapolation of the experience gained in shallow mines will not yield efficient solutions at depths. In some instances, the old experience does not even give guidance as to how a solution to the new problem should be sought. Obviously, new concepts must be evolved to suit the challenges of the new conditions.

The aim here is to introduce some of these new concepts and show that the implementation of the new ideas will require the application of high technology involving automation, remote control, and so on.

SOME PARAMETERS OF GOLD MINING

An examination of the most important physical parameters of deep gold mines is a pre-requisite to the effective discussion of the future need and employment of advanced technology in these mines.

A typical gold mine in South Africa may operate at a mean depth of about 1 700m, hoist 250 000 tons of rock per month, some 72% of which is extracted from stopes of about 1,3m in width, and produce some 1 260kg of gold per month, Salamon (1976). To achieve this it employs some 8 800 men underground and 2 600 on surface, giving a total complement of 11 400 persons. These figures reveal that the mine employs very labour intensive methods.Every month, ore is extracted over an area of some 50 000m² where the width of the gold-bearing channel is likely to be less than 0,5m. At any one time it will have available for production some 7 200m length of face. To provide the face length necessary to maintain production using current technology, the mine will have to develop some 30km of tunnels every year.

The weighted mean virgin rock temperature at the typical mine is about 38°C. In 1983 the maximum of the weighted mean virgin rock temperature for a mine was 46,8°C. During the same year the highest measured virgin rock temperature at an isolated working was 66,9°C at a depth of 3 300m. Moreover, the average mass of downcast air moved per unit mass of rock broken was 11. Also the total installed refrigeration capacity of the industry has exceeded 1,0 GW(R) in 1984. These numbers give some indication of the scale of a gold mine and of the industry as a whole.

PROBLEMS OF MINING TABULAR
DEPOSITS AT DEPTH

In the light of the statistics concerning the
typical mine, it is not necessary to go far to
identify the major physical factors that are
responsible for the problems of deep level gold
mining.

Rock Pressure

Some 56% of all fatal accidents in the gold mining
industry are due to rockfalls and/or rockbursts,
Salamon (1983). Mining efficiency also suffers
because of the interruptions of operations caused
by the same major events.

As a result of the high virgin rock stresses,
virtually all excavations in a gold mine are
surrounded by fractured rock which fact, of
course, gives rise to severe strata control
problems. Although the difficulties are serious,
miners would be able to handle most support
problems resulting from heavy rock pressure. The
predicament becomes especially challenging when a
mine, in addition to the ordinary strata control
difficulties, begins to experience rockbursts as
well. Rockbursts are violent events which result
in sudden damage or even in the destruction of
mine openings. These events are regarded by many
as the phenomenon which ultimately will limit the
expansion of mining to greater depths.

Rockbursts constitute a small subset of seismic
events, which are caused either by a sudden slip
along a pre-existing geological discontinuity,
such as a fault, or by a violent failure of solid
rock. Both of these mechanisms are induced by
changes in stresses which disturb an unstable
equilibrium. The source of the kinetic energy
liberated during an event is the strain energy
stored in the rock surrounding the focus of the
event. The change in stress disturbing the
equilibrium is induced by mining. The stored
energy results from the straining of the rock
caused partly by the virgin stresses due to
gravity and partly by the stresses induced by
mining and/or by tectonic forces, Salamon (1983).
The seismic events resemble in many respects
natural earthquakes and their magnitudes have been
known to be as high as 5,2 on the Richter scale.

Theoretical and field studies have indicated that
one of the most promising means of alleviating
both the rock pressure and rockburst hazards is
based on the control of the convergence volume in
the stopes, Cook et al (1966), Salamon
(1974,1983). To illustrate the validity of this
statement a simple limiting case can be postulated. Assume that some support material is
available which can be made to replace immediately
the mined-out portion of the reef and that this
material expands sufficiently after installation
to eliminate all roof and floor convergence which
might have taken place before backfilling. If such
an effective backfilling could be performed
continuously then a working stope would reduce to
a mere tunnel which is being moved parallel to its
long axis in the course of mining. If such utopian
mining were to be possible, then most adverse
effects of rock pressure and rockbursts would be
eliminated.

In reality no material exists which would provide
such a perfect backfill. Nevertheless, backfilling
with real materials, such as waste rock, does
reduce the severity of strata control and rockburst problems. This is a conclusion which must be
learned by mining engineers who are involved in
deep level mining.

Heat

In 1974 some 58% of all rock hoisted from the gold
mines were broken at depths at which the virgin
rock temperature exceeded 38°C. The implication of
this observation is that about two-thirds of the
working places are created in rock the original
temperature of which is higher than the normal
human body temperature. This means that most
miners in gold mines are working in a man-
controlled environment. Not only noxious gases,
fumes etc., have to be diluted below threshhold
levels but heat must be removed to reduce
temperatures to acceptable values.

At first, the approach adopted was the obvious one
of circulating greater and greater volumes of air
with the view of using the return air as the means
of heat rejection. However, as the depth of mining
was gradually increasing,the volume of circulating
air necessary to achieve the required cooling
started to increase alarmingly. It was soon
realised that with increasing depth auto-
compression heating of the descending air becomes
really significant; therefore, at a depth of some
2 500m air becomes an inefficient cooling medium.
This is so even if the heat pick-up from the rock
is not taken into account. Hence, many gold mines
require artificial cooling or refrigeration.

Refrigeration plants are installed either underground or on the surface or both underground and
on the surface. Although the purpose of this paper
is not to describe ventilation and refrigeration
practices, it is worth noting that significant
technological advances are being made which have
opened up new avenues for development in this
field.

The cost of providing acceptable temperature
conditions in the stopes is increasing at an
appalling rate. In 1984 this cost had reached more
than 10 per cent of the total working cost of a
moderately deep mine.

Productivity and logistics

The technology of underground mining of coal seams
and of massive ore bodies has advanced very
rapidly in the last several decades. The technology of stoping narrow reefs at considerable
depths in hard, abrasive rock has not progressed
at anywhere near the same pace.

Reefs and coal seams are both tabular deposits. In
view of this geometrical similarity, it is natural
to ask whether developments in coal mining could
be adapted to suit stopes in gold mines. However,
much of the coal related development makes use of
non-explosive rock breaking. Obviously, coal
extraction is a considerably easier task than the
breaking of hard rock in a similar manner. Thus,
advances in coal breaking cannot be transformed
readily into usable techniques in hard rock.

Massive hard rock ore-bodies are mined usually by
blasting. To achieve high productivity, large,
often self-propelled, mechanical equipment is
employed in drilling and rock handling. Space
limitations prevent the application of these
machines in narrow stopes.

In view of these and other problems, the stoping
technology employed in the industry has serious
short-comings. It is very labour intensive and of
low productivity. The overall labour productivity
ranges from 200 to 300 tons/employee/year, which
is several times lower than productivities
achieved in most other types of underground
mining. Low labour productivity imposes a limit on
earning capacity of the men involved.

Another shortcoming in the present stoping system is the low rate of face advance, or in other words, the low utilization of the stope face. To illustrate this point note that in 1983 there was available a total face length of some 275 000m in the industry and this face was advanced, on average, at a rate of some 7 m/month. To express the significance of this in different terms observe that the ore production in the industry averaged at 30 ton/month for each metre of face and each panel of the face was blasted on average about seven times per month.

Until relatively recently little significance was attributed to the shortcomings of the conventional stoping system. Inappropriate accounting procedures and even the taxation system have tended to blur the financial significance of the under utilization of the stope faces. Today mine managements have become more aware of this financial burden, as well as the problems in supervision, communication, and logistics which emanate from the spread out nature of their operations.

R & D to Alleviate Mining Problems

Perhaps the greatest impediment which has hindered advances has been the lack of world-wide interest in a modern technology for hard rock mining in narrow stopes at depth. While there are mines in various parts of the world which work narrow tabular deposits, they tend to be isolated and lack financial resources to initiate expensive and protracted programmes of research and development which would be required to evolve new, more efficient methods of mining. The South African gold mining industry might be the only exception. Recognizing this, the industry adopted an appropriate R & D policy in the mid 1960's

Currently, the resulting R & D programme is focused on five major aspects of mining and of miners. Three of these have been discussed in this section, while the remaining two are outside the scope of this paper.

Good progress has been made during the last two decades in combating the rock pressure and heat problems and in laying down the foundation for advances in stoping technology.

SOME THOUGHTS ON DEEP LEVEL MINING

The preceding discussion has revealed that the South African gold mining industry, if it wishes to have a greater control of its own destiny, will have to alleviate the problems which arise from the depth and geometry of mining and from the hard and abrasive nature of rock. To achieve this it will be essential to accept and introduce innovative ideas and solutions.

The lifeline of an underground mine is its shaft system. Before a mine can start production its initial shaft system must be established. In deep level mining the capital cost of such a system is tremendous and the period of shaft sinking is measured in years. The success of the venture requires optimal design and efficient utilization of the shaft system.

Shafts are used to transport men, rock (ore and waste rock), material, air, pollutants (heat, gases, dust, etc.), and services (water, compressed air, power, etc). Usually, requirements of rock hoisting and air movement determine the number and sizes of shafts. These two tasks of transportation, together with men riding and water pumping, are responsible for most of the running cost of operating a shaft system. Thus, a tight control over the quantities of ore hoisted and of

air pumped can contribute appreciably to the economic well-being of a deep mine.

The rapid yielding hydraulic face support which was developed in the 1970's has enhanced considerably the available defences against rockbursts. Developments in refrigeration provide the means of controlling the heat environment in stopes. Unfortunately current under-utilization of stope faces renders the intensive application of these new methods of improvement economically unattractive and even impractical in some circumstances.

Concentration of activities usually has favourable attributes in underground mining. This is especially so where the costs of the preparation, equipping and maintenance of service excavations and stope faces are high. Deep level mining obviously falls into this category. Concentration of operations can result either from a better spatial distribution of workings or from improved utilization of faces or, of course, from taking both of these steps simultaneously. To gain the full benefits of concentration when mining at depth, it is preferable, if and when possible, to improve both the layout of the mine and the utilization of faces.

Industrial research and development should contribute to the well-being of the industry it is supposed to serve. The work of the Research Organization will be regarded as successful only if the advances achieved, when taken together, improve the profitability of the gold mines. Naturally, profitability is affected by factors other than those which can be influenced by researchers. Some of these factors seem to be unpredictable, e.g. gold price, others can be anticipated, e.g. labour cost increases. A well balanced R & D programme should aim at changes which will lead to relative improvement of the profitability of the venture, in spite of the predictable and unpredictable changes.

AUTOMATION AT DEPTHS

General Comments

Safety and profitability of gold mines will deteriorate unless the hazards arising from rock pressure (or rockburst) and heat are combatted effectively. As the depth of mining increases the burden of transportation between the workings and the surface becomes progressively more severe. The outmoded and labour-intensive stoping technology leads to serious under-utilization of the faces which, in turn, results in excessive length of face scattered throughout the mine.

The Research Organization is attempting to tackle most of the problems arising from these facts. In this paper only those projects which relate to automation in a broad sense and have an element of originality are discussed. The aim is to highlight their importance with no attempt being made to detail the technical aspects of the various projects.

Controlled Recirculation of Air

Once-through ventilation systems are used traditionally to ventilate mines. This means a fixed quantity of air is passed from surface through the working areas and then rejected from the mine. As the depth of mining increases it becomes less and less attractive to employ this approach.

Take a simple hypothetical example and assume that the exposed surfaces of all cavities in a mine are perfectly insulated so as to ensure that no heat flows from the rock mass into the ventilating air. As the depth of mining increases, the temperature rise of the air between surface and the workings becomes progressively higher because of auto-compression. Eventually a depth is reached, somewhere below 2 500m, at which the ventilation air enters the deep workings at wet-bulb temper-atures higher than those generally recognized as reject values. Obviously, in such a case bringing more surface air into the mine only adds to the already serious heat problem. In these circum-stances controlled recirculation provides the additional quantities of air required to achieve better cooling and more effective distribution of refrigeration within the working areas. This is achieved without the need for higher quantities of downcast air, Burton et al (1984).

Controlled recirculation can be applied in a section of a mine where heat is a major problem. However, its full benefits will be realized only in new sections of existing mines or, even more importantly, in whole new mines that have been designed with this new approach to ventilation in mind.

Recirculation involves returning some of the air flowing out of a section of a mine (or a whole mine) back into the intake airway of the same working area. Hence, it increases the air flow in the working areas and, therefore, results in higher air velocities. The increased velocities enhance the cooling power of the air. Perhaps more importantly, recirculation provides the means of achieving more uniform temperature conditions throughout the workings and permits cooling to be effected at relatively low cost using centralized bulk air coolers instead of a network of low capacity cooling coils.

The numerous advantages of recirculation include, Research Organization (1984a):

(i)   a potential saving in the number of ventilation shafts and airways,
(ii)   overcoming the effects of auto-compression with less refrigeration capacity,
(iii)   smaller and lower cost main ventilation fans, and
(iv)   less electrical power consumed by the ventilation and cooling systems of the mine.

Recirculation, of course, is not a new idea. Until recently, however, mining men were suspicious of its application, largely because of the possibil-ity of noxious fumes arising from a fire in the circuit being forced by the system to enter into the fresh intake airway. For recirculation to be viable in practice it must be devised so as to ensure without doubt that it is as safe as the once-through system. Developments in recent years in telemetering and automation have provided the basis for giving such a safety assurance.

To ensure the safety of men in workings where the ventilation is assisted by recirculation, it is essential that the fans forcing air recirculation are switched off as soon as a fire occurs in the circuit and the ventilation reverts immediately to a once-through system. Two technological advances are necessary to accomplish this. It is necessary to have firstly, an effective and reliable method to detect the occurrence and approximate location of a fire and, secondly, in case of a fire the fan motors must be stopped remotely using the signals generated by the fire detectors.

After many years of development and testing an effective fire detection system, based on the so called 'becon' sensors, is in wide-spread use in South African gold mines, van der Walt (1980). Experience has shown that the system is reliable and can be adapted to become the basis of a

recirculation system, Burton et al (1984). Mines in fact have realised the potential benefits of recirculation and already several schemes, in-volving some 10% of the total volume of air circulated in the industry, have been or are being commissioned.

Combatting the Threat of Rockbursts

There are at least two aspects of the fight against rockbursts in which the technology of automation plays major, or even vital, roles.

Mining-induced seismic activity resembles closely the phenomenon of natural earthquakes. Since the early 1960's seismic instrumentation, after suit-able adaptation to the requirements of using a three-dimensional network of geophones, has been used on an increasing scale. This has been done to gain an insight into the mechanism of seismic events and to determine, virtually instantan-eously, the location of rockbursts for the guidance of possible rescue operations. These networks have become more and more sophisticated, incorporating improved seismometers, hard-wire and radio transmission of signals, on-line analysis, advance data storage, and so on.

Perhaps the most exciting possibility involves the timeous and reliable prediction of an imminent seismic event. This ambitious goal of researchers has remained elusive for some decades, in spite of world-wide interest in the matter. Early attempts were based on counting the frequency of noise, later this was expanded to incorporate the estimation of seismic energy release as a function of time. In spite of optimistic reports which emanated from various field trials from time to time, the technology employed did not permit the development of an acceptable prediction system. Recent advances in data capturing, handling, and analysis hold out the promise of augmenting the system so as to obtain both time and space related distributions of the seismic events induced by the rock fracturing process. It is possible that the so enriched information may prove adequate to form the basis of a usable prediction method which will locate, in time and in space, an imminent major event.

Another modern application of computer and automa-tion technologies involves the utilization of rock mechanics principles in mine planning. It has been known for some time that the geometrical rate of energy release is a reasonably good predictor of the rockburst hazard and, generally, of rock conditions, Cook et al (1966).

This application has reached a level of refinement where the plans of some existing mines are held in a data base in digital form. At the start of each planning period, various mining options (i.e. possible changes in mine layout) are drawn, digitized, and analysed using one of the advanced programs designed to compute the stresses and displacements induced by the mining of tabular deposits. This process facilitates the selection of that option which yields the most favourable distribution of energy release rate.

Automatic Ore Sorting Underground

It was argued earlier that one of the most effective means of diminishing the threat of rockburst is the employment of backfill. Material used for this purpose can be material from old slime dams, or the waste product of gold recovery plants, or the waste rock obtained from off-reef developments, or the reject rock of some under-ground process of concentration. These substances can be subdivided into two categories: comminuted rock imported from the surface and rock acquired underground

Material in the latter category, which may include development waste or the reject of a concentration process, ordinarily would be hoisted to the surface. Its disposal underground liberates a part of the hoisting capacity of the shaft for the transportation of ore. Thus, it provides an opportunity for an increase in the gold output of an existing mine. Of course, the application of a concentration process also upgrades the ore which reaches the recovery plant, hence represents a further chance for the augmentation of metal production. In the case of a new mine, especially if concentration is applied in conjunction with recirculation, the shaft system and other services can be designed so as to take into account the reduced rock hoisting load and diminished air requirement.

There are, therefore, powerful incentives for the utilization of waste rock obtained underground for backfilling.

Underground upgrading of ore can be achieved either by employing a reasonably conventional metallurgical process or by practising some form of sorting. After several years of research, the idea of using metallurgical means of concentration underground had to be abandoned because it did not appear feasible to produce a rejectable tailings.

Hand sorting is well known to the industry. For some years automated sorting on the surface, based on photometric or radiometric principles, has been used with some success by a number of mines. There can be little doubt that a new deep level mining system, which incorporates sorting, must be based on an automated concept. An extensive study is in progress to establish the economic and technical feasibilities of using underground sorting on a large scale. It will be appreciated that this job is formidable in the case of ores containing precious metals with grade measured in a few parts per million.

The task of separating the reject rock from the ore involves the establishment and proving of a viable physical basis for the sorting, on line recognition of the discriminating feature, and the automatic classification of particles, with a high degree of reliability, into the two required classes. This is a formidable challenge, especially when the unfavourable environmental conditions are taken into consideration.

Hydro-power

Two developments have taken place in recent years giving inspiration to the concept of hydro-power, in which chilled water is piped directly from surface into deep level stopes to provide cooling and to power stoping machinery from its hydrostatic head.

The first of these developments was the increasing use of water, chilled on the surface for mine cooling, Research Organization (1984b). The second advance was the development of hydraulic drills powered by fluids containing 95% water, Research Organization (1984b). The success of this development of the drills and other related advances, suggests that it will be feasible to have machines which are powered with reasonably good quality water only. It was recognized that there was scope for combining cooling and powering in a common system by using the hydrostatic head in the chilled water to provide a convenient source of power. Calculations have shown that the flow rate of chilled water needed to cool the workings of a typical mine is adequate to power machinery in those workings and that at depths between 1 500m to 2,000m the pressure is sufficient to drive the machines.

A comprehensive investigation has shown that the hydro-power system is technically feasible. A full scale pilot system has been designed and will be tested during 1985 in one of the mines.

For hydro-power to be used effectively and safely, both the static and dynamic behaviour of such a system must be understood thoroughly. A computer model of the steady state and transient characteristics of a hydro-power system has been developed and used to simulate the pilot system which is to be tested shortly.

Obviously, a complex hydro-power system can only function under the command of a well designed, remotely operated control system. Very strict safety standards must be observed and the system must have a 'fail safe' feature.

CONCLUSIONS

Deep level mining of narrow tabular deposits in hard rock has its special difficulties which include: excessive rock pressure which is often associated with rockbursts, high rock temperature leading to serious environmental problems underground, and problems of low productivity due to outmoded stoping technology.

Modern research, conducted by the Research Organization on behalf of the mining industry, has indicated that progress can be made in the alleviation of these problems, but this progress has been made feasible only by recent advances in science and engineering.

More particularly, modern technology will play a decisive role in the alleviation of the rockburst hazard and of the heat problem and in the implementation of more efficient production methods. For example, seismic work, recirculation of air, backfilling, automatic sorting and the hydro-power concept cannot be applied safely and effectively unless their implementation makes use of ideas of automation, automatic control, and remote control.

REFERENCES

Cook, N.G.W., E. Hoek, J.P.G. Pretorius, W.D. Ortlepp, and M.D.G. Salamon (1966). Rock mechanics applied to the study of rockbursts. J.S. Afr. Inst. Min. Metall., 66, 435-528.

Research Organization (1984a). Controlled recirculation of ventilation air. Application report 4, Chamber of Mines of South Africa.

Research Organization (1984b). Annual Report 1983, Chamber of Mines of South Africa.

Salamon, M.D.G. (1974), Rock mechanics of underground excavations. In Advances in Rock Mechanics. Proc. of the 3rd Congress, Int. Soc for Rock Mech., Denver, Colo., 1974. Vol. I, Part B, 951-1099, Washington, National Academy of Sciences.

Salamon, M.D.G. (1976). The role of research and development in the South African gold mining industry. (Presidential Address) J.S. Afr. Inst. Min. Metall. 77, 64-75.

Salamon, M.D.G. (1983). Rockburst hazard and the fight for it's alleviation in South African gold mines. In Rockbursts : Prediction and Control. Inst. Min. Metall.

Van der Walt, N.T. B.J. Bout, T Anderson and J.Newington (1980). An all analogue fire detection system for South African gold mines. J. Mine Vent Soc. S.Afr. 33, 2-13.

The Shape of Hard Rock Mining in the 21st Century

P.W. HICKSON

Manager, Production Coordination, AM&S Mining, Broken Hill

D.W. McMAHON

General Manager, Resource Planning, AM&S Mining, Collins Street, Melbourne

M.A. SCHAPPER, Ph.D

General Manager, Technological Innovations, CRA Ltd, Collins Street, Melbourne

Abstract Hard rock underground mining has been wedded to the drill-blast
method of breaking rock. Current practice and technology have focused on
greater efficiencies in the blasting and handling of ore, on the basis that the
lower costs that come with larger scale operations compensate for the waste
dilution associated with the bulk methods. However, it is suggested that new
technology is available or is on the horizon that could completely change the
shape of hard rock mining in the 21st century. Hard-rock continuous-mining
machines, possibly assisted by high pressure water jets, coupled with automated
sorting and materials handling systems, could re-orient the miner's focus to
winning metal rather than mining tonnes of ore. More continuous mining
processes could accelerate the penetration by high technology in the hard rock
underground mine, with the potential of greater manpower productivities and
lower costs per tonne of metal produced. This new technology will be important
in maintaining a competitive supply of metals, given the falling headgrades of
the orebodies remaining for development and the increasing market threat from
manufactured substitutes.

Keywords. Control engineering; computer applications; conveyors; crushing;
man-machine systems; materials handling; mining; natural resources;
technological forecasting.

INTRODUCTION

Mining is a practical art that generally responds
to, if not initiates, technological development
in its quest to win metal profitably. However,
an obsession with "things that work", driven by
the need to minimise the unknowns in an already
high risk business, tends to limit the miners'
technological advancement to incremental or
evolutionary steps. Consequently, the same
conceptual mining technique outlined in that
seminal book "De Re Metallica", written by
Agricola in 1556, is still being used by today's
underground hard rock miner. Admittedly there
have been many modifications to the process, for
instance, the rock breaking agent has moved from
fire to black powder, to dynamite, and to the
cheap Ammonium Nitrate and Fuel-Oil mixtures of
today, but essentially the hard rock mining
process is still dominated by the sequential
routine of preparation, blasting and mucking.

Many would argue that, despite its ancient
conception, the drill-blast-haul routine is far
from its technological limit, and point to many
potential advances that can be made to equipment,
manpower practice and mine design. However, the
directions of these improvements tend toward
increased scale of underground operations and
bigger volumes of material moved. The
multi-activity cycle, the variable size of the
blasted product, and the ground control problems
remain. It is these features of the drill-blast
mining methods that constrain the application of
the new technologies of the microchip, which are
sprouting in the above surface factories, offices
and laboratories.

The rapid advances in these new technologies are
providing the possibilities of new tools and
techniques that can discriminate between the
mineral and the waste and automate the
underground mining processes. On surface, they
are impacting on the world's metal consumption
habits as it learns to do more with less. These
changes challenge the miner to think afresh and
to think laterally about his ancient problem of
winning metal in a profitable way.

THE TYRANNY OF THE DRILL-BLAST CYCLE

The drill-blast cycle is probably at its most
restrictive in the development stage of mining,
that is, in the establishment of access for men
and machinery to the orebody. The full
development drill and blast cycle involves the
following discrete steps: drilling, charging,
evacuation, blasting, ventilation, re-entering,
making safe, loading and transport, and the
provision of services. "It is possible to take
all of the elements and calculate a theoretical
time to complete a full cycle. However, as all
miners will tell the optimistic scheduling
engineer, each cycle is punctuated loudly and
distinctly by the blast which accompanies it. If
the round can only be fired at a particular time,
such as the end of the shift, then so the cycle
must be adapted to this". (Price, 1983). The
development problem highlights the overall
condition of underground mining, in that it

relies on the interaction and scheduling of many discrete activities which centre on the drill-blast cycle.

Methods of mining the actual orebody range from those akin to mine development, such as cut and fill, to the massive bulk mining from long hole open stopes. Cut and fill methods tend to be more selective, have a lower productivity and higher cost, and generally are used in higher grade orebodies with weak wall rocks. The optimum scheduling, in a cut and fill stope, of the drilling, blasting, mucking, filling functions and the provision of services and ground support again is problematic because of the large number of activities involved. The large scale open stope methods tend to be used in lower grade orebodies in stronger ground conditions. Although the cyclical constraint of drilling and blasting is less disruptive in the long hole methods, the blasting process still causes problems, in both the variability in the size of the blasted product and in the aggravation of ground control problems. The ore is diluted with waste rock, both from the side walls of the stope and with sub-economic material within the design confines of the stope. Although both rock oversize and side wall damage from blasting can theoretically be minimised by optimal design and practice, the end result is dependent on the skill and the care of the operator and the mine designer, and on the vagaries of the structural features of the ground.

The variability and discontinuity of the multiple activities associated with the drill-blast mining methods make underground mining a labour-intensive activity. Fig. 1 illustrates the relativity of mining cost and its components for Australia's major underground copper, lead-zinc mines. It can be seen that underground mining costs are over 50% of the total on site mining and milling costs and are approximately 25% of the total in situ value of the ore in this sample of mines. Manpower constitutes 50% of the total on site costs and is in fact an even higher percentage of the direct mining costs. Similar profiles are found in other developed countries such as Canada.

It seems that the mining process needs to become far more structured before it can move away from its traditional labour-intensive methods. The key to more structured and hence more automated systems seems to lie in the ability to continuously mine material in a systematic and predictable way, which is not provided by cyclical, multi-activity, and unpredictable drill and blast based mine methods.

WINNING METAL IS THE NAME OF THE GAME

Recovering the maximum value of the metal in the ground has long been the focus of mining. The actual selection of the appropriate mining method often revolves around the grade and shape of the orebody and the costs and the metal recoveries associated with the mining method. In planning the mining of a particular orebody, highly selective but expensive techniques can be considered if the recoverable metal content of the ore can support the costs; less selective, higher volume, less expensive techniques can be considered if the ore is of lower grade and of sufficient quantity.

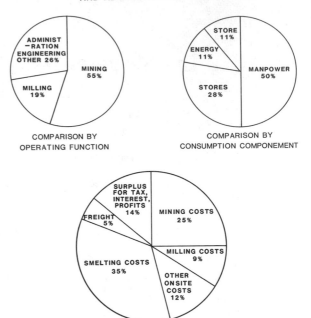

TOTAL ONSITE MINING, MILLING, ENGINEERING AND ADMINISTRATION COSTS

COMPARISON BY OPERATING FUNCTION

COMPARISON BY CONSUMPTION COMPONENET

DISTRIBUTION OF IN-SITE METAL VALUES AT 1983 PRICES

(BASED ON 7 LARGE AUSTRALIAN UNDERGROUND LEAD ZINC COPPER MINES IN 1983)

Fig. 1. Perspective on mining costs.

Regardless of the method used, some dilution of the ore with surrounding waste rock is accepted as a fact of mining life, despite the costs caused by mining and treating barren material. The reason for this is that nature generally distributes mineralisation in very irregular patterns, as illustrated in an ore grade contour plan from the CSA mine at Cobar, shown in Fig. 2, which highlights the small scale irregularity of ore grade distribution within an orebody.

Ore outlines by which bodies are represented in plan form are sometimes best considered as little more than envelopes drawn to average and rationalise often erratically mineralised areas into sensible manageable shapes.

On the other hand, the large tonnage underground mining methods require reasonably regular stope outlines. The long blast holes must be drilled straight, they cannot be bent to follow the ore outlines.
The actual long hole stope outline often includes both dilution from the internal non-mineralised material and the waste rock from the walls of the stope.

C.S.A. MINE

■   HIGH GRADE COPPER ORE (>2%Cu)
▨   MEDIUM GRADE COPPER ORE (1-2% Cu)
▦   LOW GRADE COPPER ORE (0.5%-1%)

SCALE
0   10   20   30   40
METRES

Fig. 2.   Typical mineral distribution at the
           CSA Mine at Cobar, NSW.

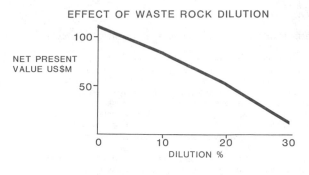

EFFECT OF WASTE ROCK DILUTION

NET PRESENT
VALUE US$M

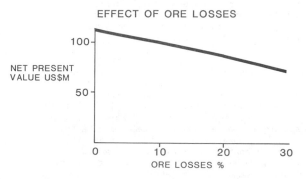

EFFECT OF ORE LOSSES

NET PRESENT
VALUE US$M

BASED ON THE FOLLOWING PROJECT CRITERIA:

ORE RESERVES UNDERGROUND DILUTED	12 MILLION mt
INVESTMENTS OVER THREE YEARS	$52 M
PRODUCTION PER YEAR	1 MILLION mt
CONCENTRATES PER TON UNDILUTED ORE	0.2 mt
PRODUCT PRICE F.O.B. SMELTER	$260/tconc
PRODUCTION COSTS F.O.B. SMELTER	$30/t
ANNUAL INFLATION	15%
INTEREST	20%

Fig. 3. Effect of dilution and ore loss on the
          valuation of a Swedish metal mine
          (White, 1984)

The bulk mining methods, however, can produce
large tonnages of ore at relatively low cost, and
it becomes a matter of weighing the overall
economics of metal recovery to determine whether
the tradeoff between mining cost, tonnage
productivity and ore grade dilution is a winning
formula for bulk mining methods. Obviously, the
grade or the value of the ore is a major
determinant, as is the relative mining costs of
the various techniques available. Recent
arguments (White, 1984) suggest, that because of
reductions in ore dilution, and increases in the
percent of ore extracted, cut and fill stoping
systems may produce the cheapest tonne of
contained metal in a far wider range of ore types
than previously thought. This is despite the
inherently high unit mining costs of the cut and
fill systems. Fig. 3 illustrates the effects of
varying degrees of waste rock dilution and ore
losses on the overall value of a typical Swedish
base metal project at 1 million tonnes per annum
of ore production capacity. These figures show
that 20% dilution can reduce the net present
value of this project by over a half. A 20% ore
loss, through poor mining discrimination, would
reduce the project's net present value by a
quarter, if all other costs remained static.

In fact, average dilution from surrounding waste
rock can rise as high as 30%, depending on rock
conditions, and the dilution in a particular
stope can be much higher; furthermore submarginal
or low grade inclusions within a mining block can
contribute to the lowering of the average mine
grade by up to 50%, depending on the variability
of the ore and the size of the stope. Therefore,
if a particular mine had a net metal value of ore
of $100/t, the reduction in value of a tonne of
ore by the inclusion of both waste and
submarginal material could be up to $40-$50 per
tonne.

Current mining technology and practice do not
help in this problem of ore dilution, for, as
one gets closer to the actual mining operation,
the immediate operating measures of performance
tend to be more in terms of tonnes of ore, rather
than tonnes of metal. Part of the reason for the
subtle but significant difference is that, having
drilled and blasted an ore block, the
discrimination in the stope between high grade,
low grade or waste diluent is difficult if not
impossible in the actual underground working
environment. Besides, the operator's concern
generally becomes one of moving the tonnes of ore
as cheaply and effectively as possible. At the
operating face, the definition of ore and waste
becomes dependent on what hole it comes out of or
goes into; current technology just does not
provide the miner with the easy and immediate
feedback necessary to measure performance other

AMR-B

than by tonnes moved.

THE PROMISE OF CONTINUOUS MINING IN HARD ROCK

Continuous mining methods are well established
and proven in many underground soft rock mines.
The longwall coal mining operation is a good
example of the automation achievable in a
continuous mine system when the cutting and
materials handling technologies become
available. It might be claimed that it was
continuous longwall mining that saved the
underground coal mines from the onslaught of
increased low cost open pit coal in tight world
markets. Underground hard rock mines currently
face similar challenges to their costs and
markets and it appears that continuous and
automated hard rock mining systems hold the
promise of future viability.

A continuous mining system could yield a broken
ore product that was smaller and more consistent
in size than the blasted product and therefore
more amenable to automated materials handling.
Pneumatic or hydraulic means of materials
handling emerge as possible technologies for
moving the material broken by a continuous
miner. A continuous mining system would
eliminate damage to walls and roofs caused by
blasting and provide a better and safer work
environment. With the elimination of the damage
caused by blasting, it may not be necessary for
the rock to be supported. If the rock still
requires support, the potential exists for this
to be automatically placed by the continuous
mining system. When coupled with automated ore
discrimination and ore sorting facilities, the
continuous miner would have the potential to
reduce costs caused by the handling and treatment
of diluent and marginal grade materials.

A greater recovery of the total ore resource
could be expected with a more positive
identification of ore in the face or immediately
after the ore has been broken. Such early stage
identification would permit the separation of the
ore from the waste in much smaller size lots than
is at present possible. An automated continuous
mining system holds the promise of a significant
reduction in manpower costs.

The cost of breaking a tonne of ore by this
system in a stope may be significantly higher
than by the high volume tonnage, efficient, long
hole methods currently used. But compared on the
basis of the total cost per tonne of metal
produced as distinct from the cost per tonne of
ore produced, the continuous mining system, with
its potential to utilise technologies that reduce
dilution, enhance ore-waste discrimination and
allow automation, has the promise of lower
production cost in some if not many future
orebodies.

ELEMENTS OF THE NEW MINING SYSTEM

The Continuous Miner

Although continuous miners are in place at many
soft rock mines, the cutter costs and the
associated physical problems associated seem to
limit their current use in hard rock mining. The
Robbins mobile miner is currently doing battle
with rock with an average compressive strength of
about 180 MPa, and perhaps represents the latest
in mechanical hard rock cutting design.

This type of machine is designed to break hard
rock but its size and the potential cutter costs
suggest that the economically viable mechanical
hard rock continuous miner may require some
assistance in either breaking and/or
preconditioning the face to be broken. The
problem of size is due partly to the squares and
cubes rule of mechanical design, as pointed out
by R.J. Robbins. This leads to the size and cost
of a machine increasing more rapidly than its
output payload, such that at some point the
increase in production capacity is more than
offset by an increase in capital and maintenance
costs. (Robbins, 1984).

One possibility is the assistance of the
mechanical cutting action with high pressure
water jets. Considerable experimentation has
been carried out over the last 10 years using
moderate pressure water jets (less than 60MPa) to
assist mechanical tools, notably drag bits. This
research has shown that the efficiency of cutting
medium to soft rock could be doubled with the
assistance of water jets. (Hood, 1984; Ropchan
and colleagues, 1980). The benefits for waterjet
assisted cutting includes a reduction of forces
acting on the tool, in particular the normal
forces, a reduction of toolwear and created
dust. The actual mechanisms, however, that yield
these benefits is still not well understood.

The main operating variables that need to be
considered in the design of waterjet assisted
cutting are the water pressure, the nossle
diameter producing the jet, the speed of traverse
of the cutting system, the stand-off distance of
the waterjet and the waterjet-cutting tool
configuration. There is a high degree of
interaction among these variables and optimum
performance depends on achieving the right
combination of these variables for the particular
rock type being cut. It has been suggested
(Hood, 1984) that the most critical factor is the
location of the waterjet in relation to the
cutting tool. This is based on the premise that
the dominant mechanism by which moderate pressure
waterjet assists the rock cutting process is by
effective rock chip clearance from the region
ahead of the cutting tool. Hood suggests that at
these pressures, the waterjet should be
positioned within 1 or 2 mm of the leading face
of the tool. Not only does this allow the
clearance of rock chips but also aids greater
interaction of the waterjet and the tool in rock
crack propagation ahead of the cutting tool. If
the waterjet is directed too far ahead of the
leading tool face, it strikes one of the plate
like particles that form ahead of the tool and
dissipates its jet energy. On the other hand,
others (Ropchan and colleagues, 1980) have found
that best results for waterjet assisted cutting
at moderate water pressures are achieved with the
jet positioned behind the bit. Fig.4 illustrates
these alternative configurations.

Most experimentation with waterjet assisted rock
cutting has been confined to moderate water
pressures because of the equipment cost and
reliability reasons. Recent developments,
however, by several manufacturers, particularly
Flow Industries of Seattle Washington, have
provided equipment capable of reliably supplying
waterjets with pressures up to 340 MPa.
Other hardware necessary for the reticulation of
high presure waterjets, such as timing and swivel
valves and flexible hosing capable of carrying

such high pressure water is also being rapidly
developed.

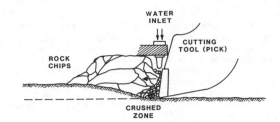

(i) WATER JET IN FRONT OF CUTTING PICK

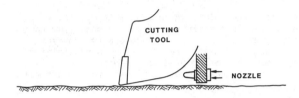

(ii) WATER JET BEHIND THE CUTTING PICK

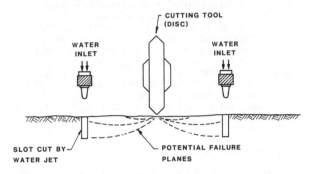

(iii) WATER JET WITH CUTTING DISC

Fig. 4. Some waterjet-cutting-tool
         configurations.

The effect of high pressure waterjets on rock is
significantly different from that of moderate
pressure waterjets. Fig.5 illustrates this as
the rock cutting effect from waterjets alone
varies from occasional spalling of the rock at
pressures approximately 0.7 of the compressive
strength of the rock to deep, clean cuts at the
high pressures of 340 mpa. The range of these
cutting effects depends on the rock strength, or
rock cuttability. Compressive strength of the
rock is a reasonable but not a comprehensive
index of cuttability. Rock porosity and cleavage
also affect the cutting performance of waterjets.
(Roxborough, 1984).

The speed of traverse of the cutting system is
another critical variable that has not been fully
explored to date. Most experimentation has been
limited to cutting speeds of 250 mm/sec.
Currently operating roadheaders and tunnel boring
machines, however, have cutting tools moving well
in excess of 1000 mm/sec.

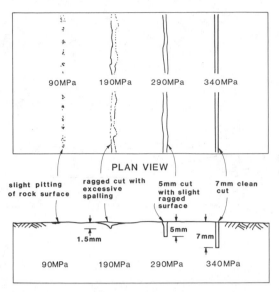

The above illustration is based on a sample of silicified
shale cut at 250mm/sec with a 0.23mm waterjet.The
sample's compressive strength was approximately 150MPa

Fig. 5. Impact of increasing water pressure
         in rock cutting.

The experimental data for cutting speeds up to
250 mm/sec show a diminishing of the cutting
performance as the speed increases as shown in
Fig.6. The extrapolation of this decrease in
performance for cutting speeds in excess of 1000
mm/sec is problematic and there is a need to
fully study the effects of waterjet cutting at
these higher speeds to ensure adequate design of
waterjet installation on production model rock
cutting machines.

The most effective configuration of waterjets and
cutting tools at higher speeds maybe completely
different from what was found at the common
experimental speeds and moderate pressures.
Possibly one successful configuration with high
pressure jets may involve the cutting of a slot
in the rock by the waterjet some distance from
the cutting tool which would effectively provide
a free face for the tensile stresses developed in
the rock by the normal forces of the cutting
tool. This configuration is illustrated in Fig.4
(iii). The lateral distance, depth of cut and
hence optimum waterpressure in this configuration
is very dependent on the rock being cut. This
suggests that considerable research work is
required for a variety of rocks to develop
optimum waterjet assisted rock cutting systems.

At this stage of the research it appears that the
area looks promising. High pressure waterjets
are offering the potential to apply high energy
inputs to the rock face without the size
constraints of purely mechanical cutting devices.

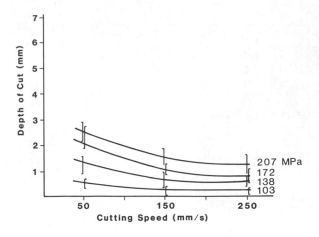

Fig. 6. Relationship between cutting speed and
depth. (Roxborough, 1984).

The technology of waterjets cutting itself is
developing rapidly. It has been shown that
cavitation and pulsing improve the cutting effect
of the waterjet. (Conn, 1984; Nekeber, 1984).
Similarly chemical additives can also yield
cutting improvements by reducing dispersion in
the waterjet. These developments may provide
further scope in the matching of cutting speed,
water pressure and waterjet-cutting tool
configuration over a greater range of rock types.

Ore Discrimination and Sorting

Technology is also advancing rapidly in this
area. The reliability and the miniaturisation of
microcircuits have given the industry the ability
to perform very complex functions, such as ore
discrimination and sorting, which only a few
years back required complex, purpose-
designed computer hardware.

The range of sensors available has expanded
considerably, allowing ore waste discrimination
to be carried out using a wide selection of
detection mechanisms such as electromagnetic
phenomena, natural or applied gamma sensing,
X-ray absorption and fluorescence, and neutron
activation fluorescence. (Schapper, 1977).

On-stream analysers are operative in milling
plants and neutron scanners, and down the hole
loggers are used to analyse solid rock masses for
some metals. The development of this technology
could conceivably see devices capable of scanning
a rock face and then automatically controlling
the cutting head of the continuous miner, on the
basis of the computed mineral content of the
cut.
Alternatively, ore-waste recognition could be
performed on the conveyor just beyond the
continuous miner. Sorting technology, such as
the RTZ Ore Sorters Model 18 Radiometric Sorter,
can discriminate parcels of ore as small as 1 kg

on a V-loaded conveyor. The Outokumpu's "Precon"
sorter, based on gamma absorption, currently can
sort a range of base metal ores up to about 60
tonnes per hour per sorting line.

Automatic switching of the material flow could be
controlled on the basis of the computed ore and
waste discrimination. The combination of a
continuous miner and an automatic ore
discriminator could dramatically reduce the
minimum mining block to tens of kilograms,
whereas reasonably selective drill and blast
techniques have minimum mining parcels of 100 to
200 tonnes.

Automated Materials Handling, Roof Support Systems

Existing technology in this area is currently
available in a variety of applications, from the
slurry transport of pelletised ore to conveyor
handling of crushed ore. Continuous coal mining
systems could be adapted to the hard rock
environment once a continuous hard rock miner was
in place.

The uniform and small size of the broken rock
would allow consideration of direct transport by
pipeline slurry from the mining face to the
grinding mills of the surface concentrating
plant. This would eliminate the need for the
underground crushing plant, the expensive haulage
shaft, and the considerable manpower required to
move the ore from the stope to the surface.
Waste rock mined either in development or in the
stope could be immediately directed from the
continuous miner to a stope needing fill or to
the back of the stope being mined.

SOME PROBLEMS TO BE FACED

The development of the integrated technologies of
continuous mining, ore sorting and materials
handling will not be easy. The mining
environment is difficult and changing according
to the vagaries of rock type, orebody
configuration and ground conditions. Much
development of the technology will have to be in
underground conditions to ensure the highest
reliability of the new mining system.

The development work will be expensive and there
may be some reluctance to invest in new
technology in what many consider to be a mature
if not declining industry. This may be an
accurate view of the industry if it persists with
current mining technology. However, the
introduction of continuous and automated mining
possibilities that allow considerable upgrading
of the insitu resource as well as the elimination
of some processing steps holds the promise of
reducing the cost of metal produced.
Furthermore, it will make mining costs far more
controllable. This in turn could ensure that
mining's metal products maintain if not increase
their market share against manufactured
substitutes.
Despite some reluctance from the traditional
mining industry, there is considerable support
emerging for fundamental research in rapid
excavation from newer areas such as military oil
shale development and the development of
underground nuclear waste repositories.
Developments in these areas could well flow into
the mining industry in time.

As with the introduction of all new technology, continuous and automated mining systems will require the retraining and redeployment of human skills.

Because of the nature of established mining practices in existing mines, it is envisaged that new automated systems would be mainly used in the development of new orebodies, including some previously considered uneconomic when using traditional mining methods. It is envisaged that the new mining machines would be essentially tools which still require the ingenuity and skill of the miner and his engineer for them to be used successfully in the unsympathetic and changing environment of the underground hard rock mine.

REFERENCES

Conn, A.F., Johnson, V.E., Lindenmoth, W.T., Chahine, G.L., (1984). Some unusual applications for cavitating water jets. 7th International Symposium On Jet Cutting Technology, Ottawa June 1984.

Hood, M. (1984). Waterjet assisted rock cutting systems, the present state of the art. University of California, Berkeley.

Knickmeyer, W., Bowman, L. (1983). High pressure water jet assisted tunnelling technique. Bergbau-Forschung GmbH, Federal Republic of Germany.

Mathews, K.E. (1977). Opportunities for technological innovation in development mining. Unpublished paper to Board on Mineral Resources for the National Research Council, Washington, D.C.

McMahon, D.W. (1982). The Cobar stoping system present and future. CRA internal paper.

Nerbeker, E.B. (1984). Potential and problems of rapidly pulsing waterjets. 7th International Symposium on Jet Cutting Technology, Ottawa June 1984.

Price, I.L. (1983). The Problem with Development. AMIRA Technical Meeting.

Robbins, R.J. (1984). The future of mechanical excavation underground mining. Mining Engineering, June 1984.

Ropchan, D., Wang, F.D., Wolgomott J. (1980). Application of water jet assisted drag bit and pick cutter for cutting of coal measure rocks. Colorado School of Mines.

Roxborough, F.F., Hagan P. (1984). Unpublished Papers. School of Mining Engineering, University of New South Wales.

Schapper, M.A. (1977). Beneficiation of large particle size using photometric sorting techniques. Australian Mining, April 1977 Reprint.

White, L. (1984). Boliden cut and fill : Not just a means of mining weak ground. EMJ, August 1984. pp. 30-36.

Automation, Mining and the Environment: Where are we Heading?

TOM P. FARRELL

Technical and Commercial Services CRA Limited, Melbourne

Abstract. The role of automation in the environmental management of the various
stages of mining projects is discussed. Remote sensing can be useful in the
exploration phase, but tends not to have sufficient resolution for individual
operations. Automatic samplers, whether for meteorology, air or water, are widely
used, but can have operational problems. The environmental scientist, the electronics
engineer and the computer programmer must combine in the future to develop products
which are applicable, robust and simple to use.

Keywords. Computer software; data acquisition; data reduction and analysis;
environment control; mining.

INTRODUCTION

Less than twenty years ago, mineral exploration,
and the subsequent development and processing of
the discovered ore, were the almost exclusive
realm of the geologist, mining engineer and
metallurgist. Other professionals, from
accountants to engineers, provided support for
those whose task was seen to be the discovery
and ultimate exploitation of a mineral
resource. Indeed, some people may even have
regarded their role as dominant over all
others. Subsequently, however, we have seen the
upsurge of other professions which vie with
those "classical" professions for important
roles in the industry. Among these are two that
are particularly relevant to this Symposium –
the electronics engineer and the environmental
specialist.

It is the aim of this paper to examine the often
complementary roles of these specialists in the
different phases of a mining operation, and to
examine deficient or problem areas where their
specialities may be used in the future.

THE EXPLORATION PHASE

Not so long ago, people searching for minerals
picked up clues simply by walking over the
ground and examining it visually. Rocks stained
a particular colour, sudden changes in
vegetation, material thrown up by burrowing
animals – all these could provide evidence of a
mineral deposit. Nowadays, the exploration work
often starts with images of the ground obtained
by an orbiting satellite such as Landsat. With
computer techniques, a researcher can analyse
the image to look for changes in structures or
base geology which may indicate the present of a
deposit.

Similarly, in the past, scientists studying the
natural environment relied on covering the
ground on foot or by vehicle, recording the
presence or absence of plants and animals and
predicting the inter-relationships both within
them and between them and the external
environment. As the geologists were hampered by
the vastness of the Australian continent, so too
were the natural scientists. It was only in the
early 1950's, for example, that the Northern
Australian landscape was extensively studied,
but again the problem of scale meant that very
general techniques had to be developed
(Christian and Stewart, 1953). Even today,
there are vast areas of the continent that have
had barely a glance by natural scientists.

The result of this is, of course, that when
confronted by the need to report on the natural
environment of a prospective minesite, the
environmental scientist is frequently hampered
by the lack of information available about the
area.

In an attempt to overcome this problem, some
have turned to satellite imagery to give at
least a small scale data set on which to base
larger scale interpretation. One of the most
successful examples of such studies has been the
seven volume publication "The environments of
South Australia" (Laut and others, 1977), which
contains the environmental information that was
prepared for a study entitled "A feasibility
study for an ecological survey of Australia".
Because of the enormous area involved, extensive
use was made of Landsat imagery for base
mapping. Nevertheless, this is still no
substitute for detailed on-the-ground
inspection; where it is useful is in enabling
an initial assessment of the relative importance
of a particular landscape or, less conclusively,
vegetation type.

Despite the large advances over recent years in
the resolution of satellite imagery, and
subsequent computer enhancement and numerical
analysis, ground truthing is still necessary.
This important fact is unfortunately sometimes
forgotten in the rush to produce elegantly
presented publications, or reports for

clamouring regulatory authorities. Two major problem areas immediately spring to mind; firstly, it can easily be forgotten that satellite imagery is comprised of electronic signals and is not photography in the commonly accepted form. Colours are assigned during processing on the basis of the energy reflected from the earth's surface, and the spectral signature of vegetation can vary greatly depending on the season, water availability, or whether it has recently been subjected to stress, for example burning. Thus differences observed on the image may be artefacts and may not represent actual differences on the ground. The second problem area is the scale of such imagery, and consequently the poor resolution obtained. While the most recent Landsat satellite has a resolution of around 70 metres, this still means that small scale features of interest to the environmental scientist cannot be discerned. Although new satellites are being prepared with much better resolution, in the order of 10 metres, satellite imagery will probably still remain best suited for broad scale repetitive inventory, as described by Foran and Pickup (1984).

THE FEASIBILITY STUDY PHASE

As the exploration phase passes into the feasibility study phase, environmental studies become more and more site specific, and begin to concentrate on the actual and perceived impacts of a potential mining or mineral processing operation. Two important factors need to be considered at this stage: what data need to be collected and how best to collect them; and how best to organize the data so obtained in order to minimize impacts while, at the same time, monitoring these impacts. It is also at this stage that an Environment Impact Study is produced.

In the early stages of feasibility studies, especially for a deposit in a remote location, personnel are not on-site full time, and therefore data collection needs to be organized on a campaign basis or be automated in some way. For basic physical information, e.g. meteorology, this most likely mean the installation of an automatic data aquisition system. For some time small "package" weather stations, which record a variety of meteorological information on paper charts, have been available in Australia. Such systems have proved in the past to be extremely robust and reliable, with the major problems appearing to lie with the timing mechanism or with pens that become blocked or dry out. The latter problem has been overcome to some extent by the introduction of felt-tip pen cartridges. In some high humidity tropical areas there have been problems with charts tearing after becoming saturated with moisture. In addition, chart changing has been a skill which many technicians have lacked.

The greatest difficulty with these systems has been the processing of the often voluminous charts obtained from them. Although chart digitising services are becoming more and more available, these are expensive and time consuming. Thus, the "second generation" of recording devices – electronic data loggers – has appeared. These, powered by batteries and backed up by solar cells, frequently scan a variety of recording instruments and write raw or processed information onto magnetic tape or, more recently, into electronic data storage

modules. Although potentially more reliable than the mechanical systems, many of those currently in use in Australia have been plagued with problems (see for example John, 1982). The most common fault appears to be the sensitivity of electronic systems to environmental extremes. This unreliability of such equipment, coupled with the comparative state-of-the-art nature of it (e.g. most serial numbers are below 100), has led to its use being abandoned at some remote sites. Nevertheless, if these problems can be overcome, such equipment would appear to be the simplest to install and use. Already, a third generation of these automatic weather stations is becoming available – those that transmit signals in real time through geostationary satellites to ground stations, where they are recorded in computer storage for subsequent processing (Allison and Morrissy, 1983).

Once data have been recorded, it should be a comparatively simple matter to read it off the recording medium into more accessible storage in a micro- or mainframe computer. However, in my experience, this is complicated by the absence of a strict industry standard in communication devices, and the appalling unreadability of most documentation. The concept of "user friendly" has certainly not progressed to this level.

Although strictly outside the scope of this paper, the subsequent storage and accession of any such data must be important in environmental programs. Until recently, the operation and use of data-base management systems has been outside the scope of many scientists' training. It is no use having a user-friendly system if the potential user has no idea of how to make friends with the system. It is only with adequate instruction that the environmental scientist can collect his data in such a way as to make maximum use of such systems. As a result of frustration with existing systems, some environmental scientists have developed specialised information retrieval systems to handle the data being obtained (Allen, 1983).

The complexity of environmental data types has for many years prevented their full use in land use planning and resource management. Perhaps the first successful attempt at this on a regional basis was the approach of McHarg (1971), which used successive overlays of maps detailing environmental features to highlight areas suitable for different types of development.

This approach has been further refined with the advent of rapid computer processing, and the ability now to handle data based on a variety of points, lines, polygons or rectangles (see, for example, Gates, McCown and Butler, 1977; Kessell, Good and Potter, 1982). The integration of these systems with sophisticated graphics packages can provide the environmental planner or manager with an extremely powerful tool. However, until now, the relative high cost of these systems has precluded their use in all but the largest environmental programs, even though their benefits are obvious to the practising environmental scientist.

THE OPERATING PHASE

Once a mining or mineral processing operation gets underway, a major function of the environmental scientist is to ensure compliance with legislative and licence conditions. In order to achieve this, extensive and sometimes quite sophisticated monitoring systems need to be established. For example, automatic water samplers, working on either a time- or event-based procedure, are commonly found around sites which have a potential to contaminate water resources. However the high apparent cost of such systems often preclude their use in many situations.

A further step in automation has been achieved by the introduction of computer controlled real time analytical systems, such as that installed for sulphur dioxide control at Mount Isa, Queensland (Jones, 1980). In this system a network of sulphur dioxide analysers throughout the city continuously analyses the sulphur dioxide concentration in the air. These analysers are regularly interrogated by a central computer, which processes and summarizes the data, presenting it to an operator in such a way as to provide him with a constant picture of sulphur dioxide movement throughout the city. Coupled with real time meteorological and emission data, this enables the operator to control operation of the smelters so that statutory limits are not exceeded. The system also produces summaries for management and government reports.

The introduction of similar systems at other smelting complexes for air pollutants other than sulphur dioxide has been hampered to date by the lack of reliable real time analysers for the substances in question. For example, lead-in-air has become a matter for some concern both inside and around lead smelters, but monitoring systems such as that developed at Mount Isa for sulphur dioxide have not been installed. Only recently has a rapid analyser for lead-in-air been developed, but there is still a considerable delay in reporting relatively low levels of lead (but levels which are above the statutory limits) so that the analysers are of limited use.

Similarly, fugitive or non-point source air emissions are now regarded as a major problem in smelters, as most other emission sources can be readily identified and controlled. It is still difficult, if not impossible, to easily identify the source of fugitive emissions, although there have been attempts in recent years to use newly developed analytical techniques to do this. Bradley and others (1981) used a combination of scanning electron microscopy and X-ray energy-dispersive and wavelength-dispersive spectrometry to investigate the chemical compositions of airborne particles sampled around a copper smelter. Four types of particles were identified, which could be uniquely associated with different stages of the smelting process, thus allowing calculation of the contribution of each stage of the process to any particulate sample.

Harrison and others (1981) examined particle size and chemical composition (by x-ray powder diffraction spectrometry) of workplace dust samples within a pyrometallurgical zinc-lead smelter. These factors were found to be correlated with the smelting operations occurring close to each sampling site and in at least one case a chemical phase specific to a single operation was identified. This was an elemental lead aerosol at the bullion floor of

an Imperial Smelting Furnace plant.

However, neither of these techniques are yet able to be used to easily identify the relative contribution of a number of fugitive emission sources to a particulate sample collected around a metallurgical plant.

In some open pit mines, particularly those located close to communities, control of blasting noise and vibration is of major concern. At Alcoa of Australia's bauxite mining operations in the Darling Range south east of Perth, an extensive blast noise monitoring program has been established (Delaney, 1982). The development of a computerised blast noise prediction model and the establishment of a radiosonde weather balloon facility have reduced blast induced complaints by a factor of ten.

Similar systems are under development at other minesites, but again the lack of robust equipment suitable for extreme Australian conditions limits their efficient utilization.

Recent investigations into airborne dust generated by intensive open cut coal mining in the Hunter Valley of New South Wales have been hampered by the complexity of the meteorology and topography, and the difficulty in defining the dust load from any one mine. The State Pollution Control Commission, the principal regulatory authority, has resorted to a computer model in order to predict the dust generated by new mines in the Valley (Ferrari and Ross, 1983). Although numerical models may appear to be very useful to the controlling authorities, care must be taken that their application is backed up by vigorous testing before being used in a predictive sense. In addition, although seemingly cost efficient through the reduction in labour and analytical costs, models such as this have frequently proved to be prohibitively expensive in terms of computer processing time. Ideally, what we need are simple cost effective methods of monitoring and predicting air pollution which are capable of being widely applied, to both other sites and other mine types. Unfortunately, such tools have still to be developed.

In the design of a waste disposal facility, for example a tailing dam, waste products that are toxic, highly reactive or otherwise represent a significant threat to public health, safety and welfare must be isolated from the environment. Complete isolation of these wastes from groundwater resources is not always practicable, particularly when they are disposed of as liquids or in slurry form. The extent of groundwater contamination can be determined by establishing monitoring wells around the disposal facility, and sampling and analysing the groundwater (Humenick, Turk and Colchin, 1980). Mathematical models can then be used to predict the transport of contaminants, and the effectiveness of various control measures. These models can also be used to predict the total length of time that contaminated leachate or seepage will be generated after a disposal facility has been completed, or after rehabilitation has taken place. However, the comments above regarding air pollution models also apply in this situation.

THE SHUTDOWN OR REHABILITATION PHASE

In the rehabilitation phase of mines, the surface drainage pattern after mining is coming increasingly under scrutiny. In the United States, the Surface Mining and Reclamation Act of 1977 requires that mined sites be regraded back to the original contour and that the hydrologic integrity of surface streams is preserved. This is generally taken to mean that stream networks be returned to as close to a pre-mining configuration as practicable.

Post-mining topography can be modelled by using Geographic Information Systems, resulting in a topography that will achieve the design for stream gradients and the concept of original contour (Thames and Rasmussen, 1982). Appropriate stream gradients are computed using runoff-stream flow models.

Although requirements in Australia are not as strict as those in the United States, advanced modelling techniques such as these are increasingly being used in mine and waste emplacement design.

As the resources available for a rehabilitation program decline after shutdown of a mining operation, remote sensing and automation can be used to maintain an acceptable monitoring program (e.g. Anderson and Tanner, 1978). All of the techniques discussed above can play an important role in ensuring the effectiveness of rehabilitation while at the same time keeping costs to a minimum. However, the faults and problems outlined also add to the uncertainty as to the quality of the information gained.

CONCLUSIONS

Poor quality control is probably the greatest obstacle to wider use of automation in environmental control of mining and mineral processing. Second to this is the lack of simple yet elegant tools which can be used by the environmental scientist employed in the mining industry. Until this situation is improved, scientists will be loath to drop their tedious but effective manual methods, in favour of the new technology. It must be the role of the electronics engineer, the computer programmer and other specialists to take note of the environnmental scientists' problems, if their products are to achieve the market penetration they desire.

REFERENCES

Allen, N.T. (1983). Worsley Alumina project fauna studies, principles and practice. In Papers of the North Australian Rehabilitation Workshop, Bowen, June 1983. pp. 78-101.

Allison, I. and Morrissy, J.V. (1983). Automatic weather stations in the Antarctic. Aust. Met. Mag., 31, 71-76.

Anderson, J.E. and Tanner, C.E. (1978). Remote Monitoring of Coal Strip Mine Rehabilitation. U.S. Environmental Protection Agency, Report No. EPA-600/7-78-149.

Bradley, J.P., Goodman, P., Chan, I.Y. and Buseck, P.R. (1981). Structure and evolution of fugitive particles from a copper smelter. Environ. Sci. Technol., 15, 1208-1212.

Christian, C.S. and Stewart, G.A. (1953). General report on survey of Katherine-Darwin region, 1946. Land Research Series No. 1. CSIRO, Melbourne.

Delaney, W. (1982). Meteorological forecasting of blast noise from the Del Park minesite. Proc. The Aus. I.M.M. Conference, Melbourne, Vic., August 1982. pp. 221-231.

Ferrari, L. M. and Ross, I.B. (1983). Monitoring and modelling air quality around mining, smelting and processing operations. In Papers of the Australian Mining Industry Council Environmental Workshop, Hunter Valley, September 1983. pp. 80-99.

Foran, B. and Pickup, G. (1984). Relationship of aircraft radiometric measurements to bare ground on semi-desert landscapes in central Australia. Aust. Rangel. J. 6, 59-68.

Gates, W.A., McCown, B.H. and Butler, K.S. (1977). A resource data management system, GRASP: Description and documentation of software. IES Report 90. Institute for Environmental Studies, University of Wisconsin-Madison.

Harrison, R.M., Williams, C.R. and O'Neill, I.K. (1981). Characterization of airborne heavy metals within a primary zinc-lead smelting works. Environ. Sci. Technol., 15, 1197-1204.

Humenick, M.J., Turk, L.J. and Colchin, M.P. (1980). Methodology for monitoring ground water at uranium solution mines. Ground Water, 18, 262-273.

John, C.D. (1982). The MRI Weather Wizard climatological system. In Papers of the Australian Mining Industry Council Environmental Workshop, Darwin, September 1982. pp. 163-167.

Jones, D.G. (1980). Meteorological aspects of air quality control at Mount Isa. In Papers of the Australian Mining Industry Council Environmental Workshop, Rockhampton, September 1980. pp. 313-336.

Kessell, S.R., Good, R.B. and Potter, M.W. (1982). Computer Modelling in Natural Area Management. Special Publication 9. Australian National Parks and Wildlife Service, Canberra.

Laut, P., Heyligers, P.C., Keig, G., Loffler, E., Margules, C. and Scott, R.M. (1977). Environments of South Australia Handbook. CSIRO, Canberra.

McHarg, I.L. (1971). Design with Nature. Doubleday & Co., New York.

Thames, J.L. and Rasmussen, W.O. (1982). Design of post mine topography to maintain hydrologic integrity of surface streams. Min. Congr. J., Jan. 1982, 46-50.

Abrasive Jet Cutting in Flammable Atmospheres — Potential Applications for Mining

R.A. ELVIN and R.M. FAIRHURST

British Hydromechanics Research Association, Cranfield, Bedford, UK

Abstract. High pressure abrasive water jets have been shown capable of cutting a wide range of materials including steel, stone, rock and concrete. Acceptable performance can be achieved using relatively low pressures and expendable abrasives. The cutting head can be made small and abrasive feed systems have been developed that allow reliable pumping of slurries over long distances to supply such cutting heads. These features make abrasive water jets ideally suited to automated control in remote areas.

Another major advantage of abrasive entraining water jet cutting heads is that they may be operated safely in potentially explosive environments due to their cold cutting ability. Therefore there are many possible applications for such equipment in mining.

This paper gives an introduction to the abrasive water jet process and includes results of trials carried out at BHRA on a range of hard materials. Work is also described which investigated the operational safety of abrasive water jets in explosive environments. In a special environmental chamber, cuts were made on steel and sandstone with atmospheres of methane/air and hydrogen/air, without causing ignition.

Keywords. Jet cutting; abrasives; flammable atmospheres; mining; environment control; concrete.

INTRODUCTION

The use of hydraulic cutters for mineral extraction and tunnelling is now an established technique worldwide. In many instances water jets are used in combination with mechanical cutters to optimise performance. Typical applications include coal cutting (Krämer, 1982) and roadheaders (Plumpton and Tomlin, 1982).

In other industries, equipment for cutting materials such as reinforced concrete and steel with abrasive laden water jets is currently operating. Such systems run at pressures typically in the range 600 to 1000 bar and utilise inexpensive and readily available abrasive materials. Trials at BHRA have demonstrated that abrasive water jets are effective on a wide range of hard materials encountered in winning minerals, including various rocks and ores.

This paper sets out to explain the principles and advantages of abrasive jet cutting and discusses potential applications of the technique for mineral extraction.

EQUIPMENT

A complete abrasive jet cutting system comprises a high pressure water pump and delivery hose together with an independent abrasive feed system connected to the cutting head containing the high pressure nozzle.

Several systems are currently available, many of which are tailored to suit particular applications. One such example is the emergency jet cutting rig developed for British Petroleum by BHRA (Fig. 1). The unit is specifically designed for operating in hazardous atmospheres and in particular, for gaining access in an emergency situation to confined areas on North Sea oil platforms. It is a self-contained system consisting of a high pressure pump and abrasive feed housed in a transportable skid-mounted steel container.

High pressure pumps are generally of the reciprocating plunger type, which were originally developed for plain water jet cleaning duties (boiler descaling, marine growth removal, etc). They are equally suitable for abrasive jet cutting applications. A typical pump delivers flow rates of approximately 50 to 70 1/min at maximum pressures of 700 to 1000 bar.

In order to maintain a steady flow of abrasive to the cutting head a feed system is necessary. In a typical system the abrasive is loaded manually into a supply hopper from which it is fed and mixed into a water supply at a suitably metered rate. It is than pumped in a slurry form at a relatively low pressure (2 to 10 bar) to the remotely located cutting head. Complete abrasive feed systems of the type described are now commercially available as transportable units.

It is also possible to dry feed abrasive to the cutting head. With this technique, the abrasive particles are conveyed pneumatically rather than hydraulically. Dry fed abrasive has been observed to produce sparks during cutting however.

The key element in an abrasive cutting system is the cutting head (Fig. 2). High pressure water is accelerated to high velocities through a simple nozzle, and thence through a mixing chamber and outlet tube where the abrasive slurry is entrained and accelerated by the water jet. The chamber and outlet tube are designed to ensure thorough mixing and even distribution of abrasive particles within the jet before being discharged from the head. This arrangement minimises pumping problems and avoids high rates of nozzle wear. Other components within the cutting head are inevitably exposed to the abrasive material and are therefore designed for ease of replacement. However, to counteract wear problems, resistant materials such as tungsten carbide are employed thereby ensuring prolonged reliable operation.

The choice of abrasive depends upon the target material. For cutting steel, copper slag has proved to be particularly effective. It is readily available, gives good quality cut surfaces and is sufficiently inexpensive to be considered expendable. It has also been used successfully for cutting concrete ~ both plain and steel reinforced.

For some rocks and brittle materials, sand is a more effective abrasive, and is also inexpensive. In many cases further improved cutting performance would be obtained from more expensive abrasives such as silicon carbide and aluminium oxide. The high cost of these materials would normally demand that they be recycled, making them more suitable for fixed installations.

The selection of abrasive for each application is therefore dependent on a number of factors including target material properties, type and location of the cutting system, and also cost and availability of suitable abrasive. In mining applications, one material that is generally available in large quantities is the stope material itself. In crushed and graded form this could possibly be utilised as the abrasive in the cutting process, and has the advantange of being virtually self-generating.

CUTTING TRIALS

Cutting trials with abrasive water jets have been conducted at BHRA on behalf of numerous industrial organisations and also as part of a general research programme in recent years.

In this time cutting has been performed on materials as diverse in nature as cast iron, sandstone and quartzite rock. However, a high proportion of the work has been concerned with cutting steel and concrete sections (both plain and reinforced). As an illustration of the results obtained, a reinforced concrete block 700 mm deep (Fig. 3) was cut through using a multi-pass technique, whilst 13 mm mild steel plate was similarly cut through in a single-pass at 200 mm/min. Fig. 4 demonstrates that good quality cut surfaces are obtained on metal sections.

In order to investigate the behaviour of abrasive water jets in explosive environments, cutting trials on steel and sandstone in flammable atmospheres were conducted jointly by BHRA and the Health and Safety Executive on behalf of the National Coal Board (UK). The tests were conducted in controlled atmospheres of hydrogen and air (17% to 20% hydrogen), and methane and air (10% to 12% methane) in a 1.7 m^3 chamber of 2 m diameter. A total of 79 cuts were carried out (equivalent to approximately 26 minutes of continuous cutting operation) in explosive mixtures of various compositions without ignition occurring.

The safety aspect of this cutting technique is attributed to the drenching action of the jet which restricts the generation of sparks (particularly if a slurry fed abrasive is employed), creates a damp working atmosphere and suppresses dust. Furthermore, the cut surface is continuously cooled by the water.

DISCUSSION

From the above text, is is apparent that abrasive jet cutting equipment and techniques, which were originally developed for use in other fields of technology, could be successfully adapted for operation in the mining industry. Potential applications might include the direct extraction of ore and mineral deposits or alternatively, employment as a cutting tool for emergency rescue.

Abrasive water jets could also be used in combination with mechanical cutters or explosives in a similar manner to conventional hydraulic cutters. However, the cutting capability of an abrasive water jet greatly exceeds that of a plain water jet. Therefore existing equipment for combined cutting operation is unlikely to be suitable for this purpose.

Abrasive jet cutting is particularly advantageous in environments where risk of explosion exists, such as the flammable or dust-laden atmospheres frequently encountered in mines. The actual cutting head is dimensionally small and can be used in confined or otherwise awkward areas where access for conventional equipment is not possible. In addition the reaction force from the head is comparitively low. It is also constant and independent of both target material and distance from target. The cutting head is thus highly suited to control by automatic means.

A further advantage is that the power unit and abrasive feed system can be located remotely from the cutting head. The previously mentioned BP emergency jet cutting unit is equipped with 150 m high pressure water and abrasive slurry hoses, enabling cutting to be carried out at a considerable distance from the rig. Thus, if a similar arrangement for a mining operation were adopted, supply equipment could be kept well clear of the hostile working environment at the face.

In some circumstances, it may be feasible to utilise finer particles of the stope material as the actual abrasive in the cutting process. This could only be applicable to mining operations in which some pre-treatment, such as crushing, of the mineral is required prior to transportation. It is particularly attractive, however, since the cost of supplying conventional abrasive and its subsequent separation from the mineral can be avoided

CONCLUSIONS

The technique of cutting with abrasive water jets could be effectively employed in mineral extraction. It has two main advantages. Firstly, it can be used in atmospheres where risk of explosion through spark ignition exists. Secondly, it is ideally suited to control by remote or automatic means, since the actual cutting head is physically small, manoeuvreable, and produces a low and constant reaction force. Thus it could be used in areas where access for conventional cutting equipment is likely to be difficult, or alternatively, in conjunction with mechanical cutters in order to improve upon the performance of existing combined hydraulic/mechanical cutting equipment.

There is a definite need for the performance of abrasive water jets to be assessed in an authentic mining environment before any further conclusions on the subject can be drawn.

REFERENCES

Barton, R.E.P. and D.H.Saunders. (1981). Jet
  cutting of reinforced concrete. RR 1679, BHRA
  Fluid Engineering, Cranfield, UK

Krämer, T. (1982). Jet miner surface trials run
  at Bergbau-Forschung, Germany. Proc. 6th Int.
  Symp. on Jet Cutting Technology (Guildford,
  U.K. 6-8 April 1982), BHRA Fluid Engineering,
  Cranfield, U.K.

Plumpton, N.A. and M.G. Tomlin. (1982). The
  development of a water jet system to improve

the performance of a boom-type roadheader.
  Proc. 6th Int. Symp. on Jet Cutting Technology
  (Guildford, U.K. 6-8 April 1982), BHRA Fluid
  Engineering, Cranfield, U.K.

Saunders, D.H., N.J. Griffiths and K. Moodie.
  (1980). Water abrasive cutting in flammable
  atmospheres. RR 1608, BHRA Fluid Engineering,
  Cranfield, U.K.

Saunders, D.H. (1982). A safe method of cutting
  steel and rock. Proc. 6th Int. Symp. on Jet
  Cutting Technology, (Guildford, U.K. 6-8 April
  1982), BHRA Fluid Engineering, Cranfield, U.K.

Fig. 1.   General view of abrasive jet
           cutting rig built for
           British Petroleum

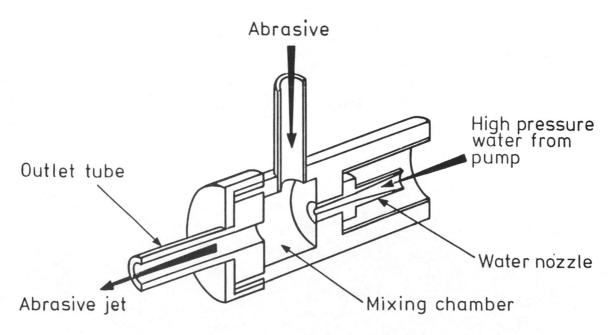

Fig. 2.   Schematic view of cutting head

Fig. 3.   Reinforced concrete cut by
          abrasive water jet

Fig. 4.   Mild steel plate, steel I-beam
          and cast iron cut by abrasive
          water jet

Robots and Mining: The Present State of Thought and Some Likely Developments

M. KASSLER

Michael Kassler and Associates Pty Limited, Suite 2, 2 West Crescent Street, McMahons Point NSW 2060, Australia

Abstract. Two recently published studies have considered whether and how robots, which have been designed for manufacturing applications, could find use in mining. The conclusions of both studies, if not identical, are compatible. Automation of mining activities will continue. Robots, at least in the short term, will have little application in mining. However, robotic concepts, both hardware and software, are of considerable relevance to mining automation, and can be anticipated to diffuse gradually into mine environments. The present paper is intended to convey the current state of thought on this subject to IFAC Symposium participants who may be unfamiliar with the referenced studies, and is not intended to report previously unpublished research.

Keywords. Robots; mining; control engineering computer applications; pattern recognition; telecontrol; man-machine systems.

INTRODUCTION

A robot may be defined as a device that possesses some form of mobility, is capable of being programmed to carry out a wide variety of tasks, and operates automatically after it has been programmed. Since a general-purpose computer satisfies the second and third of these characteristics, it is fair (although the definition displeases some mechanical engineers) to regard a robot as a computer enhanced by some form of mobility.

We know how computers--which are limited to processing data or information--have penetrated virtually every field of human activity in the developed world, generally with beneficial results. Robots, which not only can process data or information but can carry out actual physical work as well, by analogy could be expected to find application in virtually all industries. Although the number of installed robots is comparatively small--last year perhaps there were 40,000 in the Western world, including 16,500 in Japan--the growth rate is high: in Australia the growth rate has been about 43% per year (Kassler 1984b).

Although the initial applications of robots have been in manufacturing industry, accomplishing tasks such as welding of automobiles, spray painting of appliances, and transferring objects from one machine or conveyor to another, robot applications in primary industry (e.g., to shear sheep) and in the service sector (e.g., to clean buildings) are currently the subject of research. It is reasonable, therefore, that some consideration has been given to the possible applications of robots to mining.

The subject has been considered by a number of people. We have knowledge, however, of only two substantial studies of the subject that have found their way into print. I am responsible for one of these studies, which was sponsored by the Australian Science and Technology Council, who in 1982 asked for a report on the implications of robot technology for the Australian mining industry in the 1980's. Our report to ASTEC contained information of a confidential nature, but an article conveying the essential disclosable findings of our study has recently been published (Kassler, 1984a). A separate investigation of potential applications of robots to mining has been carried out by the Mining Research and Development Establishment of the U.K. National Coal Board and reported by Tregelles (1982). Since our background and experience is in the area of computers and robotics, and Tregelles's background and experience is in research and development of new technology for coal mining, the two reports jointly constitute we think a good overview of the current state of knowledge, at least as of a few years ago.

THE STUDY FOR ASTEC

One recurrent problem besetting those who work in robotics is that journalists and others frequently use the word 'robot' to refer to a device that by our definition should not be so called. A reasonably thorough literature survey uncovered a number of reports that purported to be about robots and mining. Upon analysis, all of these reports turned out to be either speculative about some future circumstances where robots might be used for mining or descriptive of technology that was but should not have been described as a robot.

A device controlled by a human operator, no matter how remotely situated he or she may be, is not a robot because the device does not operate automatically after it has been programmed. Although an unmanned submersible in principle might be a robot, in every case where we were able to check unmanned submersibles used in mining applications in fact were controlled by people located on the surface in boats. Remotely controlled devices--the technical term for them is 'teleoperators'--are currently being used in Australian and overseas mines, but unless their human operators can be entirely replaced by computer programs they do not count as robots.

A computer-controlled drilling jumbo, no matter how automated it may be, also is not a robot, for it is special-purpose in the sense in which a robot is general-purpose. A robot, by change of program and gripper mechanisms, can accomplish one task such as welding between 9 and 10 a.m. and a completely different task such as materials handling between 10.30 and 11.30. A drilling jumbo, like a numerically controlled machine tool, lacks this versatility.

As the literature search and several other avenues of investigation led to the conclusion that robots were not in operational use in mines anywhere at the time of our study, we were able to report this conclusion to ASTEC with some confidence. However, ASTEC was interested to know not only the then current state of the art but also how the Australian mining industry might beneficially exploit robot technology in the 1980's. Part of our work therefore involved examination of relevant research and development activities and how their outcome might be transferred to Australian industry in this time frame.

We found that the Australian mining industry is extremely conservative with regard to the adoption of new electronic mining technology. The word 'conservative' is descriptive, not pejorative. The main reasons for this conservatism are as follows.

First, despite the importance of mining to the Australian economy, most mining technology is imported, and the path by which new technology usually is transferred to Australian mines involves prior successful adoption of the technology elsewhere. Secondly, most electronic and electromechanical equipment, including robots, is not designed to withstand the dust, damp, vibration, heat and other environmental attributes of mines. Although conceivably robots could be 'hardened' to work reliably in mine environments, this inevitably imposes substantial additional cost and lessens the economic attractiveness of robots. Thirdly, many Australian mines are located remote from capital cities, which imposes a substantial added cost burden to keep complex technology maintained.

For all these and other reasons, the Australian mining industry's stance with regard to new technology such as robots is understandably unfavourable to fast introduction of innovation. And since there has been virtually no R&D specifically targeted at robotic mining--well-publicised initiatives such as that of the Carnegie-Mellon University's Robotics Institute never undertaken because funding was never obtained--we felt confident in advising ASTEC that robots as such would have negligible impact upon the Australian mining industry in the 1980's.

Our study gave special consideration to Professor M. W. Thring's concept of teleoperated (or, as he calls it, 'telechiric') mining, as a possible step toward robotic mining. In Thring's scenario, coal is won by people who from the surface remotely manipulate mine machinery inside the mine. In place of, for instance, a human driver in an underground mine vehicle there would be a human-sized device that was electronically linked to the operator on the surface who would see on a display the images acquired by the device's cameras and who could command the device to take particular actions. The U.K. National Coal Board, as mentioned below, has indeed evaluated this concept.

If it ever becomes possible to reduce the set of actions that such a human operator should take in every conceivable circumstance to a sequence of explicit, step-by-step rules, then indeed underground mining could be done by robots driving mining equipment with no supervision from people on the surface. But no one anywhere

appears to be carrying out research with this goal in mind. The route to mining automation that mining engineers are preferring to take involves design of new mining machinery rather than design of robots to operate existing mine machinery.

Our report to ASTEC, although concluding that robots had no substantial place in mining automation this decade, was not, I think, a pessimistic report. For although robots themselves will have little application in the short term, what we called 'robotic concepts' are having and appear likely to continue to have considerable impact upon mining automation.

Such techniques as software methods to process 'intelligently' data simultaneously arriving from multiple sensors, software methods to adapt resolution levels to changing content of sensed data, improved algorithms for automatic analysis of real-world scenes, and advances in vehicle locomotion, touch sensing, and understanding when and how to use non-standard sensors, all are being developed by researchers in robotics or in such allied or parent disciplines as artificial intelligence and mechanical and electrical engineering. Their relevance to solving particular problems in mining is clear, and the advantages of including robotics researchers along with mining engineers in an interdisciplinary approach to solving these problems seems clear.

Our report to ASTEC identified specific projects in mining automation where robotic concepts seemed applicable, and finally identified some needs of the Australian mining industry for systems that could benefit from inclusion of robotic concepts.

THE STUDY FOR THE NATIONAL COAL BOARD

Tregelles notes that the falling capital cost of a robot used in manufacturing industry is rapidly approaching the increasing cost of employing a human miner. He accordingly envisages opportunities for the economic use of robots in mining, and considers what the first step should be in introducing robots to the mining industry.

Tregelles says (1982, p. 360) that 'the opportunity will initially be found in those single and repetitive tasks which are performed by men in a clearly prescribed area and such that when applied the job can be declared redundant. Ideally the first application will be on the surface...'.

From these constraints, Tregelles selects as an appropriate initial robot application to be evaluated the automatic removal of 'cakes' of tailings from filter presses employed in washery plants for dewatering tailings. Experiments have begun at the Mining Research and Development Establishment 'to assess the feasibility of this application'.

Tregelles also reports on the MRDE's investigation of Professor Thring's proposals for telechiric mining. This investigation concluded that there is 'no fundamental impossibility in the concept of establishing and operating a longwall face remotely by means of telechirs controlled from the surface' but that 'substantial problems' exist that would have to be solved before such a system became technically practicable. It seems that 'the timescale must lie somewhere between fifteen and twenty five years' before telechiric mining could have a significant impact upon the coal industry, and the National Coal Board is not pursuing the development of telechiric mining by any vigorous programme of research.

CONCLUSIONS

In the short term, that is to say the next five years, one may anticipate a very limited use of robots by the mining industry, probably to automate tasks such as drill bit sharpening that can be accomplished without extensive modification of commercially available robots. Such initial applications will serve also to acquaint the mining industry, in comparatively congenial circumstances, with the capabilities and limitations of robots now on the market.

What we have called robotic concepts will play an important role in advancing mining automation, not by replacing conventional mining machinery with myriads of mining robots nor by replacing existing human operators of mining machinery with teleoperators controlled from some distance away. Robotic concepts will aid mining automation by advancing pattern-recognition methods so that mining machinery can better adjust in real time to changing seam thicknesses, by advancing image-processing techniques so that mobile devices can better detect objects in dusty or underwater environments, by advancing the capabilities of automated vehicles to travel efficiently without collisions, by providing methods for effective on-stream analysis, sorting and grading of the product, utilising current and novel sensors, often in parallel.

To date, mining engineers and roboticists have worked largely in isolation from each other. The time has come to remove such an unproductive barrier.

REFERENCES

Kassler, M. (1984a). Robots and mining: the implications for Australian industry in the 1980's. Proc. Nat. Conf. & Exhibition on Robotics, Melbourne, August 1984 [published by Institution of Engineers, Australia] pp. 80-85. Accepted for publication in Robotica, a journal published by Cambridge University Press.

Kassler, M. (1984b). The Australian Robot Marketplace; A Comprehensive Independent Guide to Robots in Australia. Michael Kassler and Associates Pty Limited, McMahons Point NSW.

Tregelles, P. (1982). An appreciation of the potential for robots and telechirs in the mining industry. Mining Technology, August, pp. 359-363.

Fragmentation Control and Underground Automation

DR ING. KAI NIELSEN

Division of Mining Engineering, Norwegian Institute of Technology, 7034 Trondheim, Norway

Abstract. Compared with modern longwall coal mining systems, automation
has not been developed very far in the underground metal mining
industry. One of the main reasons for this is that present hard
rock mining systems do not produce ore with a controlled degree
of fragmentation. The often poor fragmentation from the stopes
causes large extra costs and precludes a continuous flow of
material which lends itself to automation.

The paper investigates the possibility of adapting coal mining
technology for automatic loading and transport of stope ore by
conveyor belt to a shaft dump pocket. With this concept, it will be
necessary to control the maximum fragment size by careful blasting
design. The degree of fragmentation will have to be finer than is
common to-day. The study shows that the belt conveyor mining concept with controlled fragmentation may be economically competitive
with conventional cyclic loading and transport, even for an annual
production as low as 0.5 mill. tonnes. Fragmentation design and
cost calculations have been based on a combination of the
Kuznetsov relationship between blasting parameters and mean fragment size, and the Rosin-Rammler formula for fragment size
distribution.

Keywords. Fragmentation control; drilling and blasting design;
underground mining; conveyor transport; mining automation.

INTRODUCTION

In some of the coal mining industry, there
has been a development of new equipment
and operating techniques, leading towards
a highly automated underground production
of coal.

The same development has not taken place
in the metal mining industry. In most of
to-day's hard rock underground mines, only
the hoisting cycle has been successfully
automated. The rest of the production system is organized in the traditional manner,
with cyclic operations and frequent rehandling of ore. This leads to high costs
for ore loading and transport, even if
most mines are highly mechanized.

One of the main reasons for this is that
present hard rock stoping systems do not
produce ore with a controlled degree of
fragmentation. The loading and hauling
equipment has to handle a varying proportion of coarse muck and even large boulders, which leads to ineffective utilization of costly mining equipment, and
extra costs and delays caused by secondary
blasting. These extra costs may amount to
10% or more of total mine operation costs.

Thus, the often coarse fragmentation of
ore from the stopes causes large extra
costs, and makes it impossible to establish an undisturbed flow of material which
lends itself to further automation.

The initial fragmentation in hard rock mining will still have to be done by drilling
and blasting. To-day there is not any economic technology available for continuous
breakage of material at the face.

If the material flows in hard rock, underground mines are to be automated, the ore
fragmentation has to be carefully planned
and controlled by proper blasting design.

MINING CONCEPT

In order to evaluate the feasibility of
adapting coal production technology for
ore loading and transport, the following
mining concept will be studied:

The orebody has the form of a steeply dipp-
ing, thin plate, suitable for sublevel
open stoping. It is assumed that the orebody extends for a considerable distance,
both along the strike and down the dip. The
ore from the stopes will be delivered by
chutes into mobile grizzly feeders loading
onto a belt conveyor. The conveyor trans-
ports the ore horizontally out to a dump
station by the shaft. The ore will then
be hoisted without any underground crushing
being necessary.

This concept may be illustrated as shown

in Fig. 1, and the handling and transportation of ore from the stope to the shaft,
and up, may be fully automated.

The investment costs reflect the Norwegian cost level of middle 1984, with 1 U$=
8.50 NOK. It is obvious that the conveyor
width will be dictated by material sizes,
and not by capacity demands.

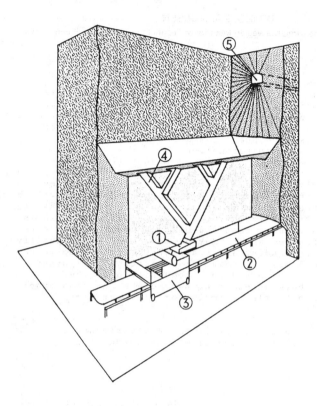

LEGEND:

1. CHUTE
2. CONVEYOR
3. GRIZZLY FEEDER
4. FINGER RAISES
5. DRILLING DRIFT

Fig. 1. Mining lay-out with conveyor loading and transport.

In order to use conveyor transport, the
fragmentation will have to be finer than in
most mines to-day. The blasting design must
be such as to virtually eliminate all over-
sized material. The grizzly feeders will
serve as an additional control measure to
protect the conveyor belt. The conveyor
system is assumed to be 1000 meter long.

CONVEYOR TRANSPORT

For the purpose of this study, standard
belt conveyors of 1000-2000 mm widths will
be considered. Some operational and economic data for these conveyors are given in
Table 1.

The conveyor capacity data have been based
on the following assumptions:

    Density   : 4.0 tonnes/m^3, massive ore
                 2.4 tonnes/m^3, broken ore
    Belt speed: 1 m/sec
    Operating time: 3500 hrs/year

BLASTING SYSTEM

A typical blasting design with fan drilling in sublevel stoping, may be described
as follows:

    Hole diameter   : 51 mm
    Hole length (max): 15-20 m
    Burden x spacing : 2.0 x 2.4 m
    Loading density : 2 kg $ANFO_3$ pr. meter
    Powder factor   : 0.41 kg/m^3
    Specific drilling: 2.7 m^3/m drilled

In fan drilling, the coarsest fragmentation will be at the outer perimeter of the
fan, where the spacing is largest. Poor
fragmentation caused by drill hole deviation may also be expected to increase with
hole lengths.

The expected fragmentation may be calculated by using the "Kuz-Ram" model as described by Cunningham (1983). This model
combines the Kuznetsov relationship between blasting parameters and mean frag-

TABLE 1   Operational and economic data for belt conveyors

Conveyor width mm	Max lump size mm	Capacity		Investment costs	
		tonnes/hr	mill. tonnes pr. year	NOK pr. meter	mill. NOK 1000 m length
1000	430	924	3.2	5500	5.5
1200	530	1356	4.7	6050	6.1
1400	680	1874	6.6	6600	6.6
1600	800	2479	8.7	7150	7.2
1800	900	3162	11.1	7700	7.7
2000	1020	3939	13.8	8250	8.3

Table data compiled from Mechanical Handling Engineers' Assoc. (1977) and Nielsen (1982)

ment size, and the Rosin-Rammler curve describing fragment size distribution:

The Kuznetsov equation is:

$$\bar{x} = A \cdot \left(\frac{V_0}{Q}\right)^{0.8} \cdot Q^{0.167} \qquad (1)$$

where: $\bar{x}$ = the mean (50% passing) fragment size (cm)

V_0 = volume of rock broken pr. hole (m^3)

Q = the mass of TNT of equivalent energy to the explosive in one borehole (kg)
1 kg ANFO = 0.87 kg TNT (weight strength)

A = rock factor which is
A = 7 for medium rock
A = 10 for hard, fissured rock
A = 13 for hard, competent rock

In this study, a rock factor $A = 10$ will be used.

The Rosin-Rammler formula is:

$$R = e^{-\left(\frac{x}{x_c}\right)^n} \qquad (2)$$

where: R = proportion of material retained on screen

x = screen size

x_c = characteristic size parameter

n = size-distribution constant

In competent rock, the constant n will normally have a value of 1.7-1.8. A high value of n indicates a uniform fragmentation without excessive amounts of either fines or oversize. For the purpose of this study, a size constant of 1.6 will be chosen, in order to also allow for some drill hole deviation.

The "Kuz-Ram" model may be used to calculate the drilling pattern that will give the fragmentation necessary for direct loading on to the various belt conveyors listed in Table 1. However, the maximum lump sizes in Table 1 are the largest allowable single dimensions, whereas the sizes calculated by the "Kuz-Ram" model refer to screen openings. To allow for irregular fragments, the maximum fragmentation size should be 75% of the maximum lump dimension as given in Table 1.

Looking only at the outer 1 m of the drilling fan, the calculations will be as follows:

First, the Rosin-Rammler formula is used to calculate the characteristic size parameter

x_c by using x = max.fragment size for each conveyor and $R = 0.001$, which is a very small proportion of the total blast volume. Then using $R = 0.5$, the mean fragment size $\bar{x}$ may be calculated,to be used in the Kuznetsov equation. Since the amount of explosive in each hole, Q, is constant, the volume and area to be broken by each hole can be calculated. The results of these calculations are shown in Table 2.

FRAGMENTATION COSTS

In fan drilling, the specific drilling (m^3/meter drilled) will be substantially higher than when drilling parallel holes. In this study, specific drilling will be assumed to be 80% higher for fan drilling with a maximum hole length of 15-20 m.

The fragmentation costs will be calculated from the following cost factors:

Drilling costs   :   25 NOK/meter
Explosives (ANFO):   5 NOK/kg

These costs include depreciation of drilling equipment and loading costs for the explosives.

Based on the drilling patterns and powder factors listed in Table 2, the fragmentation costs may be calculated as shown in Table 3.

Table 3 shows clearly that the drilling costs will increase dramatically as the designed fragmentation becomes finer. The drilling costs per m^3 can be reduced by increasing the drill hole diameter, but in order to maintain the fragmentation, the powder factor will have to be increased also. This optimization process is discussed by Nilsson (1983), but will not be pursued here. However, the fragmentation costs in Table 3 can undoubtedly be reduced.

CONVEYOR COSTS

The conveyor system will use modules, each with a length of 100 m. These modules can be dismantled and re-erected in another production area when the ore is exhausted.

A well designed stationary conveyor system will have an economic lifetime of 15-20 years. The annual operating costs will be in the order of 2-3% of investment costs.

In this study, however, the conveyor

TABLE 2   Drilling patterns and fragmentation for conveyor belt loading and transport

| Belt width mm | Fragment size, mm | | Drilling pattern | | Powder factor |
	mean	max.	Area m^2	Burden x spacing m	kg/m^2
1000	76	320	1.1	0.95 x 1.15	1.82
1200	95	400	1.5	1.15 x 1.30	1.33
1400	121	510	2.0	1.30 x 1.50	1.00
1600	143	600	2.4	1.40 x 1.70	0.83
1800	162	680	2.8	1.55 x 1.80	0.71
2000	183	970	3.3	1.70 x 1.95	0.61

TABLE 3   Fragmentation costs

| Belt width mm | Drilling pattern B x S | Specific drilling m^3/meter | | Powder factor kg/m^3 | Fragmentation costs NOK | |
		parallel	fan		m^3	tonne
1000	0.95 x 1.15	1.1	0.6	1.82	50.77	12.69
1200	1.15 x 1.30	1.5	0.8	1.33	37.90	9.48
1400	1.30 x 1.50	2.0	1.1	1.00	27.73	6.93
1600	1.40 x 1.70	2.4	1.3	0.83	23.38	5.85
1800	1.55 x 1.80	2.8	1.6	0.71	19.18	4.79
2000	1.70 x 1.95	3.3	1.8	0.61	16.93	4.23

TABLE 4   Annual conveyor loading and hauling costs

| Belt width mm | Investments cost mill. NOK Feeders and conveyor | | Operating costs mill. NOK | Annual 0 & 0 costs mill. NOK |
	Total	Annual		
1000	11.0	3.28	0.55	3.83
1200	12.2	3.64	0.61	4.25
1400	13.2	3.94	0.66	4.60
1600	14.4	4.30	0.72	5.02
1800	15.4	4.59	0.77	5.36
2000	16.6	4.95	0.83	5.78

TABLE 5   Unit costs for mining with conveyors

| Belt width mm | 0.5 mill. tonnes NOK/tonne | | | 1.0 mill. tonnes NOK/tonne | | |
	Blasting	Hauling	Total	Blasting	Hauling	Total
1000	12.69	7.66	20.35	12.69	3.83	16.52
1200	9.48	8.50	17.98	9.48	4.25	13.73
1400	6.93	9.20	16.13	6.93	4.60	11.53
1600	5.85	10.04	15.89	5.85	5.02	10.87
1800	4.79	10.72	15.51	4.79	5.36	10.15
2000	4.23	11.56	15.79	4.23	5.78	10.01

modules will probably have a shorter life-time. The cost calculations will be based on a lifetime of 5 years and an interest rate of 15% p.a. It is also assumed that the conveyor costs will be 50% of the total investment costs for the loading and transportation system.

The annual operating costs are assumed to be 5% of investment costs, independent of the tonnage, which will be much lower than the conveyor's nominal capacity.

Table 4 shows the annual loading and hauling costs for the various conveyors.

COST COMPARISON

Based on Norwegian experience, conventional open stoping with trackless equipment, will cost:

Drilling & blasting	:	2.80 NOK/tonne
Loading	:	3.50 NOK/tonne
Transport	:	4.00 NOK/tonne
Secondary blasting	:	3.50 NOK/tonne
Crushing (extra)	:	2.00 NOK/tonne
Direct mining costs	:	15.80 NOK/tonne

These costs include equipment depreciation. The crushing costs in this case, cover the extra expenses due to the operation of a large underground primary crusher. If the stoping fragmentation is reduced, all crushing costs will be lower.

To compare the conveyor mining concept with conventional mining, two cases with an annual production of 0.5 and 1.0 mill. tonnes will be considered. The mining costs for conveyor loading and transport will then be as shown in Table 5.

Table 5 shows that the optimal conveyor size will be 1800 mm for an annual production of 0.5 mill. tonnes, and probably 2000 mm for a production of 1.0 mill. tonnes pr. year.

CONCLUSION

The simple study performed here indicates that the conveyor mining concept with controlled fragmentation may be economically competitive with conventional open stope mining using trackless equipment. The annual capital costs for conveyor loading and hauling have been estimated conservatively with a life time of only 5 years. Since the conveyor width will be dictated by the size of material, conveyor utilization will be far below the nominal capacity. By careful design, capital costs may therefore be reduced compared to the con-

ventional conveyor costs used in this study. In addition, the costs for fragmentation can be reduced by optimizing the blast hole diameter, balancing costs for drilling and blasting.

The handling and transport of ore from the stopes may be fully automated with the conveyor mining concept. The organization and supervision of the ore production will then be much simpler, and more effective.

Since the feeder and conveyor equipment is much more reliable than trackless machinery, ore quality control by blending can also be improved.

It may also be expected that the amount of underground maintenance work will be reduced.

The conveyor mining system with controlled fragmentation will therefore have potential for cost and efficiency improvements, in addition to possible savings in direct mining costs.

REFERENCES

Mechanical Handling Engineers Assoc. (1977). Recommended Practice for Troughed Belt Conveyors. Harrow, Middx. England.

Nielsen, K. (1982). Mine project evaluation. (In Norwegian). Div. of Mining Engineering, Norwegian Institute of Technology, Trondheim, Norway.

Cunningham, C. (1983). The Kuz-Ram Model for Prediction of Fragmentation from Blasting. Proc. First Int. Symposium on Rock Fragmentation by Blasting. Vol. 2, pp. 439-453. Luleå, Sweden.

Nilsson, D. (1983). Optimum Fragmentation in Underground Mining. (In Swedish). Swedish Mining Research Foundation. Report F 8334. Kiruna, Sweden.

Blast Design Considerations for In-Situ Retorting of Oil Shale

G. HARRIES & G.G. PAINE

ICI Australia Operations Pty Ltd, ICI House, PO Box 4311, Melbourne

Abstract. The in-situ retorting of shallow oil shale deposits is an attractive
alternative to the material handling problems of conventional open pit mining
techniques, which involve removing overburden, fragmenting and transporting the shale
to a fixed retort, and finally disposing of the hot spent shale.

For in-situ retorting to be successful, the oil shale has to be fractured and the
fragmented rock expanded into a void, to allow controlled propagation of a flame
front through the in-situ retort. The principles of cratering with explosives have
been used to develop a process for the in-situ fracturing of a thin oil shale deposit
overlain by a formation of porous rock. The overlying porous rock formation is
first compressed by explosive charges to create voids in the porous rock. The oil
shale is then fragmented and pushed in to the voids with separate explosive charges.
This paper discusses methods of creating voids in porous rocks with minimal surface
disturbance, and then methods of obtaining the most uniform fragmentation of the
shale. The whole process can be completed, in practice, in one blast. The results
of an experimental blast are presented, and possible modifications are then discussed.

Keywords. Computer-aided design, mining in-situ, geology, blasting.

INTRODUCTION

An experimental in-situ retort was implemented at
the Julia Creek Oil Shale Project of CSR Limited
during 1981 (Marcovich, 1983; Paine, 1983).

The major design considerations for the blasting
were

1. The properties of the oil shale and the
overlying Coquina (limestone).

2. The location and mass of explosive charges to
compress the overlying porous Coquina, forming a
chamber into which the broken oil shale can expand
to give a swell of 20%.

3. The explosive charge configuration to fracture
and expand the oil shale into the void created by
blasting the overlying coquina.

4. Blast induced cracking should not extend to the
surface.

Computer modelling was used to predict the blast
parameters, such as blasthole pattern and explosive
charge mass and location for the experimental
in-situ retort.

GEOLOGY

The Toolebuc oil shale formation occurs as an
extensive but relatively thin deposit over much of
Northern and Central Queensland. The geological
seqence at the site of the experimental in-situ
retort is:

Depth (m)	Rock Unit
0-15	Allaru Mudstone
15-21	Coquina (limestone)
21-28	Oil shale
>28	Runmoor Mudstone

The relevant rock properties required for the
blast design are:

	Coquina	Oil Shale
Density (g.cm^{-3})	1.8	1.8
Porosity %	30	18
Young's Modulus (GPa)	17.5	7.8
Poisson's Ratio	0.44	0.28
Compressive Strength (MPa)	4.1	30

Thin, near surface deposits of oil shale have
been exploited by Geokinetics at Kamp Kerogin
using the in-situ retorting technique.

This deposit is overlain with stronger rock types
than those that occur at the Julia Creek site.

To produce a retort, the whole deposit including
the overburden is heaved up in large blocks.
This gives well fragmented oil shale with
sufficient swell to allow propagation of a flame
front.

It can be seen from the above rock properties
that the shale and the overlying Coquina are
porous. From experience in blasting porous

limestones similar to the Coquina, it is known that
small explosive charges chamber holes rather than
crack the surrounding rock. Extensive experience
with cratering also suggested that the Coquina could
be chambered without causing any surface disturbance.

EXPLOSIVES CRATERING

When short (quasi-spherical) charges of explosives
are buried at a series of depths, as shown in Fig. 1,
the charges buried deeply cause no disturbance at
the surface. As the depth of burial of the explosive
charge is decreased, small radial cracks are observed
at the surface. This depth of burial is known as
the critical depth. As the depth of burial is
further decreased, craters are formed, and it
has been found that a depth known as the optimum
depth a crater with the greatest volume is formed.
With depths of burial less than the optimum, the
volume of the crater decreases and more and more
material is ejected from the crater.

If the blasting can be carried out at depths greater
than the critical depth, there will be little or no
surface movement and little or no loss of vapour
from the retort. To ensure that a flame front can
propagate through the retort, the volume of the
fractured shale has to be increased by about 20%.
Voids then have to be made in the porous Coquina to
accommodate this extra volume. A schematic diagram
of the in-situ retort is shown in Fig. 2.

BLAST DESIGN

Explosives cratering principles can be applied to
the process of chambering the Coquina to determine
the limiting depth of blast induced cracking. In
the absence of a suitable test site, the Harries
Blasting Model (Harries, 1973, 1977) has been used
to predict the depths of blast induced cracking
for differing explosive types, charge lengths
and diameters.

The following limits of blast induced cracking were
predicted for ANFO explosives (ammonium nitrate,
fuel oil mixture).

Blasthole Diameter (mm)	Explosive Charge Length(m)	Depth Overburden [1] Required to Ensure No Surface Cracking(m)
200	1.0	17.2
152	1.0	13.3

[1] Assumes that the bottom of the explosive charge
is located at the base of the Coquina.

Blasting to yield uniform cracking in the oil shale
was simulated by means of the Harries Blasting
Model. The blasthole pattern selected for the
experimental in-situ retort was 6m equilateral
triangular with a 152mm diameter blasthole. To
blast a retort 18m in length and 10.4m in width, 11
blastholes drilled to the bottom of the oil shale
were required (Fig. 3). For the chambering of the
Coquina, another 14 blastholes, drilled to the base
of the Coquina, were considered necessary in
addition to the 11 blastholes drilled to the base
of the oil shale (Fig. 3). All Coquina explosive
charges need to be detonated instantaneously some
200 to 300 milliseconds before the individually
detonated explosive charges in the oil shale.
The design depths of blastholes were progressively
increased from the blast initiation end of the
retort with a view to providing a sloping base
to facilitate the collection of liquid

Factors that could affect the efficiency of
collection of liquid hydrocarbons were recognised
as

1. The permeability of the oil shale.

2. The viscosity of the liquid hydrocarbons.

3. The permeability of the underlying mudstone,
particularly after blasting.

EXPERIMENTAL IN-SITU RETORT BLAST

A small scale test blast was fired initially to
investigate the predicted explosives performance.
Evaluation of increased fracturing in the oil
shale by re-entry drilling (cored through oil
shale) and pressure test for permeability proved
inconclusive. The lack of surface disturbance
showed that the explosive performance in design
work had not been understated.

The small scale test blast allowed the following
modifications to the in-situ retort blast design.

1. The first firing blasthole could be increased
to 200mm diameter.

2. All other blastholes could be increased to
171mm diameter.

For the experimental in-situ retort blast the
general blasthole charging procedure was to charge
the blasthole in the oil shale (numbers 1 to 11,
Fig. 3) with explosive to approximately 2m from
the Coquina, then place 2m of stemming
(screenings) prior to placement of the explosive
charge in the Coquina, and finally stemming
to the surface.

Explosive charge weights varied from 58 to 192 kg
in the oil shale and 17 to 50 kg in the Coquina.
The overall explosives powder factor for the
blast was 1.37 kg.m^{-3} of oil shale.

All explosive charges in the Coquina were
initiated with a No. 1 'L' Series millisecond
delay detonator (nominal firing time 25
milliseconds) while explosive charges in the oil
shale (see Fig. 2) were initiated with Nos 15 to
25 'L' Series millisecond delay detonators
(nominal firing times 395 to 695 milliseconds).

After the blast the surface expression was shown
in two ways.

1. Surface heave of approximately 30cm over the
top of the retort.

2. Minor surface cracking 3 to 4m outside the
confines of the blasthole.

Subsequent re-entry drilling showed that the oil
shale was well fractured but relatively
undisturbed.

The Coquina was not significantly compressed,
however the underlying Ranmoor mudstone showed
the effects of compression from the blast above,
exhibiting crushed zones and opening of bedding
planes.

CONCLUSIONS

The experimental in-situ retort blast and subsequent retort commissioning has shown

1. The oil shale can be satisfactorily fractured in-situ with explosives.

2. The relatively undisturbed nature of the blasted oil shale and the subsequent problems with maintenance of combustion in the retort indicate that the explosive charges in the Coquina did not provide sufficient expansion void.

3. The damage to the underlying mudstone is most likely due to insufficient expansion void for the blasted oil shale, however consideration should also be given to explosive charge placement in the oil shale.

4. Blast design modifications are required to increase the explosives distribution in the Coquina for an increased expansion void to provide better permeability in the oil shale and reduced damage to the underlying mudstone.

5. The properties of the underlying mudstone should be considered in more detail.

ACKNOWLEDGEMENTS

The authors extend their thanks to ICI Australia Operations Pty Ltd for permission to publish this paper.

REFERENCES

Marcovich, B.J. (1983). In-situ Retorting of Julia Creek Oil Shale. Proceedings of the First Australian Workshop on Oil Shale Lucas Heights. pp. 47.

Paine, G.G. (1983). Explosives Fracturing of Oil Shale for an Experimental In-Situ Retort at the Julia Creek Project. Proceedings of the First Australian Workshop on Oil Shale Lucas Heights. pp. 51.

Harries, G. (1973). A Mathematical Model of Cratering and Blasting Proc. Nat. Symp. On Rock Fragmentation (Aust. Geomech. Soc.) Adelaide. pp. 41.

Harries, G. (1977). The Calculation of the Fragmentation of Rock from Cratering, Proc. 15th APCOM Symp, Brisbane. pp. 325.

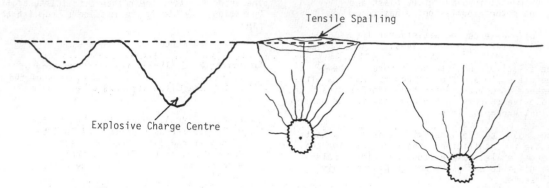

Fig. 1. Effect of Depth of Burial on Cratering

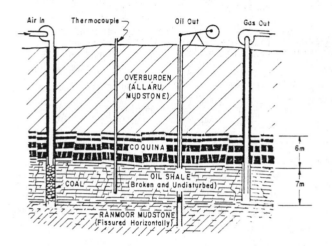

Fig. 2. Schematic of Field In-Situ
Retort (Marovitch 1983)

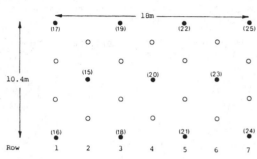

o   blasthole drilled to bottom of oil shale

●   blasthole drilled to bottom of coquina

(15) millisecond delay numbers for oil shale
charges

Fig. 3. Blasthole Pattern for Experimental
In-Situ Retort Blast

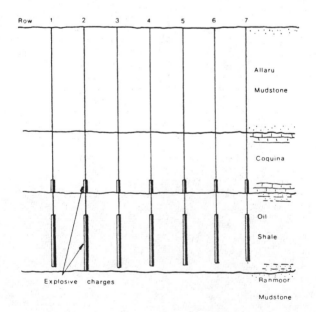

Fig. 4. Longitudinal Section of Experimental In-Situ Retort Blast

The Rotary Multi-Plate Valve as an Integrated Control System of a Longwall Roof Support

W.S. SZARUGA

Research and Development Department, Sentrol Systems Ltd., 4401 Steeles Ave., West, Toronto, M3N 2S4, Canada

Abstract. An automatic control system of a longwall roof support is discussed. Hydraulic automatic control systems for longwall roof support have not met all requirements for such systems. Utilization of standard hydraulic elements to form a system that would comply with all demands leads to a very complicated structure with a large number of hydraulic hoses within and between adjacent supports. So, only simplified systems have been made. This paper describes the multifunctional valve that permits realization of any desired logic functions, while maintaining a compact form, and analyses its application as a fully hydraulic automatic control system of a roof support.

Keywords. Hydraulic systems; longwall roof support; Boolean functions; switching theory; control system analysis.

INTRODUCTION

An automatic control system of a longwall roof support should meet certain requirements in the domain of controllability, structure complexity, reliability, safety of operation, monitoring, and total costs. Controllability requirements include the following:

- self re-setting of a chock if it cannot be fully advanced;
- lowering of a chock only if adjacent chocks are re-set;
- minimization of re-set forces while advancing a chock;
- remote manual control of adjacent chocks.

Realization of a hydraulic system with hitherto existing elements that would comply with the above requirements is very difficult. Their fulfilment leads to a very complicated structure and costly systems. That is why only simplified systems have been made. None of the existing hydraulic control systems of a roof support utilizes the minimization of re-set forces while advancing a chock, instead continuous lowering is used. This causes roof decompression over the chock, and results in a fully lowered chock position when the advance cannot be completed. Also, these systems are not provided with automatic re-set of a chock if it cannot be fully advanced. This blocks the rest of the chocks in the longwall system. Simplification of the system algorithm by not taking into account all the controllability requirements has not simplified the system structure itself. It has remained complicated, with a large number of hydraulic connections within and between adjacent chocks.

The multi-plate valve described herein permits realization of any desired logic functions while maintaining a compact form.

THE ROTARY MULTI-PLATE VALVE

The rotary multi-plate valve shown in Fig. 1 consists of a number of bi-positional rotary plates (1) that can be swiveled by a single- (2) or double-acting cylinder (3), and divided by stationary plates (6). Every plate has at least one port (4) and/or passageways (5) associated with the ports.

The ports are hydraulicly connected to supply and outflow channels or to control inputs, and together with passageways form a flow channel that can be opened or closed by each of the rotary plates. This creates a hydraulic output that can be described by Boolean functions with values of 1 or 0 meaning connections with supply or outlet channels respectively.

The last stationary plate is provided with a hydraulic jack (7) to prevent leakage in the system.

Figure 2 shows a basic element of the system with its truth table, i.e. one rotary plate activated by a single-acting cylinder, with marked inputs and outputs. The logic functions performed by this element are listed in Table 1.

TABLE 1

Inputs	Function
x_2, x_3	$Y = x_1 x_3 + \bar{x}_1 x_2$
$0, x_3$	$Y = x_1 x_3$
$1, x_3$	$Y = \bar{x}_1 + x_3$
$x_2, 0$	$Y = \bar{x}_1 x_2$
$x_2, 1$	$Y = x_1 + x_2$
$0, 1$	$Y = x_1$
$1, 0$	$Y = \bar{x}_1$

The function that is actually performed
depends on inputs x_2 and x_3, i.e. where
they are connected to. If both x_2 and x_3
are the control inputs, the output Y is
a logic function of three variables; if
one of them is connected to the supply or
the outflow channel the element produces
functions of two variables, i.e. AND, OR,
implication; if inputs x_2 and x_3 are connected to the supply and the outlet channels the element performs NOT-operation
or repetition x_1.

The same element activated by a double-
acting cylinder is shown in Fig. 3. The
logic function performed by this element
is a flip-flop function. Assembling the
above elements in series permits one to
realize any desired logic function of a
multivariable system while maintaining its
compact form.

THE CONTROL SYSTEM OF A ROOF SUPPORT

The connection diagram, with marked inputs
and outputs, is shown in Fig. 4. The meaning of them is as follows:

x_f - input from the limit switch of conveyor flitting;

x_d - input from the limit switch of chock
advance;

x_m - input from the re-set pressure switch;

x_a - input from the advance pressure
switch;

x_l - input from the left adjacent chock
to block lowering;

x_p - input from the right adjacent chock
to block lowering;

x_s - input from the minimum pressure
switch;

Y_r - re-set;

Y_l - lower;

Y_f - conveyor flitting;

Y_a - advance;

Y_b - output to the adjacent chocks to
block lowering.

During normal operation (see Fig. 5), the
conveyor is pressed continuously against
the working face and flitting is performed
behind the combined cutter loader.
The limit switch (1) of the chock's conveyor flitting, whose advance ram is fully
extended, activates the closed control system of the chock (see Fig. 4). If the
adjacent chocks are re-set, the operations
of lower, advance and re-set take place
automaticly. In an early phase, lower and
advance are requested together. The motion
of the chock, detected by the advance pressure switch (2), turns off lowering and
only chock advance is performed afterwards.
This feature meets one of the controllability requirements, i.e. minimization of
re-set forces while advancing a chock.
Once advance is completed, the limit switch of chock advance (3) activates re-set
that is finished when the re-set pressure
switch (4) indicates the working pressure
of the chock.
In case the chock cannot be fully advanced,
the advance pressure switch (2) detects
the maximum pressure and releases additional lowering of the chock. If the chock
is lowered too much, the minimum pressure

switch (5) indicates the minimum allowed
pressure and activates emergency re-set
that is finished after the re-set pressure switch (4) detects the working pressure of the chock.
During the above sequences, the adjacent
chocks are blocked to prevent lowering.
The blocking of the adjacent chocks is
turned off after re-set is finished.
The above algorithm leads to the following
logic fumctions of the system:

$$Q_1 = x_f + \bar{x}_d q_1 \qquad (1)$$

$$Q_2 = x_s + \bar{x}_f q_2 \qquad (2)$$

$$Y_r = \bar{x}_m(x_s + x_d + q_2) \qquad (3)$$

$$Y_l = \bar{q}_2 x_a \bar{x}_l \bar{x}_p \bar{x}_s \qquad (4)$$

$$Y_f = x_m(x_d + \bar{x}_f \bar{q}_1) \qquad (5)$$

$$Y_a = x_f + \bar{x}_d q_1 \qquad (6)$$

$$Y_b = \bar{x}_m \qquad (7)$$

where Q_1 and Q_2 are the states of the
system.
Realization of these functions requires
use of five rotary plates. Two of them are
of the flip-flop type. To meet all controllability requirements, two additional
rotary plates must be added to perform
bi-directional remote manual control of
the adjacent chocks.

SUMMARY

The rotary multi-plate valve can be designed as a power system that performs direct control of power cylinders or as
a logic device to provide control inputs
to the valves of the associated cylinders.
The latter permits miniaturization of the
system since the required flow section is
low. Instead of using return springs for
single-acting cylinders of rotary plates,
the main pressure of the system can be
supplied.

REFERENCES

Kruszecki, L., and W. Szaruga (1980). Possibilities of application of hydraulic
elements in automatic control systems
for longwall roof supports (in Polish).
Proceedings of The 6th International
Mining Automation Conference, Katowice,
Poland, Section 4, 87-92.

Szaruga, W. (1978). Analysis and synthesis
of the hydraulic automatic control
system of a longwall roof support (in
Polish). Ph.D. Thesis, University of
Mining and Metallurgy, Krakow, Poland.

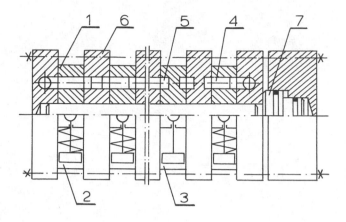

Fig. 1. Multi-plate valve

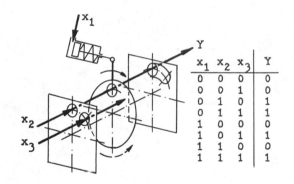

Fig. 2. Basic element

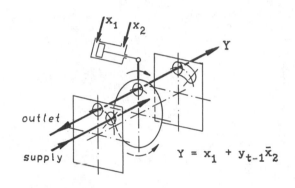

Fig. 3. Flip-flop

$$Y = x_1 + y_{t-1}\bar{x}_2$$

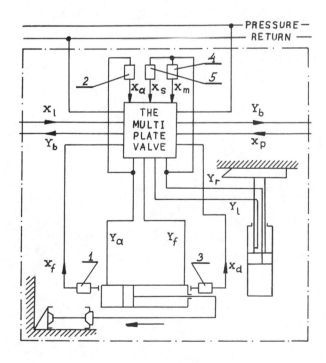

Fig. 4. Connection diagram

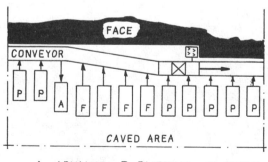

A - ADVANCE; F-FLITTING; P-PUSH;

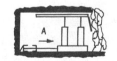

Fig. 5. Longwall roof support

Identification of Mining Shearer Loader Parameters Suitable for Automation Control

DR ANDRZEJ PODSIADLO, PhD.(Engr)

Institute of Mining and Dressing Machines and Automation, University of Mining and Metallurgy, Cracow, Poland

The paper presents two methods for periodical identification of the dependence between haulage speed and loading torque of the shearer for periodical corrections of control algorithms. These algorithms are useful for working with small microprocessor-based systems. The paper is concluded with results of identification using above methods and with discussion about their merits and demerits.

INTRODUCTION

The longwall mining system is the basic exploitation system used in coal mining in Poland. The quantity of coal produced by this system, using a shearer loader as a cutter-loader machine is estimated at 170 million tones per year. For this reason scientific research concerning these machines is of major practical significance.
Shearer construction and the shearer control system should achieve:

- maximum output represented by speed of advance $\dot{x}$, m/min;
- minimization of unitary energy of cutting E_{jv}, kWh/m^3.

There are many limitations on shearer design. The most important include the following:

- limited motor power supplying the shearer circuits;
- limited overall dimensions of the shearer due to excavation dimensions and other machines working with it;
- necessity for securing long shearer life.

The major topics for research connected with the construction and automation of shearer relate to the limitation of dynamic loads on machine elements such as cutting drums. These can be interpreted from the proportion of variable components of load and controlled on the basis of their average values.
Mining shearer loaders are complex electromechanical units. Forces arising from the cutting process are not only external to the system. Their influnce on the cutting drum is represented as a three-dimensional force system. The main vector and the main torque are not sufficiently described in the literature.
These forces cause feedback between drum torque and drum movement, because forces affecting the picks or cutting disks also affect the resisting force of shearer movement.
On the other hand, shearer movement affects drum torque. It is possible to represent the cutting shearer loader diagramatically, as is shown in Fig. 1. Mathematical descriptions of units NZG and NZC are relatively simple and are sufficiently described in the literature (Podsiadlo, Gawelek, Jurkiewicz, 1978; Dokukin, Krasnikov, Khurgin, 1978). The description of blocks bounded in Fig. 1 by the dashed line is a real problem.
Cuttability of the virgin coal and the magnitude of friction forces as represented by functions describing these blocks as random variable coefficients to allow for variations in the shearer working environment. One should also emphasize the empirical method of describing by means of mathematical functions of forces affecting picks. This affects the description of blocks O" and OP.

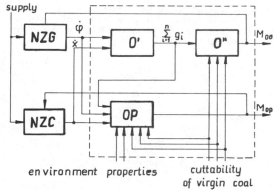

supply

Σg_i – total thickness of cutting picks in contact with coal
$\dot{\varphi}_0$ – angular speed of drum M_{OG} – drum torque
M_{OP} – moment of movement resistance forces
$\dot{x}$ – shearer travel speed
NZG – drive of shearer drum
NZC – drive system haulage
O',O" – units describing drum's work
O P – unit of movement resistance

Fig. 1. Structural of shearer loader diagram

The structural scheme shows feedback lines between marked variables. Because current systems for stabilisation of the torque or drum load M_{og} use $\dot{x}$ as a control variable the process is shown only as the function $M_{og} = f(\dot{x})$.
Equations for the shearer movement are described in another of the author's publications (Podsiadlo, Kwasniewski, 1975).
The majority of identification methods, particularly the correlation function method, requires the fulfilment of the assumption that the object input, for example $\dot{x}$, is not correlated with the disturbance affecting the object (variability of coal properties). Howether, the travel speed $\dot{x}$ of the shearer loader depends, during the cutting process, on forces affecting the picks or disks thorough movement resistance. In the literature (Byrka, 1972), author has not considered what decrease in value of the given relations were obtained by means of the correlation function method. In the following, two methods to mathematically describe the dependence of torque or drum load on travel speed are present-

ed; methods of parameter identification for these descriptions are also outlined.

ANALYTICAL MATHEMATICAL MODEL OF THE RELATION $M_{og} = f(\dot{x})$

The following variables combine to give the total shearer torque load: torque of the cutting forces M_{os} and torque of the coal loading process on the face conveyor M_{ol} , $M_{og} = M_{os} + M_{ol}$ (1)

The proportion of the loading torque in the total torque on the drum is not high and may be omitted (Podsiadlo, 1976). Analysis of a drum with a pick system, as shown in Fig. 2, is described as follows.

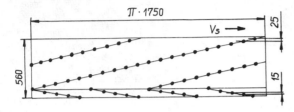

Fig. 2. Drum with a pick system

The cutting loading torque on the drum is the sum of the torque produced by the forces of picks on the cutting shield (M_{osT}) and picks on the transport worms (M_{oss}):

$$M_{og} = M_{osT} + M_{oss}$$

and

$$M_{osT} = \sum_{i=n}^{n+k/2} P_{zi} R$$

(2)

$$M_{oss} = \sum_{j=n}^{n+1/2} P_{zj} R$$

where:

P_{zi} - tangent force of the i-th pick on the cutting shield;
P_{zj} - tangent force of the j-th pick on the transport worms;
R - radius of cutting drum;
k - number of picks on the cutting drum;
l - number of picks on the transport worms;
i = n - the pick on the cutting drum starting the process of cutting;
i = n + k/2 - the pick on the cutting drum ending the process of cutting;
j = n - the pick on the transport worm starting process of cutting;
j = n + 1/2 - the pick on the transport worm ending the process of cutting.

The tangent cutting forces P_{zi} and P_{zj} depend on: pick parameters, coal compressive strength, friction coefficient of coal against steel, average thickness of cutting, side angle of coal crumbling, etc. The large number of these variables has made it difficult to devise a convenient expression to describe the relationships.

Soviets authors propose an empirical expression (Byrka, 1972; Dokukin, Krasnikov, Khurgin, 1978) but it is very complicated and it is difficult to describe torque load by means of such an equation. Therefore a nonlinear regression function was applied to aproximate the forces P_{zi} and P_{zj} in equations of the form (Podsiadlo, Szaruga, 1976)

$$P_{zi} = (A_{01T} g_i - A_{02T} g_i^2) K_{odi}$$

(3)

$$P_{zj} = (A_{01s} g_j - A_{02s} g_j^2) K_{odj}$$

where:

A_{01T}, A_{02T} - coefficients describing coal properties pick parameters on the cutting drum and the effects of picks from adjoining cutting lines;
A_{01s}, A_{02s} - coefficients describing coal properties, parameters of picks on the transport worms, and the effects of picks from adjoining cutting lines;
K_{odi} - coefficient of coal recovery for the i-th pick on the transport worms;
g_i - current thickness of cutting of the i-th pick on the cutting drum;
g_j - current thickness of cutting of the j-th pick on the transport worm

When we assume that the coefficient of coal recovery $K_{odi} = 1$ for picks on the cutting shield and we omit the link with the second power of cutting because of its low influence value on the tangent force of picks on the cutting drum (Podsiadlo, 1976) then the relation for P_{zi} will reduce to the form:

$$P_{zi} = A_{01T} g_i$$

(4)

The current cutting thickness for the i-th pick of the cutting shield and the j-th pick on the transport worms are available from the formulae:

$$g_i = \sin \omega_0 t_i \int_{t_i - \tau_T}^{t_i} \dot{x}\, dt$$

(5)

$$g_j = \sin \omega_0 t_j \int_{t_j - \tau_s}^{t_j} \dot{x}\, dt$$

where:

ω_0 - angular speed of drum;
$\dot{x}$ - shearer travel speed;
τ - time of cutting of the pick in one cutting line described by the formula:

$$\tau_T = \frac{2\pi}{\omega_0 m_T}$$ - for picks of a cutting drum;

$$\tau_s = \frac{2\pi}{\omega_0 m_s}$$ - for picks on transport worms;

m_T - number of picks in cutting line for a cutting drum;
m_s - number of picks in cutting line on transport worms;
t - time period described by relations:

$$t_i = \frac{\pi}{\omega_0} - \frac{\pi(n + k/2) - i}{\omega_0 \, k/2}$$

$$t_j = \frac{\pi}{\omega_0} - \frac{\pi(n + 1/2) - j}{\omega_0 \, 1/2}$$

Substitution in equations (5), (4), (3) and (2) gives the integral equation of drum torque load, for a drum with two picks in the cutting line, in the form:

$$M_{og}(t) = R \left(A_{01T} \sum_{i=n}^{n+k/2} \sin \omega_0 t_i \int_{t_i - \tau_T}^{t_i} \dot{x}\, dt \right.$$

$$+ A_{01s} \sum_{j=n}^{n+1/2} (K_{od1j} + K_{od2j}) \sin \omega_0 t_j \int_{t_j - \tau_s}^{t_j} \dot{x}\, dt$$

$$\left. - A_{02s} \sum_{j=n}^{n+1/2} (K_{od1j} + K_{od2j}) \sin^2 \omega_0 t_j \left(\int_{t_j - \tau_s}^{t_j} \dot{x}\, dt \right)^2 \right)$$

(6)

where:

K_{od1j} - coefficient of coal recovery for the
j-th pick on the first worm;
K_{od2j} - coefficient of coal recovery for the
j-th pick on the second worm.

It is assumed that the relationship between coefficients of coal recovery K_{od1}, K_{od2} and web depth are linear. This requires an analysis of the configuration of picks on the worms and a consideration of the variation of these coefficient values as a function of angle of drum rotation. This is illustrated in Fig. 3.

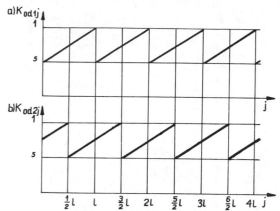

a) K_{od1j} - for the j-th pick on the first worm

b) K_{od2j} - for the j-th pick on the second worm

Fig. 3. Process of coefficient of
coal recovery

These processes are written in the form:

$$K_{od1j} = \frac{1 - s}{l} (j - l(h - 1)) + s \qquad (7a)$$

$$K_{od2j} = \frac{1 - s}{l} (j + 1/2 - l(h_1 - 1)) + s \qquad (7b)$$

where:

s - minimum value of the coefficient of coal
recovery;
h = 1, 2, 3, ... - interval of existence of j,
for example:
for $0 < j \le l$ h = 1
for $l < j \le 2l$ h = 2   and so on
$h_1 = h + 1$ for $j > l(h - 1/2)$
$h_1 = h$ for $j \le l(h - 1/2)$

and it is assumed for discontinuous points:

$$K_{od1j} = K_{od2j} = \frac{s + 1}{2} = K_{od\ av} \qquad (8)$$

as well as for the one pick in a cutting line $m_s = 1$, $K_{od2j} = 0$ and for $s = 1$, $K_{od1j} + K_{od2j} = 1$.

The above mathematical description of the relation $M_{og} = f(\dot{x})$ includes all essential relations existing during the cutting process. Using the above relations it is possible to optimize the pick configuration on a drum by considering the dynamic load for assumed coal properties (s, A_{01T}, A_{01s} and A_{02s}). This description has been used by the author in simulation tests to adapt a system of control for the shearer loader using the simplified model and the relation $M_{og} = f(\dot{x})$ with the algorithm for continuous estimation of the parameters.
This is given in the next section of the paper.
The method of parameter identification in the equations is described in the literature (Podsiadlo, Szaruga, 1976).

In summary, it is possible to show it in the following way: - intoducing auxiliary variables

$$U_1 = R \sum_{i=n}^{n+k/2} \sin \omega_0 t_i \int_{t_i - \tau_T}^{t_i} \dot{x}\, dt$$

$$U_2 = R \sum_{j=n}^{n+1/2} (K_{od1j} + K_{od2j}) \sin \omega_0 t_j \int_{t_j - \tau_s}^{t_j} \dot{x}\, dt$$

$$U_3 = R \sum_{j=n}^{n+1/2} (K_{od1j} + K_{od2j}) \sin^2 \omega_0 t_j \left(\int_{t_j - \tau_s}^{t_j} \dot{x}\, dt \right)^2$$

- one can write equation (6) in the form of a
linear regression function

$$\hat{M}_{og} = \hat{A}_{01T} U_1 + \hat{A}_{01s} U_2 + \hat{A}_{02s} U_3 \ ;$$

- parameters $\hat{A}_{01T}$, $\hat{A}_{01s}$ and $\hat{A}_{02s}$ are estimated by means of the least squares deviations method;
- one can estimate approximately parameter s from diagrams given in the publicatio by Podsiadlo and Szaruga (1976) regarding the dynamics of variations of $M_{og}(t)$ and $\dot{x}$. Fig. 4 shows an example of process M_{og}, real and computational, with parameters of equation (6) obtained by means of the method given above.
Values of parameters were equal: s = 0.02; A_{01T} = 44.15; A_{01s} = 787.8 and A_{02s} = 101.2 for M_{og} in Nm.

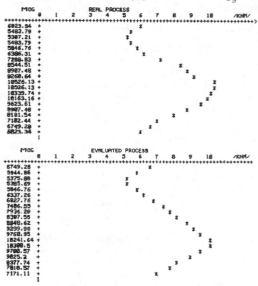

Fig. 4. Real and evaluated processes

DISCRETE MODEL OF RELATION
$M_{og} = f(\dot{x})$ USEFUL FOR CONTROL

The analytical mathematical model of the relation between drum load and travel speed given in formula (6) is insufficient for control because of the complexity and nonlinear character implicit in the relation M_{og} of $\dot{x}$.
It was therefore decided to describe this relation by use of the discrete, linear model in a general form (Manczak, Nahorski, 1983):

$$Y_n = - A(z^{-1}) Y_n + B(z^{-1}) U_n + V_n \qquad (9)$$

where:

z^{-1} - operator of delay;

$$A(z^{-1}) = a_1 z^{-1} + \dots a_R z^{-R} \ ;$$

$$B(z^{-1}) = b_0 + b_1 z^{-1} + \dots b_S z^{-S} \ ;$$

V_n - disturbances in the form of discrete, stochastic processes of autoregression:

$$(1 + C(z^{-1})) V_n = e_n ,$$

e_n - white noise.

As far as relation $M_{og} = f(\dot{x})$ is considered the formula (9) will be:

$$M_{ogn} = -\hat{A}(z^{-1}) M_{ogn} + \hat{B}(z^{-1}) \dot{x} - \hat{C}(z^{-1}) \hat{V}_n = e \quad (10)$$

and error vector $\hat{V}_n = M_{ogn} - \hat{M}_{ogn}$

where $\hat{M}_{ogn}$ - value of model output.

Choice of model rank at the first stage of parameter estimation was made regarding the formula (6). In this formula there are times of delay τ_T and τ_s resulting from cutting of the coal with a drum and pick configuration shown in Fig. 1.
It was possible to expect high values of the estimated parameters b_n

for $n = \frac{\tau_T}{\Delta t}$ and $n = \frac{\tau_s}{\Delta t}$,

which represent the time of sampling of signals M_{og} and $\dot{x}$.
Because of certain asymetry of the pick configuration on the drum there should exist a parameter b_n that relates to

$$n = \frac{\tau}{\Delta t} ,$$

where τ - time of one drum rotation.
The frequency spectrum of signals M_{og} and $\dot{x}$ gain high values in the frequency interval from 0 to 5 Hz.
Because of the large moment of inertia of the drum and the large shearer mass, the maximum time of signal sampling $\Delta t_{max} = 100$ ms. Regarding angular speed of the drum (complete rotation in 1.28 s), it was determined that time of sampling $\Delta t = 80$ ms (16 data parts for complete rotation).
For parameter estimation in model (9), variable components of M_{og} and $\dot{x}$ were used to gain better clarity in the distribution of model parameters. With the time of signal sampling $\Delta t = 80$ ms, 512 data points were used to estimate the model parameters for 32 drum rotations.
At first stage model identification was in the form:

$$M_{ogn} = B(z^{-1}) \dot{x}_n - C(z^{-1}) V_n + e .$$

Results of parameter estimation obtained by means of the generalized least squares method are shown in columns 1 and 2 in the table for different model ranks s, for the rank of polynomial equal 1. Column 3 shows results of identification obtained by means of the instrumental variable method while during the last iteration the method of direct coefficient correlation (Manczak, Nahorski, 1983) was used for model (10) with r = 1, s = 9.
As the results in the table show, the highest values of parameter b exist for delays $\tau_T/\Delta t$, $\tau_s/\Delta t$ and $\tau/\Delta t$. The precision of the model based on root-mean-square error is sufficient for practical use with the limitation of rank of model as s = 9.
Analysis of model (10) with one coefficient a_1 improves the precision of the model.
The next increase of polynomial rank $A(z^{-1})$ did not influence the precision of the model.
Use of the instrumental variable method (column 3 of the table) for parameter estimation is recommended by the author as the optimum one for adaptation of the system to the control of the shearer drum.

CONCLUSION

The presented mathematical description of the relation $M_{og} = f(\dot{x})$ has different practical applications. The analytical model can be used for:

$M_{ogn} = -A(z^{-1}) M_{ogn} + B(z^{-1}) - C(z^{-1}) V_n$; $M_{og}\{kNm\}$; $\dot{x}\{m/min\}$			
Parameters	1	2	3
a_1	-	-	- 0.455
b_0	0.064	0.069	0.066
b_1	0.121	0.166	0.112
b_2	0.276	0.228	0.173
b_3	0.187	0.187	0.157
b_4	0.199	0.090	0.076
b_5	0.053	0.064	- 0.039
b_6	0.001	0.087	- 0.073
b_7	0.156	0.018	0.077
b_8	- 0.157	0.181	0.110
b_9	- 0.113	- 0.280	- 0.241
b_{10}	- 0.050	-	-
b_{11}	0.000	-	-
b_{12}	0.001	-	-
b_{13}	0.001	-	-
b_{14}	0.001	-	-
b_{15}	0.110	-	-
b_{16}	0.000	-	-
b_{17}	0.000	-	-
c_1	0.601	0.627	0.295
Error mean-square of the model	0.187	0.211	0.163
Correlation coefficient of error	0.094	0.128	0.047

- optimization of drum construction and drum drives;
- simulation tests of other simple models and control systems.

The discrete, linear model, which ranks results from the analytical model, is a convenient method of control for the shearer. This is sufficiently precise for practical application in the field.

REFERENCES

Podsiadlo, A., H. Gawelek and A. Jurkiewicz (1978). Analiza dynamiki reduktora organu urabiająco-ładującego kombajnu węglowego. Zeszyty Naukowe AGH, No. 58, pp. 71-80 (In Polish).
Dokukin, A., Y. Krasnikov and Z. Khurgin (1978). Staticheskaya dinamika ghornykh mashin. Mashinostroyenye, Moskva (In Russian).
Podsiadlo, A. and J. Kwasniewski (1975). Ogólna postać równania opisującego ruch posuwowy węglowego kombajnu frezującego. Zeszyty Naukowe AGH, No. 526, pp. 79-93 (In Polish).
Podsiadlo, A. (1976). Moment sił skrawania frezującego kombajnu węglowego. Mechanizacja i Automatyzacja Górnictwa, No. 3/88, pp. 27-31 (In Polish).
Podsiadlo, A. and W. Szaruga (1976). Metoda modelowa określania własności urabianej calizny organem frezującym. Mechanizacja i Automatyzacja Górnictwa, No. 3/88, pp. 14-20 (In Polish).
Manczak, K. and Z. Nahorski (1983). Komputerowa identyfikacja obiektów dynamicznych. PWN, Warszawa (In Polish).
Byrka, W. (1972). Identikatsya yghlyedobyvayushchey mashiny s ryeghuliruyemnymi skorostyami ryezanya y podachy. Ghorny Zhurnal, No. 8.

Computer Assisted Remote Control and Automation Systems for Longwall Shearers

M. BOUTONNAT and P. VILLENEUVE DE JANTI

Electronic Engineering Group, Centre d'Etudes et Recherches de Charbonnages de France Verneuil-en-Halatte, France

Abstract. The automation and centralized monitoring of longwall type underground mining is presently an important subject matter of research in France. The purposes are :

. to improve safety and working conditions
. to increase productivity.
. to improve efficiency in machine operation through continuous "health monitoring".
. to make winning possible in difficult seams.

For these purposes, CERCHAR has developed adequate equipment for shearer monitoring and remote control :

. a direct view remote control system, TELSAFE, which allows one or two machine-operators to control the shearer from a remote place (typically 30-40 feet) safe from dust and falling blocks. This system is of general use in French coal mines.
. a high rate teletransmission system transmitting analog and digital data from the shearer to a remote station through the machine power cable.
. a remote control system to drive the shearer from a remote station also through the power cable.

These two systems operating through the power cable were mounted on a shearer working in a very steep seam (almost vertical) in the Lorraine Collieries. The driver's station is located a few hundred meters from the face at the end of the panel. Data Collected from the sensors on the shearer is transmitted to a micro-computer which then displays on a graphic color terminal a mimic diagram of the machine, of its position and motion in the face, including different values of typical operating parameters and automatic default warning. Research is in progress to build algorithms for real-time optimization of shearer command.

Keywords. Mining ; longwall ; shearer ; robots ; telecontrol ; computer application ; telecommunication.

INTRODUCTION

Most underground coal mines are operating with longwall techniques. Three major types of equipment are used in these workings :
. the shearer
. the powered roof support
. the armoured conveyor

Automation of these equipments is of great interest for the following purposes :

To improve safety and working conditions. Among these harsh conditions are :
. important dust nuisance
. high temperature and moisture content
. little heights (down to 70 to 80 cm in thin seams)
. dips (30° or more) in steep and semi-steep seams
. falling blocks

Any solution to move workers away from these areas will contribute to safety improvement.

To reduce machine downtime (hence increase productivity).

To permit profitable coal getting in seams where it wouldn't be possible with usual techniques.

Automation in mining has been investigated using different guidelines in some countries for several years. The purpose of this paper is to describe Charbonnages de France achievements in shearers control and Cerchar studies in progress concerning their automation.

DIRECT VIEW REMOTE CONTROLLED SHEARER OPERATION

Cuttermen are equipped with radio transmitters to command the shearer from the optimum place safe from falling blocks and other nuisances. Safety improvements and higher average cutting time are obtained, thanks to these equipments.

A new device, TELSAFE-DV, benefits by a long experience of CERCHAR in shearer remote control and incorporates the latest techniques in electronics and microprocessors. Actually, the first generation developped by Cerchar has been introduced in French mines in 1968 (Reinhard, 1973). We are giving now a brief description of TELSAFE-DV which is the second generation to be used in French mines.

TELSAFE-DV comprises one or two portable emitters and a receiver mounted in the shearer.

The emitter (fig. 1) is intrinsically safe and features :
. an ergonomic keyboard with 16 on/off commands
. a rotary switch to command shearer speed and direction (proportional order)

. an emergency stop switch.

The on/off commands are generally used for the
following actions :
. drums up/down
. deflector rocking
. protection shields up/down
. electric motors start/stop
. control of spraying, crushing....

A small size, long lasting (more than 8 hours)
demountable battery pack is fitted on the emitter.

TELSAFE-DV can operate on two transmission channels thus permitting double drum shearer remote
control. In this configuration a cutterman and
his assistant can work together on each side of
the shearer for better cutting optimization. The
cutterman keeps control of the speed and emergency stop is effective from both emitters.

The mounted receiver can operate in highly difficult conditions. For example :
. temperature between - 20° C and 100° C
. supply voltage variations from - 50 % to + 30 %

This unit is easily accessible and interchangeable. Its modular design allows easy adaptation to
a wide range of machine control requirements.

The commands sent by the emitter are digitally
coded and transmitted by a carrier wave whose
frequency lies between 155 and 165 MHz. The experience of the last 15 years has led us to implement
the following general features :
. the transmission system is continuously operating, carrying either orders or "watch" sequences
  when no order is sent.
. default stop occurs when no signal (order or
  "watch") has been received during a preset time.
. emergency stop has priority over all other commands. Both default and emergency stop cause a
  complete stop of the machine.
. a "remote/manual" switch allows quick return to
  manual operation to keep the machine working in
  case of remote control component failure.
. redundancy is used in the coding/decoding process
  to avoid burst errors due to electromagnetic
  noise.

The first TELSAFE-DV was put into service in
January 1982. Now ten machines are equipped and
are working satisfactorily in French mines
(Fig. 2).

AUTOMATION APPLIED TO SHEARERS

Studies in automation applied to shearers were
initiated in France by the development of a new
longwall winning method for very steep seams
(60°) in Lorraine Collieries. There is fundamental
interest in automation in that case because :
. working conditions are extremely difficult
. safety demands that the shearer operates
  behind protection shields which reduce
  the driver's field of view.

Automatic control implies continuous monitoring
of the shearer's position (and it's jibs
positions) relative to the coal seam with
good accuracy. The only way to solve this complex
problem is to use a powerful computer interfaced
with numerous sensors mounted in the shearer.
Considering the lack of volume available on
board and the environmental conditions, it appeared to us that the computer had to be placed
outside in a safe area underground, or even
better at the surface. For that purpose we
needed first a reliable transmission system.
So our automation program started with the
development of such a system. This development
has been successfully completed and the equipment

named TELSAFE-CA.

A system for data transmission between a shearer and a remote station : TELSAFE-CA

TELSAFE-CA uses the three phases of the shearer
power cable as the transmission medium. It can
be considered as two separate equipments
on their own or together :
. a remote control unit to drive the machine
  from a distant place
. a remote monitoring unit to collect data
  on the machine and transmit it to a distant
  place.

These two equipments transmit simultaneously
full-duplex, digital coded information through
the power cable (lenght 3000ft.) using a
single coupler. The full duplex link can
be extended to the surface using simple twisted
pairs. The general structure of the system
is shown on the diagram of Fig. 3.

The capacity of the system presently working
is for remote monitoring :
. 24 "on/off" informations
. 16 analog informations
. usable bit rate : 2,4 Kbits/s

for remote control :
. one proportional and 16 "on/off" orders
  used as in TELSAFE-DV
. one emergency stop order
. usable bit rate : 700 bits/s.

The modular concept of TELSAFE-CA allows
increase (or decrease) in transmission capacity.
The development of TELSAFE-CA took great
advantage of our experience in direct view
remote control systems. Consequently,
performances in reliability, maintainability and environmental tolerance are the
same as those detailed in TELSAFE-DV.

A TELSAFE-CA system has been working in a
steep seam longwall working in Lorraine
Collieries since October 1984. It is used
to operate the shearer from a remote station
located in the top road a few hundred meters
away from the face. Besides we noticed that
the collection of numerous informations from
the machine was of great interest for mining
people. It potentially provides :
. better face monitoring
. reduction of costly troubleshooting time
. continuous "health monitoring" to improve
  machine maintenance and breakdown prevention.

A second shearer is currently equipped for
a high production level seam working.

Computer assisted command

The information transmitted by TELSAFE-CA
were first displayed on a classical synoptic
panel in the remote station. The driver was
at the same time talking on a radiophone
with an assistant following the machine in
the face to give him complementary information.
To ease the driver's work a microcomputer
has been interfaced to the TELSAFE-CA system.
It acquires all information (position in
the face, speed, direction, jibs positions,
tilt, etc...) and processes them to display
on a color graphics CRT :
. a mimic diagram of the shearer in the face
  (Fig. 4)
. a set of information to help the driver
  to optimize machine command.

This software is currently developed in
the above mentioned steep seam working. We

still face problems concerning sensors.

When this development is completed, we shall be able to reach the final step by coupling the microcomputer to the shearer commands and achieve complete automation ; yet we think that human supervision will still be necessary.

CONCLUSIONS

The studies and developments carried out in France these last years have provided tools to ease shearers control and maintenance and to make automation a sound prospect :
. the "TELSAFE-DV" radio remote control system for direct view command by one or two drivers following the machine in the face
. a teletransmission system to centralize data in a remote station and command the machine from the same place.
. software for computer assisted command and "health monitoring".

These achievements couldn't have been possible without the constant contribution of all the mining people in Lorraine Collieries. We want to thank them for their help.

This work is part of a national development program in robotics and is supported by the Research and Technology Agency.

REFERENCES

Reinhard, A. (1973). Les Télécommandes radio des Machines du Fond. Industrie Minérale, Vol. 55.

AMR-C*

M. Boutonnat and P. Villeneuve de Janti

Fig. 1 TELSAFE-DV emitter

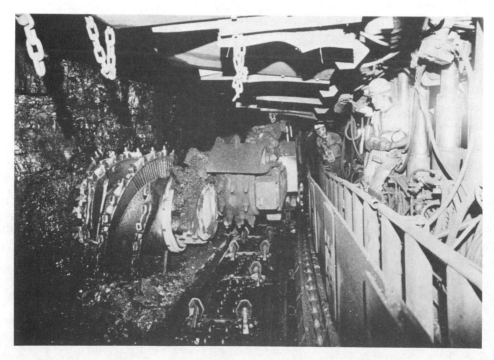

Fig. 2 TELSAFE-DV (with two emitters)
          in a longwall working (Lorraine Collieries)

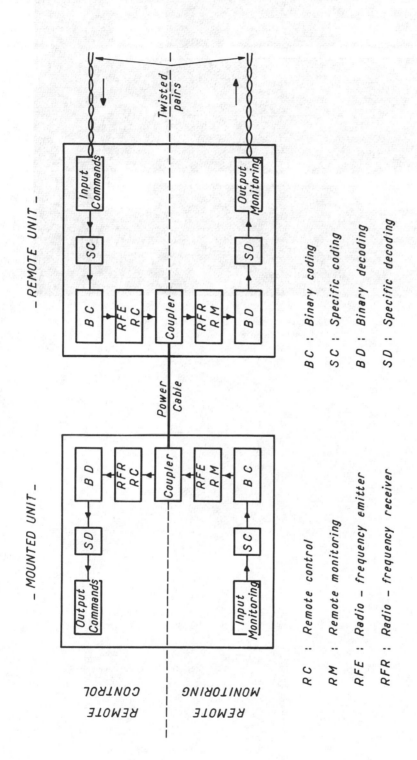

Fig. 3   SCHEMATIC DIAGRAM OF TELSAFE _ CA

BC : Binary coding   SC : Specific coding

BD : Binary decoding   SD : Specific decoding

RC : Remote control

RM : Remote monitoring

RFE : Radio – frequency emitter

RFR : Radio – frequency receiver

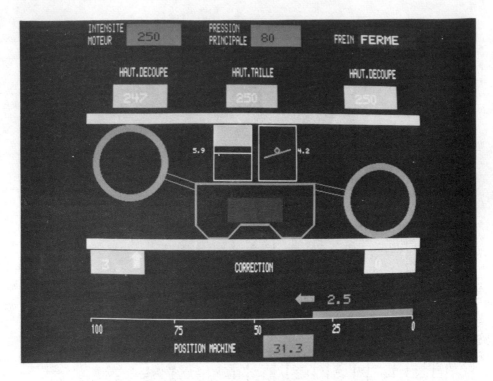

Fig. 4   Mimic diagram of the shearer
         in the face.

Computer Monitoring of Environmental Conditions and Production Processes (WCC)

R.J. FRASER, D.T. EAGER, R.J. MILLER and R.D. LAMA

Kembla Coal & Coke Pty. Limited, Wollongong, NSW, Australia

Abstract. For safe and efficient mining operations, it is essential that environment and equipment status be monitored. A system that gives quick access to status at any given time, and also permits quick analysis of their history, is required. Such a computer based monitoring system has been installed at West Cliff Colliery.

The system is based on 2 super mini computers. Three major sub-systems monitor conditions in the longwall operations, environmental conditions in the intake and return roadways, status of the underground and surface installations, and mining and transport equipment.

The environmental and gas drainage monitoring system is based upon that developed by Charbonage De France Research Institute (CERCHAR), and the longwall support and equipment monitoring system is a development based upon Mines Research Development Establishment U.K. (MRDE). Equipment monitoring system and computer software has been modified and expanded within the KCC Group. All systems operate through central stations, over telephone lines, to the computer system. Information received on the surface is processed and displayed on colour monitoring stations. Information from one or more channels can be retrieved onto the monitoring station screen at any time. The system incorporates alarm signals, which are activated if a given parameter is reached.

Keywords. Computer hardware; computer software; mining environment control; mining production control; pressure measurement; threshold elements; speech recognition.

INTRODUCTION

West Cliff Colliery is one of the deepest and gassiest mines on the South Coast of New South Wales. The Colliery mines the Bulli Seam at a depth of 465-500m using high capacity longwall mining and Wongawilli bord and pillar extraction systems. The thickness of the seam varies between 2.2-3.5m. The mine roof and floor strata consist of massive sandstones and mudstone requiring high load bearing capacity support systems. The mine has been developed using continuous miners and shuttle cars with chemical anchors, roof bolts and "W" straps as the method of roof support in the development headings of 5.5m width.

In the design of the mine layout, consideration has been given to the virgin stress field and geological aspects of the roof and floor, and as such, the mine has been laid out with the major development axis placed North-South and East-West.

The mine faces severe gas emission and gas and coal outburst problems. Over the period of 8 years, 128 coal and gas outbursts have occurred, with the largest outburst throwing 160 tonnes of coal.

The seam gas pressures are of the order of 3000 KPa and the gas content of the Bulli Seam has been estimated to be about 13-15 cubic metres per tonne of coal. Estimates of gas content of lower lying seams have shown that gas emission in the longwall panels is of the order of 36 cubic metres per tonne. High gas emissions in the development headings result in interruptions to mining.

As a result, the mine introduced full scale gas drainage systems in March 1980 (first in Australia). Both pre-drainage (advance drainage using lateral holes) and post-drainage (floor holes, roof holes and goaf drainage) are practiced.

A high capacity longwall system rated at 1250 tonnes per hour was introduced in June 1982, with a face length of 137.5m and 4 leg shield chocks with a bearing capacity of 900 tonnes each (the largest in the world at that time).

Some of the basic problems that occur in the mine are as follows:-

1.   High gas emissions in the
        development headings, 4-6 cubic
        metres per tonne.

2.   High gas emissions in the longwall,
        35-45 cubic metres per tonne.

3.   Major floor heave in the longwall
        panel gate roads.

4.   Massive sandstone roof in the major
        area of the Colliery requiring high
        capacity chocks.

5.   Maintenance of the integrity of the
        roof when mining under shale areas.

6.   Outbursts of gas and coal.

REQUIREMENTS OF A COMPUTER MONITORING SYSTEM

Computer monitoring for safety of mining
environments and production processes has been
used in mines for over 10 years.   Developments
in micro-electronics and data transmission
without distortion over long distances and with
high reliability have lead to the introduction
of a number of data collection, on-site
processing and display systems in mines
(McPherson, 1979; Morris and Gray, 1977;
MSA, 1984; Ribour and Stenning, 1983;
Schuermann, 1983a, 1983b, 1984; Thomas and
Chandler, 1978; Tregelles, 1984 and Welsh,
1983.)

The computer monitoring system designed and
developed at West Cliff Colliery has two basic
components:-

1.   Environmental Monitoring - System
        Requirements

2.   Monitoring of Production Processes -
        System Monitoring

Environmental Monitoring - System Requirements

The environmental monitoring is required to
determine air flow rate and percentages of
methane, carbon dioxide, carbon monoxide and
oxygen at various selected points in the mine
atmosphere, flow rate vacuum and percentage of
methane in the methane drainage pipe ranges
from various districts, ventilation pressure
differences at the stoppings, and activation of
pre-determined alarm signals when a selected
parameter is exceeded.   The frequency of
sampling is dependent upon the importance of a
particular location.   The percentage of
methane in the ventilation network must be
monitored at least every 15 minutes, but the
main methane drainage range requires continuous
monitoring to avoid any low methane mixture
passing into the drainage lines.   At the same
time, the monitoring should be independent of
the mine power supply so that continuous
monitoring can be done even when the power
supply to the mine or a part of it is
completely cut off.

Other requirements of the system should be the
possibility of switching over from automatic to
manual control with monitoring of a single
location continuously, on-line graphic display,
automatic and manual control of data
transmission lines in case of any defect (short
circuiting or line break) in the functioning of
the system, provision of the data transmission
line also for use as a telephone communication
system from underground to a central control
point on the surface.

There are currently two basic systems of
on-line monitoring of mine environments,
namely:-

1.   On-the-spot sensing

2.   Tube-bundle system

In the present application, on-the-spot sensing
has been selected because of inherent
advantages of the system, though this system is
more costly to install and requires more
stringent monitoring.

A number of instruments are available for the
detection of gases and monitoring of
velocities, pressure differentials etc.   The
two main types, which have undergone extensive
testing, are those that either utilise thermal
conductivity or infra-red sensors for detecting
the percentage of various gases.   Quantity
measuring instruments either make use of a
pitot tube or an orifice plate with a pressure
differential sensor head or a thermister for
direct measurement of velocity.

The system chosen uses a thermal conductivity
sensor with a range of 0-100% for the
monitoring of methane gas in the pipe, a
catalytic sensor with a range of 0-5% for the
monitoring of methane percentage in the
roadways, a thermister sensor for the
monitoring of air velocity in roadways, a
pressure differential coupled to a pitot tube
for monitoring velocity in methane gas ranges
and across stoppings and a pressure transmitter
for monitoring vacuum in gas ranges.   The
range of the sensors is given below.

0-5% CH_4 --- $\pm$ 0.1% up to 2%

0-100% CH_4 --- $\pm$ 4% from 10%

Air velocity --- 0-25 m/s in roadways
0-8 m/s $\pm$ 8% and 0-25 m/s --- $\pm$ 8%

20% - 0% O_2

0-300 ppm CO

0 - 2% CO_2

0-70m Bar pressure differential $\pm$ 0.25% for
measuring pressure across stoppings

0-500 Pa pressure differential of monitoring
velocity pressure in pipe ranges

0-100 KPa for monitoring vacuum in pipe ranges

A schematic diagram of the monitoring system for the methane gas drainage lines and environmental monitoring underground is shown in Fig. 1.

The central cabinet is located on the surface in a non-hazardous area and supplies the current for charging of the batteries for the sensors placed at different points underground, as well as for interrogation of the sensors and graphical recording of data as obtained from all the sensors. The underground system consists of a network of telephone lines and sensors which are connected to the surface through a barrier cabinet.

The speed of the chart recorders is 12.7mm/hour and a 24 hour record is visible through the front window. Visual and audio alarms are set when the percentage of CH_4 exceeds 1.25% and 2.5% in the returns, or is below 40% in the gas drainage lines. Any cut or defect in the line is also signalled by a visual and audio alarm.

Monitoring of Production Processes - System Monitoring

The monitoring of the various elements of the production processes has three basic requirements. Firstly, to monitor equipment conditions to facilitate scheduling of changes for maintenance purposes; secondly, monitoring to pinpoint the location and type of breakdown of the element; and thirdly, to inform the maintenance personnel of the steps to be taken. The system requires monitoring of a large number of elements forming the production chain. The frequency of monitoring required varies, depending upon the elements. In the case of fixed equipment belts, face conveyors etc., it means a constant check on the status (on/off) and a log of the reason for the failure. In the case of moving equipment, it is required to locate its position, as well as to monitor its functioning (on/off) and the status of the various components by a system basically designed for health monitoring. Health monitoring is also coupled to a computer programme designed to initiate, execute and record all maintenance and scheduling of equipment and its history, prepare work orders, spares inventory, resources and execute management reporting.

Presently, the production process monitoring system on the longwall includes the district belts, shearer, crusher, tail gate and main gate drives. The district belts are monitored for on/off, belt tension, and temperatures of drives and bearings. The total length of the belts has been divided into a large number of zones of approximately 200m length and the information is individually monitored. This helps locate the fault and facilitates rectification by underground personnel. Facilities exist for automatic belt start, though this is not utilised.

The face conveyor is monitored for load current, supply voltage, chain speed, motor temperatures and tension, as well as on/off. The shearer is monitored for temperature, power input, drum speed and on/off.

The longwall face support system is monitored for chock pressures in front and back legs and convergence and strains in the chock structure. Figure 2 shows the schematic diagram of the monitoring of the each bank on the longwall face.

The main surface substation, the service facility at the pit bottom, the men and materials hoisting system and the parameters of the main fan are also monitored.

Telemetering System

Three independent telemetering systems are used in the mine. These are:-

1. Transmitton TM204 for monitoring 48 analogue channels from the longwall support system.

2. Oldham CTT 63/40U (CERCHER) for 80 gas drainage and ventilation monitoring channels.

3. GEC SRC 1 for monitoring 800 digital and 60 analogue channels of the various production equipment.

Each of the systems has a stand-alone capability.

The Transmitton TM204 has a capacity to scan 48 analogue points every 2.6 seconds and communication is carried out using two telephone lines, one for address down and the other for data back. Four outstations are combined into one cabinet. The transducers (Kratus, UK) provide 4-20 mA current loop output. All underground electronics have at least 12 hours of battery backup.

The Oldham CCT 63/40U system with 80 channels interrogates in a 40 minute cycle through a pair of telephone lines which carry the battery charging current and the telemetry signal, as well as allowing a plug-in telephone set to be used during maintenance. The analogue transducer signal is converted into a variable frequency a.c. signal in the range 6-12 kHz by a voltage to frequency convertor and then returned to its original form on the surface by a frequency to voltage convertor. Signals are recorded on charts and also fed to a computer.

The GEC SRC 1 uses one of 25 audio tones per channel. A channel is time multiplexed up to 120 digital points, each point represents one contact operation, and 8 or 16 points are combined to form one analogue point. The modulation method is a three level frequency shift of $\pm$ 30 Hz. Up to 20 channels can be carried in one radio telephone circuit. The telemetry signal is superimposed on the existing working telephone lines by selecting the six tone frequencies above the useful voice range ($\sim$ 2.4 KHz) and incorporating suitable low pass filters and barriers to isolate the telemetry tones from the telephone equipment.

A general layout of the Telemetering System and the computer interfaces is given in Fig. 3.

Computer Hardware and Software

The general criteria in selecting a computer monitoring system is as follows:-

1.   It should provide an accurate (acceptable) real time picture of the environmental and production processes occurring in the mine.

2.   The system should have the facility to record the input and permit analysis of trend.

3.   It should provide alarm signals and automatic shutdown, stop and start of the individual units.

The information from the three telemetering systems relating to equipment and environmental monitoring is fed into two DEC 11/750 Vax computers using RS 232 and IEEE 488 interfaces. One of the Vax 11/750 computers is used for environmental monitoring and the other is used for backup, programme development and as a general purpose computer. Four RM 80 124 megabyte disc drives with one TU 77 tape drive units are used for mass storage. The control room has a number of black and white and colour monitors, one LXY 11 printer and two LA 120 printers.

Figure 4 shows the control room and the video monitors, communication system and environmental recording cabinets.

Unified software has been developed, which permits scanning of any channel or the monitoring of a set of channels on the video-monitor simultaneously. Although the programme is capable of handling almost any conceivable parameter, input data about a number of limited operations is also put in manually because of the excessive costs involved in sensors. These include presently, the values of each gas drainage hole, maintenance of mining machines, coal quality etc.

The data from any of the channels can be continuously monitored on the video while being simultaneously stored. The data trends can be monitored over 24 hours, one week or on a monthly basis.

Figure 5 shows a record of the monitoring of a gas drainage pipeline. Figure 6 also shows a record of monitoring of 4 transducers in a roadway showing methane percentage, velocity, quantity of air flowing and the differential pressure. The percentage of porosity, vacuum and flow rates are logged as a function of time. Figure 6 shows the ventilation district with the location of the various sensors.

Figure 7 shows an example of monitoring the status of the various conveyor belts in the mine.

The information from the computer is available at various points on the surface and underground through black and white and colour monitors, as well as to the Head Office of the parent company. In the case of a breakdown, maintenance personnel can dial into the system and a voice modulaton system will inform them of the failure mode and location of the fault in the equipment. This facilitates quick attention and rectification.

A number of daily and weekly reports are generated for attention of the supervisory personnel. These include:-

1.   Ventilation status of the various districts.

2.   Gas drainage line status of the various districts, including number of holes on line/off-line and flow rates etc.

3.   Status of the main transport belts, skip hoisting, including times the belts were stopped, number of stoppages and lost time, reasons for stoppages, coal transported.

4.   Status of mining equipment, health history, tonnages mined, maintenance scheduling etc.

5.   Coal washery throughput, coal quality etc.

CONCLUSION

The mine monitoring system introduced at West Cliff Colliery has proved successful.
The system gives reliable information on the safety of the mine environments and status of equipment. The system has greatly facilitated data collection, report generation, analysis of various factors that influence production and maintenance of high standards of safety and environments in a very gassy underground mine.

REFERENCES

McPherson, M. J. (1979). Mine environmental monitoring and control, computer methods for the 80's in the mineral industry. In A. Weiss (Ed.), A.I.M.E., New York. pp. 705-116.

Morris, I. H. and G. W. Gray, (1977). Environmental monitoring and control in the United Kingdom. Proc. Int. Conf. Remote Control and Monitoring in Mining, NCB, London. pp. 10.1-16.

MSA, "DAN" (April 1984). Data acquisition network for mine-wide monitoring. MSA.

Ribout, J. and G. Stenning, (1983). Mine monitoring system provides instant information. Canadian Mining Journal, Nov, 91-94.

Schuermann, F. (1983a). Entwicklung einer ferngestauerten Vielsteller Messtechnik zur Fernuberwachung von Betriebsuorgangen. Gluckauf, 119, 684-90.

Schuermann, F. (1983b). Messgeber fur die prozessorgesteurte Vielstellen-Messtechnik zur Fernuberwachung von Betriebsvorgangen uber und unter Tage. Gluckauf, 119, 935-41.

Schuermann, F. (1984). Fernuberwachung und
    Fernlenkung Langsam ablaufender
    Betriebsvorgange in Abbanbereich. Gluckauf,
    120, 590-99.
Thomas, V. M. and K. W. Chandler, (1978).
    Monitoring and remote control progress
    towards automation in British coal mines.
    Mining Engng (London), 136,
    No. 206, pp. 251-62.
Tregelles, P. G. (1984). Microelectronics
    provides new coal production techniques.
    Aust. Coal Miner, 6(6), 7-11.
Welsh, J. H. (1983). Computerised, remote
    monitoring systems for underground coal
    mines. U.S.B.M., I.C. 8944.

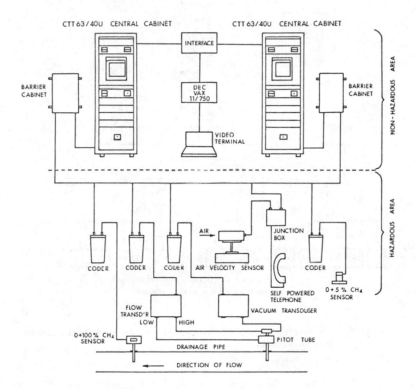

Fig. 1.   A schematic diagram of the monitoring system for the
          methane gas drainage lines and environmental monitoring
          underground.

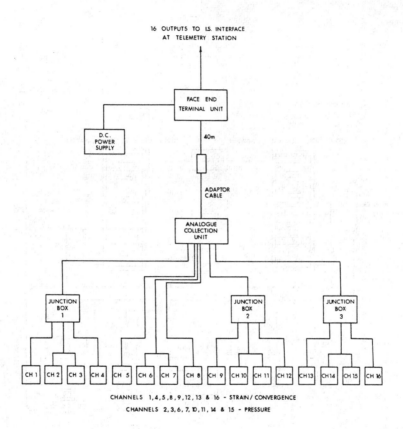

Fig. 2.   Schematic diagram of the monitoring of each bank on the
          longwall face.

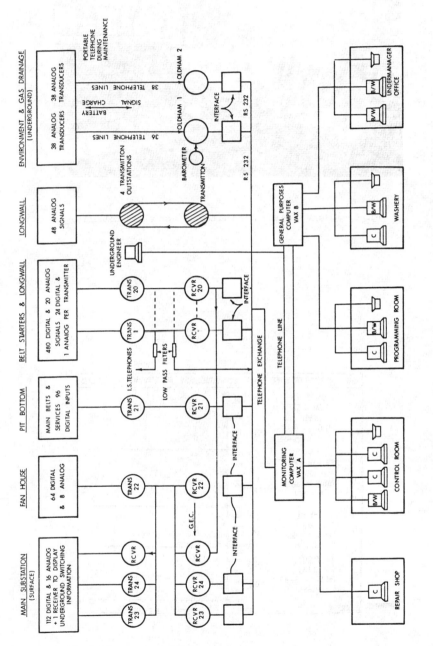

Fig. 3. General layout of the Telemetering System and the computer interfaces.

Fig. 4.   Computer Monitoring Room at West Cliff Colliery.

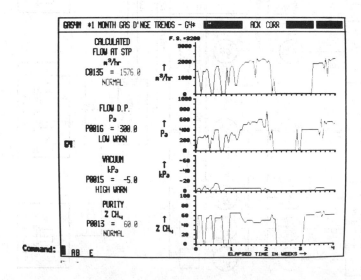

Fig. 5.   A record of the monitoring of a gas drainage
          pipeline and monitoring of 4 transducers.

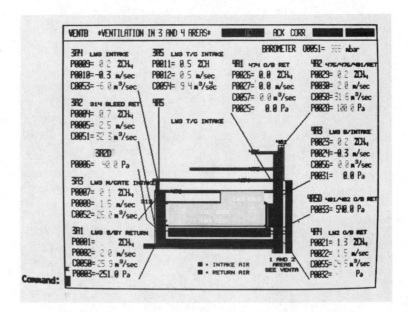

Fig. 6. The ventilation district with the location of the various sensors

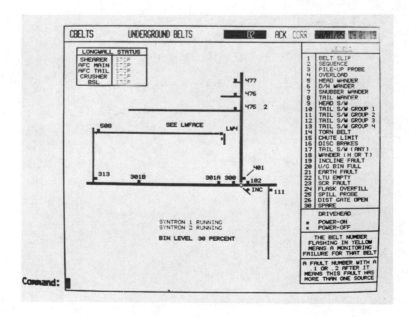

Fig. 7. An example of monitoring the status of the various conveyor belts in the mine.

A Microprocessor Based Information and Control System for Essential Services on a South African Gold Mine

H.C. BAKER DULY

Assistant Consulting Mechanical and Electrical Engineer Anglovaal Limited, Johannesburg, South Africa

Abstract. The application of cheap robust microprocessor based units to achieve monitoring and control in an underground mining environment is described. The functional layout of the programmable controller and the methods of multiplexing and processing information from local and remote sources are examined as are features for refrigeration plant control.

High noise immunity communication to a remote micro computer uses existing telephone lines and enables display and print of essential data. The development of this system then leads to applications in monitoring compressed air, ore transport and underground fires.

In conclusion, the problems of data presentation to different levels of management and the future philosophy of a central multi-reporting system are discussed.

Keywords. Microprocessors; distributed control; analog-digital conversion; PID control; hierarchical systems; refrigeration; temperature control.

INTRODUCTION

During the past decade the South African Gold Mining Industry has witnessed a concerted drive towards higher production, improved environmental conditions and greater safety. The rapidly increasing cost of unskilled labour and the installation of large expensive equipment has meant that management have had to tighten up their whole concept of man-power utilisation. The result has been a need for a Management Information System where none existed before.

Ventilation equipment consumes approximately 33% of a gold mine's electricity demand, and while fans may be seen as essential to production, refrigeration plant is frequently forced to take second place as regards maintenance when compared with ore handling equipment. As Hartebeestfontein Gold Mine was in the midst of a R20m addition to its refrigeration, it seemed logical to install some sort of monitoring and control system with certain basic abilities. Simultaneous accurate monitoring of all vital information was to be carried out in each plant room and the status of peripheral equipment such as pumps, bulk coolers, pipe networks and other external equipment should also be indicated. At the same time, a reliable control centre was required for water distribution, air temperature control and the provision of

sophisticated interlocking and load control on actual refrigeration machines. However, to accurately fulfil the function of a Management Information System, this information had to be available where senior management could see it without taking an extensive underground journey. Although the requirement was to centralise information at an intelligent device, the temptation to eliminate the original idea of distributed control was resisted for obvious safety reasons.

The original selection of equipment took place in 1979 and after much discussion it was decided to opt for a programmable controller, mainly because the complexity of the problem meant that a hard wired solution would be immediately outdated by new requirements. Although such a decision would seem obvious today, at that time electronic equipment had a particularly poor reputation underground and the programmable controller was looked on as a production line tool rather than a rugged data collection device. For these reasons, the selection of the toughest, most robust unit on the market was essential. Physical attributes such as 60°C cabinet temperatures with 90% humidity, dust proof, shock and vibration proof were to be allied with internal features such as diagnostics, battery back up, communication option, easily accessible I/O modules and easy ladder programming. The only candidate to fulfil those requirements was the Gould Modicon 484.

For minimal remote presentation of information, a VDU driven by a micro-computer included storage, higher calculation facility, and printer drive in one package. A simple inexpensive monochrome device (Motorola Exorset 100) was chosen as an interim, mainly because of its expander facilities and robust keyboard. The sheer volume of information and attendant processing required was however not envisaged, nor were the demands by management for more sophisticated presentation.

PROGRAMMABLE CONTROLLER

Referring to Fig 1 it can be seen that the 484 is a single, sealed box unit containing the microprocesser CPU, ROM, diagnostics, Comms interface, battery back up and 84 user RAM. The unit is attached by a multi pin plug to a TTL parallel data bus feeding a maximum of 256 I/O in 8 housings. The end of the bus may be terminated in a hot standby 484 or an expander module to allow a further 256 I/O. Digital signals are addressed by switches on the housing and by the module position in the housing while the on/off status is indicated by a single L.E.D. The number of analogues, addressed by switches on each module, is limited to 32 in each 256 so that it has been necessary to multiplex the analogue input signals using spare digital outputs to switch reed relays every half second to allow de-bounce time. There are two LED's on each analogue input, one to indicate a blown fuse and the other for 'on' status. The multiplexing operation can be carried out in single or parallel banks of 4 or 8 signals depending on access time and numbers of inputs.

Modules usually have an output of 24 V D.C., but are available in AC and other voltages. Bus communication is 0-5V and the PC acts as an analogue to digital converter storing the signals as 0-999 in RAM which gives an accuracy of 0,01%. When a display is required, this raw data is multiplied by an engineering unit factor and appears in a 3 digit window. Modules are polled by the CPU as dictated by the programme, usually approximately every 20m sec. The programme contains an elementary alarm and/or trip routine which drives warning lights at the front of the panel. The trip functions as first hold until accepted, to ease diagnostics.

The P.C. is programmed by a portable programmer which contains a VDU and on which is displayed the ladder network which is in the form of an electrical circuit. When the programme is running, those parts of the network which are passing current are brightly illuminated, making fault finding easy. The programmer can be plugged in locally at the plant room or used remotely via the communications module, thus saving an underground visit.

The control of a refrigeration machine, apart from obvious switching operations, requires several PID loops. These are not available on 484 as part of the programme package and therefore had to be programmed into the software. The requirement (See Fig 2) was for vane opening to be adjusted via an electro-pneumatic converter, to achieve specific outgoing water temperature while at the same time maintaining motor kW within a certain range and also not allowing the compressor differential pressure to exceed a preset value, to avoid surge. An overriding manual control was to be provided for water temperature only. Each PID loop contained three time constants which depended on individual machine characteristics and were set on site. Whichever function required the vane to be closed naturally had preference and was located by feeding all PID outputs into a Low Signal Detector. As a set point is crossed and control passes from one variable to another it is necessary to latch the two signals for a short time period to ensure a bumpless transfer.

The measurement of flow presented some problems as the P.C. had only limited arithmetic ability. Square root extraction had to be accomplished by means of a convergent series and then finally the previous nine readings were averaged to provide damping and reasonably consistent readings. When calling up a reading, the slight delay is not however noticeable.

In order to receive groups of signals from remote installations over a long distance, a teleterm squeezer unit was used. This is an electronic multiplexer that receives 32 analogue or digital signals and digitizes these signals from 4-20 mA to 0-1024. This code number is then sent by a Serial Current Loop Communication to a demultiplexer standing beside the P.C. and feeding 32 analogue or digital input modules. The main advantage is an enormous saving on cable and elimination of loss of accuracy due to voltage drop and line noise.

By far the greatest expense of the whole system was caused by the probes. The temperature, pressure and differential pressure probes were fully sealed stainless steel units, also supplied by Gould. These were adjustable for zero and range by two magnetic plugs operating through the casing to corresponding magnetic adjusters inside. A calibration problem occurred initially as, after setting, the external magnets were removed and vibration allowed the internal adjustors to move. Calibration is now done on an exchange basis every six months in order to maintain the 0,1% accuracy demanded by refrigeration efficiency calculations. The method of calibration is to extract the digital 0-999 reading from the P.C. for the particular probe and check this against an independent instrument. Air flow is measured at points of known cross section by vortex velocity meters which although frequently obscured with dust and dampness maintain surprising accuracy.

COMMUNICATIONS

Having established a number of plant rooms with their own self-contained P.C. systems embodying monitoring and local control of machines and water distribution, it was necessary to bring this information to surface (see Fig.3). Existing twisted pair telephone cables were available throughout the mine but were usually run next to main 6,6 kV power cables with consequent noise pollution.

The method of communicating with the surface micro-computer was by means of the standard FSK modems available for the Modicon P.C. but at non-standard frequencies to give very high noise immunity. Binary messages were sent on 50kHz mark and 80kHz space on a 4 wire half duplex system. That means that two-way communication is carried out, but not at the same time, as there is very bad capacitance between the pairs and, indeed, the signals in the four wires are virtually indistinguishable, and only the computer knows which direction is being transmitted. The chosen transmission rate is 2 400 Bauds although 9 600 is quite possible but unnecessarily fast for the rest of the system. Maximum transmission distance is 5 000m which is sufficient for any particular shaft system.

Communication is initiated by the micro computer by means of a 'header' followed by an address which is recognised by a particular P.C. Protocol allows message codes as follows: 1 start, 8 data, 1 stop + CRC + 1 Parity. Security is provided by the CRC or Cyclic Redundancy Check in that the total binary message is divided by a particular polynominal expression and the remainder put at the end. The receiving device performs the same calculation and if the remainder differs logs a CRC failure and rejects the information. A further check is provided by Parity which means that all the binary marks or ones are added up and if the total is even a zero is attached and, if odd, a one.

These diagnostic operations are performed rather crudely by means of logical shift but at a speed of more than 1 000 operations per second so that transmission is not delayed. The computer polls each P.C. in turn extracting all the digital readings from its registers and also giving such commands as it is allowed as well as clock synchronisation in the event of power failure.

MICRO COMPUTER

The Exorset 100 micro computer has the appearance of a simple mono-chrome unit but is in fact a highly flexible machine as can be seen in Fig. 4

Externally the computer has a bus expansion system which allows the connection of a real time clock card which includes 1 k of battery backed RAM. A number of A/D modules may also be fitted for the collection of local data which is unprocessed by a P.C. Another option which is used is a communications board for hierarchical passing of information to a similar computer using a line driven UART system. Two 5¼ inch 165k floppy discs are used, one to hold the executive programme and the other for the long-term data storage. The computer is able to carry out multi-task operations except when the disc drives are operating, when it is dedicated and information is stored but not processed. The EPROM monitor has been modified to accept the special programme disc rather than the standard MDOS so that the normal calculation options are not available and the machine is dedicated to the monitoring and control function. The monitor now also immediately reloads after power failure and re-sets the time from the clock module.

Apart from these modifications, careful programming techniques have enabled everything to be put directly in assembler language with the added advantage of speed of execution. Data processing is done by Microprocessor Pascal (MPL). To carry out certain enthalpy calculations on different refrigerants as well as air and water, it has been necessary to store complete sections of tables and interpolate two dimensionally between them. If space becomes critical, it will be possible to burn these into EPROM.

The programme also contains factors for conversion to engineering units for, as mentioned previously, accuracy has dictated that readings are transmitted directly from the digital registers of the PC. This information is retained on No. 2 disc for 24 hours on a continuous update basis and may be called upon in graphical form at any time. A sophisticated alarm system is applied to each reading as it is received, it being allocated into four inclusive brackets; normal operating range, out of range and advise daily, alarm conditions advise immediately, reading error - advise daily and discount readings.

Because of the method of programming, it has been found necessary to provide certain easy to use routines for the changing of the value of certain variables such as alarm ranges, engineering units etc. and to control the type, format and time of daily print out. A short manual giving access codes has been given, but it excludes codes to the programme proper or the setting of the real time clock.

In order to overcome the problem of having both a printer and a plotter, a printer with pin addressable graphics was used and driven through an ASCII interface via an 8 - bit data highway. Using hexa decimal code, 0-1F is kept for special symbols while 20-7F is used for normal alfa numerical and 80-FF is available for graphics. In order to do graphics with this system the form feed is taken away from the printer and a line feed supplied from the micro-software.

The higher order calculation ability of the micro as compared with the P.C. has been well utilised in performing several routines on incoming data. Performance analysis of a refrigeration plant includes general heat balance, test for non-condensibles, instantaneous utilisation, compressor efficiency, Carnot cycle efficiency. For a spray chamber or condenser chamber, the factor of merit is calculated as well as water and air side efficiencies and duty. For a refrigeration system, rates of heat transfer are calculated for service water cooling, air cooling and balanced against the thermal capacity of the make up water system and total refrigeration effect. At one particular shaft, which has a complex water distribution system, it enables the amount of water in process to be known, and when it is expected to be returned to main settlers, and what quantity of make up is required from surface.

FURTHER APPLICATIONS

Having developed a system giving centralised information and yet distributed control, it was easier to apply it to other essential services on the mine.

The next to benefit was the compressed air reticulation, which is constructed in a series of main feeds linked by surface and underground ring mains. Compressed air is an extremely expensive commodity, and it was clear that there was an unaccountable base load of more than 50% of maximum demand. The existing refrigeration monitoring infrastructure was used and orifice and pressure tappings put in all main feeds in the vicinity of a PC or Micro. The micro at each shaft then was able to do a PV calculation and give a reasonable indication of volume flow at standard pressure which could then be compared with known compressor performance, assuming similar temperatures. This information is then fed across to a micro at the compressor house, which collects data from all over the mine and is able to fill in the figures on a mine-wide network and estimate flows in unmeasured connections, and consequently pinpoint with some accuracy sudden unusual usage or pipe breakage.

The transport of ore or waste has always lead to anomalies in the quantities claimed, ranging from underground hopper transport by the hopper, to hoisting by the skip and then final run of mine crushing figures. Clearly the major figures can be obtained by measuring quantities of hoisted rock (see Fig. 5). The best place was found to be in the loading flask at shaft bottom where a dual function can be performed. The radial door supplying the flask is operated by air pressure and can be controlled by the output signal from a small P.C. which is working on the input from two load cells. These latter are in the form of KIS beams from which is hung the loading flask and are driven by an exciter giving 5mV output. Skip position is monitored by micro switches and is then filled by the correctly loaded flask on a signal from the P.C. Each skip is therefore neither overloaded nor underloaded and a record of hoisting and quantity is available for transmission to surface or display to the hoist driver although most rock hoists are automatic.

In shaft systems with different mining sections, it is sometimes important to know how much ore is coming from a particular section. This is most easily done by using load cells in a particular section of conveyor in conjunction with a roller to record belt speed. Another method using nuclear bombardment to form X-rays is more suited to coal conveyors.

The use of the micro computer to record and display information and thus displace the chart recorder enabled considerable refinements to be made to the fire detection system. As received by the mine the MkVIII system consisted of 64 beacon heads each producing a signal of 0-1 volt. This information was transmitted from underground on 64 discrete frequencies in the 10kHz band, the 0-1v being modulated as 5-25Hz on the carrier.

On surface the signals were decoded and fed to 32 dual pen chart recorders, each with mechanical alarms to indicate falling voltage. The flexibility of the micro was such to allow the frequency modulated signal to be fed directly, without modem, onto two 32 channel A/D boards, similar to those used to collect compressed air data. Once stored by the computer a 24Hr graph can be displayed, and if necessary printed, for each beacon. Also the fourteen previous 24 Hr graphs are retained.

The alarm threshold is individually adjustable for each point and may be altered to allow for age of the beacon or the continuous proximity of diesel exhaust. To prevent spurious alarming, which was a feature of the old system, a fixed level alarm and a tracking alarm are used. In addition, an alarm is only flashed if the signal has remained in a so called window for a pre-determined length of time. There are both short windows (0 - 60m) for normal operation and long windows (0 - 180m), which are activated during blasting time. Given that beacons operate in the range 0,1 to 0,9 V, anything outside is indicated as a head fault and the threshold is normally set for readings below 0,7 V.

DATA PRESENTATION

All the systems described in the text were designed and constructed by highly technical personnel to provide every bit of information necessary to carry out a complete performance and diagnostic check on remotely situated equipment. However, to the person who was not a refrigeration engineer, the austere monochrome tables of figures and even trend graphs of particular variables were not meaningful. It was also intended that as little paper should be generated by the computers as possible and that interested personnel would use the access code book and carefully choose only essential information for printing. It was expected that interested maintenance personnel would report malfunctioning probes to the technician and repair poor system operation as reported by the computer. Resistance to the new technology was to be expected and manifested itself as apathy for several years. An improved method of data presentation could possibly have shortened this time but was unfortunaely not possible with the small micro computer originally thought to be adequate. The solution would have been the implementation of central computers with hard disc storage and advanced colour graphics and processing. This project was unfortunately postponed due to the economic recession but would have the following features: information storage for a complete year to provide annual trends, colour mimics of complete systems and their more important sub-components with information displayed on the mimic, more advanced calculation ability for impending fault diagnosis and the ability to take over system and plant control in the event of the withdrawal of underground

attendants, multiple reporting facility that would offer senior management a single page with system performances and major faults during the previous 24 hours and, at the other end of the scale, offer the fitter a terse bulletin listing faults and problems to be rectified in a particular plant.

The obvious configuration would be to send all information directly to the central computer and then use the existing micros as monitors. The weakness is that a fault in the central unit would shut down the whole mine's information system so that the less attractive method of operating through the micros may have to be used.

CONCLUSION

Without the help of an in-house laboratory and staff it would not have been possible to have developed the necessary flexible tailored system which is now installed. One of the spin offs of the installation has been that several years of pre-occupation with the process and the accuracy of readings has meant that everybody concerned has learnt a great deal and is therefore better able to execute the maintenance or management function.

Acknowledgement

This work was carried out as part of technical development by Anglovaal Limited. The author wishes to thank the Executive Director (Mines) for permission to publish this paper.

REFERENCES

1. Bateman Process Instrumentation Ltd. Distributed control using programmable controllers. Paper presented at South African Council for Automation and Computation User Forum '84 Pretoria, February 1984.

2. Coetzee R.P. The practical implementation of programmable controllers in the mining industry - a case history. Certificated Engineer, 53, 1983.

3. Baker-Duly, H.C.L. Control of refrigeration and ventilation systems in a South African Gold Mine. Paper presentated at I.M.M. Third International Mine Ventilation Congress, Harrogate 1984.

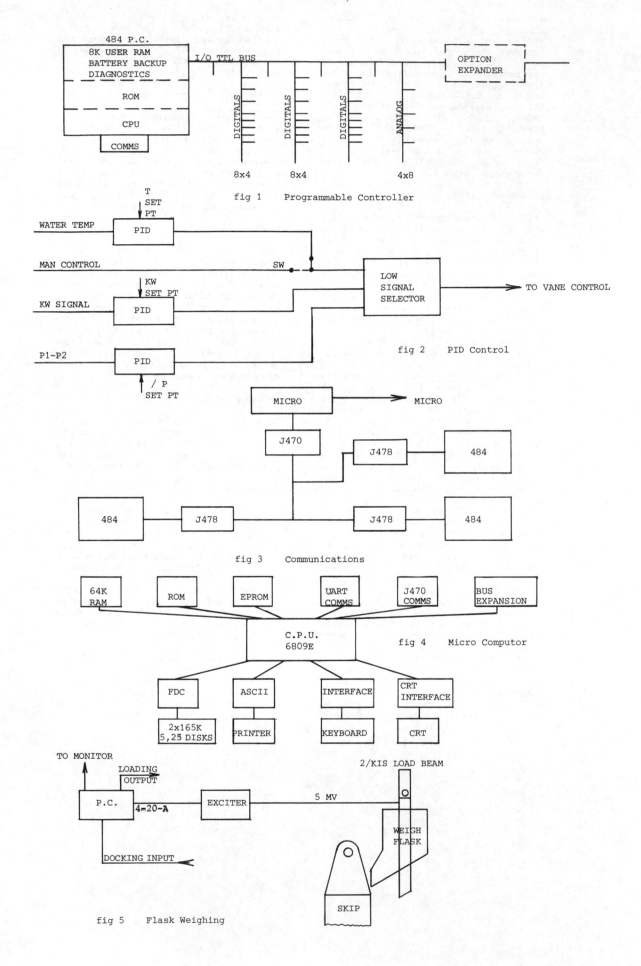

fig 1   Programmable Controller

fig 2   PID Control

fig 3   Communications

fig 4   Micro Computor

fig 5   Flask Weighing

Experiences with Computer-Based Monitoring Systems in Polish Coal Mines

S. CIERPISZ

Technical University of Silesia, Gliwice, Research and Manufacturing Centre, EMAG, Katowice, Poland

Abstract. A modular computer-based monitoring system for mines is discussed. The system consists of the following modules : production monitoring system HADES, methane monitoring system CMC, seismoacoustic and seismometric systems SAK and SYLOK, communication systems. The system enables one to compose flexible, economic structures according to the specification of a mine. Application of microcomputer-based monitoring and control systems in coal preparation plants also is discussed.

Keywords. Computer application; mining.

INTRODUCTION

Production of a mine as well as the safety depend on a well designed centralized monitoring system, the equipment used, and the methods of controlling the production process. Initially, in the early sixties, the dispatcher supervision was limited to a communication system only. Over the years, the control methods and equipment have been systematically improved, and now in some mines there are management centers based on computer monitoring and control systems.

Digital technique was introduced with caution in mining, and the obtained results did not always meet the requirements. Negative results were mostly due to poor reliability of the hardware and conceptual errors.

Since the very beginning, two trends in application of computers emerged. First, was the use of one computer, sufficiently large to cope with monitoring and data processing concerning all mining processes. The second trend was to use small digital systems, each to solve one problem.

Because of easier implementation and greater reliability of small systems, the second solution was adopted, whereby a minicomputer links with a set of sensors and a transmission system and has typical and flexible software. These specialized "modules" allow one to build flexible monitoring and control systems according to the specification of a mine.

A set of such "modules", covering a wide range of the needs of mines, was developed in the Research and Production Centre EMAG. All modules are built around one type of minicomputer PRS-4 designed and produced in EMAG .

GENERAL CONCEPT OF THE SYSTEM

The general concept of the modular monitoring system is as follows :

- The system should enable : failure and emergency signalling, monitoring of the production process and work safety parameters, current reporting, supporting the dispatcher in case of rescue operations, possible control of selected installations or technological links.

- The system should be designed as a uniform construction based on a typical central unit, the structure of the system should be modular in hardware and software, enabling easy economical

configurations according to the current
requirements of a mine.
- The system has to be designed for stage-
wise construction and operation in a
mine.

In line with the adopted concept, the
following sub-systems ("modules") were
designed and applied in mines:

- Monitoring of the production process-
providing monitoring of the main winning haulage process and production
balance - "HADES".

- Monitoring of methane concentration,
providing also shut-down of power sup-
ly in the hazardous areas- "CMC".

- Bumps hazard evaluation by means of
seismoacoustic methods - "SAK".

- Localization of seismic events by means
of microseismological methods -"SYLOK".

- Rock and gasses outbursts hazard evaluation, basing on the measurement of the
volume of the emitted gasses and employing seismoacoustic methods- "SAK-SG".

The system is completed with :
- Alarm communication system "AUD-80",
- Overall mine communication "UDG",
which are not based actually on the digital technique.
The structure of the modular monitoring
system is shown in Fig.1.

During 1985-90, it is planned to redesign
these systems using microprocessors and
to widen the structure, including early
detection of fires and ventilation, which
are under field tests.

Use of microcomputers as controllers on
the first level of the system should
provide the system with more functions
of direct control. First experiences with
the industrial application of microprocessors were gained in the coal preparation
processes, which is presented in the last
part of the paper.

PRODUCTION PROCESS MONITORING SYSTEM
"HADES"

The leading objective of our development
efforts was to design a system for general supervision of a mine operation and
the monitoring of all important data

regarding: longwall face shearer loaders
and other winning machines, armoured face
conveyors, belt conveyors, loading bunkers,
shafts, main fans and booster fans, pumps.

The system shown in Fig.2 covers the
following functions:

- Data acquisition from sensors installed
on machines.
- Data processing, which includes:detect-
ion of state changes of the machines,
keeping the timing card of the machine,
filtering out stppages longer than a
given time interval (e.g. 10 min), calculating the balance of the output by
counting the cars and skips at loading
points.
- Data presentation in the form of messages and reports automatically displayed
on VDU.

For monitoring the operation of machines
and underground equipment, the CP-10 two-
state current sensors are used, which are
mounted on the power supplying cables.
As the sensing element, the contacts of
the switching circuitry may also be used.

For data transmission, an intrinsically
safe multiple frequency transmission system CTT-32 is used.

An exemplary installation of the "HADES"
system in mine "Moszczenica" has the
following configuration :

Transducers
- monitoring of shearers in 20 faces -
20 CP-10 sensors,
- monitoring of face conveyors and other
face equipment - 80 CP-10 sensors,
- monitoring of shaft operation - 8 contacts,
- monitoring of face headings - 16 CP-10,
- monitoring of ventilation fans - 80
CP-10 sensors.

Transmission

Multiple frequency transmission system
CTT-32 with 12 units - each to transmit
by a single pair of wires 22 binary data
at a distance of max. 10 km.

Computing facilities

PRS-4 mincomputer with 16k memory,
Set of application programs including
modules for:

- face operation analysis,
- shaft transport balance,
- mine areas output balance,
- monitoring of main haulage,
- monitoring of main and booster fans.

METHANE MONITORING SYSTEM "CMC"

This system, shown in Fig.3, is provided for continuous monitoring of methane concentration in mines with a great number of controlled sensors. The system is built-up of underground, intrinsically safe equipment comprising:
- Zener protection barriers,
- Interface modules BOL,
- Two PRS-4 minicomputers working in parallel to increase the reliability of the system, handling 128 telemetry signals of low and high concentration of methane and air flow and 128 power switching-off devices.

Interface BOL blocks perform the following main functions:
- supplying all lines with a charging current for the batteries in the coders,
- measurement of the noise in the selected telemetry line,
- line break checking,
- simplex audio communication for maintenance purposes.

The PRS-4 minicomputer provides for:
- selection of sensors for measurement according to the given repetition time,
- comparing the measured value with the preset warning and alarm levels,
- activating the generator of switching-off pulses sent to selected lines associated with the sensor where the alarm level was exceeded,
- periodical printouts of the processed data such as: alarms, mean value of CH_4 concentration and its changes, duration of alarm and fault states.

The "CMC" computer-based system is expected to replace most of the current CTT-63/40Up and CMM-20 methanometric systems.

MINING SEISMOACOUSTIC AND SEISMOMETRIC SYSTEMS

The mining computer-based seismoacoustic system for the evaluation of the degree of bumps hazards is designed for continuous recording of the state of stresses in strata. The system analyzes the monitored parameters in order to determine seismoacoustic activity related to the "conventional" energy of seismic events. It also analyzes the distribution of the energy with time and its deviations from the mean value. The structure of the system is shown in Fig.4.

Output signals from max.16 geophones are amplified and transmitted by the TSA-32 transmission system to the inputs of an analog memory, where the signal amplitudes are recorded. Output signals from analog memory are transmitted to comparators. When these signals exceed the threshold value (0.5V), the comparators change their state and the interrupting signals are generated. The interruption addresses are then detected, the values of stored signal amplitudes are measured, and analog memory is cleared. The measurements are made every 10 msec. Additionally, the state of four signals from mining machines are monitored. During the operation of machines, the results of recording are stored in separate counters.

The computer system performs the following functions:
acquisition of data such as:
- exact moments of signals occurence in each channel,
- maximal signal amplitude in each channel,
- duration of the signal in each channel,
- the operation of mining machines (on-off),

data processing :
- evaluation of time differences of the signals occuring in each channel,
- classification of signals by the value of their amplitude,
- calculation of the "conventional" energy of seismic events.

Another system, based on the PRS-4 minicomputer ("SYLOK") is designed for continuous monitoring of microseismic signals and for automatic localization of the seismic events, determination of their occurence and their energy. The system is recommended as standard equipment in mine stations for seismic investigations.

The system "SYLOK" permitsone to determine the absolute seismic activity in the
mine and adjacent areas. It operates on
low frequencies, which are little attenuated by the rock, and thus its range is
several kilometers from the sensor
(seismometer). The system performs the
following functions:

- acquisition of data from 16 SPI seisme-
  meters, determination of first entry
  times in each channel, measurement of
  maximal amplitude of the cross wave in
  each channel, measurement of time of
  signal termination in each channel,
- classification of pulses with respect
  to amplitudes,
- calculation of "conventional" energy
  during one shift,
- determination of energy deviation from
  the mean energy,
- monitoring of operation of winning machines that are the source of noise,
- collecting of parameters necessary for
  more accurate interpretation of data,
  e.g. face advance, value of small- diameter drillings,
- determination of differences of entry
  times in order to perform simplified
  localization of the centres of shocks
  within the face.

Particularly useful is the look-back
facility, whereby, in given time intervals
(e.g. after shooting) and with storage
up to 44 shifts, a display of the data
permits one to trace the rate of changes.

COMMUNICATION SYSTEMS

In Polish coal mines, besides technological loud-speaking and radio communication,
two central communication systems are
used, which usually complete the whole
monitoring and control centre of a mine:
AUD -80 alarm-broadcasting system and
overall mine telephone UDG system.

The AUD alarm-broadcasting system is used
for central rescue operations in case of
fire, water, methane and other hazards.
In such situations, withdrawal of the
miners is necessary.

The system consists of:
- audible alarm signalling devices ASA -
  max.64,

- alarm telephones ATA - max.16,
- Zener protective barriers SBO,
- ACA exchange ,
- AUR tape recorder block.

The dispatcher can communicate,choosing
individual subscribers or transmiting
evacuation signals to a maximum of 12
areas, using 4 programmable or 2 permanent
warning communications. All rescue operations are recorded in the tape recorder
block.

EXPERIENCES

About 60 computer-based monitoring systems
("modules"),as described above, have been
designed and installed in Polish coal
mines and abroad. The number of installations is given in Table 1.

TABLE 1   Computer-based monitoring
           installations in Polish coal
           mines   1984

System	Poland	Other countries
Production monitor- ing "HADES"	10	6
Methane monitoring "CMC"	4	3
Seismoacoustic sys- tem "SAK"	22	5
Seismometric system "SYLOK"	4	5
Early detection of fires (experimental)	1	–
Alarm-broadcasting system "AUD"	35	6 not com- puter - based
Overall mine tele- phone system "UDG"	48	–

During eight years of designing, implementation and operation of computerized
monitoring systems, the following conclusions can be withdrawn:

- A modular monitoring system based on a
  relatively small central unit (mini or
  microprocessor) appears to be a reliable, economic solution for the deversi-
  fied needs of mines. Such a system
  meets most of the functions demanded by
  the staff of a mine.
- However, the necessity of standardiza-

tion (uniformity) of hardware and soft- ware, from the point of view of design, implementation, and maintenance of the system, is obvious. It is very import- ant to cooperate with experienced staff of the mine on the final form of reports to meet particular needs for processed information.

- Information systems that are limited to a mine find almost full acceptance by the managing staff. It is much more difficult to convince people of the value of hierarchical information sys- tems for groups of mines (e.g. areas, concern, branch, etc.).

- The Production Process Monitoring Sys- tem is used mainly by the staff of a mine for evaluation of the advance of winning works, finding weak points in the technological links, and for plan- ning of maintenance works.

- The Central Digital Methanometry System has all the features of conventional systems and allows better control of the ventilation network of a mine.

- The benefits from seismoacoustic and seismometric systems for evaluation of the energy of bumps and their localiza- tion are not so evident, however, in several cases it was possible to pre- dict bumps which occured. In these sys- tems, which are based on imprecise knowledge of the relation between "acti- vity" of the strata and a generated acoustic signal, other phenomena should be studied also. However, all monitor- ing systems that give some information, even if not very precise, on bumps ha- zards are welcomed in mines.

- As regards the discussed modular moni- toring system, minicomputers used in "modules" will be replaced by micropro- cessors. The use of microprocessors as controllers on the first level of the system should provide more functions of direct control. From the point of view of required reliability in direct cont- rol, it seems desirable to separate the functions of control from data **proces- sing** and presenting results.

COAL PREPARATION PROCESSES MONITORING AND CONTROL

In recent years, substantial progress has been made in the automation of coal pre- paration processes as a result of long - term research in many countries. One can expect that future coal preparation plants will use more microprocessors in various sections of the plant.

In EMAG Centre, microcomputer-based cont- rol systems have been designed and appli- ed to such processes as flotation, loading of coal into wagons, monitoring of calo- rific value of coal and of ash content in coal slurry. Control systems for sepa- ration in heavy media and in jigs are also under investigation. It is planned to design a modular distributed-control system for a typical microcomputer cont - roller applied to separate sections of the plant. The structure of the system is shown in Fig.5.

Control of the flotation process

In this system, a microprocessor control- ler for dosing the reagents into flota- tion cells is used. The value of control signal I_d for reagent dosing is calcula- ted by procedures of the application pro- gram using relation:

$$I_d = k \cdot V_n \cdot C_n$$

where : V_n - quantity of flow of the feed,

C_n - density of the feed

The flow of solids in the feed to the flotation process, as well as the total mass of the solids, are also calculated. Analogue input signals to the controller are generated by the densitometer, elec- tromagnetic flowmeter and the setting unit.

Monitoring of calorific value of coal

The first microcomputer-based monitoring system was applied to continuous measure- ment of coal quality, namely its calori- fic value. The system is shown in Fig.6. Basic coal quality parameters, i.e. ash content and moisture, are measured con- tinuously on the belt conveyor. The measurements are performed by the G-3

AMR-D

radiometric ash monitor and WILMAG microwave moisture meter.

The calorific value of coal is calculated
by the equation:

$$Q = Q_o - k_1 \cdot A - k_2 \cdot W$$

where: k_1, k_2 - coefficients determined
experimentally,

A, W - ash and moisture content,

Q_o - calorific value of ashless
and free of moisture coal.

The signals from monitors are processed
in the microprocessor to determine, using
regression analysis, average values of ash
and moisture content and calorific value
of coal. The results are printed out and
projected on a monitor together with the
number of a loaded wagon.

Coal loading into wagons

Automation of a coal loading station
with buffer bunkers was the other problem that was solved by application of the
microprocessor controller. The control
system performs the following functions:

- storage of load capacity of individual
  wagons,
- generation of steering signals for controlling the coal stream to batch bunkers,
- monitoring of the load of wagons,
- signalling the phases of the loading
  cycle,
- testing the correctness of the system
  operation.

The system consists of EWZ-10 hopper
scales, PSP-2 controller, operator's desk,
and interface equipment.

Heavy medium separation process

Designed and installed in 1983, the control system for the heavy medium separation
process is an example of the application
of the microprocessor controller, in
which its advantage over conventional
control system can be demonstrated.

The process control actually includes:
- stabilization of the density of heavy
  media,
- control of the levels of slurries in
  tanks.

Experiments are being made with control
of the preparation of fresh heavy medium,

monitoring of the state of machines,
start-up of drives in a sequence.

Coal separation in jigs

The reason for automation of jigs is to
increase their separation efficiency, to
obtain high and uniform quality of a concentrate, better capacity of the machine,
and lower losses of coal in refuse. This
can be achieved by:
- better stability of the feed to a jig,
- better loosenning of the bed in a jig,
- better accuracy of refuse removal from
  a jig,
- control of water flow in the machine.

Two microprocessor control systems have
been proposed for the separation process
in jigs.

The first one is assigned for the control
of the cycle of air pulsations in a jig.
The system allows one to choose easily
the shape of a cycle, the moment of its
beginning and the end according to mean
characteristic of the coal being processed.

The second system is assigned for refuse
removal from a jig and is based on a conventional float as a sensor of the thickness of the bed. The use of a microprocessor allows one to measure the thickness of the bed in a chosen time interval
in each cycle of air/water pulsation.
This procedure is expected to increase
the accuracy of the control system in
comparison with the apparatus being used
now.

REFERENCES

Firganek,B.,and Żymełka,K. (1984). System
for monitoring of production and safety in the Polish coal industry. Procee
dings of ICAMC-84,Budapest.

Cierpisz,S.,and Mironowicz,W. (1984).Trends
of automation in mining in Poland.Pro-
ceedings of ICAMC-84,Budapest.

Cierpisz,S.,et al.(1984). Microcomputer-
based control systems for coal preparation processes. Proceedings of ICAMC-
84, Budapest.

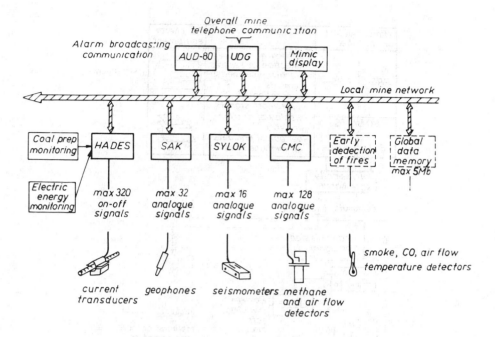

Fig.1. Modular computer-based monitoring system for mines

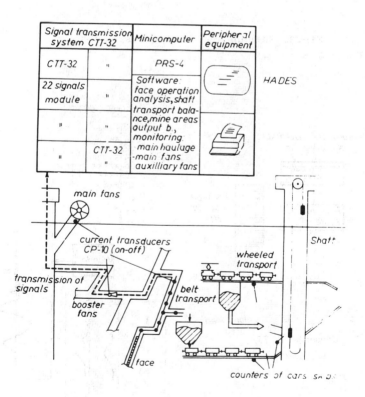

Fig.2. Production process monitoring system HADES

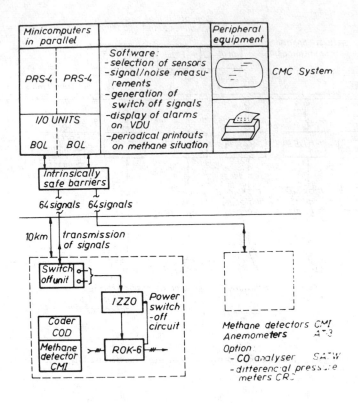

Fig.3. Methane monitoring system CMC

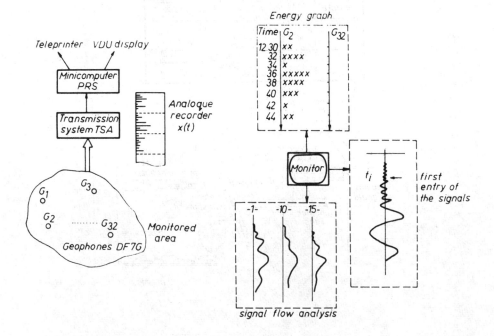

Fig.4. Mining seismoacoustic system SAK

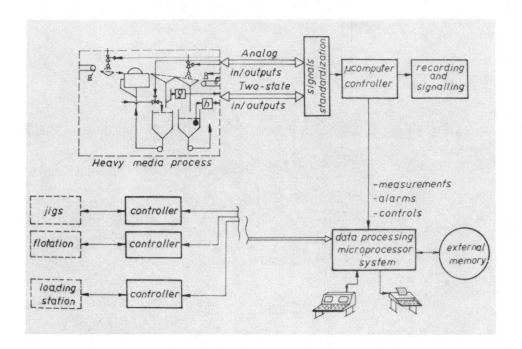

Fig. 5 Distribution-control system for coal preparation plants.

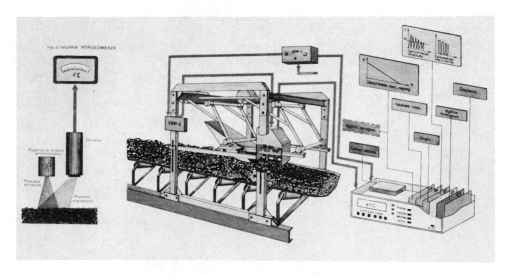

Fig. 6 Radiometric ashmonitor G-3.

Development of Monitoring and Control Techniques of Coal Mines in China

PAN SHANGDA

Deputy Director, Changzhou Automation Research Institute, Central Coal Mining Research Institute,
Changzhou, Jiangsu Province, People's Republic of China

Abstract. The application and development of monitoring and control
techniques of coal mines in China is mainly described in the paper.
The techniques of mine control and monitoring system have been developed
as an important means of production dispatch, command and management
with ever-increasing coal production and gradual raising of the level of
mechanization and centralization of continuous mining. The continuous
stage monitoring and control system in China coal mines has been
developed from the initial single fixed plant monitoring, and the
technique of computers has also been applied in the system for
productive and environmental monitoring, featuring a new type of mine
control and monitoring equipment of China.

Keywords. Control and monitoring of coal mine production and environmental parameters; telecommunication; telecontrol; information transmission; pulse modulation; data processing; time division transmission.

INTRODUCTION

Coal industry in China has been developed
rapidly since the establishment of people's Republic of China. Coal output was
only 32.43 million tons in 1949, but increased to 715 million tons annually by
1983. With the increase of coal output,
the level of mechanization of mining,
drivage and transport has been raised
gradually and coal mine equipment have
been renewed uninterruptedly.

Coal, being approximately 70% of components of consumed fuel, is the primary
energy in China, possessing an important
position of energy sources under the unchanged current composition of energy
sources.

To produce coal is a complex process with
its multiprocessing and interacting on
each other, and furthermore, there are
still such unsafe factors as water flood,
fire disaster, methane or dust explosion,
roof pressure, etc. It is imaginable
that monitoring and control techniques of
coal mines will surely play an important
role in ensuring safe production, smooth
running of various equipment, giving
effective play to equipment and resulting
in an increase of the productivity.

In order to meet the requirement of coal
production, centralized and continuous
monitoring and stage control systems have
been developed in China coal mines from
the individual point and fixed plant
monitoring, and the technique of computers
has also been applied in the system for
production. Thus, a modernized monitoring
and control system for mine production
featuring China coal mines is being
constructed.

CURRENT STATUS OF MONITORING AND CONTROL TECHNIQUES FOR COAL MINE PRODUCTION IN CHINA

In China, the research work has been engaged in coal mine monitoring and control
techniques since the middle age of the
sixties. The monitoring of the environmental parameters underground has simultaneously been developed.

The transportation line is known to be
one of the important links in coal production process. A great many of labour
power are needed to achieve the local
control check and maintenance of local
fixed plants in the system, the reliability of which will be directly related to
the continuity of production in the face.
Since the sixties, China has begun to
obsolete the coal cutters with their low
power and inferior performance and various
types of shearer-loaders and ploughs were
used in the face instead. Consequently,
the promotion of renewing transport equipment resulted from the raising of the face
productivity. In the meantime, development of the centralized control system of
conveyors was paid more attention to.
However, the belt conveyor automatic control system Type PZK-1 installed and commissioned in "Victory" Mine of Fushun Coal
Administration then was operated in the
control mode with polar combination based
on relays. The maximum number of belt
conveyors controlled was 20 and the magnetic amplifiers were used in the control
circuit in order to display on the console
the number of conveyors running practically. Stoppage according to the direction
of coal flow in the system was realized
and speed sensors and piling sensors were
provided to make the plants run smoothly.
During the early period of the seventies,
the system Type SYK in mine use for remote
signalling, telemetry and remote control

at a transmission distance of 10 km, was
successfully developed with the progress-
ing electronics technique. The system, in
which HTL circuit was selected to achieve
data multiplex transmission in time-division system, was applicable for monitoring and controlling the dispersed mine
equipment. The system consisted of three
parts: dispatching terminal in the centralized control station, actuating terminal
at the equipment controlled or where signals were centralized and cables for transmission. Distributors X and Y were provided at both the dispatching terminal and
the actuating terminal. Synchronization
operation of the whole system was ensured
by the positive "group synchronization"
pulse and the negative "step synchronization" pulse. The units for call, response
and check of selecting addresses were supplied with in the system in order that the
information exchanging could only be carried out by one actuating terminal with
the dispatching terminal at a time. And
the information transmission was thus more
reliable with the protective modes of constant ratio code and repeated checking.
During the middle age of the seventies,
the efficient long distance cable-belt
conveyor system with high power and large
capacity was started to be used in main
belt slopes and haulageways usually with
multi-feeding points. In order to give
full play to the efficiency of the transport equipment for achieving the optimum
control, the computerized control system
should be used and the perfect devices for
protection be provided. The system Type
KJD-J1 nowadays operating in No.2 Slope,
No.3 Mine in Yangquan is the first computer-based monitoring and control system
for mine belt conveyors, the core of which
is a China-made computer Type JS-10A using
the technique of data multiplex transmission in time-division system to achieve
the telecommunication, telemetry, remote
control and remote regulation to the conveyors. It can operate with up to 36
flameproof outstations in maximum, each
with nineteen switch inputs and five analogue inputs, and with three pairs of
switch outputs in the mode of non-potential contacts and one 4-20 mV analogue
output. The information is transmitted by
4-core cables at the maximum distance of
10 km. The mimic display is provided in
the central dispatching room on the surface for dispatching the corresponding
signals of each switch input or output,
and signals of coal level in bunkers, voltage and current of the electric motor and
various faults occurred can also be displayed in the room. A visual-audible
alarm   will display on the console as
well should faults appear in the equipment
controlled or telemetry variables exceed
the limit. In addition to the various
protection devices for belt alignment,
off-trough of wire ropes, exceeding temperature of the electric motor, excitation
undervoltage, etc. necessary for the whole
set of the system, speed protection device
Type SB-1 and the signalling device for
the electric protection Type BX-1 are also
set up.

There are four operating modes for the
system KJD-J1: on-line automatic control,
on-line manual centralized control, off-
line manual centralized control and local
control. Printing types as timed, automa-

tic and manual calling can be accomplished when operated with the on-line mode.

As for the monitoring of environmental
parameters in mines, the portable instrument was mainly used in the past in China
for interrupted detecting. Nevertheless,
it took a long time space and operated
still beyond the limit, and so it could
not give an assurance to safety. The
successful development of fixed methane
detecting instrument with an alarm used
in automatically cutting off electrical
devices basically solved the problem in
existing manual detecting. But it was
impossible for such kind of instrument
to meet the requirements of dispatching
the ventilation on the surface, for its
signals could not be sent to the surface.

The progress of mine carrier wave technique created a favourable condition for
developing the environmental parameter
monitoring system with the function of
telemetry. The methane monitoring devices Types ABD-21, AYJ-1, AYJ-2, MJC-100,
etc. have been developed successively
since 1968, and microcomputers have been
put up in certain coal pits.

Signals from the methane detector Type
ABD-21 developed on the basis of the fixed uninterrupted alarm can be transmitted
to the surface with the carrier channel.

The telemetric methane alarm Type AYJ can
be applied to transmit the signals to the
surface with the techniques of pulse modulation and of frequency division
multiplexing (FDM).

Signals can be transmitted to the surface
by the methane detectors Type AYJ monitoring the methane content of five points
underground continuously in a long period.
The detector can give a visual-audible
alarming signal underground and on the
surface respectively when the methane
content at the detected point exceeds the
limit and the power supply of the hazardous area can instantaneously be cut off.
The instrument is set up with a recorder
for continuously operating in the central
dispatching room.

The telemetric distance of Type AYJ is 10
km with the carrier frequencies of 17,
20.5, 25, 30 and 36 kc for transmitting
signals.

The monitoring system Type MJC-100 is
composed of fixed methane detectors with
alarms, carrier transmitters and power
boxes underground and the surface main
computer, printer, etc., the maximum monitoring capacity of which is 100 detected
points.

DEVELOPMENT OF MONITORING
AND CONTROL TECHNIQUE IN
MINE PRODUCTION

Since last decade, centralization and
continuity in coal mine production has
come to be developed in China. Levels,
mining sections and coal faces have been
reasonably lengthened in the layout of
developing, and centralized and combined
arrangement of roadways has been carried
out. As for the means of winning and
drivage, the comprehensive mechanized

mining faces have been taken place instead of the conventional mechanized coal faces. The comprehensive mechanized winning and drivage has been developed actively, as a result, the output of coal faces has been increased and the level of mechanized mining continuously raised. The increase in coal output pushed forward the development of electro-mechanical equipment with high efficiency and large capacity. In the meantime, a large amount of gas burst will inevitably be increased and the problem of safety in mine will be prominent day by day with the increment of mining intensity in the coal face and the ever-increasing deepening of the mining levels. Obviously, it is impossible to make the mine production well coordinated and managed, neither can the safe production be realized.

The development of current micro-electronic and computer techniques has put up an excellent condition for developing the modernized monitoring and control system in mine production. Nowadays, the research work of the microcomputer-based monitoring system has been engaged in instead of that of the main computer with Types AYJ-1, ABD-21 and MJC-100 methane detectors and a good result has been obtained. This new system will be installed and operated in various kinds of coal mines in China.

The monitoring system Type KJ-1 consists of a computer-based information processing center in the central dispatching station, the information transmission device from underground to the surface and detectors and sensors for operating mode and environmental parameters. The micro-computer is the core of the central dispatching station on the surface supplied with the devices such as the kinescope display with Chinese characters, mimic display of operating mode of production, the printer with Chinese characters, etc. so as to perform the storage of information, data processing, various types of display, automatic alarm for exceeding the limit and various printout records. The dispatcher can talk with the microprocessor through the function key with Chinese characters on the console to call out various kinds of data required.

In the system, the outstations of information transmission are equipped according to different processes underground and the information areas are respectively divided. The function of outstations is as follows: data acquisition, data conversion, coding, transmission and pre-processing of information. Realtime processing is carried out by the output with non-potential contacts to cut off the power supply in the hazardous area when the methane detected exceeds the limit, and the information resulted from the processing can be transmitted to the surface.

Methane concentration, wind velocity, CO, temperature, underpressure, output of coal sections, coal level in bunkers, location of the miner in the coal face and the station of status of start/stop of main electro-mechanical equipment underground can all be monitored by various detectors and sensors.

The KJ-1 system is of TDM (Time-Division Multiplexing) using 4-core cables at a maximum transmission distance of 15 km with 16 monitoring outstations, each of which consists of microcomputers and interfaces with 12 analogue and switch inputs and 4 switch outputs. The outstations are intrinsically safe with a floating power supply. The unified signal system is supplied for the input in the outstation for serialization and standardization.

Error detecting techniques, such as constant ratio code, odd-even check, repetition check, feedback check, etc. and the antijamming measures with an optoisolator have been applied with the result of a great increase of reliability of the system KJ-1.

The system KJ-1 will be installed and used in large- and medium-scaled coal mines in China.

At present, with the development of the above monitoring system, application of microprocessors in the belt conveyor monitoring system, the system of pit rails with centralized signals and the subsystems of electric dispatching and underground drainage has also been developed. The above mentioned subsystems, designed in accordance with unified signal systems, interfaces and modes of transmission, can be used as an independent system each as well as a part of the monitoring system for the whole mine. Such kind of stage control system is flexible and convenient in combination, favourable in changing and expanding and applicable to different coal mines.

CONCLUSION

The research work has been engaged in coal mine monitoring and control techniques together with the computer technique in China since the middle age of the sixties, and consequently a modernized monitoring and control system for mine production and environmental parameters featuring Chinese characteristics has gradually been constructed, by which not only the environmental parameters underground and the information of operating status of fixed plants can continuously be collected with accuracy in time so that natural disasters and hidden troubles (dangers) underground can be forecasted through the storage and analysis of collected information, but also the control and adjustment of all links of production can be carried out to make the equipment and system run in the optimum operating condition. The construction of overall mine monitoring system coordinated with the communication dispatching system throughout the whole mine will surely make the technical face and safe status of China's coal mines further changed and so a new era of coal mines in China will appear with the raising level of management and ever increasing of techno-economic quota.

Health Monitoring of Underground Mining Machines

E. SCHELLENBERG

Dip.E1.Ing. ETH, Department of Mining Automation and Electronics, VOEST-ALPINE AG Linz, Austria

G. STEINBRUCKER

Dipl.Ing. for Mechanical Engineering, Department of Mining Technology, VOEST-ALPINE AG, Zeltweg, Austria

Abstract. Improvements in mining by the application of electronics necessitates analysis of the objectives and problems. A detailed analysis has shown that increased tunel-
ling performance is one of the most important objectives. In mining, the environmental
conditions for machines, their handling and servicing are very hard. Therefore a health
monitoring system for tunnelling machines in hazardous areas has been developed. This
system measures operational parameters, such as flow, pressure, temperature, etc., derives forecasts of possible failures from these values, and assists maintenance personnel. Important data are transmitted to a surface computer via telephone cables. Health
monitoring is based on the cognition that slow deterioration of characteristic parameters, e.g. fluid loss, are symptoms of wear. Sudden changes of machine parameters indicate sudden destruction. Computer technology is very advanced, but transducer technology has not been able to keep pace. Filling the yawning gap in the mining industry is a
challenge to technology and requires special techniques. Machine monitoring and automation in underground mining is still in its infancy. A lot of development and research
work is still necessary. The risks are great. The expected benefit due to increased mining performance (up to 5 %) justifies the risks in development.

Keywords. Mining; tunnelling; health condition; system analysis; microprocessor; computer; sensor.

INTRODUCTION

The objective of a mine is to produce a certain
quantity and quality of raw material for the market
at reasonable costs.

New technologies, e.g. electronics, are necessary
to fulfill the increasing market requirements.
Micro-computers and sensor technology have been introduced in several fields of the mining industry
and these allow many new opportunities in underground mining.

A precise analysis of objectives, requirements and
problems of the mine, the miner and his equipment
is a prerequisite for the successful application.

Some of the most important objectives (Steinbrucker,
1983) are:

o Humanized work environment, e.g. less dust, remove personnel out of dangerous areas, avoid
rockfall.

o Increased tunnelling performance, e.g. less downtime, avoid failures.

o Increased quality of mined product.

The more precisely the objectives and problems are
analysed, the higher the chances are of a successful application of electronics. Suitable techniques
are therefore necessary for system analysis and project management (Schellenberg, 1982).

Mining machinery used for underground mining is subject to extreme stress.

The characteristics of the raw material often
change, sometimes soft coal is cut and sometimes
bands of hard stone have to be worked.

A system analysis study carried out in a potash
mine (Steinbrucker, 1983b) shows that similar conditions can be found in other mining fields.

In comparison to other industries, mining personnel
have a lower level of education. Machines are often
incorrectly operated and overloaded; maintenance is
unsatisfactory.

Surface machinery, such as trucks, excavators, etc.,
are overhauled regularly in the maintenance and repair shops. It is practically impossible to carry
out repairs under the bad conditions underground,
so machinery is dismantled and the defective parts
transported to the surface and repaired there.

This results in long maintenance and repair downtimes. The extracting machines therefore lose, on
average, up to one hour for maintenance and repair
work per one hour's cutting time.

Appendix 1 shows a tree of objectives, how the main
objective is dissected into sub-objectives.

One of the most important sub-objectives is the
reduction of downtime of the tunnelling machines,
where the health of the machine plays a significant
role.

It is therefore necessary to analyse the machine's
problems and their causes.

PROBLEM ANALYSIS

Failures always have causes. It is far better to recognise failure causes early, and thereby deduce the tunnelling machine's health condition, rather than wait for a failure effect.

The most difficult and time consuming failure sources are found in the gearbox, hydraulic and electrical systems. Failures in these fields are analysed for their cause.

Appendix 2 shows the effect cause tree for a gearbox.

Failure causes usually lead back to

- overload of machine
- misoperation
- insufficient maintenance

which results, for example, in defects to the gears and bearings and also to contamination in the hydraulic and lubricating oils.

The electronic, and in particular the sensor technique, allows early recognition of the failure causes. Due to cramped space conditions and the harsh environment in underground mines, the machines can only accommodate a limited use of electronics.
Furthermore, an electronic system that is too complex reduces the reliability of the entire mining machine.

Using the method of Saaty (1982), the priorities of functions for the health monitoring system were determined.

MONITORING STRATEGY

Analysis of the problem leads to the following monitoring strategy:

1. Determine measurements for the prevention of downtimes. This is achieved by means of the <u>failure prognosis</u>, which is much more important than the diagnosis.

2. In cases where a failure occurs, it must be eliminated as quickly as possible. This leads to the <u>failure diagnosis</u>.

3. The responsible personnel must be informed immediateley, in order that the necessary counter-measures can be taken.

FUNCTIONS OF THE HEALTH MONITORING SYSTEM (HMS)

Indications of wear to the mechanics, hydraulics and electrics are recognisable by a slow deterioration of the characteristic parameters, such as energy consumption, performance characteristic, loss of liquids, as well as changes in the composition of the lubricating oil. Changes in the physical behaviour, e.g. the vibrating pattern of the housing, can also result from indications of wear (Broch, 1980).

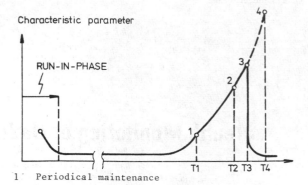

Characteristic parameter

RUN-IN-PHASE

1   Periodical maintenance

2   Health Monitoring System's alarm

3   Maintenance during the Maintenance shift due to Alarm HMS

4   Without HMS: Maintenance after breakdown

Fig. 1. Machine Health Condition: Characteristic Parameters as a function of Operating Time

Based on the measuring data, the monitoring system continuously calculates the characteristic parameters that describe the health condition of the individual machine components.

During periodical maintenance without the health monitoring system, the machine components are replaced after their operating time T1 (see Fig. 1) and are therefore not fully used.

The HMS recognises a threatening deterioration of the health condition in its early stages. The personnel are informed in time (point of time T2). The endangered machine component is repaired during the scheduled maintenance shift. The operation can continue after point of time T3.

If the health condition is not detected, and therefore not remedied, this results (in point of time T4) in an unexpected failure, which causes a shut down of operations and costs unplanned repair time.

If the characteristic parameters change rapidly, then this indicates a <u>sudden destruction of components</u>, such as a fracture in the hose, gear, supporting components, etc. The failures are diagnosed according to such information.

The Health Monitoring System (HMS) is divided into the following main functions (Fig. 2):

o   Operating values such as pressure, flow, current, debries content in oil, velocity, vibration are measured and converted into technical units. Average values are calculated (1).

o   These measured values are joined together. The characteristic health data for individual machine components are derived from these values (2). Recommendations for a careful operation method are indicated via the LCD display (8) to the machine operator.

o   Some of the measured values are checked for operational limits (4).

o   If a failure is discovered in the machine, HMS recommends actions to the machine operator (9) and, if necessary, switches off certain components (10).

o   Recommendations for maintenance are derived
    from the health data, and are only accessible
    to the maintenance and surface personnel (7).

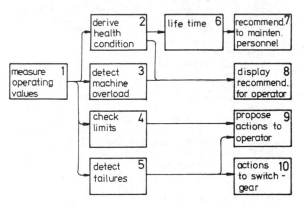

Fig. 2. Block diagram of Health Monitoring
        Functions

The evaluation of the debris content in oil is explained in more detail, as an example of measuring
values and their further processing in the microprocessor.

Many defects can be traced back to wear of the components' surface.

The relative movements of the surfaces against each
other - rolling or sliding - cause wear, which removes particles from the surface.

In oil lubricated systems, these particles are
transported away from where they are produced.
They either are deposited, are filtered off, or remain in the circulation.

These particles contain valuable information relating to the kind, location and degree of wear;
from this, conclusions can be drawn regarding the
health of the components.

The size and shape of the particles depends, to a
large extent, on the failure cause. For example:
bearing or gear wear leads to powder-like debris.
Thick, shorter particles with fracture surfaces
indicate fragmentation, such as, for example, gears
or bearings. Longer, curled debris of various thickness, that resemble lathe turnings, indicate defects
caused by abrasive impurities, such as quartz sand,
between the moving bearings.

The type of material of the particle can also give
further information relating to the origin of defects, e.g., brass particles originate from bearing
cages and synthetic particles from the sealing elements.

In addition, the debris production rate indicates
impending defects. Fig. 3 shows the debris production rate function in time (simplified).

Numerous debris particles are produced initially
from the run-in effect. The development of new
debris is minimal in a healthy gearbox with normal wear. If a defect is developing, then the
production rate will increase progressively.

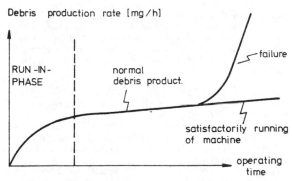

Fig. 3. Debris Production Rate as a Function
        of Operating Time

Before deciding on a certain method for measuring
the debris production rate, the following questions
have to be clarified:

1. Type of potential failure modes?
2. How critical are these failures?
3. How often can they be expected?
4. What type of typical debris relates to the fai-
    lure mode (size, frequency, material, etc.)?
5. What is the best method of determining the
    debris?
6. Can exact statements regarding the health sta-
    tus be derived from these measured values?
7. Is the method suitable for mining?

In the following, a few methods for debris sensing
will be outlined.

They have different main fields of application.
Fig. 4 shows, for instance, which methods are best
suited to the determined particle size.

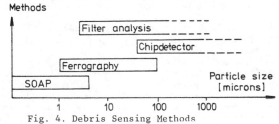

Fig. 4. Debris Sensing Methods

SOAP (SPECTROMETRIC OIL ANALYSIS PROCESS)

An oil sample is spectrometically analysed in the
laboratory. The elements in the oil are determined,
based on the spectral lines (e. g. Fe, Cu, Al, Mo,
Ag, Ti).

FERROGRAPHY

This method is based on the magnetic precipitation
of ferrous particles in a high gradient magnetic
field. The particles flow through the magnetic
field in the oil and are deposited according to
size. This method, like the SOAP method, is usually
carried out in the laboratory. Attempts to further
develop this method on-line and on-board are currently being made.

CHIPS DETECTOR

A magnet, submerged in the flowing oil, attracts
and removes the ferrous particles out of the oil.
The deposited particles are determined according to
quantity or number. Present day sensors must be
cleaned after use, and the debris is then checked
visually. Automatic detectors are currently being

developed. Only this can meet the requirements of mining for a maintenance-free operation.
A chip detector will be used in the existing system. As soon as a certain quantity of debris has settled, the sensor sends a signal to the microcomputer. This initiates a report to the responsible personnel.

The duration of the interval between these individual signals represents the measure for the debris production rate. If the time interval is shorter than a preset boundary (Fig. 5), the microprocessor draws conclusions on the health condition of the gearbox and the personnel are given recommendations for maintenance.

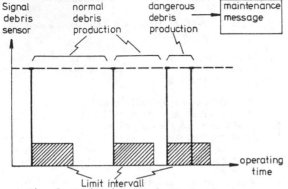

Fig. 5. Monitoring Debris Production

HARDWARE

The Health Monitoring system has been developed for application in underground coal mining. It consists of the following components, which are intrinsically safe and have been approved by the authorities:

o   Microprocessor System (underground)
o   Sensors and actuators
o   Machine display
o   Personality unit
o   Transmission system
o   Surface computer

Fig. 6 shows how these components are connected to each other.

The microprocessor system is located on the tunnelling machine. This system samples all sensors. It checks both the analogue and the digital signals for integrity and performs the essential calculations that are necessary to determine control actions and energise the interface for the actuators. This unit comprises three separate RCA microprocessors, which communicate with each other via an 8 bit interprocessor communication system.

The sensors convert process values such as pressure, temperature, flow, etc., into electric signals ranging from 5.4 to 7.0 V, which are further processed by the microprocessor.

The actuators, such as, for example, the electric motor's power switch, are controlled via the triac opto coupler. The opto coupler separates the different intrinsically safe circuits.

The machine operator receives necessary messages, such as:

    "GEAR BOX OIL VERY DIRTY"

on the machine display via a liquid crystal display (LCD). Maintenance and authorised personnel receive additional operative information such as

    "HYDRAULIC PRESSURE XX BAR".

The system's machine specific data, such as limit values and geometric values, are entered via the personality unit by means of hex switches.

The transmission system transfers condensed information via telephone cable to the surface computer.

The surface computer prepares the data and provides shift reports, faults and maintenance reports via a printer. Management heading and production information is available via a CRT display.

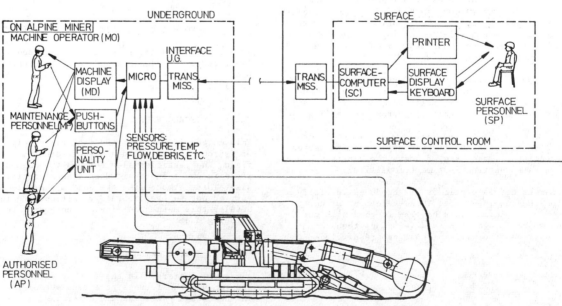

Fig. 6. Layout of the Hardware

CONCLUSION

Health monitoring on underground mining machine
with a microprocessor and computer system increases
operation time, while preventing major defects.

Maintenance work can be carried out during maintenance shifts; scheduling of personnel, operating material and spare parts are improved.

Machine monitoring and automation in underground
mining is still in its infancy. A lot of research
and development work is still necessary in this
field. The risks are high. Suitable system analysis
techniques and management methods are important
prerequisites for success.

ACKNOWLEDGEMENT

The authors would like to express their appreciation to Director Dr. Ing. Santiago Ramos for his
kind support during the system analysis in the
Potasas del Llobregat (ERT) Mine.

REFERENCES

Anand, D.K. (1974). Introduction to Control Systems
Pergamon Press Inc., Oxford.

Broch, J.T. (1980). Mechanical Vibration and Shock
Measurements. Brüel & Kjaerbooks, Danmark.

Olaf, J. (1976). Automatisierung und Fernüberwa-
chung in Bergbaubetrieben. Glückauf GmbH, Essen.

Saaty, T.L. (1982). Decision Making for Leaders.
Lifetime Learning Publications Wadsworth Inc.,
California.

SAE (1984). Guide to Oil System Monitoring in
Aircraft Gas Turbine Engines. Aerospace Infor-
mation Report, AIR 1828. Society of Automotive
Engineers Inc., Pennsylvania.

Schellenberg, E. (1981). Systemanalyse und Automation im Bergbau. Internal Report. VOEST-AL-
PINE AG, Linz.

Schellenberg, E. (1982a).
Project Management, one of the key factors of
reliable Process Computers. In E. Lauger (ed.),
Reliability in Electrical and Electronic Com-
ponents and Systems. North-Holland Publishing
Company.

Schellenberg, E. (1982b). Project Management, Key
to Successful Computer Projects. In R. Trappl
(ed), Cybernetics and Systems Research. North-
Holland Publishing Company, Amsterdam. P.567.

Steinbrucker, G., and Schellenberg, E. (1983a).
Systemanalyse Maschinenüberwachung ALPINE
MINER. Internal Report. VOEST-ALPINE AG, Linz.

Steinbrucker, G., and Schellenberg, E. (1983b).
Systemanalyse der Kaligrube Potasa del Llobregat. Studie für ERT, Linz.

Steinbrucker, G., and Schellenberg E. (1984).
Specification Monitoring System for ALPINE
MINER. Internal Report. VOEST-ALPINE AG, Linz.

Tauber, T., W.A Hudgins, and R.S. Lee (1982).
Full-Flow Debris Monitoring and Fine Filtration for Helicopter Propulsion Systems. Rotary Wing Propulsion System Specialists Meeting. American Helicopter Society, Washington D.C.

Yardley, E.D. (1979). The Use of Ferrography and
Spectrographic Oil Analysis to Monitor the
Performance of Three 90 kW Gearboxes. Wear
56, Elsevier Sequoia S.A., Lausanne.

APPENDIX 1 - Tree of Objectives

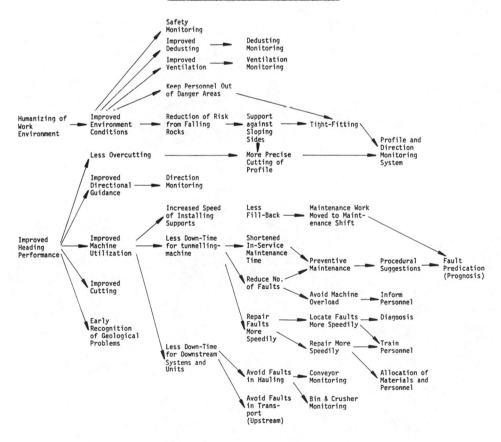

APPENDIX 2 - Failure, Effect → cause tree

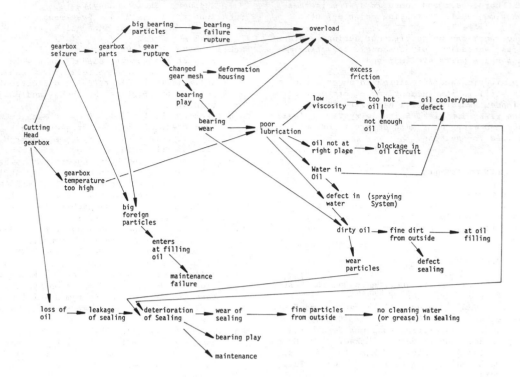

Monitoring and Control in the Recirculation of Underground Ventilation Air

J.N. MIDDLETON and R.C. BURTON

Environmental Engineering Laboratory, Chamber of Mines of South Africa Research Organization, Johannesburg, SA

K. WALKER

Engineering Services Branch, Chamber of Mines of South Africa Research Organization, Johannesburg, SA

Abstract.   An outline of the ventilation and refrigeration requirements of deep South
African gold mines is presented.   This is followed by a brief description of the history
of controlled recirculation in mines and the practical and economic benefits such schemes
can offer.   The potential hazard of a fire in a recirculation system is discussed.

The major part of the paper describes the first large-scale field trial of controlled
recirculation to be conducted in South Africa.   Emphasis is placed on the monitoring
system that was used for data acquisition during the trial, and also on a safety system
designed to shut off the recirculation fans in the event of a fire.

The paper concludes with a review of the latest development work taking place on recirculation safety systems.

Keywords.   Mine air recirculation, computer control, alarm systems, data acquisition,
environment control, sensors.

1.   INTRODUCTION

The use of controlled recirculation of air in mines
is becoming more accepted worldwide.   In order to
assess all aspects of recirculation in a deep South
African gold mine, a large-scale field trial of
controlled recirculation was commissioned at Loraine
Gold Mines Ltd during 1982.

This paper emphasizes the monitoring and control
aspects of the field trial.   Of particular importance was an automatic safety system which, in
the event of a fire in the area, would stop recirculation and return the area to normal once-
through ventilation.   The reason for incorporating such a safety system was that, with recirculation in operation, such a fire would cause
noxious fumes to be introduced into the fresh intake air.   In view of the large amount of timber
that is used in South African gold mines, a fire
is a potential hazard and makes the installation
of such a safety system essential for any recirculation scheme.

The emission of radon or methane gas could also
be potentially hazardous, although both are rarely
found in significant quantities in South African
gold mines.   Under the South African Mines and
Works Act and Regulations of 1956, certain mines
with significant methane emissions are classified
as fiery.   Recirculation is prohibited on these
mines without an exemption from the Regulations.
Loraine gold mine is not classified as fiery.
Nevertheless, throughout the field trial, a representative of the Government Mining Engineer was
involved in meetings of the recirculation project
team.

The field trial provided an important opportunity
for gathering data, and the monitoring of various
environmental parameters was undertaken to enable
validation of the theoretical effects of recirculation.   The monitoring system that was installed was experimental, and neither the instruments nor the procedures used would necessarily be
the most suitable for normal mine monitoring requirements.

This paper gives a brief introduction to the problems associated with the ventilation of deep South
African gold mines, with an outline of the potential benefits and hazards of using controlled recirculation.   This is followed by a description
of the Loraine gold mine controlled recirculation
system.   The major emphasis of the paper is
placed on a description of the monitoring and control aspects of the installation, together with
the safety system for automatic fan control in
the event of a fire.   The final section deals
with the latest developments taking place in the
design of safety systems for recirculation schemes.

2.   VENTILATION IN SOUTH AFRICAN GOLD MINES

The major environmental problem encountered in the
deep gold mines of South Africa is heat, the main
sources of which are the strata and adiabatic compression of the downcast air.   At the current
mean working depth in the industry of 1640m below
the surface, the Virgin Rock Temperature (VRT) is
38°C.   The deepest working place is at a depth
of 3582m, with a VRT of 56°C.

Cooling of the intake air is essential for providing acceptable temperatures in deep working
areas.   This has resulted in many large refrigeration installations : the total capacity in 43
gold mines in September 1983 was 930 MW(R).   A
great deal of ventilation air is required to dis-
tribute the required amounts of cooling, the
average amount for a typical South African gold
mine in September 1983 being 820 m³/s, whilst the
largest mine used over 3000 m³/s.

Greater mining depths will inevitably result in
increased refrigeration requirements and, if air
coolers are to be spaced at practical intervals,

the quantity of ventilation air within the working areas will also need to be increased. This is because the rate of air temperature rise with distance is inversely proportional to the air flow-rate.

Clearly, controlled recirculation is a means of providing the additional air within the working areas, without incurring the ever increasing costs of providing air from surface. Such costs not only include the electrical power for surface and booster fans, but also the capital costs associated with the provision of new airways and ventilation shafts. The optimum use of available air in this manner has long been an accepted practice in industrial ventilation and air-conditioning systems. However, due to the potential hazards involved, until recently recirculation was not generally considered for mining applications.

2.1 Recirculation of Mine Air

Controlled recirculation is the re-introduction of a portion of the exhaust air from a working area into the intake air of the same area. In such a system, the recirculated air is always in addition to the 'fresh' intake air normally supplied to the area.

The use of recirculation in mines is not a new concept. Several authors (Aldred and others, 1984; Bakke and others, 1964; Pickering and Aldred, 1977) have considered its application for temperature, dust and methane control since the first use of recirculation was recorded (Lawton, 1932). A paper by Allan (1983) gave a comprehensive review of the development and use of recirculation in British collieries, where 61 sites were using controlled recirculation by January, 1983. A description of the theoretical effects of recirculation as applied to gold mines may be found in a recent paper (Burton and others, 1984), which also gives details of the Loraine field trial. Until this field trial, there had been no large-scale investigation of controlled recirculation in South Africa.

There are two major practical benefits resul-

ting from the introduction of recirculation. Firstly, an increased airflow results in increased air velocities. This has two effects : slightly improved air cooling powers, and better 'scouring' of pollutants. Secondly and most important, increases in airflow combined with bulk cooling of the air and possibly dust filtration provide a convenient and economic means of controlling temperatures and dust within working areas.

The commissioning of a recirculation scheme may be rapidly instituted since development work should be minimal. The costly and time consuming development of new airways or modifications to main fans, necessary to bring more air from surface, may be avoided. Hence, the capital cost of a recirculation scheme can be very low compared with providing the same quantity of air by conventional means. A cost analysis of the introduction of the Loraine recirculation scheme and a comparison with normal ventilation is included in a paper by Fleetwood and others (1984).

It must be emphasized that it has been demonstrated on many occasions in operating mines that recirculation per se does not lead to any general increase in the concentration of contaminants in the return air. Furthermore, recirculation does not lead to a gradual build-up of any contaminant over time. These two important findings were supported by all the measurements made during the Loraine trial. The major problem associated with controlled recirculation is the hazardous situation which would arise if there were to be a fire in the area. All South African gold mines use large quantities of timber for support and, although much of this is treated by fire-retardant chemicals, fires nevertheless occur. If a fire were to occur during a working shift, then dangerous gases, such as carbon monoxide, as well as smoke would be partially introduced into the intake air, thus preventing a safe fresh air exit for the personnel in the area.

This illustrates the need for an effective and reliable system that is designed to detect a fire and automatically stop recirculation by either control of fans or ventilation doors. Such action would be designed to return the area to a

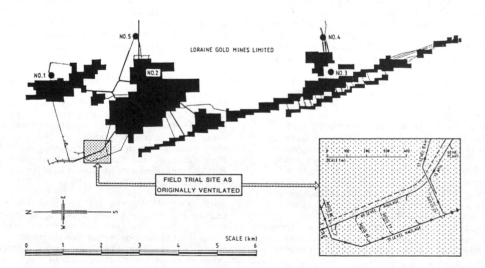

Fig. 1. Site of the recirculation field trial

conventional once-through ventilation system and
thus prevent any noxious fumes from entering the
intake air.

3.   THE LORAINE FIELD TRIAL

The field trial of controlled recirculation at
Loraine Gold Mines Ltd had three main objectives:

(i)   To demonstrate that controlled recirculation
       can be a practical, workable and effective
       aid to ventilating and cooling gold mines.

(ii)   To demonstrate that a controlled recircu-
       lation system can be operated safely.

(iii) To show that the effects of controlled re-
       circulation can be predicted by means of
       mathematical models.

To achieve the objectives, a computer controlled
monitoring system was installed to acquire data
that would enable objectives (i) and (iii) to be
fulfilled.   In addition to this, an experimental, automatic safety system was installed to ensure that the recirculation scheme would operate
safely (objective ii).   The development of such
a safety system may be considered to be the most
important part of the trial, since this would be
an essential part of any future recirculation system.   The monitoring system is described in Section 4 of this paper whilst the safety system is
described in Section 5.

Some theoretical results and conclusions gained
from the field trial are described in Section 4.2.
A description of the trial itself is fully documented in a recent paper (Burton and others, 1984).

3.1   Outline of the Field Trial

A schematic of the field trial site in the Eldorado
section of Loraine Gold Mines Ltd is shown in Fig.
1.   An important consideration when choosing the
site was that the introduction of the recirculation system would offer the mine real benefits.
Thus the site was chosen in an area that was re-

mote from existing shafts and at a depth of 1800m,
where the virgin rock temperature was 46°C.   The
fresh airflow to the area was restricted to about
35 m³/s at a wetbulb temperature of up to 28°C.
Production in the area was limited due to the poor
environmental conditions in some of the working
places, caused by high temperatures and low air
flowrates.   The refrigeration plant indicated in
Fig. 1 could not be utilized to its full capacity
of 1760 kW(R), since the intake airflow at that
location was only 15 m³/s.   A further consideration was that the 20 m³/s of intake air shown
flowing from the north was also required elsewhere
for development headings.

The recirculation scheme designed for the field
trial ventilated four stopes (working areas).
A schematic of the recirculation system is shown
in Fig. 2.   The recirculation circuit was established by developing a 1.8m diameter raisebore airway between the two operating levels of the system.
A three-stage, direct contact spray chamber was
constructed at the bottom of the raise-bore to cool
the recirculated air before it was mixed with the
intake air.   The 1760 kW(R) refrigeration plant
was dedicated to this spray chamber.   The recirculation fans (5 X 40kW) were installed in parallel
at the top of the raise-bore on 57 level to control
the recirculated airflow.   Intake air was controlled by two similar fans installed in the intake
airway on 60 level.   The design conditions for the
scheme required an intake air quantity of about
15 m³/s and a recirculated quantity of about 35m³/s
thus providing a total of 50 m³/s of air to the
area, at an intake temperature of 23°C wet-bulb
and a reject temperature from the stopes of 28°C
wet-bulb.

A specially constructed, air-conditioned instrument
room was built on 60 level to contain the bulk of
the monitoring and safety systems.   Signals from
all the remote sensing devices in the area were
routed to this room.

The installation was successfully commissioned in
October 1982.

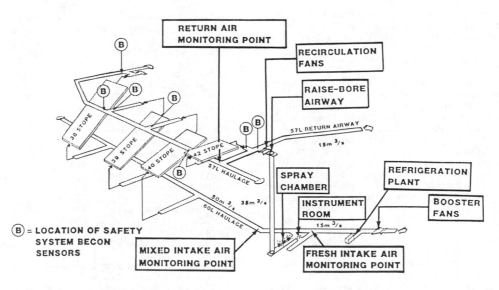

Fig. 2. Schematic of recirculation installation

4.   MONITORING SYSTEM

The main objective of the monitoring system was to acquire and store data automatically from a variety of remote sensing devices.   Various environmental parameters were studied, which enabled the effects of recirculation to be closely investigated.   A block diagram of the monitoring system, showing the remote sensors, instrumentation and control output functions is shown in Fig. 3.

The instrument room was manned continuously for a period of one month following commissioning of the recirculation system.   Following this period, it was manned one shift only on a daily basis.   In addition to the computer-based monitoring system, regular manual surveys of various environmental parameters were undertaken throughout the area.

The parameters automatically monitored by the system were as follows:

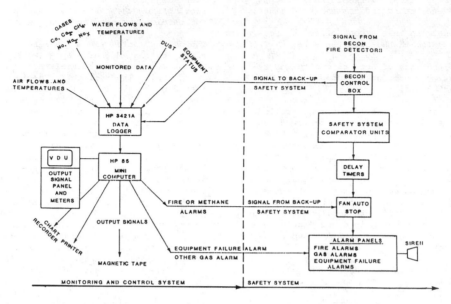

Fig. 3. Block diagram of instrumentation and safety system

4.1   Design of the Monitoring Instrumentation

The instrumentation was designed such that comprehensive monitoring could be performed on the following:

(i)   Fresh air conditions.

(ii)   Mixed-intake air conditions (downstream of the point at which the recirculated air and the fresh air are mixed).

(iii)   Return air conditions.

Data acquisition was by means of a Hewlett-Packard HP 85 desk top computer coupled to a Hewlett-Packard HP 3497A data logger.   Although the main task of these units was to monitor and store data in order to verify the predicted effects of recirculation, two further capabilities were incorporated.   First, the computer provided a daily print-out of all relevant environmental parameters and equipment performance status.   Second and more important, it provided a back-up safety system for automatic control of the recirculation fans in case there was a failure of the primary safety system. The safety system is described fully in Section 5.

Since the Loraine field trial was the first large scale test of controlled recirculation, it was considered essential that as many parameters as possible should be examined.   The quantity of measurements taken was intended to be able to allay any misconceptions about recirculation and to be able to predict with certainty the effect of recirculation under various conditions.

(i)   Gases (CO, CO_2, CH_4, NO, NO_2, NOx).

(ii)   Air temperatures (Wet- and Dry-bulb).

(iii)   Air velocities.

(iv)   Dust concentrations.

(v)   Chilled service water and Spray Chamber water temperatures and flows.

(vi)   Equipment status (Fans, Pumps, etc).

Air was sampled from the three monitoring points (fresh air, mixed-intake air and return air), by means of plastic sniffer tubes which led to gas analyzers in the instrument room on 60 level.

The gas analyzers were controlled by the computer, which sequenced the analyzer to sample from each sampling point once every six minutes.   The concentration of each gas was displayed on LED monitors on the front of the analyzers as well as being logged by the computer.

Air temperatures were monitored at the three main sampling points using specially designed psychrometer units.   These utilised two PT-100 Resistance Temperature Device (RTD) probes; one RTD measured dry-bulb temperature and the other the wet-bulb temperature.   The RTD's farthest from the instrument room utilised field transmitters, which generated a 4-20 mA signal, thus avoiding voltage drop on the long signal wires.   The RTD's nearer the instrument room used a direct 3-wire voltage signal.   The signals from all the RTD's were taken by armoured cable to the instrument room.

In the case of air quantities, it was only necessary to monitor the intake and mixed-intake air-
streams, since the return air quantity would be the
same as the intake quantity (assuming no leakage).
Initially, hot-wire anemometers were used, but
turbulence in the air-stream caused great fluctuations in the readings.   After damping circuits
were incorporated, the unit in the fresh air performed satisfactorily.   However, the extremely
humid nature of the mixed-intake air, due to the
spray chamber, caused repeated failure of the hot-
wire anemometer.   This was subsequently replaced
by a vane anemometer, which performed satisfactorily for the remainder of the trial.

Dust measurements were made in the intake, mixed-
intake and return air, by means of copper sniffer
tubes which led back to the instrument room.
Measurement of dust concentrations in each of the
samples was by means of a Tyndallometer.   Sequencing control between the three sampling points was
performed by the computer, thus allowing the concentration at each of the measuring points to be
logged once every six minutes.   Some problems were
experienced with condensation in the sniffer tubes,
but this was overcome by the use of heating tapes
and pipe insulation.   In addition to the automatic monitoring, manual dust measurements were
made in the intake, mixed-intake and return air
with an Anderson 2000 cascade impactor.

The service water temperature at the inlet to the
stopes was monitored with a single PT-100 RTD which
transmitted its signal to the instrument room.
The flow was monitored on a chart recorder which
was not connected to the computer system.   The
flow of water to the spray chamber was monitored
using a Differential Pressure cell, whilst the
temperatures at the inlet and outlet of the spray
chamber were measured with RTD's.

The operational status of the recirculation fans,
the spray chamber interstage pumps and the computer system itself were monitored.   In the event
of a failure, warning lights were illuminated on
three alarm panels (described in Section 5).   In
the case of a fan or pump failure, indication of
this was logged by the computer with appropriate
print-out.

4.2   Operation and Performance of the Monitoring System

The monitoring programme yielded a large amount of
data, which was subsequently used for validating
recirculation theory and for establishing design
parameters for future systems.   In terms of the
stated objectives (Section 3), the monitoring of
the field trial was a considerable success.   The
system was operated without any major problems for
more than a year.

The introduction of recirculation and cooling had
an immediate effect on the wet-bulb temperatures of
the air within the   area.   It must be emphasized
that these changes were not due to recirculation
per se, but to the bulk air-cooling which was incorporated in the system.   The various wet-bulb
temperatures around the recirculation circuit, both
before and after the system was commissioned, are
shown in Fig. 4.   The actual temperatures at any
point in the system showed considerable fluctuations due to a number of factors, such as the rock
production rate and the intermittent use of chilled service water.   The values shown in Fig. 4 are
therefore mean values.

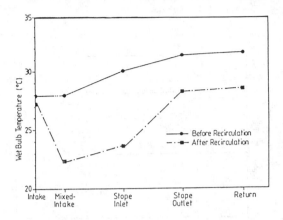

Fig. 4. Wet bulb air temperature changes
around the recirculation system

Theoretical analyses of the concentration of gaseous contaminants in any recirculation system show
that three main conclusions can be drawn:

(i)   Recirculation does not cause return air
       contaminant concentrations to increase above
       the values existing before recirculation.

(ii)   Recirculation increases the mixed-intake
       air contaminant concentrations.

(iii) Recirculation does not cause a gradual
       build up of any contaminant in an area.

One of the major requirements of the monitoring
system was to confirm these conclusions.   In order
to demonstrate the effect of varying the recirculation fraction, R (defined as the ratio of recirculated airflow to the total mixed-intake airflow),
many measurements were made at varying airflows, of
the concentrations of gaseous contaminants.   The
results for carbon dioxide are shown in Fig. 5,
which shows that return air contaminant levels do
not rise with increasing R.   In the mixed-intake
air, there is a linear rise in contaminant concentration, but the level is always below that of the
return air concentration.   The gaseous contaminant
concentrations were monitored continuously throughout the field trial.   Carbon dioxide levels were
typically 295, 425 and 465 ppm in the intake,
mixed-intake and return air respectively.   Carbon
monoxide levels at the same locations were typically 7, 9 and 10 ppm and oxides of nitrogen (NOx)
were typically 0,55, 1,23 and 1,59 ppm.   These
figures are well within the limits of 5000 ppm for
carbon dioxide, 50 ppm for carbon monoxide and 5
ppm for NOx as prescribed by the Occupational
Safety and Health Administration (OSHA).   These
figures are all for the period outside the blast.
These figures rose considerably during the blasting period
but returned below the set limits within about two
hours of the blast.   Methane levels at all times
were barely detectable.

Dust measurements were made in the intake, mixed-
intake and return air, the average values in these
locations being 0,222, 0,231 and 1,05 mg/m³ respectively.   The value for the return air was
barely acceptable.   However, it should be noted
that the concentration in the mixed intake air is
only slightly higher than that in the fresh intake

air and that the airflow has been considerably increased compared with before recirculation.   Thus,
it is clear that recirculation has not created the
high level in the return air.

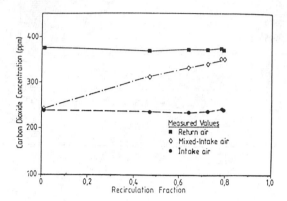

Fig. 5.   Relationship between CO_2 concentration
          and recirculation fraction

It also should be noted that rock production increased from about 6000 to 18 000 tons per month
after the introduction of recirculation and it is
this that has led to the increased dust concentrations.   If filtration were to be employed, these
levels could be significantly reduced.   It was
observed during the field trial that the bulk air
cooling spray chamber was effective at removing
the large, but not the small, dust particles.
This has led to further research, currently under
way at the Chamber of Mines Research Laboratories,
to assess the dust suppression characteristics of
sonically atomized water sprays.   A successful
laboratory test has produced promising results and
a full scale field trial is now under way.

The monitoring system was virtually maintenance
free.   Apart from weekly calibration checks on
the remote sensing devices, the only maintenance
required was cleaning of the psychrometer units,
which was done once per week.

The monitoring system was removed after 12 months.
The recirculation system is still operating
(November, 1984) and the only instrumentation re-
maining is the automatic safety system.

5.   SAFETY   SYSTEM

If a fire were to start within the area served by
the recirculation system, hazardous fumes would be
introduced into the intake.   This can be prevented
by stopping recirculation.   An automatic safety
system was therefore designed to detect a fire and
automatically shut down the recirculation fans.

5.1   Design of the Safety System

A mine-proven, 'Becon' fire detection system was
chosen for the trial.   This system was developed
during the 1970's by the Anglo American Electronics
laboratory (Van der Walt, 1983).   Since 1979, over
3 000 detectors have been installed in 23 South
African gold mines and collieries.   Such systems
have also been sold to Australia and the USA.
Normally, signals from the various Becon sensors
are telemetered to surface, where they are displayed on chart recorders which are monitored manually.
The Loraine field trial was the first instance of a
Becon system coupled to automatic fan control.

Seven Becon sensors were used for the trial, these
being installed at the stope exits and in the return airway on 57 level (Fig. 2).   The sensors detect ionized particles caused by a fire and, in the
presence of these, the output signal drops.   Transmission of the sensor output signals to the instrument room was by means of a 12-core armoured cable.

In the instrument room, the Becon signals from the
four most important sensors (at the exits from the
four stopes) were fed to the safety system control
unit.   This unit comprised four comparators,
which compared the Becon signals with pre-set alarm
levels.   If any Becon signal dropped below the
pre-set alarm level, then an alarm situation would
be created.   However, the recirculation fans were
not immediately stopped in such a situation, since
short term influences, such as exhaust emissions
from diesel locomotives or blasting fumes, could
cause the Becon signal to fall temporarily below
the pre-set alarm level.   Thus, delay timers were
built into the circuit, which would only activate
the fan trip relays after a certain time period had
elapsed.

Since a methane emission (although extremely unlikely) could also be potentially hazardous in the
recirculation system, the return air methane gas
level (Section 4.1) was connected to a fifth comparator in the safety system control unit.   In the
event of an unacceptable methane level being detected, this comparator could also shut off the recirculation fans.

The intake air booster fans were not affected by
the safety system and continued to run at all times,
thus ensuring an uninterrupted fresh air supply to
the system.   The wiring was configured to be fail-
safe in the recirculation fans, such that none of
these fans could be restarted whilst any one or
more Becons was alarming.   The fans could be restarted manually once normal status was resumed.

Once an alarm had tripped the recirculation fans,
visual alarms were given at three panels on 60
Level;   at the refrigeration plant (continuously
manned), on the outside wall of the instrument
room, and inside the instrument room.   In addition, a siren was placed in the refrigeration plant
to indicate an alarm.

These alarm panels also carried warning lights for
indication of equipment failure (see Section 4.1).

5.2   Operation of the Safety System

The system was extremely reliable, there being only
one false fire alarm, which resulted in the recirculation fans being shut down unnecessarily during
the first year of operation.

The response of the Becon sensors was tested to
determine the effect of the exhaust gases from
diesel locomotives.   This was done by purposely
operating a diesel locomotive close to a detector.
A locomotive that had recently been overhauled and
one that was due for an overhaul were continuously
shunted back and forth upstream of a detector.
The results were favourable, in that, although the
detectors reacted to the exhaust fumes for short
periods, the built-in time delay of 15 minutes
(referred to in Section 5.1) prevented the recirculation fans being shut down unnecessarily.

Distinguishing between blasting fumes and a fire
was a more difficult problem.   An analysis of
Becon responses during many blasting periods indicated that, with a 15 minute delay, the safety
system would cause the fans to be shut down at

every blast. A 'window' was therefore activated in the blasting period (13h00 to 18h00 daily), during which time a longer delay of 60 minutes was used.

Unfortunately, the most likely time for a fire to start is in the period immediately following the blast, and the early detection of such a fire is clearly desirable. Thus, the need is illustrated to be able to clearly define the difference in Becon response between a fire and a blast.

6. FUTURE DEVELOPMENTS

The latest developments by the Anglo American Electronics Laboratory incorporate computer analysis of telemetered Becon signals from underground in combination with advanced colour graphics. It would require only a small amount of additional programming to enable such a system to control fans in the event of an alarm, thus making it suitable for recirculation systems. Fan and ventilation door monitoring and control is already practised on a

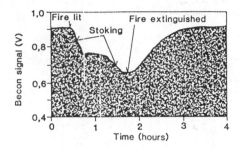

Fig. 6.1. Signal due a small test fire

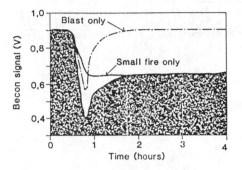

Fig. 6.3. Predicted signal due to blast and small fire

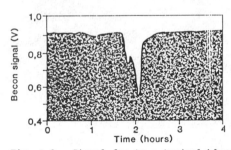

Fig. 6.2. Signal due to a typical blast

Fig. 6. Becon signal levels due to typical blast and small fire

A test was designed in which a controlled fire was lit and the Becon response closely recorded. The test was conducted on a Sunday, when no personnel were present in the area. A small wood fire was lit at the entrance to one of the stopes and, during the test, the automatic safety system was disabled. Figure 6.1 depicts the Becon response at the exit of the stope in which the fire was lit. The effect of a typical blast on a Becon signal is shown in Fig. 6.2, and the difference between these two graphs can clearly be seen. The response time for the blast is very much more rapid than for fire, both with the signal dropping and rising. The two responses are shown in combination in Fig. 6.3.

Thus it is fairly simple to devise a method of distinguishing between a fire and a blast, by monitoring the rate of change of the Becon Signal. The hardware for such a system is now being developed by the Anglo American Electronics Laboratory.

The Becon sensors and safety system operated extremely reliably during the entire period of the trial, with the only maintenance being regular cleaning and calibration of the sensors. The system, in a slightly modified configuration, was retained by the mine once the field trial was completed and is still in continuous operation.

number of South African mines.

However, for mines with no telemetering system, the need is clearly defined for a small robust safety unit that could be used underground and which would monitor Becon signals and automatically shut off recirculation fans in the event of a fire. The design of a unit is currently being investigated by the Environmental Engineering Laboratory of the Chamber of Mines.

7. ACKNOWLEGEMENT

This paper arises from work carried out as part of the research programme of the Chamber of Mines Research Organization.

The authors wish to acknowledge the assistance of many people who were involved in the field trial. In addition, thanks are due to the staff of Loraine Gold Mines Ltd and Anglovaal Ltd, for all the enthusiasm, effort and help that they gave during the operation of the first major trial of controlled recirculation in South Africa.

REFERENCES

Aldred, R., Sproston, J.H., and Pearce, R.J.(1984).

Air conditioning and recirculation of mine air
in North Nottinghamshire.
Min. Engr., 143, 601-7

Allan, J.A. (1983). A review of controlled recirculation ventilation systems. CIM Bull., 76,
83-7.

Bakke, P., Leach, S.J., and Slack, A. (1964).
Some theoretical and experimental observations
on the recirculation of mine ventilation.
Colliery Engng., 41, 471-7.

Burton R.C., Plenderleith, W., Stewart, J.M.,
Pretorius, B.C.B., and Holding, W. (1984).
Recirculation of air in the ventilation and
cooling of deep gold mines. In Third International Mine Ventilation Congress, Harrogate,
Howes, M.J. and Jones, M.J., eds (London : IMM)
291-9.

Fleetwood, B.R., Burton, R.C., Pretorius, B.C.B.
and Holding, W. (1984). Controlled recirculation of ventilation air at Loraine Gold
Mines Limited. Association of Mine Managers
of South Africa, Circular 2/84.

Lawton, B.R. (1932). Local cooling underground by
recirculation.
Trans. Instn. Min. Engrs., 85, 63-71

Pickering, A.J., and Aldred, R. (1977). Controlled recirculation of ventilation - a means of
dust control in face advance headings.
Min. Engr., 190, 329-345.

Van der Walt, N.T. (1983). Smoke without fire.
Nuclear Active, Jan 1983.

The Modelling and Control of Refrigeration Plant Used in Cooling Systems on Deep South African Gold Mines

N.T. MIDDLETON

Engineering Systems Branch, Chamber of Mines of South Africa Research Organization, Johannesburg, SA

Abstract. Many of the large refrigeration plants used in the cooling systems on deep South African gold mines are situated on surface where they are exposed to seasonally and diurnally varying climatic conditions, and to a diurnally varying demand for cooling made by the mine. It has been traditional to accommodate the seasonal variation in climate by manual control applied to the plants at appropriate times of the year, and the plants have normally been specified to conform passively to the diurnal variations. This paper introduces new strategies for the active automatic control of the plants on a diurnal cycle, and shows that attractive benefits in energy and chilled water resource management can be accrued through the use of these strategies. The results have been derived through the modelling and simulation of a representative refrigeration plant.

Keywords. Mining; environmental control; refrigeration; stochastic control; optimization.

INTRODUCTION

This paper deals with the modelling and dynamic control of the large refrigeration plants which are used extensively in the South African gold mining industry, for generating the cooling required to counteract the severe heat loads which exist in deep mines.

The deepest South African gold mines are approaching 4 000 m below surface, where the virgin rock temperature can exceed 60 °C; the average depth of workings in the industry is about 2 000 m, and at this depth the virgin rock temperature is approximately 40 °C. In the last twenty years, with mines reaching greater depths, and with the determined effort by the industry to provide comfortable environmental conditions for a work force of nearly half a million people, the capacity of installed refrigeration plant has grown from less than 50 MW in 1963, to its current value of about 850 MW.

The cooling provided by the refrigeration plants is distributed through a mine using chilled water, which, in turn, is used to cool air in heat exchangers, and to provide a widespread cooling effect in the working areas where it is used as service water.

Roughly one ton of service water is used for activities like drilling and washing down, for each ton of rock broken from the face. It has been proved (van der Walt and Whillier, 1978) that the practice of chilling the service water significantly assists the conventional air cooling methods, in deep working areas. Service water chilling has been the major stimulus to the massive growth in refrigeration installations in the gold mining industry. Approximately half of the refrigeration plants used are installed on surface at the mines, and this paper concentrates on the control issues affecting these plants, as distinct from those which are installed underground. Surface refrigeration plants are required to conform to widely varying operating conditions; not only is the demand for chilled service water variable and dependent on the users of this water, but also, the surface refrigeration plants must dissipate heat to ambient climatic conditions which vary both seasonally and diurnally.

A number of shutdown strategies for individual machines, aimed at matching overall plant characteristics to the seasonally changing climatic conditions, have been devised (de Nysschen, 1980, for example). In respect of the diurnally varying operating conditions, however, it has been traditional to specify surface refrigeration plants which can conform to these changes passively, so that the plants are only in need of occasional manual control. It is the purpose of this paper to introduce new automatic control strategies for the diurnal control of a surface refrigeration plant, and to show that considerable benefits in energy and chilled water resource management can be accrued.

In the subsequent sections to this paper, the layout of a typical surface refrigeration plant is described, and the measured response of a plant to diurnal manual control and real operating conditions is appraised, indicating a scope for an improved control strategy. A brief section on plant modelling follows, and then emphasis is given to the development and simulated performance of automatic control strategies. Detailed elaboration on the content of this paper can be found in Middleton (1984).

PLANT LAYOUT

The layout of a typical surface refrigeration plant is shown in Figure 1. Warm water returned from the mine is 'pre-cooled' in a cooling tower to near the ambient wet-bulb temperature, and then it is piped into an evaporator network which chills it to about 4 °C. A temperature and flow bypass control valve, which spans the evaporator network, serves to mix controlled quantities of chilled water with the water off the pre-cooling tower, so affecting both the cooling load on the evaporator network, and the flowrate of chilled water delivered into the reservoir, since the flowrate of water in the evaporator network is kept constant.

Heat is transferred from the evaporator network by the compression of a circulating refrigerant, which condenses at a high temperature in the condenser network, during the transfer of heat to a circulating coolant water stream. The heat carried in this stream is dissipated to the atmosphere in the condenser cooling towers.

The variable demand by the mine for chilled water affects the level of chilled water stored in the reservoir, while the changing climatic conditions, particularly the wet-bulb temperature, affect the temperature of the water off the pre-cooling tower and the temperature in the condenser coolant water stream. The efficiency of the plant depends, among other things, on the condensing temperatures and the temperature of the water entering the evaporator network.

DIURNAL MANUAL CONTROL

Figure 2 shows recorded operating conditions which cover a five day period. Figure 3 shows the recorded response of a plant to manual control over the same period. It can be seen, in Figure 3, that the chilled water delivery flowrate is changed from time to time, and this is because of manual adjustments made to the setting of the bypass control valve. Particularly evident in Figure 3 is the slow, and large, oscillation in the level in the chilled water reservoir; after some 36 hours into the period shown, when the reservoir was nearly empty, the operator made a correction to the bypass valve, which, in spite of subsequent smaller adjustments, nearly caused it to over-flow after 120 hours into the period. The operator was apparently not capable of satisfactorily matching the flowrates of chilled water delivery to the erratic demand, within the confined capacity of the reservoir.

With manual control, no attention was being given to minimizing the electrical energy consumption of the plant, and this fact is demonstrated in Figure 4, which shows the electrical power consumption profile and the coefficient of performance (COP) of the plant over a 24 hour interval in the five day period. The COP is a figure of merit for a refrigeration plant, as it indicates the ratio of the cooling power achieved in the evaporator network to the power which has to be supplied to the compressors in the plant. Cool climatic conditions lend themselves to the attainment of high COP's, and hence economical electrical energy consumption. In Figure 4, it can be seen that the COP is low when climatic conditions are cool, that is, during the night and the early hours of the morning. The cool climatic conditions were not being exploited to reduce energy consumption.

This appraisal of the typical performance of manual control indicated that an automatic control strategy should be investigated.

PLANT MODELLING

Refrigeration plant modelling and the setting up of a simulation facility was an important phase in the study which led to an automatic control strategy. Middleton, (1979 , 1980 and 1984) should be referred to for a full treatise on this work.

The general aims behind the modelling effort were to provide a means for examining refrigeration plant characteristics with respect to varying climatic conditions and controlled delivery flowrates, and to create a simulation facility which could be used in the assessment of control strategies.

An analytical, as distinct from a system identification, approach was followed in the modelling of the plant. The handling of the dynamic aspects were simplified by the omission of the dynamics of those elements in a refrigeration plant which have a very much quicker response than the diurnal variations in the operating conditions and the volumetric response of the chilled water reservoir.

There are some 115 variables in the plant model, and the simulated responses of key variables were compared to measured data from the real plant. In this case, the simulation was excited with recordings of the same operating conditions that prevailed at the time of the measurements. Figure 5 shows typical actual and simulated responses of two key variables. Calculated over all key variables and many side-by-side tests, the mean error between actual and modelled data was −0,3 per cent, while the rms error was 3,23 per cent.

AUTOMATIC CONTROL

This section discusses the objectives for the automatic control of a surface refrigeration plant, the modelling of such a plant in a form which is amenable to control theory, and the development of automatic control strategies. In the discussion on these strategies, it is pointed out that certain prior operating condition data are needed, and references are made to the forecasting algorithms used for this purpose.

Control Objectives

The objectives to be met by a suitable automatic control system are that the refrigeration plant consumes a minimum quantity of electrical energy, while simultaneously ensuring that it delivers sufficient chilled water to the reservoir to cover the variable demand made by the mine. There are constraints on the capacity of the refrigeration plant and the range of chilled water flowrates which can be delivered by the plant.

Modelling for Control

The simulation facility described above is capable of predicting refrigeration plant characteristics. One such characteristic is shown in Figure 6, which indicates the plant electrical power consumption, m, as a function of the delivery flowrate, q, the ambient wetbulb temperature, t_{wb}, and the temperature of the water off the pre-cooling tower, t_{wpco}. Regression techniques were used to express the characteristic in the quadratic form:

$$m = Y^T.E.Y + F^T.Y + g \quad kW \qquad (5.1)$$

where
$$Y = \begin{bmatrix} q \\ t_{wb} \\ t_{wpco} \end{bmatrix} \qquad (5.2)$$

and E and F are respectively a matrix and a vector of coefficients, with g being a scalar constant. Although (5.1) expresses a quadratic formulation in Y, it turns out that the coefficient of q^2 is approximately zero, so that an elimination of the q^2 term, achieved by setting that coefficient to zero, only modifies the predicted value of the power, m, by about 0,25 per cent. The power m is therefore approximately linear in the control variable, q, as endorsed by an inspection of Figure 6.

The quantity of chilled water stored in the reservoir can be approximated by

$$\ell(k+1) = \ell(k)+0,0036[q(k)-d(k)] \text{ Megalitres} \tag{5.3}$$

where k is a discrete time variable of interval 1 hour, ℓ is the stored quantity of chilled water, q is the delivery flowrate in l/s, and d is the demand flowrate, also in l/s.

By setting the coefficient of q^2 in (5.1) to zero, and rearranging (5.1) and (5.3), the refrigeration plant can be modelled dynamically by

$$X(k+1)=A.X(k)+B(k).U(k)+W(k) \tag{5.4}$$

$$\text{where} \quad X(k) = \begin{bmatrix} m(k) \\ \ell(k) \end{bmatrix} \tag{5.5}$$

$$U(k) = \begin{bmatrix} q(k) \end{bmatrix} \tag{5.6}$$

$$A = \begin{bmatrix} \epsilon & 0 \\ 0 & 1 \end{bmatrix} \tag{5.7}$$

where ϵ is a small number to preserve the non-singularity of A, and B(k) and W(k) are time-varying vectors which depend non-linearly on $t_{wb}(k)$ and $t_{wpco}(k)$.

Energy Minimization by Non-linear Programming

A constrained non-linear programming method was used to find an acceptable profile of the control variable q(k), which would result in minimum energy consumption by the plant, assuming that the variables $t_{wb}(k)$ and $t_{wpco}(k)$ were known deterministically for a future period of time. The high dimensionability of the problem (represented by the number of time intervals being considered) and the large number of inequality constraints on the control variable (two per time interval), led, however, to such excessive computation times that the non-linear programming approach seemed to be unsuitable for real-time control. This study showed, nevertheless, that energy consumption could be reduced through the use of dynamic control.

Energy Minimization by the Infinite Reservoir Solution (IRS)

With the failure of non-linear programming in giving an acceptable approach to dynamic control for minimum energy consumption, a purely mathematical path was followed, and this led to the 'infinite reservoir solution' (IRS). This solution derives from the following problem formulation.

The energy consumption, which is to be minimized, is expressed by

$$J = \sum_{k=1}^{\infty} m(k) \qquad \text{kWh} \tag{5.8}$$

while it is essential that the average delivery flowrate can be specified (and normally set to equal the average demand by the mine so that the reservoir neither overflows nor runs dry), so that

$$\bar{q} = \lim_{N \to \infty} \frac{1}{N} \sum_{k=1}^{N} q(k) \tag{5.9}$$

($\bar{q}$ is specified),

and at the same time it is necessary to constrain the derived values of q(k) to fall within practicable limits. This is done by specifying the variance around the average $\bar{q}$ which the sequence of q(k)'s can assume:

$$\text{var}(q) = \lim_{N \to \infty} \frac{1}{N} \sum_{k=1}^{N} [q^2(k)-\bar{q}^2] \tag{5.10}$$

(var(q) is specified).

The IRS arises from minimization of the Lagrangian

$$L = \sum_{k=1}^{\infty} m(k)+ \lambda_1 \lim_{N \to \infty} \frac{1}{N} \sum_{k=1}^{N} \left\{ q(k)-\bar{q} \right\}$$
$$+ \lambda_2 \lim_{N \to \infty} \frac{1}{N} \sum_{k=1}^{N} \left\{ [q^2(k)-\bar{q}^2]-\text{var}(q) \right\} \tag{5.11}$$

Solving $\frac{\partial L}{\partial q(k)} = 0$; $\frac{\partial L}{\partial \lambda_1} = 0$; $\frac{\partial L}{\partial \lambda_2} = 0$ $\tag{5.12}$

and developing the algebra yields

$$q(k) = \bar{q} - \frac{\text{sd}(q)}{\text{sd}(\text{D})} . [\bar{b} - b(k)] \tag{5.13}$$

where sd represents a standard deviation, and b is a quadratic function in the variables t_{wb} and t_{wpco}.

This analysis, which derives the delivery flowrate resulting in minimum energy consumption, does not account for the variation in the level of chilled water in the reservoir. Since the average delivery, $\bar{q}$, can be set to equal the expected average demand, the average chilled water storage in the reservoir will be maintained, but it is theoretically feasible that the level of water in the reservoir could fluctuate by large amounts around the average. It is this notion which has led to the name 'infinite reservoir solution' (IRS).

Figure 7 shows how the natural cooling of the night hours has been exploited by the IRS to reduce energy consumption. In comparison to the manual control situation of Figure 4, it can be seen that the power consumption profile is considerably reduced, the peak is shifted from day to night, and the plant COP is kept high during the cool hours. In order to meet, on average, the demand for chilled water, the IRS control strategy causes the plant to deliver high flowrates into the reservoir at night, when efficient refrigeration conditions prevail, causing the level in the reservoir to rise; this stored chilled water is used during the day to supplement the reduced quantity delivered by the plant, because of the less efficient refrigeration conditions which exist at that time.

Stochastic Receding Horizon Control (SRHC)

Deficiencies in the IRS are that there is no means for simultaneously achieving a reduced energy consumption while having some control on the level of water in the reservoir, and that the IRS is a supervisory algorithm, in that it does not specify any plant feedback. To avoid these deficiencies it is necessary to examine the control of the plant using the dynamic model given in equation (5.4).

Because of the variations which occur in $B(k)$ and $W(k)$, equation (5.4) represents a time-varying system. There are only a few ways of controlling a time-varying system, and one of these involves the use of optimal control theory and its application to the multi-variable setpoint controller problem. The development of the theory leads to a feedback controller which causes the plant power consumption and reservoir level to track time-varying setpoints with a degree of accuracy specified by weighting parameters which may vary with time. Unfortunately, the reservoir level and plant power consumption are inversely coupled, so that if the weighting is arranged for the level to closely track its setpoint, then the power consumption is likely to deviate substantially from its set-point, and vice-versa. Intermediate trade-offs between level and power setpoint tracking accuracies are feasible.

A practical difficulty which arises from the use of this theory is the need for prior knowledge of the variations in $B(k)$ and $W(k)$, for future time. Because a refrigeration plant runs continuously, this implies that an infinite amount of information is required for the solution of the controller equations. These difficulties have been surmounted through the use of forecasting and the notion of receding horizon control.

The essence of receding horizon control is the use of information on influential events which are expected to occur between the present and a future horizon in time, to contribute to an optimally calculated control action taken at the present time, which will cause improved aggregate performance by the plant. As the present time advances, the horizon recedes correspondingly, so that the information on expected future influential events always covers a fixed time period.

In the context of refrigeration plant control, the future influential events are the variations in $B(k)$ and $W(k)$, while, since climatic effects and water demand variations are on a diurnal cycle, the horizon is taken to be 24 hours into the future. Armed with the knowledge of the future behaviour of $B(k)$ and $W(k)$, it is possible to compute a delivery flowrate $q(k)$ to be set at the present time, which will optimally contribute to attaining the desired aggregate performance of the plant with respect to reservoir level and power setpoint control. After each

discrete time interval, k, the computations are repeated, resulting in a new value for $q(k)$ because of the new information about future variations in $B(k)$ and $W(k)$.

Since it is impossible to specify the future variations in $B(k)$ and $W(k)$ for 24 hours deterministically, it has been necessary to use forecasting algorithms for this purpose, so assigning future values to $B(k)$ and $W(k)$ stochastically. The combination of the stochastic nature of forecasting with receding horizon control was behind the name which has been given to this method of control, 'stochastic receding horizon control' (SRHC).

The forecasting algorithms which were used, derive from the theory of time-series analysis, which has been thoroughly dealt with by Box and Jenkins (1976). Middleton (1984), describes the development of the algorithms for this application.

A mathematical requirement in the implementation of SRHC, is the solution of a matrix Ricatti equation and a matrix prefilter equation. The solutions to these equations at the present time are used to compute the feedback (Kalman) and feedforward (prefilter) gains to be used at the present time in the automatic control system. However, to arrive at the solutions of the matrix Ricatti and prefilter equations at the present time, these equations must be solved backward in time from the future horizon to the present. Both the Ricatti and the prefilter equations depend on the vectors $B(k)$ and $W(k)$ and their forecasted variations over the period up to future horizon.

Since the errors between the actual and the forecasted values of $B(k)$ and $W(k)$ increase with lead time up to the horizon, and since the Ricatti and prefilter equations are solved backward from the horizon to the present, using the forecasts of $B(k)$ and $W(k)$, a certain amount of error is induced into the solutions of the Ricatti and pre-filter equations. A statistical analysis has shown that the error contamination of the solution at the present time to the Ricatti and prefilter equations is remarkably small, as indicated by the error at time k in Figure 8, which shows the induced errors into one element of the solution matrix to the Ricatti equation.

The impact of these small errors is that the SRHC strategy is, in this application, very nearly as good as the receding horizon control system would have been, should the forecasted values of $B(k)$ and $W(k)$ been deterministically available.

Figure 9 is a block diagram which shows the general scheme for SRHC. By adjusting the weighting parameters, an operator can select the emphasis toward either minimum power consumption, or toward close tracking of the reservoir level to a setpoint which can be specified (and varied) by the operator.

An example of the control of the delivery flowrate prescribed by SRHC, and the simulated response of the plant, is given in Figure 10. The first 24 hours are used for accumulation of operating condition history by the forecasting algorithms, and so no control or response is shown. For approximately the next 36 hours the SRHC strategy is emphasized toward minimum energy consumption, while for the remaining period the emphasis is changed toward regulation of the reservoir level to a constant value. In the first period it can be seen that, after an initial transient, the power consumption response is favourable, while the level in the reservoir varies quite substantially; in the second period it is the reservoir level which is controlled, while the power consumption varies. In practice, an intermediate weighting is desirable.

Performance of Control Strategies

Table 1 presents the salient performance features for manual control, control using the infinite reservoir solution and stochastic receding horizon control, where figures for emphasis toward both minimum energy consumption and minimum deviation from a constant level setpoint are given. The methods are compared in terms of the average daily kWh consumption, the maximum kVA consumption, and the span through which the level in the reservoir moves.

The infinite reservoir solution predicts a reduction of nearly 8 per cent in total energy consumption, and the maximum kVA consumption is reduced by a significant 20 per cent. These improvements do not have a seriously detrimental effect on the fluctuations of level in the reservoir, particularly since these are predicted to be some 55 per cent better than the fluctuations which characterize manual control.

Stochastic receding horizon control with the emphasis toward minimum energy consumption performs similarly to the infinite reservoir solution, as would be expected. However, if the emphasis is toward minimum deviation from a constant level setpoint, the penalty for maintaining the reservoir level constant is a maximum kVA consumption which is about 22 per cent higher than that experienced with manual control.

In the case of control for minimum energy consumption, the peak kVA demands occur during the night, which has a favourable effect on the load factor for the whole mine. However, the reverse is true for precise control on the level in the reservoir. In practice, therefore, the method which is advocated is SRHC with the emphasis largely toward minimum energy consumption; worthwhile energy consumption benefits will be derived, while should there be a substantial deviation in the reservoir level, there will be some degree of control action tending to correct this.

CONCLUSIONS

It has been shown that the use of an automatic control strategy will improve the performance of a surface refrigeration plant. In particular, the total energy consumption, the maximum kVA demand, and the capacity requirements for the storage reservoir can all be reduced. Additionally, the peak power demand can be made to occur during the night, which has a favourable effect in improving the load factor for the whole mine.

A significant implication of the reduced kVA demand is the correspondingly reduced refrigeration plant capacity which would be required to provide the same cooling as was provided by the manually controlled plant. This represents a capital saving.

Diurnally based automatic control systems of the type described in this paper have not been used on refrigeration plants in the South African gold mining industry. It is advocated that the strategies outlined in this paper should be seriously considered for implementation, because of the benefits which have been predicted.

ACKNOWLEDGEMENT

This work has been conducted as part of the research programme of the Research Organization of the Chamber of Mines of South Africa.

REFERENCES

Box, G.E.P. and Jenkins, G.M. (1976). Time-series Analysis : Forecasting and Control. Holden-Day, 1976.

De Nysschen, A.J. (1980). Considerations in the Design and Layout of a Surface Refrigeration System at the Southern Shaft, Buffelsfontein Gold Mine. Symposium on Refrigeration and Air Conditioning, Riverside Convention Centre, Vanderbijlpark, October, 1980. Mine Ventilation Society of South Africa.

Middleton, N.T. and Bluhm, S.J. (1979). Simulation of Refrigeration Systems for the Environmental Control of Deep Gold Mines. Symp. on Simulation of Dynamic Systems in Applied Science, Management and Engineering. South African Council for Automation and Computation, University of Cape Town, 27 August, 1979.

Middleton, N.T. and Bluhm, S.J. (1980). Simulation and Optimal Control of Refrigeration Systems used for Environmental Control in South African Gold Mines. 3rd IFAC Symposium on Automation in Mining, Mineral and Metal Processing. Montreal, Canada, August, 1980. Pergamon Press.

Middleton, N.T. (1984). The Modelling and Control of Refrigeration Plants in Deep South African Gold Mines. Ph.D. Thesis. University of the Witwatersrand, Johannesburg, October, 1984.

Van der Walt, J. and Whillier, A. (1978). The Cooling Experiment at Hartebeestfontein Gold Mine. Journal of the Mine Ventilation Society of South Africa, 31, 8, 1978, 141-147.

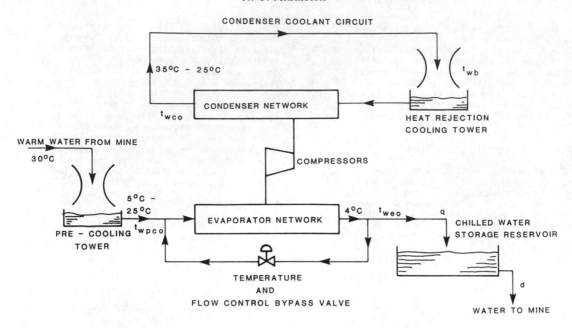

Figure 1.   Basic Layout of a Surface Refrigeration Plant

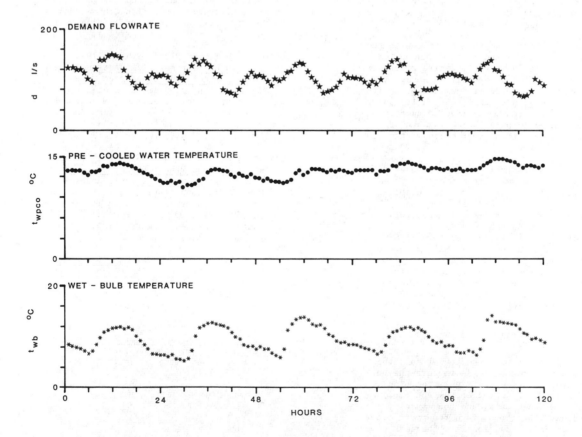

Figure 2.   Operating Conditions over a Five Day Period

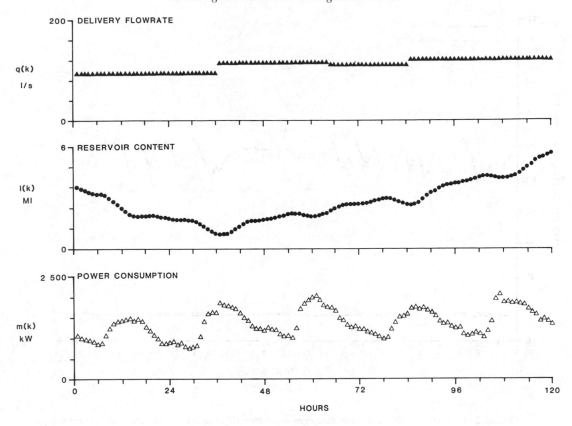

Figure 3. Manual Control over a Five Day Period

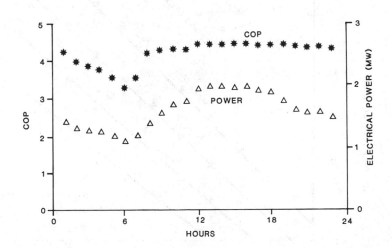

Figure 4. Response of Coefficient of Performance and Electrical Power Consumption to Manual Control over a 24 Hour Period

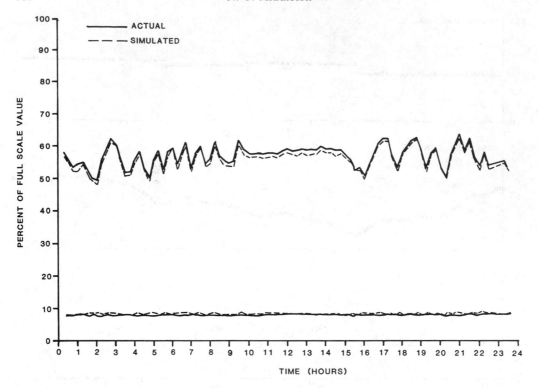

Figure 5.   Actual and Simulated Responses of Two Key Variables in the Refrigeration
             Plant and its Model

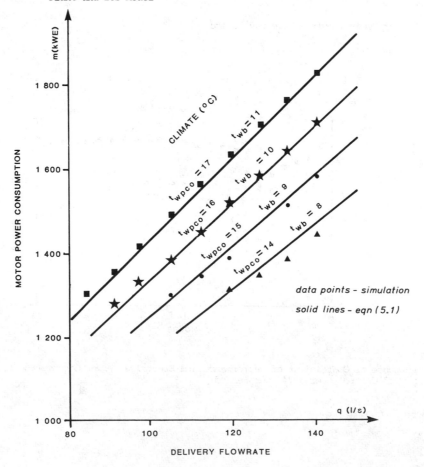

Figure 6.   Plant Electrical Power Consumption as a Function of Climate and Delivery Flowrate

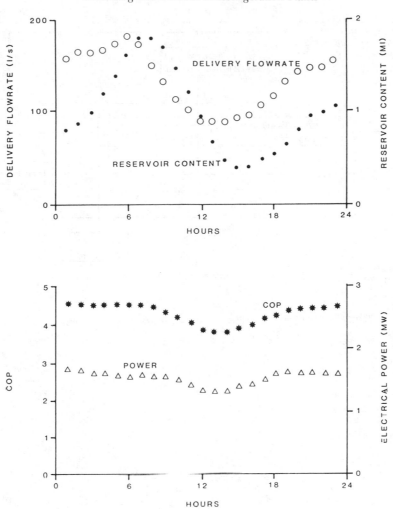

Figure 7.   Delivery Flowrate as Predicted by the Infinite
             Reservoir Solution, with Responses of Reservoir
             Content, COP and Power Consumption

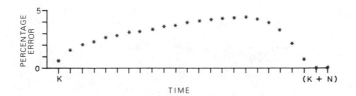

Figure 8.   Error Contamination of an Element in the Solution
             Matrix to the Ricatti Equation, as a Function of
             Lead Time to Horizon

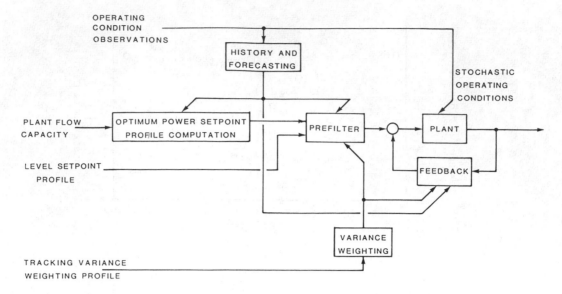

Figure 9.   The Basic Strategy for Stochastic Receding Horizon Control

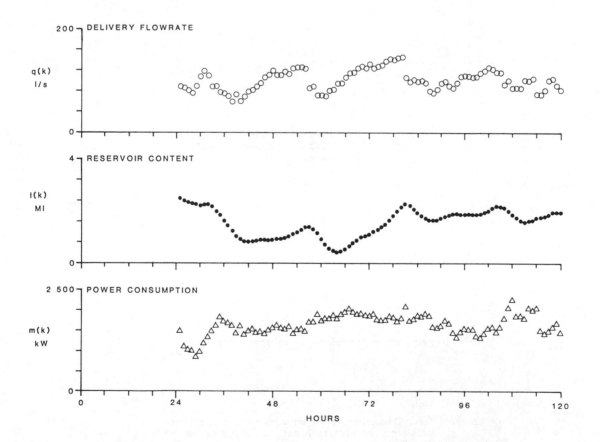

Figure 10.   Simulated Performance of the Stochastic Receding Horizon Control System

TABLE 1   Performance Comparison for Different Control Methods

Control Method		Electric Energy Consumption		Electrical Demand		Reservoir Level	
		Average daily (kWh)	Improvement over manual control (% reduction)	Maximum (kVA)	Improvement over manual control (% reduction)	Span (m)	Improvement over manual control (% reduction)
Manual		32 641	0	2 401	0	4,78	0
Infinite Reservoir Solution		30 123	7,71	1 921	19,99	2,16	54,81
Stochastic Receding Horizon Control	Minimum energy emphasis	30 618	6,20	1 925	19,83	1,51	68,41
	Minimum deviation from constant level setpoint emphasis	31 489	3,53	2 937	−22,32	0	100,0

Experiments with Air Flow Control in a Coal Mine

B. FIRGANEK MSc, PhD, Z. KRZYSTANEK MSc. PhD, S. WASILEWSKI MSc. PhD

Research and Production Centre for Mining Electrotechnics and Automation EMAG, ul. Armii Czerwonej 83a,

40-161 Katowice, Poland

Abstract. The problem of automatic control of a coal mine ventilation network is discussed. It is found that the structure of the control system should have two levels, i.e. local regulation of ventilation parameters (air flows) in faces to eliminate the effect of quick disturbances, and supervisory control of the whole ventilation process by periodical modifications to the working characteristics of the main fans. Practical and theoretical problems connected with implementation of the system in Polish mining are listed and discussed. To solve these problems a series of experiments with air flow control in three different coal faces were carried out with the aim of checking the measuring and control equipment in a real mine conditions as well as investigating the gasodynamics of the ventilation process. The results of the experiments are analysed, and used to verify the mathematical model of the gasodynamics of a longwall face.

Keywords. Control equipment; control theory; data aquisition; data reduction and analysis; environment control; identification; step response.

INTRODUCTION

The main excavation system used in Polish coal mines is longwall, with great concentration of output and high levels of mechanization in mining and transportation of the coal. Although such a system guarantees a high efficiency of production, and the possibility of full mechanization and automation of the production processes, it is very inconvenient, as far as the natural hazards are concerned. Special problems are created by firedamp in faces, where methane emission from excavated coal and goaf frequently exceeds the permissible level. This is the main limitation on increasing coal production.

In spite of widespread use of degassing systems, the basic preventive method is still the intensive ventilation of faces. Big main fans are used, with power in the range of several megawatts. Special ventilation systems are aimed to maximize dilution of the methane in faces and joined galleries. The ventilation air is directed to faces by the shortest possible routes with big cross-sections. However, the big changes in methane emission in faces results in these great masses of air being ineffectually utilised. Investigations made in Poland and in other countries showed that, in order to keep the optimal air flow from the safety and economical point of view, corrections to the ventilation network should be made every few days. This task exceeds the ability of traditional methods of control of the ventilation network, therefore research and experimental work on control of ventilation in coal mines was carried out. The main topics in those works were investigations of the effects accompanying an air flow in a ventilation network, especially in excavation faces, as well as formulation and practical revision of the mathematical model. The experiments described in this report were carried out in a Polish colliery.

CONTROL OF A MINE VENTILATION NETWORK

A present-day colliery, excavating coal by a longwall system with high concentration of output, has great and rapid changes in the parameters of the ventilation network. An average life-time of faces does not normally exceed a few months. During this comparatively short period, there are a wide range of changes in practically all ventilation parameters of faces, like: aerodynamic drag of workings, gas concentration, air velocity, as well as physical and chemical parameters of the air. Distribution of these changes is random, and therefore the ventilation network as a whole is stable, with parameters that change very slowly, but over a wide range. This is useful from the control point of view, and determines the structure of the control system, which should have two levels:

    i/ local regulation of ventilation parameters in faces to eliminate the effect of quick changes;

   ii/ supervisory control of the whole ventilation process in the mine by periodical modifications to the working characteristics of the main fans.

Use may be made of the layout of the ventilation system. The ventilation network is naturally divided into ventilation areas, the number of which is most often equal to the number of active longwall faces. The areas are ventilated by separate streams of fresh air. Consequently, they may be considered as independent control units, connected together only through the supervisory control level, assigning set values for the local systems (Krzystanek, 1981).

The above described structure has been put into practice in the Polish coal industry. The system consists of regional subsystems,,controlling the airflows, and a supervisory computer installed on the surface. Asxa controlled member in regional systems, the remotely regulated gauge door is used. Its position is determined according to the measurements of an anemometer installed in the entrance gallery. The set values for the regional control system is an amount of air necessary for maintaining the concentration of methane at a safe level. The value is calculated on the basis of a global performance index of the mine ventilation.

Implementation of the system in practice is very difficult and demands the solution of several practical and theoretical problems. Some of them are as follows.

1/ The problem of assurance of the stability and repeatability of the controlled members working in a mine environment. The construction of the doors and the manner of their installation should assure resistance to mechanical loads occurring due to natural deformation of excavations, as well as corrosion resistance. The above conditions are in great measure fulfilled by the Polish shutter gauge door of type ATWR (Fig.1). A rigid, solid construction of the door and use of a double frame, in which regulation members are positioned, assures great strength and resistance against rock pressure. One method of installation of a door in a gallery is shown in Fig.2.

2/ Problems concerning the measuring of air parameters. Operation of the system needs constant automatic measuring, mainly of the quantity or velocity of air and the methane concentration in sections. The first value is a controlled variable in a local control system, and the second is a main parameter in equations used for optimization of air flow and calculation of the set values. Air velocity measurement may be accomplished by the recently designed Polish acoustic anemometer AA-1 (Fig. 3). This instrument is without moving parts, assuring precision measurement, especially in very low velocities of airflow. To obtain constant methane monitoring, the new multi-purpose methanometer MW-1 (Fig.4) has been designed with measuring range of 0 to 5 % CH_4. Automatic measuring of other quantities such as temperature, moisture and absolute as well as differential pressure are also desirable in the system, but with the exception of temperature sensors, there is a lack

of instruments sufficiently tested in practice.

3/ The problemsof evaluation and revision of a mathematical model of the ventilation network, necessary for determination of the control and regulation algorithms. For operation of the supervisory level of the system a well known and tested steady-state model is sufficiently good, but evaluation and determining the parameters of a local regulation algorithm needs knowledge of a model of dynamic processes occurring in areas due to the interaction of disturbances and/or the control actions. Special difficulties are met when the mathematical description and identification of the methane sources in workings are attempted. This is due to the non-measurability of several parameters and the variety of gas emission patterns. In spite of several trials made to solve the problem, it is still unresolved and needs further investigation.

EXPERIMENTS WITH AIR-FLOW CONTROL

Important stages in work on ventilation control, carried out in the Research and Production Centre EMAG, were two experiments performed in Manifest Lipcowy Colliery. The colliery has an extensive and complicated ventilation network and exploits seams with high methane contents.

The aim of the experiments, apart from checking the measuring instruments and gauge doors, was the investigation of the gasodynamics of the workings by observations of transient effects produced by regulation actions with various durations and amplitudes. The investigations were carried out in three longwall faces with different outputs of coal, different systems of ventilation, and different volumes of goaf. The faces were as follows.

1/ Longwall face No.1 (Fig.5a): winning by longitudinal, retreating, caving system: the length of face was approx. 200 m. The face has been 70 % extracted and is situated adjacent to widespread goaf.

2/ Longwall face No.2 (Fig.5b): winning by transverse advance, caving system: the length of face was approx.100 m, at the beginning of exploitation, goaf rather small.

3/ Longwall face No.3 (Fig.5c): winning by longitudinal, retreating, caving system: the length of the face was approx. 180 m, 50 % exploited. Ventilated in a "Z" system; the goaf adjacent to the ventilation road sealed.

During the experiments, the following quantities were continously measured:
1/ the velocity of the inflow air stream by rotating anemometer;
2/ methane concentration at the end of the ventilation road by multipurpose methanometer MW-1;
3/ absolute pressure in the face by pressure sensor and portable instruments of a type barolux.

By means of gauge doors installed at the

entrance to each section,   step changes of the inlet air-stream were produced.
The measured data were transmitted via telemetric system to the surface and recorded. The data were used in verifying the
gasodynamic model afterwards.

RESULTS OF THE EXPERIMENTS

The characteristic curves of the measured
values of pressure (P), air velocity (v)
and methane concentration (C) in the outlet air streams, recorded during the experiments, are presented in Fig.6, 7 and
8.

The instants when regulation actions were
made, i.e. opening or closing of gauge
doors, are indicated by arrows on the time axis.

The measurements were made during non-
working shifts, therefore it may be assumed that the observed changes in methane
concentration result from superposition
of two processes:
1/ changes in methane emission from the
adjacent goaf, forming a kind of methane
reservoir, from which methane is washed out
by the air stream and/or pulled out due to
the changes in pressure produced by the
regulation action;
2/ mixing of the methane emitted from
goaf, body of coal and disintegrated coal
with ventilation air at the face.

Comparing the changes of pressure and velocity in all faces with corresponding
changes of methane concentration, the
first two may be recognized as step changes. Only in the biggest area (No.1), a
several-minute delay may be noticed in
stabilization of the air stream velocity,
but this is small in comparison with the
duration of the transient state of methane concentration, which lasted up to two
hours. The step change of pressure and
quantity of air in area No.1 produced at
the beginning a significant change in the
amplitude of methane concentration, in
the same direction as the air quantity
change, which is opposite to the expected
result of the regulation, and only later
was the slow return to the steady state
value observed. This phenomenon, characteristic of faces adjacent to extensive goaf
creates the biggest problems in formula-
ting an algorithm of the local regulation.
The effect is less disturbing (smaller
changes of amplitude and shorter transient
state periods), if changes of the regulated variable are relatively slow, which
has been confirmed during further experiments made in the area.

At section No.2, where the face is small
and so is the goaf, as is to be expected,
the complicated transient states did not
occur. The transient states are monotonic,
and the stabilization times of the measured quantities are short. The effect of
goaf is negligible, similarily, in section
No.3, where the goaf is sealed along the
ventilation road. A non-monotonic curve,
having however a different shape, smaller
amplitude, and shorter duration in comparison with section No.1, may be observed
in section No.3 only in the step change of
the quantity of air flowing through the

face. It may be assumed that the cause of
that phenomenon is the intensive ventilation, washing out the methane from ca-
verns and voids in the goaf adjacent to
the face. After the washing out is completed, the concentration of methane sta-
bilizes at a level lower than before the
regulation.

VERIFICATION OF THE GASODYNAMIC
MODEL OF A LONGWALL FACE

The transient state of flow of the composition of air and methane through the excavation containing the source of methane may be, under some simplifying assumptions, described by the equation (Trutwin, 1973):

$$V \frac{dC(t)}{dt} + q(t)C(t) = q(t)C_o(t) + q_m(t)$$
$$(1)$$

where:
V - volume of the mixing zone,
$C(t)$ - methane concentration at the end
        of ventilation road,
$q(t)$ - volume of air stream in the exca-
        vation,
$C_o(t)$ - methane concentration at the en-
        trance to the excavation,
$q_m(t)$ - volume of methane emission.

Taking into account the step changes of
the ventilation conditions, i.e.:

$$q(t) = \begin{cases} Q_1 & \text{for } t < 0 \\ Q_2 & \text{for } t \geqslant 0 \end{cases}$$

and the constant (in the majority of practical cases equal to zero) methane concentration at the entrance to the emission zone:

$$C_o(t) = C_o = const$$

equation (1) may be evaluated as follows:

$$V \frac{dC(t)}{dt} + Q_2 C(t) = Q_2 C_o + q_m(t) \qquad (2)$$

with the initial condition
$$C(0) = C_o + \frac{q_m(0)}{Q_1} \qquad (3)$$

The last variable in equations (1) and
(2), called the characteristic or function of the methane source, decides the
form of the transient states, and its
evaluation and, particularly, identification in underground conditions creates the
most difficult problems. The total quantity of methane emitted at a longwall face is the sum of gas emitted from coal,
surrounding stone, and goaf. The fraction
of the separate components is variable
and depends on several parameters that
are unmeasurable or measurable only with
great difficulty. Therefore, the more
convenient method, suitable for the purpose of automatic control, is treating
the methane emission as a whole, assuming
the general, physically justifiable model
of a source, and experimental by measuring its parameters. The model, verified
in this way, may then be utilized in de-

signing the control system, as well as in the modelling and simulation preceding the implementation of the system in practice. Such a method has been employed in this work.

The most general approach to the problem of mathematical description of the methane source in the working area is an assumption (Trutwin, 1973) that methane penetrates to the excavation through the goaf, which is treated as a reservoir of gas of a unknown volume V_z. The penetration of methane to the excavation through the filter zone may be described, like a linear process, by a relation:

$$q_m(t) = b \left[p_z(t) - p(t) \right] \qquad (4)$$

where:

b – filtration coefficient,
p – absolute pressure in excavation,
p_z – pressure in goaf.

Unmeasurable pressure in the goaf may be described by the differential equation:

$$V_z \frac{dp_z(t)}{dt} + b p(t) p_z(t) = p(t) \left[q_z(t) + b p_z(t) \right] \qquad (5)$$

For the step changes of pressure

$$p(t) = \begin{cases} p_1 & \text{for } t < 0 \\ \\ p_2 & \text{for } t \geqslant 0 \end{cases}$$

and, under the assumption that the emission of methane from the body of coal is constant during the transient state ($q_z(t) = Q_m = $ const.), equation (3), with the initial condition

$$p_z(o) = p_1 + \frac{Q_2}{b} \qquad (6)$$

has the solution:

$$p_z(t) = p_2 + \frac{Q_m}{b} - (p_2 - p_1) e^{-\frac{b p_2 t}{V_z}} \qquad (7)$$

The source function (2) has therefore the form

$$q_m(t) = Q_m - b(p_2 - p_1) e^{-\frac{b p_2 t}{V_z}} \qquad (8)$$

and the solution of equation (1) for this form of source:

$$C(t) = C_o + Q_m \left[\frac{1}{Q_2} + \left(\frac{1}{Q_1} - \frac{1}{Q_2} \right) e^{-\frac{t}{\tau_1}} \right] +$$

$$- \frac{b(p_2 - p_1)}{V \left(\frac{1}{\tau_1} - \frac{1}{\tau_2} \right)} \left(e^{-\frac{t}{\tau_2}} - e^{-\frac{t}{\tau_1}} \right) \qquad (9)$$

where:

$\tau_1 = \dfrac{V}{Q_2}$ – time constant of the mixing process in the excavation,

$\tau_1 = \dfrac{V_z}{b \cdot p_2}$ – time constant of the methane penetration from goaf.

After a small transformation, the equation in a form convenient for approximation may be obtained:

$$C(t) = K_o + K_1 e^{-\frac{t}{\tau_1}} + K_2{}^{-\frac{t}{\tau_2}} \qquad (10)$$

This equation approximates well the transient states in section No.1 (Fig.9). The calculated approximate values of parameters for this section are:
V = 31 000 m^3
Q_m = 0.117 m^3/s
τ_1 = 1090 s,
V_z = 120 000 m^3
b = 1.1 m^3/s . kPa,
τ_2 = 1700 s.

For the remaining two regions, the effect of goaf may be disregarded, and so the form of the source function may be considerably simplified. Thus, for section No. 2, model (1) has been tested, assuming constant emission of methane from surrounding rock and disintegrated coal, i.e.

$$q_m(t) = Q_m$$

yielding a good approximation for the parameters:

V = 6300 m^3
Q_m = 0.07 m^3/s,
$\tau = \tau_1$ = 625 s

For section No.3, it has been assumed that the pulse-shape emission of methane lasted for a time T

$$q_m(t) = \begin{cases} Q_{m1} & \text{for } t < 0 \\ Q_{m2} & \text{for } 0 \leqslant t < T \\ Q_{m1} & \text{for } t \geqslant T \end{cases}$$

giving parameters
V = 9025 m^3, Q_{m1} = 0.093 m^3/s, Q_{m2} = 0.121 m^3/s, T = 885 s, τ = 504 s.

CONCLUDING REMARKS

The results of experiments discussed in this report are part of research work being carried out in the Research and Production Centre EMAG in Poland, aiming at the implementation of control systems for mine ventilation in the coal industry.

The experiments have been planned to obtain the real data necessary for verification of the mathematical model of air flow dynamics and methane emission in underground excavations.

The model approximates well the transient states. The main problem is evaluation of the mathematical description of the methane source, the form of which depends on several-measurable parameters related mainly to the geological and technical conditions of a certain area.

The results obtained are utilized in modelling research and computer simulation. The control system is represented by a steady-state model of the ventilation net-

work of a whole mine and a dynamic model
of the sections being worked. The steady
state model is used for controlling the
ventilation of a whole mine, i.e. for evaluating the set values for local regulators to ensure proper levels of safety parameters while minimizing the amount of
energy consumed by the mine ventilation.
The dynamic model comprises equations describing the dynamics of transient processes and sources of quasi-random disturbances occurring during normal work, caused
for example by working machines. The model
is used for checking the quality of algorithms designed for local regulators.

Simultaneously, the hardware of a control
system is completed and investigated in

laboratory and mine conditions, according
to the concept described in section 2. The
system is to be installed shortly in a Polish colliery. At first, it will include
only the chosen working sections, and later it will be gradually extended to the
whole colliery.

REFERENCES

Krzystanek, Z.(1981). Control of a Mine
    Ventilation Process at Presence of
    Disturbances. PhD Thesis (in Polish)
Trutwin, W.(1973). Influence of Ventila-
    tion Conditions on Methane Concentra-
    tion in Mine Workings. Zeszyty Proble-
    mowe Górnictwa , Vol.11, No.2 (in Po-
    lish).

Fig. 1. Gauge door

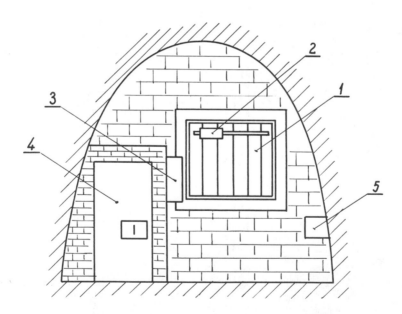

Fig.2. Gauge door mounted in a gallery

        1. shutter door   4. man-lock
        2. pnematic cylinder   5. installation hole
        3. control box

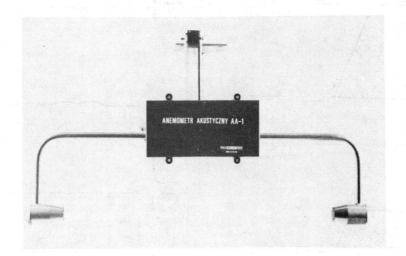

Fig.3. Acoustic anemometer

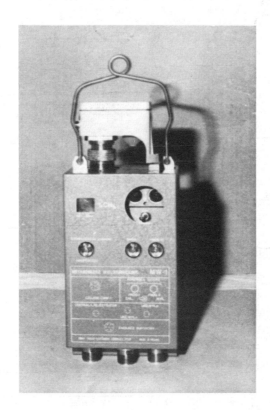

Fig.4. Multipurpose
        methanometer

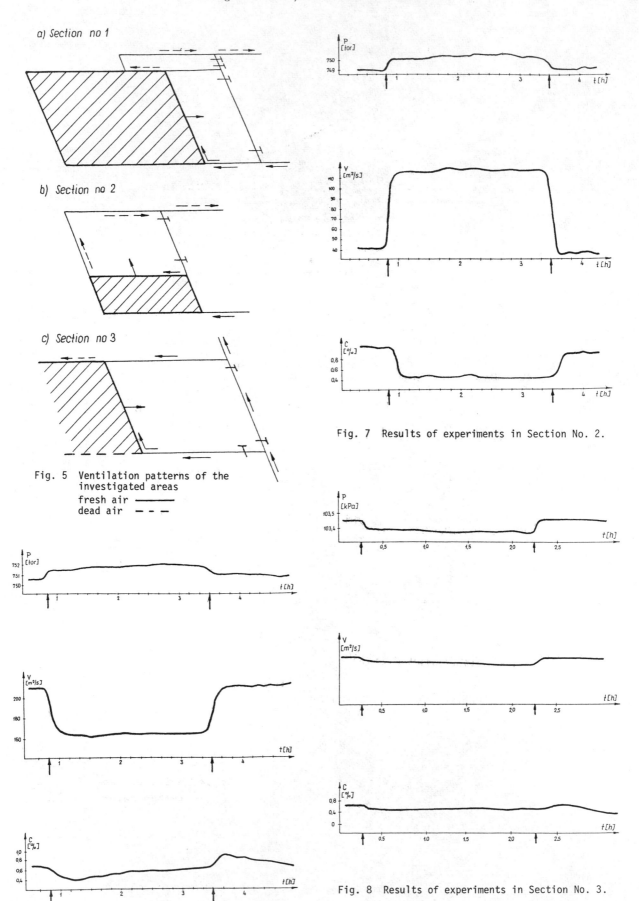

a) Section no 1

b) Section no 2

c) Section no 3

Fig. 5 Ventilation patterns of the investigated areas
fresh air ——————
dead air — — —

Fig. 6 Results of experiments in section No. 1.

Fig. 7 Results of experiments in Section No. 2.

Fig. 8 Results of experiments in Section No. 3.

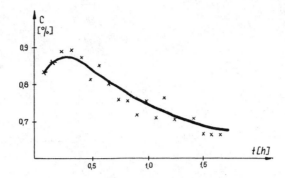

Fig. 9 Approximation of a transient process
C /t/

real results   **x**
approximation   ——

The Computer Modelling of Surface Mining Operation

V. SRAJER

Calgary Coal Research Laboratory, CANMET, Calgary, Alberta, Canada

Abstract. The Canada Centre for Mineral and Energy Technology (CANMET)/Calgary Coal Research Laboratory and the National Research Council in Ottawa established a joint working research group with industry to develop and test various computer models for application in surface coal mining operations in western Canada.

The first part of the paper deals with the problems the group experienced during the past two years especially having model specification and requirements defined and understand site management practices.

The second part of the paper deals with site specific development of general mine model for TransAlta Utilities' Highvale Mine in Alberta. Described here are the development of an interactive truck/shovel model for their three pit operation, selection and application of dragline model for tandem operation with the extended bench method, criteria for operational model of the five seams operation with selective mining approach.

Potential model application for other mining companies are also discussed. An overall structure for interactive dynamic event-based models of surface mining operation is introduced. Examples of logic diagrams for some selected events are also presented.

Keywords. Specification for computer modelling; software survey; data collection; general mine model; equipment models; mine operation model.

INTRODUCTION

A hierarchy of computer models are in various development stages at the National Research Council (NRC) in Ottawa, Ontario and at the CANMET/ Calgary Coal Research Laboratory.

This project as a joint venture of the NRC/CANMET and industry commenced in 1982. The introduction of the program was announced during the seminar "Computer Modelling Applications In Surface Mining" held in September 1982 together with the 4th Open Pit Operators Conference in Edmonton, Alberta.

During the last two years, the NRC/CANMET group cooperated with the TransAlta Utilities Corporation of Calgary, Alberta, in the development of various computer models for their Highvale Mine, located west of Edmonton in Alberta.

This paper summarizes the experience of the NRC/CANMET group in the development of computer models for surface coal mines as well as the achievements and outline of the plan for future.

BACKGROUND

Specification and Requirements

At the very start of the project, the mining company was asked to provide detailed description of their requirements that are expected from the model. A clear and comprehensive description was needed, including:
- Mine planning
- Short-term scheduling
- Selection of mining methods
- Development of data base
- Economic evaluation
- Equipment selection
- Equipment dispatching
- Special applications.

The description of the needs and identification of problems to be addressed proved to be the most difficult and time consuming task. After several trial and error runs the following steps were taken:
1. Identification of users
2. Addressing general and specific needs of each user
3. Development of computer based aids to meet the needs of users
4. Evaluation of the computer based aids.

It is estimated that actual development of the computer model and testing of the model represents only about 25% of the total time needed to complete the work. A major part of the effort was spent on identification of needs of the users. Even after several meetings the mining personnel were not aware what the computer model could do for them and how they could benefit from using the model. The most difficult task was to make the mine personnel (from pit boss to mine manager) clearly define their duties, so that they give the step-by-step instruction on the logic used by them during their daily schedule.

After several attempts to have the problem resolved and keep the project moving, more agressive approach was initiated. The following procedure was adopted and has been in use since:
1. Initial Meeting (one day) - to develop general

understanding and needs of the company, identification of users and contact personnel.
2. Consultation with Users (one week) - after
development of the work plan individual users
were contacted separately.
3. Project Meeting (one day) - to agree to work
plan, milestones and responsibilities.
4. Users Meeting (several days) - each potential
user is to be involved in the model development
during his working day (or several days).The
following personnel had to be contacted: pit
boss, supervisor (mining, mechanical, safety,
etc.), mine manager, assistant mine manager,
engineer, planner. It would not be expected of
the mine personnel to develop a structured work
program but simply to answer to the one planned
by CANMET/NRC. Only after this, the rationale
behind user's decision could be understood.
5. Presentation Meetings (several days) - each
user had to be called for any possible modification they would like. This was to be the
last meeting before the commencement of the
model development work.

Survey of Available Software

Another very important step in the development of
useable computer model was to conduct a survey of
any software available outside of the organization. This was achieved through personal
contacts with the users and developers of the
software.

After getting an idea of the possible application
and requirement of the model, an extensive survey
of software applicable to the mining industry was
initiated. Two main aspects were examined:
- software availability
- software applicability.

The available software was categorized as the one
freely accessible and obtainable from the vendors,
with source code included in the package.

The applicable software was the software of which
more than 75% was immediately useful for the
industry, well documented and compatible to the
VAX computer.

This work was carried out by CANMET through a
research contract so as to identify software
applicable to the Canadian mining industry. A
seminar was also organized with the participation
of a consulting company from United States to
identify the most useful software packages on the
American market. Several companies were later
contacted and asked for detailed presentation of
their computer software and hardware. Subsequently, the users of mining software were visited
and their experience evaluated.

After completion of the survey a very good idea of
the available software applicable to the Canadian
mining industry was obtained.

Field Investigation

The Highvale Mine in Alberta, owned by TransAlta
Utilities Corporation became the field laboratory. The mine was selected as the first mining
property for development and demonstration of the
computer models, because of the company's atti-
tude, immediate needs, availability of site and
data for model validation and testing.

The Highvale Mine is producing 8 Mt of subbitumi-
nous coal per year for the power generating
station at the mine site. The mine, one of the
largest in Canada, is operating four pits, having
draglines to remove overburden, using a truck/
shovel system for coaling operation and mobile
equipment (dozers, scrapers) to remove interburden
between the five seams. The use of bucketwheels
and conveyors for removing the overburden is under
consideration.

The field investigation consisted of well planned
detailed time studies and evaluation of previous
data obtainable from the mine site personnel.
Several weeks, at different times of the year
(spring, summer, winter), were spent at the site
to acquire knowledge of the mine operation and to
study truck, shovel and dragline cycles. The
collection technique improved with time, from
simple use of stop-watch and writing the times for
manual evaluation, to development of the recording
instrument and evaluation of data by computer
(Stuart, 1984).

At present, evaluation of the truck and loader
system and the dragline cycle study is completed.

A study is also initiated for the investigation of
different production systems used by other companies to acquire knowledge and to understand decision making process. The collected data will be
ultimately used in the development of a general
mine model that could be applied to most surface
coal mines in Western Canada.

DEVELOPMENT OF GENERAL MINE MODEL

Basic Structure

The general mine model basic structure is actually
a set of individual structural and equipment
models. These models are expected to form large
subsystems that will activate the general mine
model when the operator decides for that.

It is understood that there is definitely a limit
to the degree the model could be generalized and
applied to any other surface coal mine. It is
assumed, however, that there is a certain structure that is applicable to any surface mine in
Western Canada (use of similar equipment, mining
methods typical for mountain coal deposits or
plains coal deposits, reporting forms, etc.). The
proposed general model will have site-specific
modules attached to the general data base. Identification of these site-specific modules and the
development of the general structure is the present day priority. The basic structural tasks are
as follows:
1. Geological Modelling. It is expected to
investigate several models available at the
software market and to purchase one of them.
2. Operating Conditions Models. A description of
the mine operation is done in consultation with
the mine operators. This model will eventually
tie-up the geological model with production and
equipment models.
3. Equipment Model. Each equipment operating at
the mine is modelled, its cycle timed, use and
assignment defined. These models are closely
connected with equipment data base.
4. Mine Layout Model. Models that monitor various
mine practices and patterns typical for each
operation. Decision making practices by pit
designers are clearly spelled out in these
models. Some mining properties with similar
geological/geotechnical parameters are expected
to be similar in mine layout.
5. Production/Scheduling Models. Ultimate in use,
these models will provide answers to the mine
operator, engineer or upper management on the
problems associated, for example, with equip-

ment scheduling, transportation sequenced mine planning, etc. Finally, they could be used in the development of a dispatching system for mine trucks.

6. Cost Program Models. These are also obtainable from various software vendors. Also, each mine will have some system that would be useful in defining cost models. It is expected, the cost models will be a part of the whole package, for evaluation of productivity, financial analysis and sensitivity analysis of different decisions or policies. In general, the software and data bases for interactive computer modelling are described in Fig. 1 (Nenonen, 1981a) displaying functions of the user input, process and output reporting.

It is expected that the user will make the ultimate decision on selection of models, options requested and type of reports required in the output.

Equipment Models

The first task addressed in the project was the development of equipment models.

Each equipment model is a part of the system model designed for individual operations. For example, a model of system "A" (BWE/conveyor/bin) consists of the following individual models: Bin model, Conveyor model and Bucketwheel model. Each individual model is selected by the user from the equipment data base according to his own decision and needs. The system models could also be combined with the possibility of interchanging the models. For example, if the user is trying to have sensitivity analysis for BWE/conveyor/ stacker and BWE/hopper/truck/bin/stacker, he may use some unconventional combination of the equipment models in the system model.

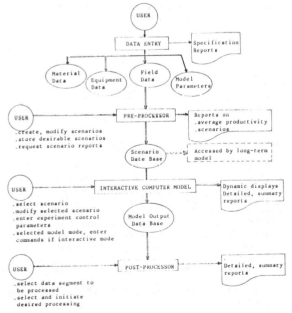

Fig. 1. Software and data bases for interactive computer modelling.

Each system model has its own typical production statistics, based mainly on the historical data specific for this type of mining operation, or projected production requirements. The production statistics models will ultimately feed into the coarse model of mining operation that will

incorporate long term dynamics and decision making models. The mining operation model may also be connected to geological or mineralization models data base. The computer model hierarchy is presented in Fig. 2 (Nenonen, 1982a).

A typical system structure that summarize operational flow is developed for each operation so that the user could choose to have the computer run the whole simulation or part of it. The typical system describing Highvale Mine is illustrated in Fig. 3. Figure 3 also illustrates several system models divided into equipment models, transportation models, etc. Each individual model has its own general structure that interacts with other structures. For example, the equipment models will interact with the models developed for maintenance facilities, fueling policies, supervisory policies, equipment specifications, mine layout, power systems, etc.

Each model has its own event flow model that simulates the parameters in detail. The event model will exercise logic for a specific task, serve as system interrupt and interpret operator commands. During "no-interruption" it will run and schedule the timer of the next event in simple interactive mode.

A show overview of equipment models under study or developed for specific application is presented in the following summary of models:

Truck/Shovel Model:
- version of truck/shovel interactive model developed for Lornex Mine in central British Columbia.
- update of the model for application to tar sands operation developed for Canstar Oil Sands project in northern Alberta.
- update of the model to include data base of the equipment used at Highvale Mine in Alberta.
- first version of the general truck/shovel model.

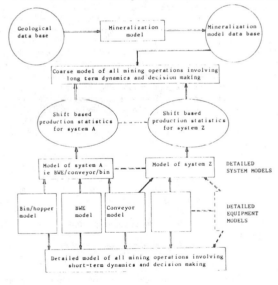

Fig. 2. Computer model hierarchy

The truck/shovel model at Highvale Mine was developed for a three-pit operation with two dumping points, several loading points and two modes of hauling (coal transportation from pits to hopper and ash removal from ash loading facilities at the power plant to pit spoil)(Srajer, 1984). Four different types of coal haulers are included for coaling and one type for ash hauling. There is

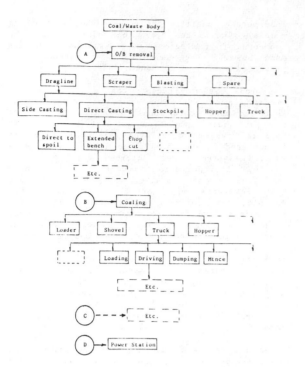

Fig. 3. Typical system flow

some provision to extend truck fleet when
different types of trucks are added. The mine
road network was digitized for the model with a
Graf-Pen Digitizer and later a Calcomp Digitizer
was used. A procedure to use road network
co-ordinates with elevation of key points is under
investigation and may eventually be adopted. The
data entry software has been developed for entry
of truck parameters as well as that of loader/
shovel. A mathematical model for loading, dumping, travel cycles has been developed and tested
for application elsewhere by Canstar for their oil
sand mining project and by one consulting company
in Calgary for coal mine project. This part of
the model was also written for IBM 3033 computer.
The multi-tasking structure was developed for the
interactive models encompassing several different
types of loading and haulage equipment that could
be activated as required by the main model.

The truck/shovel model is expected to be used
mainly:
- To determine the impact on productivity and
  operating cost of the truck fleet, various types
  of loaders, maintenance capabilities, size of
  reserve fleet, road layout and haulage maintenance, speed limit, use of tires and company
  policies.
- To evaluate truck monitoring and dispatching
  systems.
- For truck/loader productivity studies.

Dragline Model:
- Interactive model of dragline operation proposed
  for Canstar tar sands mine.
- Dragline range diagram model
- Update of the model to accommodate cycle times
  typical for chop-cut and face-cut practices at
  Highvale Mine.

The dragline model was initially developed for
Canstar tar sands mine (Graefe, 19820), later this
model became a part of the mine models package for
Highvale Mine (Nenonen, 1982b). It was used to
simulate removal and placement of overburden
material, to build up geometry of spoil piles or

to work in conjunction with some of the other
models of mining operation (Nenonen, 1982b;
Graefe, 1982). The model functions, used for
regular dragline operation, are described in Fig.
4 (Graefe, 1982). The overall procedure for
construction of dragline range diagrams is
illustrated in Fig. 5.

The dragline model at the time of writing this
report, provide the following outputs:
- Shift report with production summary,
  performance indicators (mechanical availability,
  utilization, operating efficiency), time
  statistics.
- Display (using Tektronix 4054 computer) a top
  view of the mine, mine blocks with dragline
  position, status and marking of the blocks
  mined.
- Cross section of the blocks, indicating the
  thickness and grade of the layers in the block.

The work is underway to implement the graphics
display on VS11 terminal and to include additional
modes of dragline operation. This model is now
being validated using data obtained from Highvale
Mine (Stuart, 1984).

BWE Model:
- Interactive model of bucketwheel operation,
  featuring rehandling of material from windrow in
  a tar sand mine.

The BWE model has been developed to aid equipment
selection and for mine planning. The model is
dependent on input parameters, stochastic time
variables, and geological data. At the present
stage, the model operates on event-by-event basis,
so the types of excavation cuts must be analysed
separately. The slice is the smallest increment
of the cycle. Output consists of the initial
condition, production summaries, performance
indicators, cycle time statistics, time histories
and graphic display. Work is presently underway
to validate the BWE model using actual production
data.

Conveyor/Bin/Hopper Model:
- Interactive model of conveyor/hopper system
  designed for tar sands mine.
- An updated version of the system to include
  features typical to TransAlta's surface coal
  mine.

The conveyor model is actually a simulator of the
material handling links between supply points
(BWE's, bins) and demand points (dump, hopper,
etc.) The model simulates fluctuations of the
feed rate, the flow of material through the conveyor network and the feed of the material at the
preparation/extraction plant (Chan, 1982). The
model provides summary reports on production,
utilization, energy consumption and other performance indicators. A periodically updated alphanumeric display is used for monitoring on-going
conveyor activities, the type of conveyor functions the particular section is engaged in, a
summary of the material transported. The utilization of feeding equipment (BWE's) is also
provided.

The simulation experiments were carried out on the
VAX11/780 computer at a rate of ten shifts of
operation in approximately ten minutes for two
bucketwheels and eight conveyor sections (Chan,
1982). The graphics display is available using
Tektronic 4054 computer at the present time.

All models operate on a PDP-11/60 computer with
DEC VT-11 graphic terminal, VAX11/750 computer
with DEC VS-11 graphic, or using Tectronix

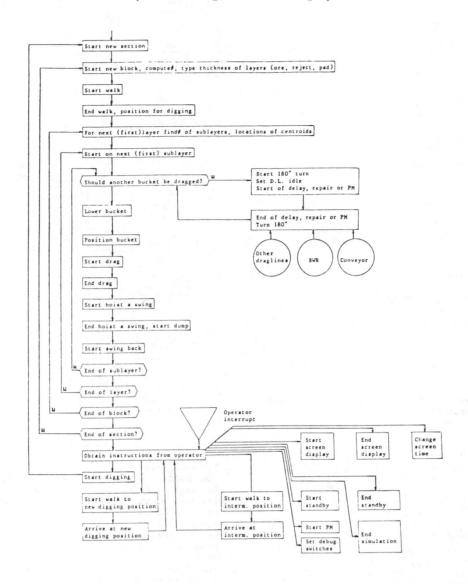

Fig. 4. Dragline model diagram

graphics terminals. Each model requires around 250K words of computer storage and is divided into sub-tasks to miminize storage requirements. The models could be implemented on a small computer system with limited high speed memory capacity. Two terminals are needed to display alphanumeric and dynamic graphics operation. The models consist of a set of programs based on the discrete event simulation language GASP II, enhanced and modified by the National Research Council of Canada (Chan, 1982; Collins, 1982; Nenonen, 1982c; Graefe, 1982) to facilitate the interactive modelling of processes and operations.

Mine Operation Model

Each equipment model developed will form a part of the mine operation model. Figure 6 shows the sequence of mining operations at Highvale Mine and the flow diagram of events (Srajer, 1984). Each model will simulate any system of operation so the central flow of events describing mining practices will be sequenced as desired. In example on Fig. 6, the selective mining of coal seams and partings

will be simulated, based on standard mining practices at Highvale Mine. The flow diagram could be different for each mining property, but individual models will fit into the existing structure without any substantial change. The diagram of an overall structure for interactive dynamic event - based models of surface mining operation is in Fig. 7 (Nenonen, 1982b). Scheduling and data flow are described here. The flow diagram will also include a certain provision for holding actual and monitored data through performance analysis.

FUTURE RESEARCH

The work on the development of computer models for simulation of surface mining operations was initiated in September 1982. It is, in effect, in the beginning as a research project that will rest on the support and cooperation between the government and the industry.

One of the immediate concerns is to develop a generalized truck/shovel model. At present, the

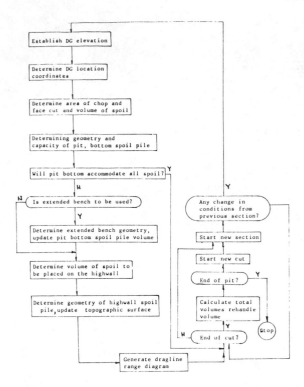

Fig. 5. Procedure for dragline range diagrams construction

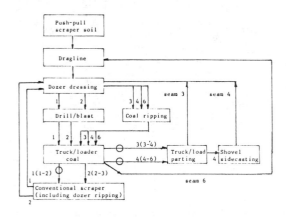

○ parting ripping   ( ) parting between seam n and n+1

Fig. 6. Sequence of operation models at Highvale Mine

work is in progress to investigate the truck selection criteria used by mining companies and truck manufacturers. This information will be used for developing the part of the model for optimization of truck haulage, sizing and truck transportation network. The testing of various processes and routines will follow using actual field data. The data base will be extended to include all the existing or planned coal haulers or dump trucks used at surface coal mines in western Canada.

Development of a general truck shovel/model will be closely followed by progress of other equipment models and validation of these models using actual field data.

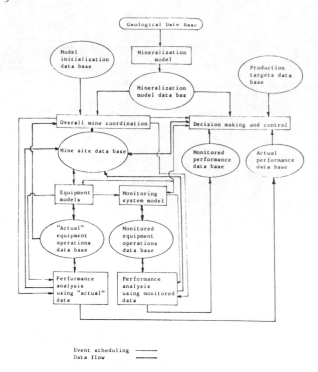

Event scheduling ─────
Data flow   ━━━━━

Fig. 7. An overall structure for interactive dynamic event-based models of surface mining operations.

ACKNOWLEDGEMENT

The work on computer modelling is a joint effort the CANMET/ Calgary Coal Research Laboratory, the Division of Mechanical Engineering of the National Research Council of Canada in Ottawa and the Canadian mining industry. Other organizations, namely Canstar, Petro-Canada, Monenco, Simons (International) were at some time or other, involved with this project and deserve some credit. The help of my colleagues, B. Das and N.J. Stuart in the preparation and review of this report is acknowledged with thanks.

REFERENCES

Chan, A.W. (1982). "Interactive computer modelling of conveyor systems"; CIM Bulletin, 75, 847, 81-83.

Collins, J.K., Nenonen, L. (1982). "Interactive computer aids for the design of oil sands operations"; Presented at the 82th CIM Conference in Quebec City, Quebec, Canada.

Graefe, U.P.W. (1982). "Interactive computer modelling of draglines"; CIM Bulletin, 75, 847, 74-80.

Nenonen, L. (1982a). Notes on project planning meeting with TransAlta Utilities, Calgary, Alberta, Canada.

Nenonen, L. (1982b). Notes on general mine model requirements definition by TransAlta Utilities, Feb. 10/82.

Nenonen, L. (1982c). "Interactive computer modelling of truck haulage syatems; CIM Bulletin, 75, 847, 84-90.

Srajer, V. (1984). "Evaluation of truck/loader system at Highvale Mine, Alberta"; Division Report of CANMET/Calgary CRL, Canada.

Stuart, N. (1984). "The use of a programmable microprocessor based data logger for widely differing monitoring tasks"; Division Report, of CANMET/Calgary CRL, Canada.

State and Development Trends of EDP-Based Operations Monitoring and Remote Control in the Coal Mining Industry of the Federal Republic of Germany

F. HEISING

Bergassessor, Geschäftsführer Steinkohlenbergbauverein/Bergbau-Forschung GmbH, Essen, Bundesrepublik Deutschland

J. STEUDEL

Dipl.-Physiker, Abteilung Fernwirktechnik Bergbau-Forschung GmbH, Essen, Bundesrepublik Deutschland

Abstract. Starting from comprehensive programme systems for monitoring tasks of any kind, this paper follows the steps of the programme system hierarchy from top to bottom. The monitoring and control centres for entire production units, the subordinate centres, i.e. district control centres, the microcomputer based autonomous control systems, and data input by mobile terminals are discussed. To an increasing extent, the main emphasis is shifted from monitoring and control of individual machinery to optimized operation of extremely complex comprehensive systems. Network displays, for instance displays of production, transport, supply- and evacuation-networks as well as of safety systems, gain increasing importance. Personnel deployment systems and planning networks as well as stocktaking and accounting are more and more closely linked with the above-mentioned networks. In order to keep development and application (purchasing) costs low, and to assure compatibility, the use of standard modules is to be assigned particular importance. New module developments are introduced, and the design of standard interfaces for computer and module coupling are discussed. For conclusion, a short view is given on the opinion of the line managements on how future hierarchic systems should be featured in terms of individual units and overall systems, and used.

Keywords. Mining, Computer applications, Data processing, On-line operation, Hierarchic systems, Standardization.

INTRODUCTION

Economic requirements in the first place, however, also safety-related, social-humanitarian, and environmental criteria, together with present-day order of magnitude of mining operations and their components, require careful and, if possible, complete monitoring of operations and the operational status. Since the correlations between the individual operational phases, between underground and surface operations as well as between the requirements of production and those of legal requests and safety regulations and, eventually, between the definite mineral quality in situ and the steadily increasing requirements of the market become more and more complex, decision-making with due consideration of all correlations and parameters to be looked at becomes difficult.

Since approx. 30 years, commercial EDP and, since the recent past, process computers and microcomputers constitute more and more a substantial aid for controlling the more and more complex correlations of individual operations phases and the interdependences within the production flow, without running the risks of human error, e.g. wrong interpretation of technical information. (Heising, 1977, 1982; Heising and Steudel, 1984).

A discussion of the details of the various systems would go beyond the time limit; so it will be discussed only one example of each of the latest developments mentioned.

PRESENT STATE

Even though EDP technology came into mining relatively late and via the simplest forms of counting and indicating monitoring (Fig. 1), present-day's highly equipped monitoring and control centres are standard in West-German coal mining. (Rauhut, 1981; Schröder, 1982).

A great number of technical and safety-related information converges on these colliery control centres for giving evidence on the momentary state of operations, for supporting the management in instantaneous decision-making, for being re-coded either manually or automatically into control signals and, eventually, for supplying the management with information serving as the basis for short-term, medium-term or long-term decision-making.

All process-relevant technical or safety-related data are stored in the computer, and can be visualized in clear text on VDU. These VDUs allow process dialogue with the control centre personnel via keyboard. The classic panoramic displays accordingly are largely replaced by a number of VDUs. The displays polled can be visualized on any selected VDU.

Besides the creation of suited electrical equipment for data gathering and transmission, the development of programmes is particularly important. Since nowadays the costs for writing programmes are generally higher than the investment costs for data processing hardware, the necessity becomes obvious to create, as far as possible, a standard system in modular design, applicable for the variety of tasks in a number of places and under various conditions. Development work in that sense was intensified particularly in recent years and resulted in a large programme package for monitoring underground operations. This software is run today already for various monitoring tasks, and is constantly complemented by additional components. The target aimed at is the availability to the users of an easily operable and adaptable system suited for any monitoring and control task.

The software is structured in a way that the individual sub-systems either assure master functions in monitoring or processing routines for confined operation sectors. All sub-system software can be run on the master computer as well as on the subordinate computer (Olaf and Steudel, 1981; Leder and Wendler, 1982).

Even though emphasis is placed differently in individual cases, the planning work for new colliery control centres and for retro-fitting process computers to existing ones reflect the standard requirement for introduction of a hierarchically structured decentralized automated system with superordinate computer.

Above all, two particular developments had to be realized for creating the basic prerequisites for that. The introduction of microcomputer based serial data transmission systems enable the high data rate required for information exchange between microcomputer control and colliery control centre to be achieved. These time multiplex systems replace, when necessary, the audio-frequency systems predominantly used up to present.

Intrinsically safe microprocessor systems in modular design are now generally approved for underground operations and, with respect to the software, can be adapted to meet various control requirements.

HIERARCHIC LEVELS OF AN AUTOMATED SYSTEM

For all collieries, introduction of the hierarchic system in modular design is strived for in order to assure adaptation to permanently increasing requirements for higher cost-effectiveness and safety as well as for ergonomic improvement.

Interchangeable hardware modules, a largely standardized software for identical application cases, and defined interfaces allow largely trouble-free extension of systems and system components. The hierarchic structure allows information on the individual hierarchic levels to be digested and transmitted to the higher information and processing levels and, by far-going decentralization, a maximum of tasks to be assigned to autonomous systems.

In this way, three hierarchic levels can be defined:

- Superordinate monitoring and control centres

- District control centres for monitoring and control of operational sub-units

- Autonomous monitoring and control of individual machine systems.

To these three hierarchic levels, a fourth application field, i.e. mobile terminals for data and information input, is intended to be added. The units concerned are supposed to be lateron connected to the data transmission network in locations on the surface or underground freely chosen according to necessity, so that their information can be used within the overall monitoring system. At present, these units are still tapped directly by autonomous terminals which process their information.

Information related to production as well as to safety converges on the superordinate monitoring and control centres.

The production-related information enables:

- Optimization of railbound and trackless transport systems as well as of belt conveyor networks, including flow monitoring over the clearance systems as well as monitoring of bunker levels and shaft winding (Behrenbeck and Schröder, 1974)

- Visual display of cable and pipe networks for utilities supply and waste disposal (electricity, compressed air, fresh water, mine water, methane drainage etc.) and monitoring in terms of voltages, flows, pressures, concentrations etc.

- Material-flow monitoring from pityard operations to the points of use underground (Kantor, 1982).

The safety-related information allows, by monitoring of complete ventilation networks and their data in terms of ventilation air-flows, air-stream velocities, CH_4- or CO-concentrations, early detection of mine fires, methane release forecasts, and escape routing which, together with the integrated telephone and alerting network, serves for giving alert to the personnel and for accelerating rescue action (Schröder, 1980; Stark et al., 1982).

In the subordinate district control centres for monitoring and control of sub-sectors of operations usually only district-relevant information is gathered for immediate control of operations within said district. Only selected production data are then supplied to the superordinate computer of the higher level.

The so-called district control centre which may be installed either underground in a place convenient from safety and human-engineering viewpoints or on the surface in the immediate vicinity of the colliery control centre - the latter configuration becomes more and more standard practice - may be quoted as a typical application example for a subordinate control centre. All production operations including those of the coal winning machine, the face conveyor, the face support, the stage loaders, and the district bunkers are visualized here and converted into instructions for manual intervention or into automaticcontrol signals. The programmed shearer loader cuts as well as the semi-automatic plough control are run from the district control centre, and, since recently, satisfactory and encouraging production results are recorded. (Kugler and Kriener, 1982; Baack et al., 1984).

On the third hierarchic level, autonomous microcomputer-based control systems running machines on their own, is dealt with. The trouble reports of these systems can be transmitted via the colliery's general computer communication network to the colliery control centre. The autonomous systems do not receive control signals from an outside control centre but from computer routines of the internal microcomputer which, on the other hand, supplies selected data and trouble reports to the higher hierarchic level. The controls of belt conveyor systems, coal ploughs, roadheading machines, central air-conditioning plants, and centralized cement distribution networks may be quoted as examples for those autonomous control systems by means of which it is tried to optimize operations of each of the above-mentioned functional units. In this field, particular progress was recorded recently.

Superordinate Monitoring and Control Centres

The colliery control centre of Lohberg colliery of Bergbau AG Niederrhein, the western area of Ruhrkohle AG, may be quoted as typical example for the present state of monitoring and control centre technology in West-German coal mining. On this control centre converges an optimized selection of information, and space requirements for this centre could be minimized (Fig. 2).

The master computer system comprises - for the sake of safety - two identical computers of the R 30 type of Siemens AG, run in parallel. Double storage of all recorded values and information assures, in case of trouble with one computer, transfer of functions to the other one within negligibly short time. For visuali-

zation, three VDUs for graphs and two for written-out information are available and, for recording, three printers and one plotter are provided. The hardware, apart from the number of peripheral units, is standard for all new computer-based colliery control centres. The implemented software, however, is different according to the priorities resulting from the operational needs of the individual collieries.

On Lohberg colliery, sub-systems are installed for monitoring measured values in ventilation, fresh water supply, and coalfaces as well as those relating to the operational status of individual machines and devices. Further program modules process data for visualizing the operational status and troubles, if any, as reported, e.g. from the subordinate controls of the centralized cement distribution system and the conveyor belt network, on coloured VDU. Another programme system automatically drafts reports on production data and down-time. All minutes and diagrams are visualized first on VDU and printed out only if required, except for ventilation data which need to be plotted permanently according to the inspectorate's regulations.

Particular importance is assigned on this colliery to the monitoring of the fresh water supply network due to its large size. Monitoring is to assure that the supply of all parts of the network is regular, and it has to be avoided that parts of the underground workings are flooded by leakage. To these ends, the flow is checked for excess and shortfall of threshold values. Various threshold values for working days and rest days are subject to re-adjustment by a computer-internal calendar automatic. Furthermore, the threshold values for working days are continuously adapted to the preset values which in turn are permanently updated by correlation with machine running time and water demand of the individual users.

The differences between the values measured by individual measuring units on the flow line is calculated for spotting leaks. Since monitoring also caters for valve positions and flow directions the computer stores the complete and regularly updated process situation. The data contained therein are processed in view of issuing synopses and detailed plots as well as for simultaneous display of curves recorded by three measuring lines, on a special VDU for graphs. The alarm fields of the flowchart give information on any irregularity. The necessary measures for re-establishing the normal operational status are not taken automatically but manually.

In a similar way, also the other sub-systems supply the necessary information for complete monitoring of production and safety.

District Control Centres

Osterfeld colliery in the western area
of Ruhrkohle AG, Bergbau AG Niederrhein,
exhibits a good example of surface-installed district control centre technology
with semiautomatic plough control (Fig. 3).

In order to assure the important personal
contact between the man on the controls
and the deputies and overmen of their
district, the control centre is arranged
immediately adjacent to the deputies' and
overmen's office. The communication with
the face crew is assured by telephone and
the coalface tannoy system.

The control centre comprises components
of a microcomputer system for data transmission and control (Simdas), a process
computer R 10 supplied by Siemens AG, and
a control panel for manual remote control.
The control panel also houses a process
control keyboard for computer operation.
Two VDUs are available for visualization
of the operational status and of operation
troubles by permanently updated displays.
For trouble analyses, sectorial displays
containing more detailed information can
be polled.

For automatic operation, coalface parameters, e.g. plough travel section, preset angles and alignment of the face
conveyor, are to be put in via VDU masks.
The coalface length is determined automatically by a manual coalface survey.
In automatic operation, plough position,
conveyor alignment, and conveyor inclination are measured, and the respective
values are transmitted to the computer.

The partial automation of coal winning
comprises a horizontal control of the
plough catering also for directional
changes in selectable positions as well
as for vertical conveyor control in the
direction of face advance.

The conveyor inclination, the guide parameter for the plough, is corrected via
the advance rams in case the actional
values diverge from the preset ones, so
that the plough is kept within the seam
horizon. If the actual face alignment
diverges from the preset one, this can
be corrected by sectionwise ploughing of
the salients in the coalface.

Besides coordination of the microcomputer
control for automatic operation, the
process computer assures processing of
data on machine running time and downtime,
i.e. it drafts VDU-displayed minutes.
A surface-installed district control
centre as set up on Osterfeld colliery
is possibly the last step of a development from the mobile control station
near the coalface to the communication
centre for the district (Fig. 4). This,
however, does not mean that the tasks
assigned to a district control centre
could not be complemented by additional
ones.

Autonomous Monitoring and Control

Out of the large variety of existing
application fields, the autonomous control
of a cement supply system for stoppings
which is standard for a number of
collieries, may be quoted as example
(Fig. 5).

This complex system optimizes the distribution of cement from the point of unloading from trucks on the pityard until
the final point of use. The cement is
conveyed pneumatically via pipelines
from the surface to the point of use
underground. Tube switches control the
material flow in the sometimes very rami-
fied systems. Due to the great distances
to be covered, relay stations are required for compensating pressure drop,
and furthermore interim storage bunkers
are needed to cater for supply bottle-
necks for the individual network branches.

Several microcomputers coupled to each
other control the proportioning of the
cement according to priorities which
change according to the consumption at
the points of use. Requests put in
manually on the sub-stations do only
entail higher priority when the cement
on hand in the appurtenant bunkers drops
below a certain threshold.

The function of the control system is
completely autonomous if the surface
microcomputer is completed by VDU and
keyboard to form a full-scale terminal
in its own right. The integration into
the hierarchically structured automation
system can be assured via a standard
interface.

The visual display of updated priorities,
of operational status, of trouble reports,
and of quantity distribution is, via
computer coupling, realized either on the
VDU of superordinate process computer
or on the one of the system's own terminal.

This example should be sufficient to show
which importance is assigned in West
Germany coal mining to intrinsically
safe micro-computer-based systems. The
flexibility of programmable control
systems contributed largely to further
development of automated underground
operations, even under difficult conditions caused by geological faulting,
environmental parameters, and discontinuous production.

Mobile Terminals

Mobile terminals allow additional dialogue with the system which lateron is
intended to be possible on each of the
hierarchic levels. An intrinsically safe
data gathering system, in a first phase
intended for works studies only, turned
out to be very useful, and meanwhile was
complemented by a new freely programmable
version. Accordingly, its application
range was widened considerably so that
this new version can also be used for
material flow monitoring.

For more ease of handling by mostly un-
trained personnel, the data gathering
unit incorporates operator's guidance
via 15 function keys, clear-text error
reporting, and a security against in-
adverted deletion of stored information.
16,000 signals can be stored and read out
by a computer on the surface. The system
is intended to be complemented by an
interface by which the unit can also be
connected to subordinate control centres
or control posts underground.

SUMMARY, PROSPECTS

To sum up, we may state, as to the present
stage of development of EDP-based operations
monitoring and remote control in West-
German coal mining, that successful use
of this technology in underground operations
and for underground operations is coupled
to an increasing extent to planning and
monitoring tasks in the administrative
field (Altena et al., 1982). Personnel
planning, working time monitoring, material
planning, development and production
planning as well as updated displays of
the deposit are now backed up by the
technical and safety-related information
for better decision-making.

Eventually, the question reads: what is
expected within West-German coal mining
from application of EDP-based operations
monitoring and remote control?

It is expected that - roughly said - the
increasing variety of more and more
complex individual tasks and influences
can be tackled to an increasing extent
more rapidly, more reliably, and with
consideration of interdependences and
interactions in view of the necessary
assessments and decision-making.

In the near future, solutions to further
individual problems are expected, e.g.
for integration of the support control
into automation of coal winning, and to
elimination of recording systems, i.e.
those for long-term recordings. The still
today officially required long-term
storage of all ventilation-related data
by recorders could be made definitely
more efficient by storage on magnetic data
carriers in connection with the data being
reduced to only safety-relevant changes.

Eventually, we should mention the first
successful tests run at present in view
of connecting the superordinate control
centres and the subordinate ones by fibre
optics.

In the long run, underground operations
require a trouble-free self-checking information basis for highly developed
programs optimizing the operations in
view of safety-related, technical,
production-related, and cost-related
criteria automatically, i.e. independently
on possible humon error. In this way it
should be possible to reduce paper work
still arising in EDP field application
in form of print-outs serving as bases
for decision-making, and to concentrate
just on few decision-relevant data.

One prerequisite therefor is the well-
scheduled introduction of a hierarchically structured decentralized information system the individual components of which, in the first phase, are
run on various collieries. This way of
proceeding is to keep the financial
stresses as well as the unavoidable
teething troubles of new systems on a
level acceptable to the individual
collieries in question.

Still today, a variety of information
now reaches the superordinate control
centres in a visualized form without
direct use being made of said information
either by the personnel of the control
centre or by the management. Incoming
signals can be ignored either voluntarily
or inadvertedly, action necessary for
safety may not be taken, and the quantity
of the possible information flow may
become confusing. In the long run, all
this can only be avoided by more and more
automated information processing and
automated triggering of necessary action.
If the envisaged automation of the EDP-
based operations control turns out to be
successful, this should mean more safety,
more amenity and higher cost-effectiveness in our collieries.

REFERENCES

Altena, H., u.a. (1982). Automatische Er-
stellung von Wetterführungsplänen mit
der elektronischen Datenverarbeitung.
Glückauf, 118, 1171 - 78.

Baack, D., u.a. (1984). Rechnergestützte
Abteilungswarte auf dem Bergwerk Osterfeld. Glückauf, 120, 953 - 960.

Behrenbeck, H.J. and L. Schröder (1979).
Steuerung und Optimierung des Fließ-
fördersystems unter Tage durch Prozeß-
rechner auf dem Verbundbergwerk Haus
Aden. Glückauf, 110, 333 - 339.

Berse, G. and E. Feistkorn (1982). Ein
eigensicheres Datenerfassungs- und
-speichergerät für die Rationalisierung
von Betriebsablaufstudien. Glückauf-
Forschungsh., 43, 38 - 42.

Heising, F. (1977). The Haus Aden Project.
Technical Paper No. 16, Intern. Conference Remote Control and Monitoring
in Mining, Birmingham.

Heising, F. (1982). Experiences in remote
monitoring and control and process
control in the Ruhrkohle AG. Paper
to The Midland Institute of Mining
Engineers, Doncaster, Translation No.
MRDE 1613.

Heising, F. (1982). Application of Electro-
nic Data Processing in Monitoring and
Control of Underground Coal Mines.
Vorgetragen auf der 6th ICCR Conference
London.

Heising, F. and J. Steudel (1984).
Computer-Based Monitoring and Control
System in Coal Mining Industries of
the Federal Republic of Germany. Vor-
getragen auf der Intern. Konferenz für
Prozeßsteuerung im Bergbau (ICAMC),
Budapest.

Kantor, J. (1982). Rechnergesteuerte
     Materialdisposition und Materialfluß-
     verfolgung. Schriften für Operations
     Research und Datenverarbeitung im
     Bergbau, 8, 215 - 38, Verlag Glückauf.

Kugler, U. and A. Kriener (1982). Vom
     Strebsteuerstand zur Revierwarte.
     Glückauf, 118, 487 - 490.

Leder, G. and G. Wendler (1982). Rechner-
     gestützte Betriebsüberwachung.
     Glückauf-Forschungsh., 43, 235 - 38.

Olaf, J. and J. Steudel (1981). Prozeß-
     rechner für den Grubenbetrieb.
     Glückauf, 117, 927 - 29.

Rauhut, F.-J. (1981). Stand der Über-
     legungen bei der Einführung von Daten-
     technik und Automatisierung bei der

Bergbau AG Niederrhein. Vortrag zum
     7. britischen Besuch des PIT (Per-
     formance Improvement Team) in Deutsch-
     land, unveröffentlicht.

Schröder, L. (1980). Brandfrüherkennung
     mit Prozeßrechnern. Schriften für
     Operations Research und Datenver-
     arbeitung im Bergbau, 7, 147 - 172.
     Essen, Verlag Glückauf.

Schröder, L. (1982). Rechnerunterstützte
     Grubenwarten. Schriften für Operations
     Research und Datenverarbeitung im
     Bergbau, 8, 264 - 50. Verlag Glückauf,
     Essen.

Stark, A., u.a. (1982). Stand und Ent-
     wicklung der Brandfrüherkennung bei
     der Bergbau AG Lippe. Glückauf, 118,
     389 - 393.

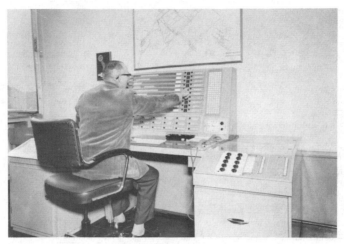

Fig. 1:   Control room, Monopol Colliery, 1960

Fig. 2:   Control room, Lohberg Colliery, 1984

Fig. 3:   District control centre, Osterfeld Colliery

Fig. 4: Mobile control station underground, 1975

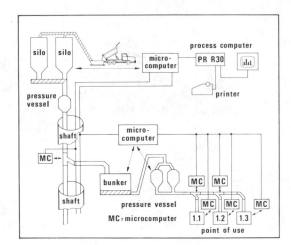

Fig. 5: Microcomputer based cement supply system

Control and Automation of Underground Operations in the UK Mining Industry

D.K. BARHAM BSc(Hons), CEng, FIMinE

National Coal Board, Mining Research and Development Establishment, Ashby Road, Stanhope Bretby, Burton on Trent,
Staffs, DE15 0QD, England

Abstract. The concentrated efforts now being directed towards the development and application of monitoring and control techniques have now resulted in their widespread use in UK mines together with evidence of benefits resulting from their application. There is little doubt that the longwall method of producing deep mined coal will continue but will be made more reliable and efficient with the aid of control technology. The major fields of research and development work are concerned with planning and design of mines and mining equipment; management and engineering information; and automation of coalface, tunnelling and transport systems. The principal achievements now in widespread use are: the standard computer system MINOS, widely applied to mine transport, coalfaces and mine environment; extensive use of machine condition monitoring leading to condition-based maintenance; underground automation of conveyor systems and guidance systems for ranging drum shearers and boom-type roadheaders are being widely introduced.

Keywords. Coal mining; automation; control equipment; environment; coalface; tunnelling; transport; robots.

INTRODUCTION

Over 92% of deep-mined coal in the UK is produced from longwall faces. The safety record of the UK industry is second to none in the world, with the accident rate now similar to that of the rest of manufacturing industry. Explosions have been reduced to very rare occurrences, and pneumoconiosis has been virtually eliminated. These remarkable achievements have been brought about by control backed up by legislation, which has set the indices by which progress can be measured. The major technical problem facing the industry is to increase machine utilisation, which will lead to higher productivity increases, decreasing costs and a greater share of the energy market.

Prior to the recent industrial relations problems the National Coal Board (NCB) operated 170 collieries with a total annual saleable output of about 100 m tonnes from 508 coalfaces. Opencast coal added a further 14-16 m tonnes and consumption was of the order of 110 m tonnes.

The rapidly changing economic situation, during which there has been a marked decrease in the expectation of growth of energy consumption, has made pre-recession forecasts appear very optimistic. Recent forecasts from various sources estimate that UK demand for the year 2000 will be as low as 75 million tonnes (Sussex University Study) to as high as 170 million tonnes (National Union of Mineworkers) with the Department of Energy's estimate in the range 100-142 million tonnes. (House of Lords, 1983). Even on the most pessimistic forecast the need to improve performance by introduction of new technology will not be diminished in importance.

The total mechanisation of the UK mining operations took place over a 20 year period up to 1974 and methods will not alter greatly in this

century. One reason for being confident in this prediction is that the results from the best UK longwall coalfaces has shown the high potential of present equipment.

It is the National Coal Board's mining R & D policy to allow Mining Research and Development Establishment to concentrate on the new and faster growing technologies. Improvements in well-established technologies, applied to equipment presently supplied by traditional manufacturers, will not be taken beyond the feasibility stage and hence manufacturers will be involved from an early stage in the development and testing work.

In addition to the requirement to introduce control and automation techniques to improve the efficient use of underground there are likely to be other changes and constraints that will necessitate increased technological advances. The replacement of inefficient and excess capacity and its replacement by new and deeper mines on the Eastern and Southern extensions of the Midlands coalfields will continue the increased concentration of output from greater depths which is likely to introduce greater environmental difficulties in terms of methane, dust and heat.

The average depth of working has increased by 100 m per decade and the exhaustion and closure of older coalfields and their replacement by the deeper Eastern coal will not halt the rate of increase of depth from its present average level of 514 m (25% of output comes from workings deeper than 750 m).

The geothermic gradient generates strata temperatures in the UK of about $40^{\circ}C$ at 1000 m in depth. A typical mine district ventilation system will have to cope with one megawatt of heat from the strata plus that from coalface plant, which itself frequently exceeds 1.5 megawatts of installed power. Hence there will be increasing need for controlled environments to produce acceptable temperatures in working areas.

The methane make in the UK is typically of the order of 25 - 60 cubic metres per tonne and tends to be higher in the newer and deeper coalfields. Ergonomic considerations limit the air velocity to 5 m per second where men are working and hence efficient techniques of methane drainage will become more vital, to maintain a high production rate, as will techniques of monitoring and suppression of ignitions.

In addition to environmental and other safety legislation frequently providing a limit to performance, legislation on noise will be introduced. A level of 90 dB at the operating position is the likely limit, which will involve considerable redesign of parts of machines, and a further limit of 85 dB will make remote and automatic control essential for most operations.

Further legislation relating to the surface environment, for example, subsidence and sulphur and chlorine emission, will place constraints on the layout and method of working of faces and the seams (and parts of seam) which can be mined.

From this brief scenario of coal mining in the near future, it is possible to describe the strategy behind the introduction of control technology and automation to the mining industry. This can be summarized as: the realisation of the potential of the underground equipment in terms of output and quality; a drive to reduce costs both of equipment and operations; the search for new markets; and last but not least the safety and health of operators.

In order to consider the input and impact of control technology, it is necessary to identify the current recent developments that are starting to be installed in many pits. These can be identified into discrete but related fields of work: planning and design; management and engineering information; and automation.

PLANNING AND DESIGN

As a means of reducing the 36% of faces that finish prematurely due to meeting unpredicted geological disturbances, considerable efforts have been devoted to developments that provide better detailed information of geological structures. The efforts have resulted in an in-seam seismics technique, which is being widely exploited and which can detect faults with a throw greater than seam thickness. A typical range of detection is 300 m although greater ranges have been achieved. A parallel development of automatically guided longhole drilling to increase the range and accuracy of seam disturbance detection has resulted in an in-seam drilling technique that uses the difference between natural-gamma radiation in the roof and floor of a seam and that in the coal to locate the position of the drill bit across the thickness of the seam. A prototype instrument has been developed, which includes inclination and azimuth measurements. Several holes over 500 m long have been drilled in seam at an outcrop trial site and the equipment has now been successfully demonstrated at several collieries. A production sensor is now available from Salford Electrical Instruments Ltd and measurement-

whilst-drilling equipment is under development which will, within the next two years, enable continuous drilling and sensing to be carried out.

The advances in new computer technology as applied to testing and design techniques have been introduced to the NCB and its manufacturers and have enabled the reliability of machines to be predicted and made it possible to indicate where improved design is required both for new and existing underground machinery. Computer-aided design techniques have aided not only the detailed design of reliable coalface equipment, but, for example, the layout of picks on cutting drums to give efficient cutting. Three-dimensional layouts of face systems, showing interaction of equipment, enable the selection and design of armoured face conveyors (AFCs), supports and shearer loaders, which allow for production standards with the minimum of operational delays. More recent developments have shown CAD to be a major contributor to colliery planning.

The end point of the design and planning phase is not only well designed and reliable equipment and face systems, but also the specification of the operating conditions and limits of the machine, which will guide the designer and installer of automation schemes.

MANAGEMENT AND ENGINEERING INFORMATION SYSTEMS

Machine Condition Monitoring

Machines that have been designed and manufactured to meet a specific duty have to be commissioned and maintained to meet the performance capabilities required by the production engineers. MRDE has developed a package of portable instruments for this task, which, by measurement of hydraulic and electrical circuits, provides instant information concerning the condition of our underground equipment. By connection into data processing and display systems at the mine surface, the information can be used to ensure that key control parameters or indices are maintained within acceptable standards during operation.

These instrumentation packages have been used at 28 collieries, and a detailed investigation at one site has shown a reduction in face delays due to power loader, AFC and powered supports breakdown of 24 minutes per shift. The packages of equipment are supplied by Davis of Derby Ltd.

The development of further instrumentation has been carried out to include measurement of debris particles in oil-lubricated gearboxes, and to give a reliable indication of oil level, since the majority of gearbox reliability problems will be cured by keeping the correct quantity and quality of oil in our machines. Following the introduction of oil debris monitoring in one colliery, the delays during coal production due to face machine breakdowns from failed gearboxes have been reduced to nil over a period of six months in comparison with over 100 hours delay in the previous six months.

The introduction of these techniques has provided evidence that condition-based (rather than time-based) maintenance can give considerable cost benefits in reduced down time of plant.

The analysis of data by desk-top computers, or by the mine central computing system at the surface, has already provided a data base on which the design and testing of manufacturers' existing and future machines can be based. A major role of MRDE now includes the design appraisal and testing of manufacturers' machines, and the extensive and growing use of machine condition monitoring by the NCB will be an essential part of improving the reliability of face machinery for successful automation.

Systems Performance Monitoring

It was recognised in the early 1970's that the development of the minicomputer would enable cost effective monitoring of colliery operations. The complex nature of software and its growing cost (in contrast to hardware costs, which have rapidly decreased) showed that it was essential to design to an industry standard known as MINOS (MINe Operating System). This uses standard hardware and hence allows for standard software, which has remained under the control of the NCB, although several companies supply the specified system to collieries. Colliery personnel can configure the programs to meet their own requirements. It was first used in conjunction with a conveyor network control system and, in addition to providing on-line monitoring, soon showed the benefit of long-term data analysis. The systems provide reports on performance indices as well as reliability trends for different plant. These reports are commonly used as the basis of planning the engineering and maintenance work. MINOS conveyor monitoring and control systems are installed at 43 collieries and work is continuing to provide stone clearance management to decrease pollution of coal by mine development in rock.

A major extension to the system has been that of environmental monitoring. The development of a reliable set of instruments to measure air flow and mine gases (methane, carbon monoxide and oxygen) has provided information to the monitoring system, which produces on-line displays and stores data for long-term analysis. The system is applicable to both general monitoring and methane drainage. Work has currently started on detection of hydrogen and oxides of nitrogen and on a software system to provide early warning of incipient fires from processing of data indicating carbon monoxide increase and oxygen deficiency. The MINOS environmental system is installed in 19 collieries and a detailed study of one of these has shown a cost reduction of £6,000/week/colliery in salaries of supervisory staff needed to inspect the mine at weekends.

The MINOS system has been further extended to the coalface and operational monitoring systems have been installed on 90 faces at 28 collieries. The running state of major face plant including coalface machines, face conveyor and stage loader is monitored, in addition to the shearer cutting state and position. The resulting information is transmitted to the mine surface and used to provide an instant pictorial representation of the state of face operations. Combined with a manual input of the causes of delays, the face delays and their causes can be displayed on-line and converted into shift and weekly reports of machine running time and major delays.

The information is available for the managers' and engineers' meetings and relieves the frustration which often exists due to unreliable information. It has highlighted problem areas which were acted on quickly and contributed towards face design.

A detailed study was made in the South Yorkshire Area over a three year period of the performance of over 40 coal faces (Barham, Hartley and Rason, 1985). Those faces with MINOS face monitoring systems (varied from 2 to 6) showed 10% improvement in output per shift in comparison with non-monitored faces after other factors were eliminated.

AUTOMATION

Coalface

A start was made in the mid 1960s towards coalface automation, which, with hindsight, did not receive widespread application due to lack of reliable plant, electronic technology and demand from collieries. It took until 1974 before an R & D policy directed towards longwall coalface automation was again pursued. Although fully automated systems are not expected until the late 1980s, a step-by-step approach to the programme is giving some benefits in the shorter term.

The automatic sequence control of roof supports has been developed by the major UK manufacturers, Dowty and Gullick Dobson, on the principles established in the 1960s. They allow for various levels of control, from manual control of adjacent individual or sets of supports to automatic sequence control of conveyors and support advance initiated by the position and direction of the shearer. More than 20 systems have been installed in UK collieries, including Selby Mine, and 16 installations have been purchased overseas, by Australia, USA and South Africa.

The main problem, for which solutions are beginning to emerge, is that of machine and face guidance to ensure that the machines cut in the seam, faces are kept straight and advanced at the optimum rate.

Vertical guidance of the main coal-getting machine, the ranging drum shearer, required extensive control system design studies, while major sensor developments have been made. The principal development has been a natural radiation detector, which will measure the roof layer of coal normally left in UK mines because of the weak overlying shale. Several hundred of these instruments have been sold in the UK and throughout the world by the manufacturers, Salford Electrical Instruments, and widespread use has shown that their practical range of measurement of coal thickness is between 10 mm and 500 mm.

In order to meet conditions where the natural radiation sensor cannot be used an instrumented pick sensor has been developed. The varying force pattern on the pick as it cuts through the seam is computed to give a seam signature against which the force patterns from succeeding drum revolutions are compared. The drum is steered until current and seam signatures coincide. Reliable data transmission and processing systems have been developed and the prototype machine is installed underground and under-going trials.

The requirements of the guidance system and the environment on a coalface machine dictated that intrinsically safe microcomputer system had to be developed, capable of receiving and processing sensor data, carrying out control functions, indicating faults, and transmitting data to the gate and surface. An experimental version became available in 1979 and has had extensive use.

However, operational limitations and obsolescent electronics have made it necessary for a redesigned production system with improved display, reliability and operational features.

Eleven installations of the prototype natural radiation system have been made and studies on several faces have enabled economic benefits to be assessed. These studies have only been possible because of the surface monitoring, which has recorded the utilisation of the system and the accuracy of the steering system. One study compared two identical faces except that the steering system was installed on only one of the faces. There was an increased production rate of the order of 40% on the automatically steered face, which was the equivalent of generating an extra £1m per year per face due to generation of better face conditions. At another colliery, which had difficulty in excluding soft floor dirt in the product, the proportion of saleable coal from the face increased by 26 tonnes per shear, equivalent to approximately £1.5 m/year proceeds. This level of payback has been repeated on most of the installations of prototype equipment.

The enormous potential from a system costing £150,000, with the possibility of application to about 60% of faces, has created a demand which will be met by the rapid development of the production versions, which are now available from the UK shearer manufacturers. The first machine is installed in a South Yorkshire Colliery. Shearer manufacturers (Anderson Strathclyde and British Jeffrey Diamond) have equipment available and rapid expansion is expected.

The second main requirement for full automation is the control of advance and alignment of the face. The principal aim is to measure the advance of face, either from a number of points and processing the data to generate the plan of the face, or by continuously surveying the face using the machine as a surveying platform. The former system relies on an unreeling ball of nylon cord, the length of which is measured as the face advances. The cord transducer is mounted on supports at intervals along the face. A successful full face trial of 40 instruments provided face management with a gate end display of the face line and this information led to a considerable improvement in face straightness and production versions are under manufacture. The second is an infra-red optical system, the information from which is transmitted to the gate, where the face line and advance is computed. Roof support manufacturers have been contracted to produce the advance control system using a link to the alignment information computer. Two installations have started and again these are linked to the surface so that the key information on performance and benefit is recorded away from the face.

A major problem at face ends is the multiplicity of plant and manufacturers. 80% of faces are advancing and 60% of rips are shotfired and hence there is still great concentration of labour at face ends. MRDE control technology development has concentrated on development of ripping machines capable of semi automatic operation, which will cope with hard strata and allow access to the face.

At two collieries, automatic roadside packing systems are under development, with transport from the surface of either cement or washery dirt directly into the packhole, with both economical and environmental benefits.

Tunnelling and Mine Development

The performance of tunnelling machines is equally subject to system delays. The latest generation of circular tunnelling machines cost up to £4 m. each and roadheaders cost approximately £500,000, and hence their life cost is several times this figure. Again the major control problem is the guidance of the total machine or of the cutting head. Tunnelling guidance systems have only recently been applied to coal mines using techniques known in civil engineering. Automatic profile control for boom-type roadheaders, as supplied by Dosco and Anderson Strathclyde, has reached underground trials. The system uses a laser beam as a reference and, together with measurement of boom angles, gives information which is processed to give roll, slew and pitch of machine, and hence the two dimensional positioning of the cutting head can be controlled. Another drawback of the boom roadheader is that the associated roof support operations cannot be carried out concurrently with cutting unless a shield is used and work on an integrated design to enable continuous operation of roadheaders now showing some success with trials of roadheaders within a shield now taking place.

The coalface guidance system has recently been applied to the In Seam Miner coalface development machine, and it is likely to become the basis for a full automatic shortwall coal producing system, with application where there are surface support or geological problems.

The applied power to boom tunnelling machine cutting heads can exceed 250 kW and, since cutting performance depends on the high inertia, the protection of power transmission equipment to improve reliability has indicated the need for sophisticated load control systems. These are currently under development and trials of a prototype system on the MRDE circular tunnelling machine have already led to an improvement to gearbox life.

Transport

The majority of mineral transport in the UK is by conveyor and the automation of conveyor systems commenced in the 1960s and was finally established in the mid 1970s when MINOS became available. It has enabled a substantial reduction of delay time plus a reduction of operators. For examples there has been a reduction in 1745 transfer point attendants nationally since 1977, from the introduction of remote and automatic control of conveyor transfer points, (Moses, 1984). Payback of the cost of these schemes which averaged £4,000/job saved varies between one to two years.

There are approximately 129 collieries using locomotive systems, and there are over 5,000 rope haulages for vehicles on rails. Outbye operations are most resistant to improvement in performance, in spite of high expenditure on mechanical equipment. A start has been made on utilizing coded transponders to identify position and type of vehicle and process the information. This will provide management with information required to control performance of the transport system and assist management to identify the areas in which an improvement can be made.

FUTURE DEVELOPMENTS

A recent paper (P G Tregelles 1984), outlined the

effect of future developments, which he considered would come as a consequence of two pressures: firstly, those from operational management because of enduring and basic problems in the present techniques, secondly, the pressure from R & D to apply new technology.

Aspects of new control technology which are likely to have a major impact on the mining industry are communications, computing, automation and robotics.

Communications and Computing

It is in this area that some of the major changes will be seen. Currently the standard digital communications used for monitoring and control in UK mines run at 600 baud, although experimental systems use 9600 baud. Developments for the immediate future will run at 64000 baud, but the electrically noisy and potentially explosive environment plus expansion of data will necessitate application of fibre optic techniques and the introduction of complex communication networks. Increasing integration of digital circuits will undoubtedly assist in introducing computing equipment into space restricted machines, although the complexity of cabling and connection remain as a difficulty. A likely source of novel mining control techniques will be the application of intelligent knowledge based systems (expert systems). Available data covering all aspects of operations is ever growing and many decisions are made on empirical knowledge. Experimental application of expert systems has commenced at MRDE as a diagnostic aid for decision making, related to strata control decisions and mining system performance analysis.

Automation, Telechirics and Robotics

It has been proposed that many of the environmental, social and economic problems of the future could be overcome by use of telechirs, ie a machine equipped with sensory equipment linked to the surface, from where it would be manually operated. The argument for this was that the equipment needed is of less complexity than that required for automation, the experience and intelligence of the operators could be utilized, and ultimately the lack of underground operators eliminates the need for ventilation and large access roadways. An 18 month study has recently been concluded which identified many difficulties of the use of telechirs to operate a longwall face remotely. Primarily, there are operations that a man cannot accurately do, whatever his sensory ability, eg he cannot see the boundary of the coal seam and rock until exposed, and the speed at which telechirs can operate is slower than human activity. An investigation has been carried out into a remotely operated vehicle to advance into hazardous areas after an explosion, flood or fire. The difficulty of predicting the mine roadway environment and the cost and possible utilization of the vehicle has led to the project being shelved. What is far more interesting is the introduction of automatic and computer controlled machines, which in many cases are reprogrammable and designed to manipulate tools for the performance of manufacturing tasks, and hence they could be defined as robots.

Although mining robots are unlikely to resemble those existing in manufacturing industry, mainly because of the non-repetitive nature of coal mining, the vertical guidance system on the coalface shearer, the profile control of boom road heading machines, and the automatic equipment for setting of supports can be regarded as primitive robots. The advance of computer systems will undoubtedly lead to certain machines becoming advanced robots, although it is not visualised that human supervision, installation and maintenance will be replaced in the unpredictable environment of the underground coal mine.

CONCLUSIONS

The immediate application of control technology and automation will be to improve the performance of existing mining equipment to more closely meet its potential and reduce costs of production. It is only in this way that a share of the estimated energy market will be retained, the increasing social aspirations will be met, and environmental difficulties will be overcome.

The application in the UK of currently known techniques and equipment and systems, both in development and in demonstration, have shown their effectiveness. The immediate task for mine management is to introduce the new technology on an ever wider scale to an established industry so that it will have a major impact on the efficiency of the industry and hence on its costs.

ACKNOWLEDGEMENTS

The author would like to thank his associates at Mining Research and Development Establishment for their assistance in preparing this paper. The opinions expressed are those of the author and not necessarily those of the National Coal Board.

REFERENCES

House of Lords: Select Committee on the European Communities European Community Coal Policy 10th Report, Page 180. 29 November 1983. HMSO.

Barham, Hartley and Rason, (1985). Development and Application of Monitoring and Control Systems for Coalface Operations. Institution of Mining Engineers Conference, 'Mining 85', June 1985, Birmingham. (Not yet published).

Rason, B. F. System 70000, The Mining Engineer, March 1984, 143 No 270, pp 417-424.

Moses, K (1984). Design for Safety in Underground Transport. ABMEC/NCB Conference on Designing for Safety. p S25, 25 November 1984, Colliery Guardian.

Tregelles, P. G. (1984). Future Production Techniques - Proceedings of the Australian Coal Association Conference 1984, Brisbane, Australia. (Not yet published).

The Control of Hydro Power Distribution Systems in Deep Mines

R.W. SCOTT and D.G. WYMER

Engineering Systems Branch, Chamber of Mines Research Organization, Johannesburg, SA

Abstract. The concept of hydro power in deep mines is based on the exploitation of the
hydrostatic head that can be gained as refrigerated cooling water descends in shaft pipe
columns from the surface, to provide hydraulic power for machines at the working face.
This paper describes the work done on two important aspects of the controlled distribution of the high pressure water within a mine, namely the control of pressure under
normal operating conditions, and the control of flow in an emergency situation following
a pipe breakage. Suitable methods of control have been determined using theoretical
techniques to model the steady and unsteady flow behaviour of the water in the pipeline,
and taking into consideration the environmental constraints on the types of control
devices that can be used. The preferred approach is one in which the high pressure
water itself is used as the control medium. The control valves themselves can be fairly
conventional in design, but in the case of pipe break protection it is shown that a less
conventional approach based on the use of a hydraulic fuse gives improved control
performance.

Keywords. Pipeline systems; mining; power system control; control valves; pressure
control; flow control; computer-aided system design; modelling.

INTRODUCTION

In South African gold mines, the mining activities
at the face are labour intensive and represent a
high proportion of the total cost of mining.
Mechanization of these activities can bring about
an improvement in this situation through an increase in productivity, and the Research Organization of the Chamber of Mines has been involved for
many years in the development of various types of
hydraulically powered mining equipment for this
purpose. Because of the harsh conditions prevail-
ing at the working faces, the only hydraulic fluid
considered sufficiently economic and practical for
use with the equipment in the long term is water
(Wymer, 1984). Meanwhile, high pressure water
jets are being used to an increasing extent for
cleaning broken rock from the working faces
(O'Beirne, Gibbs and Siderer, 1982). Thus, there
is a developing need generally for water-hydraulic
power in South African gold mines.

Many such mines operate at depths of 3 000 m and
more, and require large quantities of refrigerated
water for maintaining acceptable ambient temperatures underground. The preferred arrangement is
to chill the water on the surface, and then to
transport it to the underground workings in
pipes. The depths of the workings are such that
substantial water pressures (typically 10 MPa to
20 MPa) can be developed as the water descends,
and many mines recover the hydrostatic energy
through Pelton turbines to generate electrical
power. This also minimizes the temperature rise
of the water. If, however, the water is maintained at high pressure and is distributed in this
form to the working areas, it can fulfil a dual
function of providing both cooling and hydraulic
power to the face.

There are many advantages of this 'hydro power'
approach, not least being the elimination of large
quantities of equipment demanding regular maintenance underground, such as electrical power supply
equipment and high pressure pumps, which otherwise
would be required for the generation of hydraulic
power close to the working areas. Additionally,
it was realized that the available quantities and
pressures of water, as determined by the cooling
requirements and working depths respectively, were
generally more than sufficient for the powering of
machinery. Thus, a mine could introduce hydro
power without necessarily increasing its total
water handling capacity and without installing
additional pumps.

In view of these various attractions, the concept
was investigated in great detail, to determine
more precisely the design requirements for a
viable working system. Part of this investigation
involved the development of a suitable control
system design for the high pressure distribution
network, particularly to ensure that machines were
supplied at the correct hydraulic pressures, and
to safeguard personnel and equipment in the event
of a pipe failure. This paper outlines the techniques used in this part of the investigation to
predict the behaviour of water in a mine hydro
power system under various conditions, and
describes some of the basic hydraulic control
devices that were found to be required.

ANALYTICAL TECHNIQUES

The simplest approach to the design of a control
system for a hydraulic reticulation network is
based on steady state considerations. Variables
associated with piping systems are related by
algebraic equations which account for friction
losses. A common expression for long straight
pipes is the Darcy-Weisbach equation (Streeter,
1966):

$$\Delta P = \frac{8fL\rho}{\pi^2 D^5} Q^2 \tag{1}$$

where ΔP = pressure loss along pipe length (Pa)
  f = friction factor
  L = length of pipe (m)
  ρ = density of fluid (kg/m^3)
  D = internal diameter of pipe (m)
  Q = volumetric flow rate (m^3/s)

Similar expressions are used for other pipeline components. While such expressions are very useful, they give no indication of dynamic effects which occur when the system conditions are changed in any way. Although it is possible to calculate the state of a hydraulic network before and after an event such as a pipe rupture, an equation such as (1) gives no idea as to how long it takes to reach the new state. Moveover it gives no indication of the size of pressure waves, or 'water hammer', induced by sudden changes. Such dynamic effects can be calculated only by using the following differential equations for a deformable pipeline (Streeter and Wylie, 1967):

$$\frac{dQ}{dt} = -\frac{2f}{\pi D^3} Q|Q| - \frac{\pi D^2}{4\rho} \frac{\partial P}{\partial x} \tag{2}$$

$$\frac{dP}{dt} = -\frac{4\rho v^2}{\pi D^2} \frac{\partial Q}{\partial x} \tag{3}$$

where P is the pressure and Q is the flow at a distance x along the pipeline, at a time t.

Parameters f, D and ρ are as defined previously and v is the velocity of propagation of a pressure wave in the water. Note that in the steady state ($dQ/dt = 0$, $\partial P/\partial x = -\Delta P/L$) equation (2) reduces to the Darcy-Weisbach expression (1).

Equations (2) and (3) are a set of non-linear partial differential equations and, since there is no known analytic solution, they must be solved numerically using a computer. This is normally done using the 'method of characteristics' as described by Streeter and Wiley (1967), Fox (1977) and Chaudhry (1979). Conditions at the ends of the pipes in a network may be matched using equations (or characteristic curves) describing the operation of the connecting devices. In this way the entire water reticulation system may be simulated. Such a simulation has been developed to assist in the analysis of hydro power reticulation systems and to evaluate control systems designs.

Equations (2) and (3) may be simplified by making certain approximations and considering only idealized cases. Although such simplified equations can quickly give a rough idea of system behaviour without the need for a computer, and are often used in preference by pipeline system designers, some important features are lost. This is illustrated using two examples, typical of events that may occur in a mine hydro power system. These examples are based on Fig. 1, which shows schematically a portion of such a system. It is assumed in each example that the water flow rate is initially 0,03 m^3/s.

Example 1: Sudden 'clean break' failure of the horizontal pipeline, 500 m from the shaft.

In this case, the pressure drop over the horizontal length of pipeline changes instantaneously from its normal, low operating value to 18 MPa. In order to simplify equation (2), it is assumed that the pressure varies linearly along the pipeline, that is:

$$\frac{\partial P}{\partial x} = \frac{P(out) - P(in)}{L} = -\frac{\Delta P}{L} \tag{4}$$

Then equation (2) becomes:

$$\frac{dQ}{dt} = -\frac{2f}{\pi D^3} Q|Q| + \frac{\pi D^2}{4\rho L} \Delta P \tag{5}$$

Since ΔP changes from one constant value to another, equation (5) may be solved (Boyce and DiPrima, 1965) to give:

$$Q(t) = Q_f \frac{\exp(a+bt)-1}{\exp(a+bt)+1} \tag{6}$$

where Q_f = final flow as calculated from equation (1)

$$b = \frac{4fQ_f}{\pi D^3}$$

$$a = \ln\frac{(Q_f + Q_o)}{(Q_f - Q_o)} - bt_o$$

t_o = initial time

Q_o = initial flow

For this example, taking the initial time as zero and the friction factor as 0,025:

$$Q(t) = 0,133\frac{\exp(0,459+4,237t) - 1}{\exp(0,459+4,237t) + 1} \tag{7}$$

The system response, calculated accurately using equations (2) and (3), and approximately from equation (7), is shown in Fig. 2. Although the simplified expression gives a good indication of the settling time, it does not predict accurately the initial rate of change of flow or the peak flow.

Example 2: Sudden closure of the isolating valve in the horizontal pipeline.

By making the following substitution:

$$\frac{dP}{dt} = v\frac{dP}{dx}$$

and by assuming that the pressure and flow rate change instantaneously to new steady values, equation (3) can be rewritten as

$$\Delta P = \frac{-4\rho v}{\pi D^2} \Delta Q \tag{8}$$

where ΔP is the magnitude of the pressure rise, and ΔQ is the reduction in flow rate.

The system response, again calculated accurately using equations (2) and (3), and approximately using equation (8), is shown in Fig. 3. In the case of the approximate solution, the large diameter vertical pipeline has to be ignored, with a full reflection being assumed at the connection point at the bottom of the shaft. Thus, the pressure wave is assumed to oscillate in square wave form with a periodicity of 4L/v (approximately 2,5 s in this example) as it is reflected back and forth along the horizontal pipeline.

The exact solution predicts that the pressure will rise initially to 21½ MPa, followed by a more gradual increase to 23 MPa. Successive pressure peaks are progressively attenuated because of friction in the pipeline. The approximate solution predicts a similar initial rise in pressure, but underestimates the peak value slightly, and, of course, does not show the effect of friction losses.

These two examples show that the approximate expressions (6) and (8) can be used to gain insight very quickly into certain aspects of the response of a system, but only in simple idealized situations. In particular, only single lengths of pipe with a uniform diameter can be considered. To predict completely the response of a full hydro power network, it is necessary to develop a mathematical simulation and, for each element within the network, to solve the basic equations (2) and (3).

CONTROL SYSTEM DESIGN CONSIDERATIONS

Special considerations are necessary when designing a control system for a hydro power distribution network in a deep mine. The pipework configuration is unusual, in that it is extensive in both the horizontal and vertical planes, and the environmental conditions impose constraints on the types of control equipment that can be used.

The first and most obvious consideration is the existence within the system of high pressures, typically between 10 and 20 MPa. All control valves located in the line underground must be capable of withstanding these pressures with an appropriate safety margin. In some cases, this has precluded the use of certain designs of control valve which would otherwise have been attractive for this application.

A second consideration is that, if a structural failure should occur, a runaway flow situation will develop as described in the previous section, and the affected section of the line should be shut down automatically to minimize the hazard to equipment and personnel. For instance, referring again to the schematic in Fig. 1, if the horizontal pipeline ruptured completely at a point 100 m from the shaft, equation (1) predicts that the flow rate would rise to 0,3 m^3/s, with an exit velocity of 38 m/s. The reaction force on the broken pipe would be 140 kN, which could cause the pipe to break away from its supports. If an attempt were made to stop this flow suddenly with an automatic valve, the pressure upstream of the valve would rise to nearly 60 MPa, according to equation (8). This could in turn cause a further structural failure of the pipeline.

By contrast, failure of the same pipeline pressurized to 18 MPa by a positive displacement pump would result in no increase in flow rate or velocity. The pressure in the line would fall almost to zero, the pipe reaction force would be negligible, and the sudden closure of a valve after the rupture would produce a pressure wave of less than 6 MPa.

Thirdly, there are several environmental considerations which affect the choice and location of control equipment. Close to the working faces, the atmosphere is hot, humid, corrosive and dusty, and working spaces are confined and inaccessible. Consequently, the reliability of control devices is likely to be poor, particularly in cases where external pneumatic or electrical power is required. Therefore the use of all but the simplest control devices should be avoided. In the main access tunnels, the environment is less harsh and space is less restricted, so there are fewer constraints on the types of control devices that can be used. However, most commercially available pipeline valves are controlled and actuated using

an external pneumatic or electrical power supply and such supplies can be interrupted during the day-to-day operation of a mine. Although temporary back-up systems can be considered, the most satisfactory approach is to use controllers and actuators working directly from the high pressure water in the pipeline. Few valves of this type are available commercially, but some specially-developed prototypes are being evaluated.

Finally, the quality of the water must be considered. After passing through hydraulic machines in the working areas, the water exerts a cooling effect by coming into contact with the freshly exposed rock surfaces. In doing so, it picks up large quantities of salts and fine rock particles, and becomes chemically contaminated by blasting fumes. In conventional mining, many of the suspended particles are removed by settling, but nothing is done to remove dissolved contaminants. The concept of hydro power offers a broader economic base for the justification of bulk water treatment to remove these contaminants, so the water in the high pressure section of a hydro power system could be relatively pure. However, for economic reasons, it is unlikely that the dissolved and suspended matter will be removed entirely in the treatment plant. Additionally, it is known already that large quantities of solid particles can be generated from time to time within the pipeline by corrosion, scaling, deterioration of internal coatings, and debris introduced during installation and maintenance. Therefore, all control system components must be capable of withstanding moderate quantities of impurities in the water for long periods, and large quantities for short periods. This applies particularly to water-operated controllers and actuators, since these will tend to be more susceptible to damage or malfunction. This may impose special constraints on the designs of such components, and is the subject of current investigations.

NORMAL OPERATING CONDITIONS - CONTROL OF PRESSURE

Water must be supplied to machines in the working areas at pressures compatible with machine requirements to ensure safe and efficient operation. The control of pressure must automatically take into account variations in line losses over short time periods, due to changing demands during the working shift, and over long periods when factors such as pipe scaling and extensions to the system become significant. Pressures will also vary with location in the mine, because of differences in elevation between the various working areas. The hydraulic equipment under development for operation on high pressure water tends to be capable of working over a wide range of pressures (typically 12 to 18 MPa). Studies have shown that in most cases pressure control may be carried out using pressure regulators in the main trunk lines, as shown in Fig. 1, away from the harsh environment of the working areas, with the variations in line losses in the intervening pipework not compromising unduly the level of control achieved. Besides enabling the reliability of the control components to be improved considerably, this approach introduces substantial cost savings through the use of larger, but much fewer, pressure regulators.

EMERGENCY CONDITIONS - CONTROL OF FLOW

Normally, the water flow rate is controlled automatically, according to the demands of the machines operating at the face, by the hydraulic

resistance and mechanical inertia of the machines
themselves. In an emergency situation occurring
as a result of a pipe breakage, this control is
lost, and must be replaced by some form of flow
control within the pipe system to avoid the type
of runaway flow situation discussed earlier in the
paper. Three basic types of device have been
considered for this purpose, the flow control
orifice, the safety shutdown valve and the
hydraulic fuse.

Flow Control Orifices

The pressure drop through an orifice increases as
the square of the flow rate. Thus, in a runaway
flow situation, a suitably sized orifice will
assist in dissipating the hydrostatic head, and
will restrict the final flow rate to a lower
value, as illustrated in Fig. 4. Although this
will only reduce, rather than eliminate, the
hazards associated with a pipe break, the use of
orifices can be regarded as a simple, fail-safe
back-up to more sophisticated flow control
devices.

Safety Shutdown Valves

These devices are essentially isolating valves
controlled and actuated externally, preferably
using high pressure water taken directly from the
line. The isolating valves themselves are available commercially from several manufacturers, in a
range of sizes and pressure ratings more than
adequate for this application. Suitable high
pressure water controllers and actuators for the
automatic operation of these valves are currently
under development.

A disadvantage of a safety shutdown valve is that
it tends to respond slowly, so that in a pipe
rupture situation the flow will have already
reached its full runaway value before the valve
can close. In the meantime, the high velocity
water jet issuing from the broken pipe may have
caused considerable damage, and the large reaction
force may have pulled the pipe from its supports,
causing further damage. Additionally, in view of
the greatly increased flow rate, the valve must
now be closed much more slowly to avoid excessive
water hammer, delaying even further the shutdown
process.

Referring to the system shown schematically in
Fig. 1, assume, for example, that the isolating
valve shown 250 m from the working face is now a
safety shutdown valve, and that its control system
has a delay of 0,5 s. If a pipe fails suddenly,
close to the working face, a negative pressure
wave will arrive at the valve 0,17 s later, and
after a further delay of 0,5 s the valve will
start to close. The pressure immediately upstream
of the valve will then rise in a manner dependent
on the closing time. Fig. 5 shows what happens
if the valve closing time is 1 s, according to the
predictions of a simulation based on equations (2)
and (3). The pressure drops initially as a runaway situation develops, and then rises suddenly
as the valve closes. After peaking at 27 MPa, the
pressure then oscillates with appreciable amplitude for more than 10 s. This behaviour could
easily cause further failure of the pipe system.
Fig. 6, on the other hand, shows the situation if
the valve closing time is 10 s. After the initial
delay, the pressure rises slowly and, on completion of valve closure, peaks at only 21 MPa. The
subsequent oscillations are minor, and are soon
dissipated. In this case, there should be no
chance of a further pipe failure.

Hydraulic Fuses

A schematic representation of a hydraulic fuse in
its simplest form is shown in Fig. 7. It is
essentially a globe-type valve, in which the
closure element is normally held away from its
seat by a coil spring. As the flow increases, a
pressure difference develops within the valve,
generating a force on the closure element opposing
the spring, and the valve starts to close. During
the final stages of approach, the pressure difference across the seat increases very rapidly, causing the valve to close with a 'snap' action. The
pressure upstream will then hold the valve tightly
closed until the pressure downstream is restored.
This can only be accomplished once the failed pipe
has been repaired. The advantages of this type of
device over a safety shutdown valve are that it is
simple in construction and entirely self-actuated,
it reopens automatically, and it responds extremely rapidly.

Considering the same pipe failure situation as
before, but with the valve in Fig. 1 now being a
hydraulic fuse, and assuming this hydraulic fuse
snaps shut when the flow reaches 0,06 m^3/s (twice
the normal maximum value), the system behaviour
predicted by the simulation is as shown in Fig.
8. Despite the almost instantaneous closure, the
pressure rises only to 22½ MPa because the flow
has not been allowed to reach its runaway value.

This pressure peak can be reduced even further if
complete shut-off of the hydraulic fuse is avoided. This can be achieved, for example, by simply
drilling a hole through the closure element. The
effect is illustrated in Fig. 9, which shows the
predicted behaviour when the previous pipe failure
situation is controlled by a hydraulic fuse containing a 10 mm diameter leakage hole. The maximum pressure is reduced to 21 MPa, and the subsequent oscillations are attenuated much more rapidly than in the previous case. The flow rate
through the hydraulic fuse, shown in Fig. 10,
quickly settles to 0,013 m^3/s after the event,
giving an exit velocity at the open end of the
pipe of less than 2 m/s. This is unlikely to
cause any significant damage, and can be shut off
at some later stage with a manual isolating
valve. A further advantage of having a leakage
path through the hydraulic fuse is that, following
repair of the damaged pipe, the isolated section
of line can be repressurized through the leakage
hole in a controlled manner without the need for
an external by-pass line around the hydraulic
fuse.

CONCLUSIONS

The development of a suitable control system for a
hydro power distribution network in a mine depends
on the ability to predict the steady and unsteady
flow behaviour of the water in the pipeline.
Accurate prediction necessitates the use of a
mathematical simulation technique based on the
numerical solution of a set of governing differential equations, although simplified expressions
can be used to give a rough idea of system behaviour in very simple, idealized cases. A simulation has been developed successfully for hydro
power systems in deep mines, and has been used to
formulate a suitable control systems approach.

The pipework configuration and environmental conditions associated with a mine hydro power system
affect in a major way the preferred design and
location of hydraulic control devices, resulting
in a need for new types to be developed. A particularly desirable feature is that the control devices must not depend on external pneumatic or
electrical power supplies, but must rather use the

energy in the high pressure water itself.

Under normal operating conditions, the pressures within the system must be regulated automatically, but the characteristics of the hydraulically-driven mining equipment are such that the level of control need not be very strict. This enables the number of control devices to be minimized, and avoids the necessity for them to be located close to the workings, where the environment is particularly harsh.

Using the mathematical simulation to predict flow behaviour following a pipe breakage, three types of control device have been investigated for controlling flow in the affected pipework, to avoid the damaging effects of a runaway flow situation. Simple flow restriction orifices offer some degree of control, but are viewed rather as a back-up to other more sophisticated devices. Safety shutdown valves can isolate completely the affected pipework, but do not respond rapidly enough to prevent the flow from rising to its full runaway value, and must therefore close slowly to avoid excessive water hammer. A less conventional approach uses a hydraulic fuse which is a low-inertia, self-actuated device capable of responding so rapidly that it can, quite safely, isolate the line almost instanteously before a full runaway condition has been reached.

ACKNOWLEDGEMENT

This work formed part of the research and development programme of the Research Organization, Chamber of Mines of South Africa.

REFERENCES

Boyce, E.B. and DiPrima, R.C. (1965). <u>Elementary Differential Equations and Boundary Value Problems</u>, 2nd ed. Wiley, New York.
Chaudhry, M.H. (1979). <u>Applied Hydraulic Transients</u>. Van Nostrand Rheinhold, New York.
Fox, J.A. (1977). <u>Hydraulic Analysis of Unsteady Flow in Pipe Networks</u>. MacMillan, London.
O'Beirne, D., Gibbs, L.S. and Seiderer, A. (1982). The use of high-pressure water-jetting in gold-mine stoping operations. In H.W. Glen (Ed.), <u>Proceedings, 12th CMMI Congress</u>, S. Afr. Inst. Min. Metall. (or Geol. Soc. S. Afr.), Johannesburg.
Streeter, V.L. (1966). <u>Fluid Mechanics</u>, 3rd ed. McGraw-Hill, New York.
Streeter, V.L. and Wylie, E.B. (1967). <u>Hydraulic Transients</u>, McGraw-Hill, New York.
Wymer, D.G. (1984). The use of emulsions and water in underground hydraulic systems, In <u>Proceedings, Symp. on the Current Use of Hydraulics in the Mining Industry</u>, Minemech 84, S. Afr. Inst. Mech. Engrs., Johannesburg.

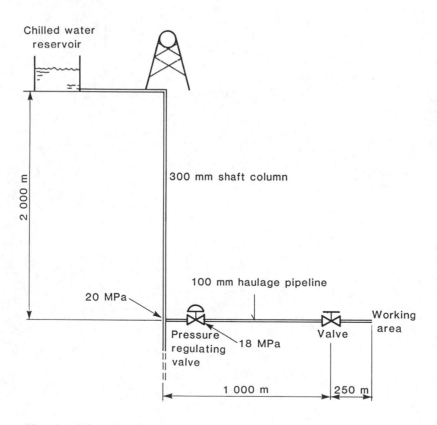

Fig. 1. Schematic of a portion of a typical mine hydro power system

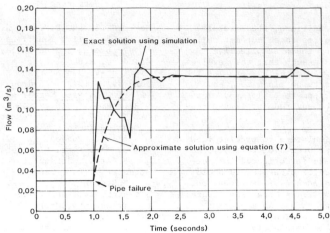

Fig. 2.   Flow behaviour following a 'clean break' pipe
          failure

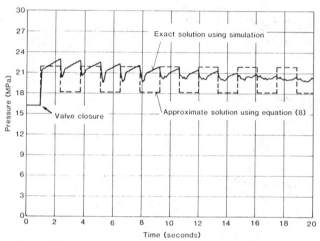

Fig. 3.   Pressure transients generated by sudden
          valve closure

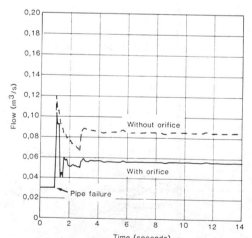

Fig. 4.   Effect of a 25mm flow restriction
          orifice following a pipe failure
          close to the working face
          (see Fig. 1).

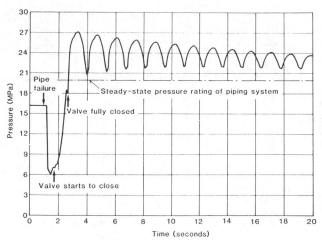

Fig. 5. Pressure transients generated by rapid closure
of a safety shutdown valve following a pipe
failure

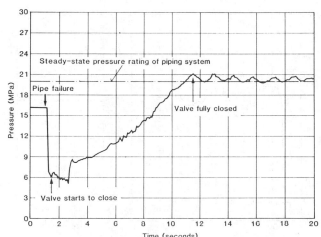

Fig. 6. Pressure transients caused by slow closure
of a safety shutdown valve following a
pipe failure

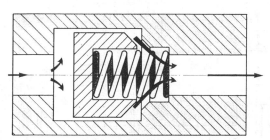

Fig. 7. Schematic of hydraulic fuse

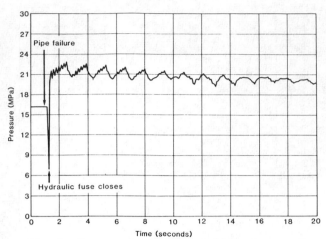

Fig. 8. Pressure transients caused by closure of a
        hydraulic fuse following a pipe failure

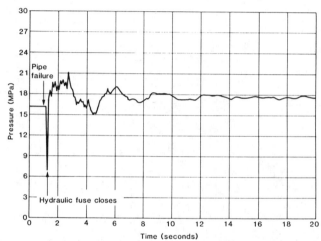

Fig. 9. Pressure transients caused by closure of a
        hydraulic fuse (with leakage hole) following
        a pipe failure

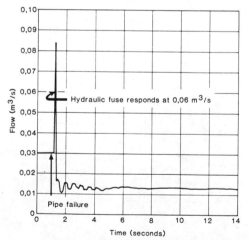

Fig. 10. Control of flow by a hydraulic
         fuse (with leakage hole)
         following a pipe failure

Experience with Operating and Controlling an In-Pit Portable Crushing and Conveying System

E.J. FREEH and C.D. ILES

Duval Corporation, 4715 East Fort Lowell Road, Tuscon, Arizona

Abstract. Portable crushing at Duval Corporation's Sierrita open pit mine has proven both the economics and operability of large, hardrock, portable systems. In-house expertise was utilized to create a system incorporating both operational and maintenance features requested by operating personnel. The monitor/control system involves over 400 channels of information and is based on a DEC LSI 11/2 microprocessor communicating with seven microprocessor-based data acquisition systems over a 4-wire, multidrop link. The year and one-half successful operation affirms the design philosophy and the hardware/software implementation.

Keywords. Monitor; control; conveying; crushing; portable; mining.

INTRODUCTION

Duval Corporation, established in 1926, is a United States mineral resource company active in exploration, mining, upgrading and refining of copper, molybdenum, gold, potash, and sulphur.

Duval's Sierrita property, located 30 miles south of Tucson, Arizona, has two adjacent open pits in a copper-molybdenum porphyry deposit of chalcopyrite-molybdenite mineralization. Mining operations will eventually produce a single excavation approximately two miles long, one mile wide, and two thousand feet deep. This is the location of Duval's pioneering in-pit portable crushing work. Sasoos (1984), an associate editor of the Engineering and Mining Journal, presents a good review of in-pit crushing and conveying.

The smaller Esperanza pit commenced mining operations in 1957; the larger Sierrita pit in 1968. Ore is processed in two separate mills, rated at 18,000 tpd and 92,000 tpd, respectively. Nominal mine production from both pits is 230,000 tpd, including 120,000 tpd waste. The ore grade averages .30-.34% copper and .033-.035% molybdenum. Janes and Johnson (1976) provide a description of the operation.

SIERRITA HAULAGE SYSTEMS

Three generations of haulage systems have been installed at the Sierrita pit since 1970. The original, or First Generation System, consisted of all-truck waste haulage and truck-crusher-conveyor ore haulage. Ore was hauled to two 60 x 89 gyratory crushers located 1,200-1,700 feet from the pit perimeter, then conveyed 2.3 miles to the mill on three 54-inch belts.

Haulage studies conducted in 1974 recommended that a new ore system be installed in the pit to reduce escalating truck costs. This Second Generation System, installed in 1976, consisted of two 60 x 89 gyratories located 350 feet down in the pit. Ore was transported to the mill over seven 60-inch conveyor belts totaling 3.4 miles length. The two belts exiting the pit were established on 16 and 14 degree slopes.

Restructuring of the haulage system also included conversion of the First Generation ore crushers to waste by increasing the crusher opening, installation of four 72-inch conveyor belts totaling 1.3 miles in length, and installation of a tripper and stacker.

The Second Generation System was effective in lowering mining cost by dramatically reducing truck haulage profiles, resulting in both decreased operating costs and truck requirements. Due to the static nature of the system, however, it was not possible to adjust to changing mining conditions, and truck costs began to escalate as the active mining areas moved away from the in-pit crushers.

With future haulage profiles projected to significantly increase in both distance and lift, a study was initiated at Sierrita in 1978 to evaluate future haulage alternatives. Portable crushing and conveying was determined to be the best haulage strategy for Sierrita. Project results showed that this Third Generation System would (1) provide significant operational and capital cost savings, (2) allow greater flexibility in mine design, (3) require minimal area for crusher set up, and (4) be technologically feasible to build and operate.

The Third Generation System currently operating at Sierrita is composed of an apron feeder, ore crusher, discharge conveyor and two drive stations, all fully portable and structurally independent. Photograph 1 is an overview of the installation in the Sierrita pit. The second photograph is a view of the crusher and the apron feeder. Support equipment includes a transporter and mobile crane. The apron feeder, 172 feet long with 10 foot wide steel pans, is hung by cables from a support tower. This design permits variable operating angles and loading configurations. It is currently being loaded by truck from the same bench as the crusher, conveying on a 20 degree slope.

The crusher is a 60 x 89 gyratory unit, housed in a toroidal, three-legged, structure. Fifty feet in diameter, the toroid houses all support equipment in enclosed rooms which are pressurized to minimize dust accumulation. The discharge conveyor, located beneath the crusher, uses a 10 foot wide rubber belt for rock removal rather than the standard apron feeder.

The portable drive stations are totally self-contained units, including transformers, motor controls, drives, and take up. The system can handle peak loads in excess of 5,000 tons per hour.

Relocation of portable units, except for the discharge conveyor, which is self-propelled, is accomplished with a 1200 ton capacity crawler transporter capable of one-half mile per hour on 12 percent grades.

The last unit in the system is a mobile crane, fitted with a retractable boom, which is used to service the crusher. Rated at 400 tons (12-foot radius), it will lift 82 tons at the 60-foot radius required for a mantle changeout.

The portable crusher currently supplies about half of the mill feed, the remainder coming from the conventional in-pit crusher. Future plans include installation of two additional portable crushers for waste crushing and conveying, completing the Third Generation System.

DESIGN AND UTILIZATION CONSIDERATIONS

The utilization strategy of portable crushers is of major importance in system design and economics. Both operating and capital costs can be significantly affected by configuration of relocation sites, time between moves, relationship to active mining areas, and changes in mine design and scheduling required for system implementation. To achieve maximum economic benefits, a portable system should be considered an integral part of the "active-mining-area" material handling system.

A portable crusher system should be relocated as necessary to maintain minimum haulage operating costs. This time increment will vary from mine to mine and within design periods for any given mine, but nevertheless the general requirement dictates equipment that is capable of rapid and inexpensive relocation. Failure to relocate to maintain minimum truck haulage profiles will result in (1) constantly increasing truck operating haulage costs, (2) increasing truck requirements and therefore capital expense, and (3) creation of unnecessary crusher landings and access roads which may be difficult or impractical to mine out.

To ensure practical equipment, Duval emphasized operational and maintenance requirements. The initial evaluation was conducted by mine engineering and operations personnel located at the mine site; people who would eventually be responsible for operation of the system. This group also specified operating requirements they believed would be necessary to ensure success in the Sierrita environment. Significant effort was spent simulating crusher-conveyor relocation schedules in order to verify that the system could be effectively implemented. Equipment was drawn to scale and located in "worst case" configurations. Conveyor routes were carefully evaluated, particularly conveyor-haulroad crossovers.

After concept approval by management, a team was established within DUTEC (Duval Technology and Engineering Corporation) with responsibility for design, construction and startup of the prototype system. Heavily oriented toward equipment maintenance, the individual members brought

expertise from underground, marine, as well as open pit environments. Guided by the previously established operating requirements, the foundation was laid for creation of the Duval system.

The Sierrita mining environment dictates portable crusher relocation an average of once per year, installation in varying site configurations, and operation in relatively limited spaces. Flexibility and portability allow the Duval system to meet these requirements.

Flexibility is achieved by separating the system into three structurally independent equipment units: apron feeder, crusher, and discharge conveyor. Truck-dumping onto the apron feeder can be performed from the same bench as the crusher or from the bench above, by adjusting the operating angle of the apron feeder truss. Duval prefers same-bench loading because it minimizes site preparation expense and does not tie up active mining areas with construction activities. The operating angle can vary from 5 to 20 degrees.

A second feature increasing system flexibility is the ability of the apron feeder to feed the crusher from opposite sides. Third, the crushed material can exit on the discharge conveyor belt from underneath the crusher in one of three directions, spaced at 120 degree intervals.

Since the portable system will be relocated every other bench as mining increases in depth (or about once per year), it is desirable to minimize move downtime. The system is designed to be relocated within a 48-hour period, measured from shutdown to startup. Site preparation is minimal, with no concrete required. All equipment units sit on steel pads which sit directly on the ground. Minimum downtime and site preparation work characterize a system both easy and inexpensive to move.

MONITOR AND CONTROL SYSTEM

General

Development of the portable crusher system by Duval Corporation required designing a system for monitoring a variety of discrete and analog variables as well as providing interlock and control facilities. The major design objectives in the developing of a monitoring and control method included considerations of system robustness, assembly/disassembly, ease of maintenance, and cost.

The system had to be relatively immune to dust, vibration, temperature variations, and electrical interference. It was necessary to allow for the access of multiple points at multiple remote locations, allow convenient reconfiguring of points at each of the locations, and permit easy disconnect/connect upon physical relocation of any segments of the crushing system. In addition, the design had to facilitate rapid field repair by electrical or instrumentation personnel, as appropriate, with a minimum amount of additional training.

There were three primary hardware alternatives for meeting the requirements: the traditional "hardwire" approach; the use of an off-the-shelf programmable controller; and the in-house development of a customized system based on a standard microprocessor.

The hardwire approach was determined to be

impractical due to a total of over 400 analog
and discrete points of interest, distances up to
4,000 feet from the control room (possible
future distances of 2 miles), and restricted
space for an operator's console.

Also, the inability to conveniently reconfigure
the system for point changes or crusher relocation and the requirement to be able to present
all this status information to a host computer,
dictated a more flexible system. Further, the
hardwire approach with the enormous amount of
wire required suffered a cost disadvantage relative to a microprocessor-based system.

The use of a standard programmable controller
system was also considered and the offerings of
several vendors reviewed in detail. This
approach was favored initially but eventually
rejected because it would not permit the desired
functionality ultimately required as design
became finalized. Further, there were numerous
benefits that accrued to implementing our own
in-house system that would not be forthcoming
with the installation of an off-the-shelf
programmable controller.

Experience gained from previous process
monitoring and control projects in the areas of
sulphur well production and control; gold production and processing; energy management of
industrial heaters and boilers; ball mill
grinding; and other areas of production and
processing, provided Duval with the 'in-house'
expertise to specify and implement all of the
monitoring and control requirements. The 'in-
house' system was eventually determined to be
the most cost effective solution while
simultaneously providing the greatest
functionality.

System Description

A schematic of the Sierrita crushing and
conveying operation is presented in Fig. 1, with
the portable crusher control room considered the
local station. The remote locations consist of
the stationary crusher and its existing overland
conveying system, portable extendable conveyor
I, portable extendable conveyor II, portable
apron feeder, portable discharge conveyor, and
the portable crusher itself.

The host computer is a VAX 11/780 located in the
engineering and maintenance area of the pit.
The so-called 'host computer' does not play an
active part in the crusher monitor/control
system. Its function is to accept data from the
portable crusher system and provide storage and
accessibility. This computer is part of a
corporatewide network permitting maintenance and
control personnel to monitor performance from
corporate operations headquarters in Tucson.
This feature was particularly valuable during
the shakedown phase.

A microprocessor-based data acquisition system
was used to provide the monitoring and control
functions at each location. Each remote station
is capable of monitoring up to 24 analog inputs,
a combination of 96 discrete I/O points, and
can provide up to 8 analog outputs. As many as
128 remote stations are possible, allowing
multiple stations to be used at the same
location if additional capacity is required.

The remote stations were designed as modular as
possible to permit "remove and replace"
maintenance techniques as a first-line approach
to system repair. The remote stations are set
up in four modular sections; power, communi-

cations, processing, and I/O. The power
section comprises three DC power supplies; the
communications section contains one line driver
modem; the processing section incorporates three
microprocessor cards; and the I/O section
consists of individual I/O modules with LED
indicators. Such modularity provides for ease
of problem detection and replacement of the
faulty section. The modular units can then be
conveniently repaired back at the shop. The
modularity also permits existing personnel to
perform the necessary field repair with a
minimum amount of training.

The system was designed to operate over a
sufficiently wide temperature range such that
air conditioning was not required. Also, all
I/O points were interfaced via solid state
devices to maximize reliability.

All the remote stations are connected back to
the operator's control console via one 4-wire,
twisted pair, "true" multidrop communications
link. Any of the remote stations can be powered
off or disconnected from the link and communications will continue with the remaining
locations. This capability was provided to
allow the remaining locations to function while
one or more stations are down. Crusher relocation is also simplified, a simple disconnect
and reconnect of the communications link and the
remote points are again accessible.

The local station, housed in the operator's
console in the control room, is used to provide
the necessary interface from the system to the
operator via the console. Also housed in the
console is the master microprocessor of the
system. The master unit is based on a DEC LSI
11/2 microprocessor with RAM and EPROM memory
and serial I/O boards to handle the communications. The master unit, with operating software, commands and polls the remote stations,
displays the information to the operator, accepts operator input, and relays the information
over the radio link to a host computer. The
program is written in assembly language and
resides in EPROM memory to provide nonvolatility
and maximum speed and reliability. This was
essential to provide real- time reporting to
assure the operator of current status
information.

Fig. 2 is a schematic of the equipment in the
control room showing both the DEC LSI 11/2
microcomputer and one of the seven data
acquisition units.

As mentioned, communication between the local
and remote stations is via a serial, 4-wire
twisted pair, multidrop link using line drivers.
The communications protocol is a packet format
with checksum and parity detection to insure
data integrity. Each station is assigned a
unique address (switch selectable) which is
embedded in the communications packet. The
properly addressed station responds to the
command issued by the master unit and the others
simply ignore the message until they are
properly addressed.

Real-time response is achieved by communicating
with each of the monitoring stations every 300
msec. All start/stop and monitoring functions
are provided for through the communications
network, except those that must be hardwired at
each station for safety purposes (e.g., pullcord
shutdown). The operator can perform sequenced
start/stop or emergency stop of the system from
the console. On startup, all the sequencing,
timing, and interlock checking is performed by
the control and monitoring system. If any

"permit-to-start" input is not active, the system will not start and a message will be presented to the operator indicating the problem.

Most of the system status information is displayed on a CRT in the console at 19.2 Kbps. Features like reverse video, blinking, bolding, and double height characters are used on a formatted screen to indicate various alarm and operating information. In addition to the screen display, a two-pen chart recorder can be used to provide a time-history plot of any two of the inputs monitored at the local or remote stations. A small number of indicator lights and an alarm horn are also present on the console.

Operator interaction is provided via a sealed alphanumeric keypad and a small number of pushbutton switches. A hardwired emergency stop button is also provided in the event of the failure of the computer system.

Also provided in the control room are means to regulate the rate at which the portable apron feeder delivers rock to the crusher. Hand controls for a jackhammer to dislodge boulders from the crusher and two closed-circuit TV monitors to observe critical points are likewise located in the control room.

Once every minute the system status is transmitted via a 1200 bps UHF radio link to the VAX host computer. The crusher control and monitoring system operate independently from the host computer used to store and facilitate analysis of the operating data. An FSK (frequency shift key) modem is used to interface the digital output from the microprocessor to the UHF radio. The radio link was chosen to provide crusher mobility and to provide ease of communications with future portable crushers. The portable crusher system data reside in a data base formatted on the host computer where they can be accessed by maintenance, operations, and management personnel to provide production reports, shift summaries, and other outputs of interest. Of particular interest to engineers and maintenance personnel is the ability to retrieve detailed historical data on system performance and conveniently analyze same.

CONCLUSION

The Duval portable crushing system is fully operational, proving both the economics and practicality of large hardrock systems. The monitor/control system, in operation for over a year and a half, has performed to expectation. The user friendliness has provided for good operator acceptance and the reliability and modularity of the system segments have given Maintenance the confidence required for acceptance of a state-of-the-art system.

Over 17,000,000 tons have been crushed as of November 1984, averaging 1,200,000 tons per month. Hourly rates average 2,800-2,900 tons per hour, varying from 2,100 to over 5,500

depending on degree of fragmentation of crusher feed.

Physical availability has exceeded 90%, closely approximating the conventional in-pit crusher. Both the apron feeder and discharge conveyor have physical availabilities in the high 90's.

The number of active haultrucks required to mine ore from the pit bottom has been reduced by seven units for a two-shovel operation, decreasing capital requirements by $9,000,000. This reduction results from an 11-13 minute decrease in truck cycle time, as shown in Table 1.

TABLE 1 Truck Cycle Time (Minutes)

| | Crusher | | | |
| Active | Portable (3,100') | Permanent (3,600') | | Percent |
Bench	Crusher	Crusher	Decrease	Reduction
3050	8.4	21.5	13.1	61
3100	7.0	20.4	13.4	66
3150	8.3	19.5	11.2	57

Utilization strategy and therefore equipment configuration is site-specific and must be carefully planned to achieve maximum benefits. The projected savings of $0.29 per ton have been realized - in Duval's Sierrita operation this will yield a net savings of forty million dollars in ten years.

Until one has experienced a development such as herein described, it is easy to underestimate the value of an in-house instrumentation and control capability versed in the application of computers and associated electronic equipment. For example, the specifications, in regard to scope alone, grew in several steps from an initial estimate of approximately twenty channels to over four hundred. Changes in system size of an order of magnitude and more require a re-thinking of the conceptual design. The advantage in such cases of systems engineers working alongside the mechanical design engineers, as the over-all design evolves, is obvious.

ACKNOWLEDGEMENT

The authors wish to acknowledge the contribution of D. M. Steines to both the preparation of this paper and the design and implementation of the subject monitor/control system.

REFERENCES

Sassos, M. P. (1984). In-Pit Crushing and Conveying Systems. E&MJ, 185, 46-59.

Janes, C. G., and Johnson, L. M. (1976). The Duval Sierrita Concentrator. Flotation., A M Gaudin Memorial, Fursteneau, M. C., AIME.

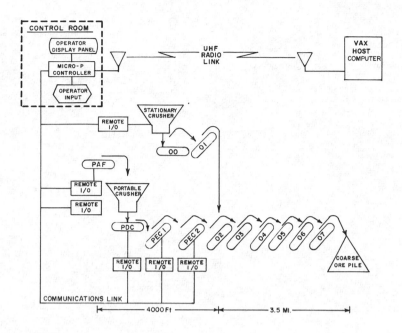

Fig. 1   Schematic Showing Portable Crusher and Auxiliaries and Its
         Interrelationship with the Stationary Crusher/Conveyor System.

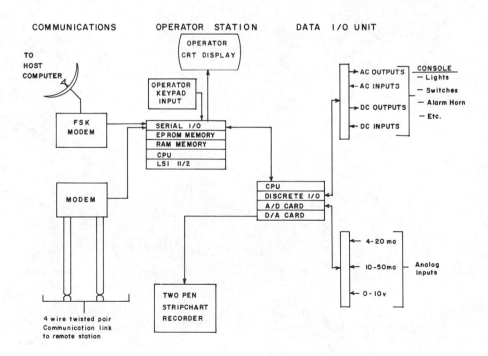

Fig. 2   Schematic of Portable Crusher Monitor/Control System. Operator
         CRT is driven by a DEC LSI 11/2. One of the seven Motorola 6802
         microprocessor-based data units is shown at right.

Photograph 1 Overview of Sierrita Pit Showing Portable Crusher,
Stationary Crusher, and Interconnecting Conveyors.

Photograph 2 Close-up of Portable Crusher and Pan Feeder.

A Review of Hydraulic Pipeline Transport and its Future Development

DR P.J. BAKER

Business Development Department, BHRA Fluid Engineering, Cranfield, Bedford, England

Abstract. There have been many developments in recent years in pipeline transport to
enable them to be more widely used in industry. Substantial tonnages of materials that
are not affected by water, such as coal and other minerals, aggregates and wastes, can
be moved over considerable distances in quite small diameter pipes. Where there are no
existing road or rail facilities, long-distance pipelines are usually cheaper to
install, and this advantage increases in difficult terrain. Pipeline operators all
over the world assert that hydraulic lines pay for themselves in a relatively short
time, in spite of the heavy capital outlay, and this is equally true for lines conveying slurries or pastes within a plant complex.

This paper reviews some of the more important investigations in the hydraulic
transport field. Although reference to systems carrying a wide range of materials is
made, the majority of recent published data centres on coal transport.

The paper looks at the application to a wide range of particle sizes, the development
of non-aqueous slurrying media and coal/water fuels. The review also looks at some of
the equipment being proposed to provide more reliable operation and in particular the
range of solids-handling pumps now available. The paper goes on to identify areas
where further research and development is required and some of the many organisations
undertaking this work.

Finally the short, medium to long term potential for their use is discussed together
with political and other factors likely to influence market growth.

Keywords. Hydraulic transport; slurry flow; mineral handling; pipelines; solids
handling pumps.

INTRODUCTION

The early history of pipelining of slurries is
somewhat obscure. There is some evidence that the
Romans used to transport ore and tailings in a
slurry form. In recent times, papers have reported
on pipelines being used in the USA to pump sand
from goldmines in 1850 and waste out of coal mines
in 1884. In this century, the earliest documented
system of significant size was operated in 1913,
conveying coal through a 200 mm. diameter 540 metre
long pipeline from barges in the River Thames in
London to Hammersmith Power Station. It is claimed
that the system operated satisfactorily for some
eleven years before it was abandoned due to
blockage.

Serious research into slurry technology began in
the early 1950's. The first methodical experimental study on fine-particle slurries leading to
empirical correlation was published by Durand and
Condolios in France in 1952. It is interesting to
note that, in spite of the tremendous amount of
research that has been undertaken since those days,
some of Durand's equations are still being used
satisfactorily in design studies for some slurry
mixtures. In the UK, work was mainly carried out
by BHRA in conjunction with the NCB on coarse
particle coal slurries.

In recent years the emphasis has increasingly been
on the development of slurry pipelines for coal
transport. However, there are many hundreds of
applications worldwide where pipelines are being
used to pump a wide range of minerals, mineral

waste and other materials in slurry form.
(Baker, 1979). Some of the more obvious
advantages of pipelines are that they are compact
and have a high carrying capacity. They can be
constructed in terrain not always possible to
other forms of transport. They are also
environmentally more acceptable, in that the
pipelines can be buried without the unpleasant
consequences of some surface transport systems.
The main disadvantages of hydraulic pipelines
are that applications are limited to materials
that can be mixed with the carrying medium
without affecting the products final use. They
are also limited to areas where the carrying
medium is economically available, and for some
mixtures there is concern about pipeline blockage. Over any distance the pipeline is a fixed
connection between two points. The relatively
high capital cost tends to limit them therefore to
applications with high throughputs operating for
several years.

A recent review undertaken by the BHRA showed that
over 50 million tons of solids are pumped each
year. If all the planned pipelines go into
operation, the quantity transported could rise to
some 300 million tons per year. These figures do
not take into account the in-plant applications,
which tend not to be published. The majority of
the proposed long distance pipelines are for coal
transport in the USA (Baker, 1983). The coal
would be finely crushed to produce a pseudo-
homogeneous mixture with water. Economic
viability of these systems have been demonstrated
by many designers. However, one of the major

cost elements in fine particle slurry systems arises through the need to dewater the fine coal (less than 2 mm) down to about 20% prior to use directly in power plants. For loading onto ships for export, the acceptable water content could be as low as 10%. Much work has therefore been going on in recent years on improving dewatering techniques and to look at the possibility of operating at higher solid to liquid concentrations and with larger particle sizes. This would have the additional advantage of reducing crushing costs.

This paper attempts to briefly review some of the recent developments, which could not only widen the range of applications for hydraulic pipelines, but also could reduce some of the operating and capital costs.

SLURRY OPERATING REGIMES

The flow regimes for slurry transport are considerably more complex than those for single-phase fluids, and slurries may be both settling or non-settling. Figure 1 identifies three main forms of mixture available to the user. Factors that will determine the regime in which the mixture is operating are flow velocity, particle size distribution, density and concentration.

FINE PARTICLE SLURRY SYSTEMS

The technology of fine particle slurries up to 1 to 2 mm. size is well established for a whole range of materials and the majority of working systems operate in this mode. In general, almost any finely ground mineral lends itself to long distance slurry pipelines. Table 1 shows a selection of some of the major pipelines built worldwide. Table 2 identifies some of the more recent fine coal pipelines proposed. It will be seen that the majority of these are for very long haul application in the USA. In addition to the USA, two other countries have been actively applying this technology to industry. These are Japan and South Africa.

A brief review of the Japanese applications show that some 37 working projects - transporting slimes, raw coal or concentrates, backfilling of mines and volcanic ash - were in operation in 1976. The longest line is claimed to be 68 kilometers, pumping mine tailings. New schemes are being designed, particularly for land reclamation.

There are also many short pipelines in South Africa, specifically for handling gold slimes and tailings. These are usually less than 20 kilometers long but have been operating successfully for many years. With recent improvements in processing costs, further pipelines are being designed and built to transport the tailings back to mine plants to extract further gold.

COARSE PARTICLE SLURRY SYSTEMS

The technology for coarse particle slurries, having particles greater than 2 mm. and usually less than 100 mm., is now becoming available. It is in this area that most research and development is being undertaken at the present time, particularly in the USA, Japan, Canada, Germany, Australia and the UK. In general, the application areas for large particle heterogeneous slurry mixtures is for short distance applications of a few kilometers.

There are a number of important advantages in pumping coarse-particle sized material, such as reduced preparation and dewatering costs. Unfortunately, these mixtures require high operating

velocities to maintain movement of the particles, which leads to both higher abrasion of the pipeline system and high specific power requirements.

A limited number of coarse slurry lines have been built, and it would appear that widest experience of this type of regime exists in the Soviet Union, although there are references to papers of applications in the UK by the National Coal Board and in the USA. Recently the NCB built a 1.2 kilometer coarse particle colliery waste slurry pipeline system for a throughput of 200 tons per hour solids, passing a 38 mm.screen. In the USA, an initial design study was undertaken in 1982 to transport coal approximately 12 kilometers across Staten Island to a port from a railway terminal for export purposes. It is understood that this project has now been abandoned because of the recent change in market conditions. Future application areas are seen for vertical hoisting of minerals from mines and the sea bed and for waste disposal schemes.

COARSE/FINE SLURRY SYSTEMS

The limitations of pumping coarse particles in pipelines, referred to above, have led to the need for research into techniques that would allow industry to move coarse materials over the long distances achieved by fine slurry systems. The possibility that coarse slurry material could be pumped at low velocities with relatively low pressure gradients was first investigated by Elliott and Gliddon (1970). They found that a heterogeneous slurry containing 13 mm. coal could be pumped, with the addition of 25% fines to give a total concentration of about 60%, at low velocities and at relatively low pressure gradients. Further in depth investigations into these mixtures are being undertaken, both in the UK at BHRA and in Australia at CSIRO.

Technical data available to the designer is still limited at the present time, however this form of regime appears to have wide applications and can be considered for both short and long distance systems. Recent work in Australia (Duckworth, Pullum and Lockyear, 1983), describes a series of tests carried out with mixtures of coarse and fine coal slurry with concentrations up to 67%. The top size of coal was some 20 mm. The researchers have examined the start up condition of the pipeline containing 67% concentration after four day shut down period. Although the pressure required was initially larger than the corresponding steady state condition, the pressure gradient dropped to the value within a very relatively short period of 30 minutes. Pipeline blockage did not occur.

In addition to coal, the BHRA have recently been applying this technology to the disposal of mine waste. Here the requirement is for the disposal of very large quantities of wide particle range material with the minimum of handling, and if possible, avoiding dewatering altogether.

The future potential for applications using this form of mixture looks promising, particularly within the coal industry. At present, the only known similar mixtures pumped at high concentration and wide particle size range are concrete and mine pump-packing systems. Both these applications are over short distances with relatively high friction losses.

NON-AQUEOUS CARRIER FLUIDS

Much effort has been put into overcoming the problem of dewatering material, an area of major concern to users. In addition to improving the performance of dewatering plants, a number of non-aqueous media have been investigated. Again, the main application has been for transporting coal. The principal candidate media were identified as crude oil, fuel oil, methanol, ethanol, and carbon dioxide.

COAL/OIL MIXTURES

These are based on heavy fuel oils loaded to a weight concentration of 40 to 50% pulverised coal, about the same as crude oil or water based slurry. They are designed for direct firing into boilers, hence require no separation.

Another end use proposed is for direct injection into a blast furnace. The rheology of COM is not yet fully established, depending on coal type, oil type, additives, temperature, particle size distribution and storage history. It would appear at the present time, for economic reasons, coal/oil research and development has been reduced.

COAL/METHANOL MIXTURES

Like crude oil, methanol owes its attractiveness as a slurry vehicle to its value as pay load. In this capacity it has been recommended as a substitute transport media for natural gas from Alaska, the gas being converted to methanol at source. Methanol coal slurries can be prepared and pumped in a similar manner to coal water slurries, but require a finer grind of coal if head losses and transitional losses are to be kept down. The research activity in this area is also small compared with conventional coal/water systems.

COAL/CARBON DIOXIDE MIXTURES

In contrast to COM and methanol, carbon dioxide owes it attractiveness as a slurry vehicle to its ease of separation from coal. This technology, which is still in its infancy, is being developed in the USA. A number of prospective advantages are claimed. It can be recovered from coal without consuming water, liquid carbon dioxide is at least 1/15th as viscous as water, allowing it to be loaded to a higher coal concentration (ie 80%) and very little CO_2 is absorbed by the coal, making it easy to separate and avoiding agglomeration or swelling problems associated with water. Carbon dioxide in large quantities could become a marketable product for such end uses as enhanced oil recovery. Separation can be achieved by simple evaporation of the carrier fluid yielding dried ground coal and purified CO_2. The researchers claim that this process can be very economic compared with other pipeline techniques, but much more work is required to confirm the technical long-term reliability and safety of such schemes.

COAL/WATER FUELS

The rise in oil prices in 1973/74 provided the incentive for research into methods to reduce energy costs. In addition to COM, CWF were investigated for direct firing in boilers. These fuels are known by a number of proprietary names and differ mainly in particle size distribution and the viscosity reducing and suspension stabilising additives employed. In general the maximum particle size is less than 300µm. with 75% to 80% less than 74µm. Concentration by weight of coal is usually about 60 - 70%.

Development work has been carried out in USA, Germany, Sweden, Holland, Japan and Italy, with the common objective of achieving a highly loaded mixture with acceptable viscosity and stability. Many full scale trials have been conducted, particularly to determine combustion performance. However, limited data seems to have been obtained and published to date on the pipeline performance of these mixtures, and this is an area where information will need to be obtained before complete systems can be designed. It is claimed that the advantage of CWF over COM is its reduced cost per unit of heat. The cost advantage is claimed to be approximately 12% at present. However, because of recent developments in the marketplace, ie lower oil prices and reduced energy consumption, the application of this technology is being limited, as boiler conversions are not being so actively carried out.

EQUIPMENT DESIGN

In parallel with developments in fluid technology, improvements in equipment design has continued. A review of papers presented at various slurry transport conferences would suggest that most elements of a slurry system have received attention, with the emphasis on improving performance and widening the range of solids handling pumps and feeders available to the designer, particularly for some of the coarse/fine particle slurry mixtures proposed. The wear characteristics of the many new pipeline materials available to the designer, now in the marketplace, have also received attention.

The choice of solids handling pumping equipment will depend on the various properties of the slurry to be pumped, as well as satisfying the normal duty and installation requirements. The initial choice lies between some type of pump that handles the slurry directly, with the attendant problems of abrasive wear and slurry damage, and a feeder system using a remote source of clean driving fluid, which may be more complex and hence costly. All have some inherent limitation on their performance or application.

The two main groups of pumps are dynamic and positive displacement (Fig. 2). Dynamic pumps tend to be cheaper in both capital and maintenance costs and can handle relatively large solids; they are normally considered for the higher flow, low pressure and less viscous applications. The converse applies generally to positive displacement types; they are also more efficient for a given output and their performance is largely unaffected by the presence of solids. The conventional roto-dynamic pump is used for most in-house applications, handling a wide range of particle size slurries over relatively short distances.

The number of pumps available to the designer for handling coarse-particle slurries in longer distance applications, where high pressures are required, is much more limited. Most positive displacement pumps have restricted flow areas at the inlet and outlet ports, which allow only the passage of fine and relatively fine particles. Some of the modern lock-hopper systems have a very high capital cost and possibly high operating costs because of the many moving parts that could be subjected to wear. Concrete pumps, which are readily available in the market, appear to offer the most promising solution at the present time to handle high concentration coarse/fine particle slurry mixtures. Most types are capable of delivering pressures >50 bar and have the additional advantage over the reciprocating pumps of the ability to handle moderately large solids, due to the 'full-way' valve design, as well as very viscous mixtures. The main disadvantage is the relatively lower output/unit, usually of the order of 50 - 100 m3/hr. Hence, several units in parallel would usually be required, compared with more conventional reciprocating pumps. Development work is being undertaken to improve reliability and phasing the stroking sequence between units in multi stage applications to avoid large pressure fluctuations when discharging into a pipeline. Fluctuation could have a detrimental effect on the slurry and the pipeline components. In recent years the manufacturers of these units have been aggressively marketing them for applications outside the concrete pumping field. An alternative design, which is now undergoing development trials, is the high-head rotary pump. This was invented in Australia for use with high concentration coarse/fine slurry mixtures for long distance applications (Boyle B. E. and L. A, 1979).

Manufacturers of roto-dynamic pumps are actively pursuing research into designs with improving operating efficiencies, longer life and easier maintenance. In addition, in order to extend their market into the long distance applications, work is being undertaken to produce units that operate at higher pressures. Extensive research has been undertaken into wear performance of pumps and pipelines because of the many factors that influence their working life.

Rubber lining has been successfully used, giving lower wear rate than metals in both pumps and pipeline applications, providing the solids are not long or sharp. Polyurethane can give very good results as a pipe lining but its use in pumps has given somewhat conflicting information. The softer grades of both natural rubber and polyurethane appear generally more resistant than the harder ones. Ceramics are very wear-resistant but can be expensive and are susceptible to brittleness and thermal shock. New developments using ceramics in small pump applications may show improvements.

Generally there is a lack of reliable quantitative data on the many factors affecting wear from which wear rates on new systems might be predicted with a reasonable certainty. Because of this the designer still has to resort to carrying out wear tests before he can finalise his design and operating cost figures.

In addition to the general advances being made in conventional dewatering of solids, a major investigation has been underway in Australia for the treatment of fine coal/water slurries (Rigby and others, 1982). The technique involves mixing the slurry with oil, which results in the small coal particles agglomerating with the oil. Spheres up to 3 - 4 mm. diameter are produced, which can easily be separated from the water by static screens. In addition, the clays washed out of the coal do not agglomerate. This integrated approach to improve the dewatering process in a slurry system and at the same time improving the quality of the product for the user could have wide application. Its economic viability depends primarily on the effective recovery of the oil at the end of the process where the user would not be prepared to accept the agglomerates containing oil for either technical or economic reasons.

FUTURE RESEARCH AND DEVELOPMENTS

Since the mid 1970's, there has been a rapid increase in research and development, particularly for applications of pipelines to the coal industry. The main emphasis has been increasingly towards establishing the technical and economic limitations of pumping coarse coal over relatively long distances. As a result of the recent down turn in the coal business, however, the level of R & D funding is now being cut back. Countries which

have been identified as having invested significant funds to this technology include USA, Japan, Germany and Canada. Many other countries have been active to a lesser degree.

Throughout this paper the author has attempted to identify some of the main areas where research and development has been undertaken. As to the future, advances will continue to be made, possibly at a lower rate until market conditions improve, into mathematical modelling of the various slurry flow regimes; equipment with greater reliability and laboratory and pilot scale pipeline tests on, in particular, the coarse/fine slurry mixtures. With limited resources available, there will be an increasing tendency for R & D to be aimed at specific application areas. To determine these areas, it is essential to look at the market opportunities for pipelines in the future. One potential growth area, particularly for Australia, is the application to moving coal and other minerals to the coast for export. The application of pipelines to the loading and unloading of ships, either in ports or off-shore from single-point mooring buoys, appears to offer many advantages both economically and environmentally. However, more demonstration work is required to establish the viability of the various schemes being suggested, the main requirement being to reduce the dewatering problem on board ships being loaded with coal slurries. Ship stability and unloading techniques need further investigation.

The advent of the computer for system monitoring and control should lead to more efficient operations. Unfortunately, the availability of a full range of suitable instruments for accurately sensoring the main slurry parameters is also very limited, despite significant efforts to date. More work is required in this direction. In the longer term, economics will force the designer to opt for more fully automated systems if he is to compete with other transport systems and deliver an acceptable product to the user. Finally, there is an increasing need to ensure designers are kept abreast of developments through training courses and other forms of informative dissemination.

There is no shortage of centres of technical excellence undertaking Research and Development studies available to designer and contractors. A worldwide review of published establishments identified some 49 in total, including 17 Universities, 8 Government laboratories and 20 industrial organisations. This total does not include many of the installations used by industry for in-house testing and are not therefore generally advertised or documented. Most pipeline test facilities are relatively short closed circuit loops having pipe diameters up to approx. 200 mm. The largest facility has recently been built in the USA and consists of 3 major loops up to 0.43m. dia. and 220 m. long. In addition, there are a number of specialist facilities, which are designed to look at both preparation and dewatering techniques.

CONCLUSIONS

The future growth, both in the short and long term, for hydraulic slurry systems will be affected by a number of factors in addition to its technical development. Interest in hydraulic transport systems has ebbed and flowed with changes in economic and political situations. We are now seeing a further down turn in activities due to the present recession in the coal industry.

Discussions with organisations actively involved

in R & D would suggest there is tremendous potential. This is a natural reaction as most are fully committed and enthusiastic about their own research. In practice there are few significant pipelines under construction or recently commissioned. It should however be stressed that most of these pipelines tend to be the major long-haul applications of typically between 10 and several hundreds of kilometers in length. It is understandable that the greatest interest is shown in these major schemes because of the large contract value to those who are involved. However, it must also be noted that, throughout the world, there continue to be many small insignificant pipelines being designed, built and successfully operated, particularly for the mining industry.

As already mentioned in this paper, most of the major long-haul fine-coal pipelines planned for construction were in the United States. Major political pressures have been brought to bear to stop these pipelines, particularly from the railway operators. In addition to obtaining Eminent Domain, there is strong opposition from farmers who are concerned with loss of their water. These factors, together with the recent industrial recession, will severely limit the number of long haul slurry pipelines to be constructed during the next five years. The number of coarse particle slurry systems to be constructed is also expected to be very small, and limited to those of a few kilometers in length, again partly due to the present weakness in the market but also due to limited technical data and experience available.

In the medium term, with growing pressures on land requirements, environmental issues, and rationalisation of various industries, a greater trend towards the application of both short and long distance pipelines for handling a whole range of minerals and mineral waste is expected. Also, unless more efficiently operated railways are available to industry, the operating cost differential will widen, which will again increase the incentive towards pipelines.

Although emphasis has been placed on the market potential for coal handling, applications for the transport of many other minerals and waste products will increase. It must however be borne in mind that alternate pipeline modes of transport, including pneumatic handling systems, will inevitably offer increasing competition in some areas of application, particularly where water availability is limited.

In predicting the long term potential for major slurry pipeline systems, both political issues such as Government's plans for railways in each country, transport infra-structure, and the rate of industrial growth will have a major influence in the final analysis.

Even without these factors, solids handling pipelines would not be expected to be as widely used as gas and oil pipelines. However, the future for slurry pipelines will be to play an increasing role in the operation of many industries.

REFERENCES

Baker, P. J., and B. E. A. Jacobs (1979). A guide to slurry pipeline systems. BHRA Publication, Cranfield, England.

Baker, P. J. (1983). Hydraulic transport by pipelines. Published Inst. of Civil Engineers, London.

Boyle, B. E., and L. A. Boyle (1979). The potential for hydraulic transportation of coal in NSW colliery department. Quarry Mine & Pit Vol. 18. No. 8 pp 10-17.

Duckworth, R. A., L. Pullum and C. F. Lockyear (1983). The hydraulic transport of coarse coal at high concentration. J. of Pipelines Vol. 3. No. 4 Elseview Science Publication, Amsterdam. pp. 251-277.

Elliott, D. E., and B. J. Gliddon (1970). Hydraulic transport of coal at high concentrates. Proc, Ist. Int. Conf. on Slurry Transportation, Hydrotransport, Warwick, U.K. Sept 1 - 4 1970. BHRA paper G.2.

Rigby, G. E., C. U. Jones., D. E. Mainwaring and A. D. Thomas (1982). Slurry pipeline studies on the BHP-BPA 30 tonne per. hr. demonstration plant. Int. proc. 8th Int. Conf. on Slurry Transport, Hydrotransport, Johannesburg, South Africa Aug 25 - 27 1982 BHRA paper D1.

REGIME	NORMAL FLOW CONDITIONS	NOMINAL PARTICLE SIZE RANGE AND CONCENTRATION BY WEIGHT	APPLICATIONS
Fine Particle Systems	Homogeneous - Pseudo - Homogeneous Flow	-2mm/50% -300µm/70%	Long haul syst. eg mine to power stations. Ship unloading. Full dewatering plant required. Coal/water fuels in-plant.
Coarse Particle Systems	Heterogeneous or Moving Bed	No top size limit but usually up to 50 mm with low proportion of -100 µm. 35-40%	Short haul syst. eg in-mine Vertical hoisting. Ship loading/unloading. Reduced dewatering stage.
Coarse/ Fine Systems	Laminar flow. Dense phase or stabilised flow.	Usually up to 50 mm. with sufficient fines present to support coarse fractions 60-80%.	Both short & long haul systems. Ship loading/unloading, in-port/offshore. Reduced dewatering stage.

Fig. 1. Different Flow Regimes with particular reference to Coal Slurries

TABLE 1   Major Systems Constructed

Slurry Materials	System or Location	Length km	Diameter mm	Annual thruput (mt/a)
Coal	Ohio, USA	174	254	1.3
	Black Mesa, USA	439	457	4.8
	Poland	200	256	N/A
	U.S.S.R	61	304	1.8
Limestone	Rugby, U.K.	91	254	1.7
Copper Concentrate	West Iran	112	114	0.3
	KBI, Turkey	61	127	1.0
Magnetite Concentrate	Tasmania, Australia	85	229	2.3
	Pena Colorado, Mexico	48	203	1.8
	Samarco, Brazil	390	508	12
	Chongin, North Korea	61	N/A	4.5
	Kudremukh, India	67	457	7.5
Gilsonite	Utah, USA	115	250	1.3
Mill Tailings	Japan	68	305	0.6
Kaolin	Sandersville	110	450	0.07

TABLE 2   Proposed Coal Slurry Pipelines

Location/Operator	Length (km)	Diameter (m)	Throughput (t/a)
Coalstream, USA	2400	0.35 to 1.2	55×10^6
ETSI, USA	2240	0.4 to 1.0	30×10^6
Pacific Bulk, USA	1040	0.66	10×10^6
San Marco, USA	1440	0.51 to 0.61	10×10^6
Powder River, USA	-	-	-
Alton, USA	262 + 30	0.56 & 0.51	9×10^6
Alton, USA *	109 + 8	0.3 & 0.23	2.5×10^6
Snake River, USA	1760	0.4 to 0.76	16×10^6
Texas Eastern, USA	2080	0.51 to 0.97	25×10^6
New York, USA *	12.8	0.76	$8 \times 10^3 \ hr^{-1}$
Virginia, USA /	640 - 744	-	5×10^6 to 25×19^6
Poland to Austria	608	0.46	5×10^6
Poland to Italy /	770	-	5×10^6 to 10×10^6
Italy to Austria	500	0.46	5.75×10^6
Jharia, India	1759	0.25 to 0.525	22×10^6
Singraulic, India	1600	0.3 to 0.525	13×10^6

* Indefinitely deferred.
/ Feasibility study only

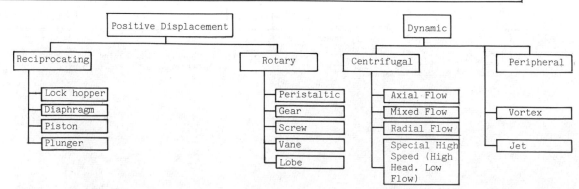

Fig. 2 Pump Classification

Future Developments in Underground Mobile Equipment

F.W. GRIGG

Department of Mechanical Engineering, University of Queensland, St Lucia, 4067, Brisbane Australia

Abstract. Recent developments in engine, transmission and control technology, likely to give rise to improved performance and efficiency of underground mobile equipment, are reviewed. Turbocharging and the electronic control of fuel pumps and injectors can improve diesel engine performance whilst minimizing exhaust emissions. Ceramic materials promise further improvements in engine efficiency. Nevertheless, electrically powered machines, equipped with trailing cables or 'travelling plugs' are expected to find increasing application where there is less need for flexibility and where the cost of ventilation and heating or cooling makes diesel engines unattractive. Voltage to frequency converters and induction motors appear to have a future in mains powered machines. Vehicles employing energy storage in the form of batteries or flywheels will find increasing application, but it is expected that they will remain a minority until there is a significant break-through in battery technology. Hydrostatic transmissions having high power density and efficiency are likely to be used more widely. Electronic control systems are likely to be used to simplify jobs and to improve performance and productivity. Enhancement of remote control systems is to be expected for work in unsafe areas, but true robot operation is considered unlikely for the moment. Attention to ergonomic factors should produce significant productivity gains.

Keywords. Mining; underground mobile equipment; diesel engines; exhaust emissions; transmissions; hydrostatic drives; electric drives; ergonomics; hydraulic systems; energy storage; control equipment.

INTRODUCTION

The economic viability of many underground mines in the developed countries is currently under challenge. Changes in consumption patterns, substitution of alternatives, and unrestrained production by the developing countries have forced commodity prices to very low levels. High labour costs combined with the costs of providing the better working conditions demanded by underground workers in the developed countries make it difficult to compete with mines where labour costs are much lower, the demands of labour are less important, and where government subsides are often very significant. To remain competitive, mines in the developed world have to find ways of reducing their production costs and, in general, this means finding ways of increasing the productivity/cost ratio of each worker underground.

During the last couple of decades, there has been a proliferation in the use of diesel powered mobile equipment underground. The flexibility and totally independent operational capabilities of this equipment has resulted in substantial productivity gains over the early mining methods. One consequence of the introduction of diesel engines underground was the need to increase ventilation so as to achieve satisfactory air quality conditions in the presence of the diesel exhaust emissions. This inevitably led to increases in the cost of ventilation.

In recent years, ventilation costs have soared as a result of the combined effects of the unit costs of fuel and power and the increased consumption of power associated with the desire to either heat or cool the ventilating air to provide a better working environment. In some cases, there has

also been an increase in the volume of air supplied in order to achieve better air quality standards.

An emission-free alternative to the diesel engine is the electric motor. This has the additional advantage that its unavailable energy is rejected at the power station and not to the underground atmosphere. The greater heat rejection of the diesel powered machine can be a significant disadvantage of its use in hot climates.

In view of the high cost of ventilation (Magnussen, 1979) and the diesel engine's significance in contributing to the need for that ventilation, it is not surprising that the future of diesel engines underground should be questioned. Indeed, in many mines, action has already been taken to replace diesel equipment with electrically powered equipment wherever possible.

Although it is easy to convert fixed equipment to electrical power, there are difficulties in converting mobile equipment, since inevitably there is the problem of how to supply the power to the vehicle as it moves around the mine. Invariably, the use of electric power on mobile machines leads to a reduction in the flexibility of their application.

Rising fuel and power costs place increasing significance on the desirability of having highly efficient energy conversion devices and transmissions. In some situations, there are opportunities for recovery of energy during downhill runs or when braking, and in these applications, energy storage devices and hybrid transmission systems can be considered. Power systems based totally on energy storage such, as batteries or flywheels, are another possibility for certain applications.

The ease with which it is now possible to install low cost, high performance control systems on mobile machines is likely to have a major impact on both the design of the machines and on the ways in which they can be employed within a mine. Significant innovations in mining methods must be expected as automatic machines, remote controlled machines, and robot machines are introduced.

It can be seen that the following are amongst the challenges facing the designers of underground mobile equipment for use in the developed countries :-

(i)   to devise ways of powering mobile machines whilst achieving efficient, reliable, flexible and emission-free operation at a low capital cost;

(ii)   to improve the performance of diesel engines and associated transmissions etc. so as to substantially improve the overall efficiency of the machines whilst providing significantly reduced emissions;

(iii)   to improve the reliability and ease of maintenance of machines so as to reduce the cost of underground maintenance;

(iv)   to apply the principles of ergonomics to the design of operator's control stations to improve the operator's working conditions whilst at the same time improving productivity; and

(v)   to exploit developments in automatic control technology so as to improve productivity and to reduce the need for making safe all areas where machines are operated underground.

Although these challenges apply to the designers of underground mobile equipment, they also apply to the designers of many other mobile machines. Consequently, there are many opportunities for the cross fertilization of ideas and the use of innovative equipment developed for other purposes. It should be borne in mind that the introduction of new technology almost invariably requires a systems approach if success is to be achieved, and this is particularly important in underground mining where so many factors interact.

Recent developments in engines, transmissions, control systems, braking systems, and steering systems are likely to contribute to new solutions to these challenges. The significance of recent improvements and likely future developments in these branches of technology are discussed in the following sections.

POWER PLANT

Electric

Electric motors are in many respects the ideal power plant for underground use. They can be made to have high efficiency, they are quiet and they do not produce dangerous emissions.

Where DC motors are employed, they are usually powered from an overhead two-wire trolley system or from batteries carried on the vehicle. In such cases, speed control is achieved by the use of Thyristor choppers. More recently, there has been a tendency to use on-board silicon controlled rectifiers (SCR's) to control DC motor speed and

to supply the power to the machine as three phase AC. This arrangement allows auxiliaries that are best operated at constant speed to be powered by relatively cheap induction motors. In most cases, where electric motors have been used to drive machines such as Load Haul Dump vehicles, the electric motor has been coupled to a torque converter on the input of a powershift gearbox connected by cardan shafts to differential planetary axle assemblies. In other words, the electric motor has been installed simply as a replacement for a diesel engine.

Although DC motors have been the traditional form of electric drive on vehicles, recent practice has tended towards the use of AC induction motors. These have the advantage of being relatively compact, light weight, cheap and maintenance free (as there are no brushes). Constant speed induction motor powered machines are usually fitted with modified torque convertor drive systems incorporating a modulating clutch, which serves to reduce the jerkinees associated with gear changes with a constant motor speed. Although such an arrangement is relatively simple, it suffers from the disadvantage that it further increases the losses that occur in the drive system to a point where overall drive efficiency is likely to be around 50%.

Recent developments of solid state devices provide significant advances in the technology of AC motor control. Several manufacturers are now offering frequency convertor systems suitable for controlling the speed of induction motors from startup to full speed. These convertors usually consist of a rectifier, a DC voltage smoothing filter and an inverter (Fig. 1). The inverter determines the frequency and size of the output voltage and hence the speed of the motor. Such drives have characteristics similar to DC systems including short duration overload tolerance. Although additional blower cooling may be required for motors operating at less than normal rated speed, the overall efficiency of such systems can be very high. There are several mainline locomotives having installed capacities of around 4.4 MW operating using this system and very many smaller industrial drives. At the moment there is a cost penalty compared with comparable DC drives, but it is expected that this will disappear in the near future. When applied to mobile equipment, such as Load Haul Dump vehicles, such drives should be capable of achieving overall efficiencies of around 70% if the motors drive directly into reduction gearboxes without the use of a torque converter or powershift gearbox.

Electrical Power Sources

Batteries. If flexibility of operation is the principal criterion, then batteries carried on the vehicle must be the most desireable source of electric energy for drive motors. Unfortunately, lead acid batteries are unable to provide sufficient energy for economical operation of machines such as Load Haul Dump vehicles. It has been found that the maximum time between recharges can seldom exceed half a shift. Much research effort is being directed to the development of new batteries and a recent report by Wearden (1984), indicates that significant progress has been made on the development of a successful nickel zinc battery. This technology offers twice the gravimetric energy density and one and a half times the volumetric energy density of lead acid batteries. Other advantages include the capability of operating over a much wider temperature range and a greatly improved charge-cycle life.

Trailing Cables. Until suitable batteries become available, electric powered vehicles will have to rely on trailing cables or similar systems for their power supply. Although trailing cables have been used on mining vehicles for many years, they are still considered to be a source of problems. Normally, cable lengths of less than 200 metres are employed. One end of the cable plugs into an appropiately located power supply outlet while the other end is wound on and off a cable reel mounted on the vehicle. Both horizontal and vertical reels have been tried and considerable effort has gone into refining the design of cable tensioning/control devices. Systems used have included constant cable tension arrangements, synchronised payout/takeup corresponding to vehicle speed, and the use of sensing arms to control the drum speed so as to maintain a constant relationship between the cable and the vehicle. None of these systems has proved to be entirely satisfactory and, almost always, the vehicle speed is limited by the cable performance, and most users experience relatively short cable lives (less than 1000 hours). The cables used usually incorporate three main conductors plus earth sensing and monitoring conductors. The cables are quite expensive and can be fairly easily damaged and typically cost in the vicinity of $10 per machine hour. Obviously, there is scope for improvement in cable and winder design and durability.

Since the range of movement of trailing cable machines is limited by the length of their cable, they are usually used at fixed locations. When the need arises to move to a new location, they either have to be towed or supplied with power from a mobile power unit, which may be trailer mounted.

Trolley Systems. To overcome the lack of flexibility associated with trailing cables, trolley systems have been used. However, these are not very attractive since they involve relatively high costs, restricted routes and there are safety problems when used in relatively low headings commonly found underground.

The 'travelling plug' concept developed by Jarvis Clark and reported by Du Russel and Muirhead (1981) is an interesting solution to the dilemma of providing flexible operation of an electric vehicle. The basic system shown in Fig. 2 consists of a trolley travelling on a steel track bolted to the roof. The track support also accomodates four conductors. Spring loaded pickup shoes mounted on the trolley are energized by three of the conductors and grounded by the fourth. Electrical connection between the trolley and the moving vehicle is via a trailing cable attached to the vehicle and the combination enables a vehicle to travel along a heading and to then divert sideways into a crosscut for a distance limited by the length of the cable. This system would seem to offer sufficient advantages to guarantee it a place in situations where greater flexibility than that available from a trailing cable is needed and where diesel engines are unacceptable.

Flywheel Drives

These drives have potential applications in vehicles requiring only intermittent mobility. Sponsored by the U.S. Bureau of Mines, Rockwell have designed an experimental system fitted to a 20 ton shuttle car (Zachary and Ginsburg, 1982). This machine uses a flywheel driven by an electric motor which can be powered from the mains every time the shuttle car unloads. The operational duration of this machine is approximately six minutes. On a later version of this unit, it is intended to use a mechanical drive system to spin up the flywheel at the shuttle car's discharge point, thus eliminating the need for an on-board electric motor. The flywheel on the experimental unit consists of 7 discs, each approx. 22 inches in diameter, operating at a top speed of approx. 14,500 rpm, and it is designed to deliver 4.5 kW-hr of usable energy.

Diesel Engines

Although air cooled diesels have been used extensively in gas-free underground mines, water cooled engines are now finding acceptance, especially in hot climates. Indirect injection has frequently been chosen where the need for relatively low exhaust emissions has been a primary consideration, however the more efficient direct injection engine is now claimed by some manufacturers to be able to achieve comparable emission levels (Salee, 1982). Catalytic converters are commonly used to reduce CO emissions. Deration is common practice to minimise smoke, and retardation of injection timing is a fairly common method of reducing the emission of oxides of nitrogen (NOx).

The need to improve the efficiency and to reduce the emissions of diesel engines is widely recognised by diesel engine manufacturers. Users of diesel engines are invariably interested in fuel economy, and there are numerous government regulations controlling emission levels in all types of applications. Competition is intense and significant improvements have been made recently in laboratory tests and more are expected. In general, the improvements arise out of work aimed at controlling the conditions of combustion.

Turbocharging and aftercooling have long been recognised as ways of significantly improving the power/weight ratio of engines. In recent years, it has been found that turbocharging has a number of other benefits and it is likely that turbochargers will become standard equipment on engines in future. Turbochargers can not only boost the power output of a given engine but can significantly improve brake specific fuel consumption and emission and noise levels. Ways are being devised to get around the problems usually associated with turbocharger delay. Aftercooling offers an enhancement of most of the benefits of turbocharging and reduces the risk of overheating in hot environments.

Electronic control of the fuel injection process is receiving considerable attention and appears to offer significant benefits (Komiyama and others, 1981; Trenne and Ives, 1982 and Day and Frank, 1982). Timing of injection and fuel quantity delivered can be independently controlled in most of the systems being developed and significant gains in efficiency are being claimed. Substantial reductions in emission levels can also be achieved by this technique (Reams and others, 1982).

Catalytic converters are widely used to control the CO and HC emissions from diesel engines. However, they are not effective in controlling the particulate or NOx emissions. In essence, the NOx emissions can be reduced by reducing the temperature in the combustion chamber and methods of achieving this include exhaust gas recirculation, the use of water/fuel emulsions, or the injection of water into the combustion air (Lawson, 1981). In recent research, using an unmodified diesel engine, emulsions consisting of 25% methanol/75% diesel fuel have been found to reduce exhaust emissions significantly (Lebel, 1983) but methanol poses a number of safety problems as an underground fuel.

Whilst bearing in mind that the solution to the emission problem will vary according to the type of engine, type of fuel (sulphur content etc) operating conditions (humidity and temperature etc), it seems likely at this stage that a very significant reduction in exhaust emissions should be achievable by using turbocharged, aftercooled engines fitted with electronically controlled fuel injection equipment and catalytic converters in association with perhaps water injection to the combustion air and the use of particulate filters in the exhaust stream. Further experimentation and development is required before the optimum solution to this problem will be achieved.

It is usual to find that pollution levels vary throughout a mine and there are some areas (such as dead headings) where good ventilation is very difficult to achieve. Pollution levels at the operator's position on machines working in such locations, can be significantly reduced by using exhaust diluters, discharging upwards and away from the operator at an angle of about 30 deg. to the horizontal (Anon, 1983).

The recent development of new ceramic materials such as partially stabilised zirconia (PSZ) appears to have potential for some exciting developments in engine technology. Bell (1984), reports the possibility of raising operating temperatures from about 700°C to around 1100°C. The US Army Tank Automotive Command (TACOM) and Cummins Engine Company have built an uncooled diesel having ceramic material on the combustion chamber, the cylinder liners and heads, the piston crowns and intake and exhaust ports. They expect to be able to demonstrate a lubrication free version of this engine, having an efficiency of 54%, this year. The ceramic insulation has meant that the radiator and 360 other parts, including the water pump, are no longer required. Ceramics are also being used for turbocharger rotors and experiements are being conducted on turbo-compounding. This means that the first true adiabatic engine is likely to be produced in the next couple of years. Such developments will undoubtedly have spinoffs into the normal production range and ensure the continued use of diesel engines underground for many years to come.

TRANSMISSIONS

The basic purpose of a transmission is to transmit the power from the engine to the wheels of a vehicle as efficiently as possible. The other functional requirements of the transmission depend on the nature of the power plant fitted to the vehicle. If the power plant is capable of providing a drive torque from zero speed, as is the case for some electric drives, then a relatively simple gear box of appropriate ratio may be all that is required in the transmission. On the other hand, if the power plant is of the type that runs continuously, such as a diesel engine or a mains powered constant speed induction motor, then a transmission offering variable speed ratios, and hence variable torque ratios, will be required.

Conventional

The familiar foot operated clutch and manually shifted gearbox is efficient and quite satisfactory on small service vehicles, but for larger vehicles it is common to employ torque converters and automatic or power shift gearboxes to achieve this function. In this role, the torque converter performs the dual functions of providing a torque multiplication effect and a fluid coupling effect which reduces the shocks associated with mismatch of speeds when changing gears.

Although torque converters are widely used, they suffer from the problem that the slip essential to their correct functioning necessarily leads to a lower efficiency than is possible with more direct types of drive. It is for this reason that in recent years such transmissions have often been equipped with 'lock up', which means that the drive through the torque converter becomes direct when the 'lock up' is engaged. At stall (a condition which can occur in many mining machines involved in pushing and loading activities), the torque converter is operating at a high slip condition and this gives rise to considerable heat generation, which can lead to overheating problems as well as excess fuel consumption.

Gearboxes, by their very nature, fail to provide the continuously variable transmission ratio necessary if the engine is to remain close to its maximum torque point as the vehicle speed increases from zero to maximum. However, if a large number of ratios is provided, it is possible to maintain the engine close to its optimum operating point. The greater the number of ratios available, the greater the effort required of the driver. In light vehicles, this problem is solved by the use of an automatic transmission, but in heavier vehicles, where the number of ratios tends to be significantly higher, there is now a tendency to use electronic controls to select the appropriate ratio for the combination of vehicle speed and engine torque concerned (Faber, 1982). Although such an approach can lead to better performance than can normally be expected from a manually controlled system, it does not eliminate the losses associated with the torque converter.

Another disadvantage of the conventional transmission is that there are only a very limited number of arrangements of the components that can be employed, since they must be joined together by mechanical drive shafts. This can have a strong influence on vehicle configuration. The relative bulkiness and noisiness of the conventional transmission system is another distinct disadvantage. Possible alternative transmissions include electric and hydrostatic systems (Bullock and others, 1981).

Electric

Apart from a few battery powered machines, the relatively direct drive arrangement usually found with DC traction motors employed in locomotive drive applications has seldom been used on underground mobile equipment. However, fully electric transmissions have been applied to very large off-highway surface vehicles, typically of 100 tons capacity or more. Companies such as General Electric, Reliance Electric and Marathon LeTorneau have all developed wheel motors suitable for such machines.

Although electric transmissions provide a desirable measure of freedom in the relationship between the various components, and they offer the potential for regenerative braking, they suffer from the problem that the power density of electric machines and control equipment tends to be very poor when compared with that of hydraulic machines. This bulkiness, together with the need for cooling of motors operating at sustained high torques at low speeds, and the availability, comparable efficiency and relatively low cost of alternatives, have doubtless mitigated against the widespread use of electric transmissions on vehicles of more moderate size.

As indicated above, this situation could change with the development of low cost solid state rectifiers and inverters so that induction wheelmotors can be used.

Hydrostatic

Hydrostatic transmissions have undergone considerable development during the last decade and the variety and range of sizes available has increased dramatically. Since they have very high power densities, hydrostatic drives are particularly attractive on underground machines where space is almost always severely restricted. Furthermore, hydrostatic transmissions have the potential of allowing the vehicle manufacturer to offer a constant speed induction motor powered unit as a simple alternative to a diesel powered unit without suffering the penalties in transmission efficiency associated with modulating clutches applied to conventional transmissions. Retarding or engine braking is another attractive feature of hydrostatic drives, since it can eliminate the need for service brakes, which are expensive and which tend to be high maintenance cost items (Emergency/parking brakes are still required).

Significant recent improvements in hydrostatic transmissions include : increases in the efficiency and operating pressures, reduction in noise levels and the development of a number of control systems to provide various operating characteristics. When variable displacement axial piston motors are used in conjunction with variable displacement axial piston pumps, it is possible to achieve a continuously variable transmission capable of operating over a wide speed range. Fig. 3 shows a comparison between the efficiencies of a torque converter four speed gearbox transmission and that of a continuously variable hydrostatic transmission. It can be seen that the hydrostatic has efficiency advantages over all except the high speed part of the operating range. The difference in efficiency is particularly significant at very low speeds for vehicles such as loaders, which simultaneously have to produce wheel torque and lifting power for bucket operation during mucking.

To get the maximum benefit from hydrostatic drives, it is desirable, if possible, for the hydraulic motor to be located entirely within the wheel hub. This provides the maximum space saving on the vehicle. Commercially available wheel motors come in two basic types: (i) Low Speed High Torque (LSHT) type and (ii) High Speed Low Torque (HSLT).

The low speed high torque (LSHT) motor is of simple construction, it has a high starting torque and a high efficiency and has the advantage of not requiring any reduction gearbox. Such units, however, are only available in a limited range of sizes and, although dual displacement and free-wheeling versions are available, continuously variable displacement types are not.

HSLT motors, on the other hand, require a substantial gear reduction for application as wheel motors, and this is usually achieved by using an epicyclic hub reduction unit. The efficiency and starting torque of these units tend to be lower than their LSHT counterparts, but they are available with continuously variable displacement and numerous control options, including electric. For some applications, the flexibility and operating range of the variable displacement HSLT motor is not required, and in such circumstances, the fixed displacement bent axis motor provides a somewhat more economical solution (Tunley and Preston, 1981).

There are many ways in which hydrostatic drives can be configured to achieve particular ends as a vehicle transmission system. Some of these have been reviewed by Bullock and others (1981). Hydrostatic drives have the potential for being applied to situations requiring regenerative braking, with energy storage being provided in the form of either a flywheel (Evans and Karlsson, 1981) or a hydraulic accumulator (Regar, 1981). The use of such techniques reduces the total energy consumption of the vehicle and it also provides an opportunity for having emission-free operation for short periods.

Significant as these benefits might be in some circumstances, there are few opportunities for worthwile energy recovery in most underground mining operations, and there is seldom sufficient room on underground vehicles to enable the installation of energy storage devices of the size required.

Other

The integrated engine transmission concept described by Wallace, Tarabad and Howard (1983) appears to have potential for future heavy vehicles, especially if success can be achieved in producing an adiabatic engine. A schematic layout of this system is shown in Fig. 4.

ERGONOMIC FACTORS

Most existing underground machines provide little in the way of creature comforts for their operators. In many cases, one gets the impression that the driver's visibility, control layout and comfort have been very low on the list of priorities of the machine designer. This situation is now changing, at least in the developed countries, where there is a demand for better working conditions and also a recognition that they can lead to increased productivity.

Although significant gains can be made by modifying existing machines, the best approach is undoubtedly to take ergonomic factors into account during the initial design stage. By doing this, it is possible to develop an integrated system offering ease of maintenance with minimum complexity. Noise reduction is best tackled early, since significant gains can be achieved by careful selection, location and installation of engine and transmission components.

Many underground machines have scope for significant improvements in the ease of operation of controls, the comfort of their seating, the reduction of their noise levels and the provision of clean air for the driver to breathe.

In the last few years there has been a proliferation in the range of controls available for the remote operation of equipment, either by means of electric cables or hydraulic hoses. In general, these controls have been designed to require only small operating forces, and they lend themselves to installation in carefully planned control layouts likely to produce effective man/machine interaction. In many cases, they are far less bulky than their predecessors and they thereby permit better accessibility to the driver's seat whilst also providing potential for better visibility. A further advantage of these remote controls is that they divorce the positioning of the control lever from that of the device being controlled, and this can often significantly improve the service accessibility of both the control lever and the controlled device.

The provision of an enclosed cab is one way of
achieving many improvements in operator comfort
and safety. Cabs can provide protection against
falling objects and against overturning; they can
reduce noise levels; and they can, if pressurized
or airconditioned, exclude dust and provide the
operator with a clean, dry, quiet work
environment. By the use of suitable remote
control levers and steering systems, it becomes
feasible to mount the cab on isolators so as to
reduce the transmission of noise and vibration
from the frame of the vehicle.

A number of manufacturer's now offer quite
sophisticated seat suspensions, which can greatly
improve operator comfort. One problem that
usually remains is that the amplitude of
oscillation of these suspended seats can be
extreme and this can lead to significant
excursions of the operator relative to his control
levers unless the controls are mounted on the
seat. The use of suitable remote controls makes
this feasible.

CONTROL SYSTEMS

The ready availability of pumps, motors, valves,
etc, equipped for electronic control, provides
almost unlimited scope for innovation in the
automation of many aspects of underground machine
operation. It is now possible to have not only
electronic controls on engines or transmissions
but also to link the controls of various
subsystems so as to either perform various
functions automatically or to reduce the number of
functions that the driver has to control manually.

One of the problems in developing such control
systems is the availability of suitable
tranducers. Fortunately, this need has been
recognised and considerable research is now being
done on the development of highly stable, robust
transducers suitable for use in such control
systems. In some cases, these transducers are
designed for use with digital controls, but many
are basically analogue in form. Although it is to
be expected that microprocessors will usually do
the actual controlling, at the moment it appears
that they will continue to comunicate with their
transducers and actuators via A to D and D to A
links for some time into the future, and in some
cases analogue controllers are quite able to cope,
thus eliminating the need for the A to D and D to
A conversions.

Although there is little doubt that electronic
equipment can survive the operating conditions
that occur on mobile machinery underground,
considerable care will be required if its
introduction is to be successful from the start.
In essence, this means considerable attention to
detail. Semiconductor junctions must be kept
cooler than 160°C and ambient conditions above 70°
C are not recommended for large scale integrated
circuits. Digital systems are prone to
interference from electromagnetic radiation, and
it is common for maintenance operations on an
underground vehicle to involve steam cleaning,
oxycutting and electric arc welding. This means
that, if an electronic control system is to
survive, the transducers, cables and the control
system all need to be made of suitable materials
and appropriate precautions should be taken to
protect them from their inhospitable environment.
Similar, though probably less severe problems
arise in surface vehicles and there is presently a
considerable effort being made to develop fibre
optic communication systems for surface vehicles
(Wearden, 1984).

Another problem associated with the introduction of
electronics underground is that of troubleshooting.
Many mines currently have little or no electronic
equipment underground, and hence have virtually no
suitably qualified technicians available to service
such equipment. In these circumstances, it is
desirable that equipment fitted to vehicles should
be designed to provide a high level of self
diagnostic capability. It is anticipated that this
should include testing of transducers, circuit
board functions and overall system performance
functions but should not go to the component level
on the circuit board. In other words, the
serviceman should be able to repair the machine by
replacing transducers or complete circuit boards.

A potential side benefit of the introduction of
electronic control is that it makes possible the
automatic keeping of detailed performance records
of each individual machine. This can be a useful
management tool (Flick and Salinger 1982).

Wheel slip control is an area where automatic
controls can provide significant benefits. By
sensing wheel speed and controlling wheel torque,
it is possible to maximize wheel thrust whilst
minimizing wheel slip provided there is a suitable
reference speed available. This reference could be
from a free rolling or undriven wheel or
alternatively from a true ground speed sensor such
as that reported by Tsuka and others (1982). In
view of the high cost of tyres used on underground
equipment, such a system has the potential for
substantial savings.

OTHER SYSTEMS

Braking

Traditionally, underground vehicles have been
fitted with air operated starting and braking
systems. These have necessitated the carrying of
quite large air receivers, which have tended to
obstruct maintenance access to the machines. With
the increasing use of hydraulic systems on
vehicles, and the decrease in the availability of
air supplies distributed throughout mines, there is
a decreasing need to continue with air operated
systems. Fully hydraulic power braking systems are
now available, which can operate from auxiliary
hydraulic pumps likely to be found on most mining
machines. Although a hydraulic accumulator is
necessary to provide fail safe operation, these
systems provide a much less bulky and simpler
braking system than with air.

Steering Systems

Several manufacturers have recently released rapid-
response swash-plate type axial piston pumps fitted
with controls capable of providing constant flow on
almost instant demand with virtually instant
compensation to zero flow when the demand ceases.
These pumps are intended for use in steering
systems where it is required to have a constant
response regardless of engine speed. Use of such
pumps will not only save energy but will also
eliminate the common problem of steering response
falling as engine speed falls.

Other steering developments include the
availability of flow amplifiers suitable for use in
the steering systems of large vehicles, which have
hitherto been poorly catered for. Electric
proportional control valves are also now available
in sizes suitable for applications of this type,
facilitating the use of electric joystick controls.

UNMANNED AND AUTOMATIC OPERATION.

At present, a considerable amount of time and money is spent on making safe the access and the areas actually being mined underground. It is not hard to imagine the development of mining robots capable of performing many of the functions currently performed by men. However, there are many potential problems to be overcome before such ideas can become reality. The need to recover petroleum and other minerals from the ocean bed has stimulated the development of much specialized equipment capable of remote operation. As far as is known, virtually none of this equipment is entirely robotic in operation. This seems to be principally because few of the operations are either repetitive or sufficiently uniform to be readily able to be carried out by a computer controlled system. Very significant advances in machine vision systems and in associated 'intelligent' computers will be required before truly robot mining machines will be developed. In the meantime, there appears to be considerable scope for the development of machines that can be operated safely by humans at a remote location.

Remote control obviously has the potential to solve most of the ergonomic and safety problems associated with the operator riding on the machine, but it does introduce a number of new ones; principally those relating to the operator's knowing what the machine is doing. At present there are several Load Haul Dump vehicles being operated via radio links at a range of about 100 metres. These systems have proved to be very successful in economic terms, as they have enabled the recovery of ore from unsafe areas where it was previously irretrievable. The major disadvantage of the present systems is that they do not provide the driver with any visual information as to precisely what the machine is doing. This tends to severely limit productivity.

In many respects, it appears at this stage that there is much more to be gained by automating jobs rather than trying to produce robot machines. In other words, it appears that significant benefits can be had by removing the tedious tasks from the operator whilst leaving him with the overall control of the system. An example of this is the automatic blade levelling control system used on laser land levelling systems. This type of system increases productivity because it relieves the operator of a very demanding task, and it tends to improve the accuracy with which the task is performed by eliminating its dependence on operator skill. Automation of at least the location and direction aspects of the drilling of holes for blasting is an example of this.

The use of electronic control to provide automatic adaptation or compensation for wear or changing working conditions is also likely to be worthwile. This could involve nothing more complex than automatic adjustment of the working speed of a machine so as to maintain a constant demand on the power plant.

Automatic vehicle guidance systems offer another form of driverless remote control. Numerous techniques have been used, however they usually consist of an induction cable with a low frequency current passing through it, laid along the proposed route so that vehicle mounted pickup coils can detect the presence of the cable and control the vehicle's steering so as to minimize the deviation. Control over the speed and other operations of the vehicle can be effected by signals sent through the cable system, thus providing fully automatic control (Chiba, 1982). The need to lay the cable along the desired route

eliminates wire following as a method of guidance in unsafe areas unless a method can be devised to lay the wire in the unsafe area under remote control. Possible alternatives include the use of a laser beam or perhaps microwave units to determine the position of the vehicle in the unsafe area. By using such a guidance system in conjunction with a TV camera, it may be possible to significantly increase the productivity of remote controlled machines since the operator would be relieved of the difficult task of steering the vehicle.

CONCLUSION

Substantial gains in efficiency and productivity of underground mobile equipment can be obtained by applying existing advanced technology. Recently published research results give reason to expect further substantial improvements in the near future, but these are unlikely to be realised unless a broad systems approach is taken to future designs. Significant improvements in the ergonomics of vehicles are within reach and substantial job simplification can be expected with more widespread application of automatic control. Although the advances in diesel engine technology have been dramatic, a break-through in battery technology would undoubtedly lead to a revolution in the design of underground mobile equipment.

REFERENCES

Anon. (1983). Modifying Exhaust Outlet Reduces Vehicle Operator Exposure to Diesel Exhaust. Technology News. US Bureau of Mines, 168, Feb.

Bell, J. (1984). Diesels lose their cool. New Scientist. Jan. 18.

Blevins, J.R., & Stephan, J.J. (1982). Truck Electronic Engine Controls. SAE Paper 820907. (TRW)

Bullock, K.J., Bandopadhayay, P.C., Daniel, P.C. and Staib, K. (1981). Drive Train Alternatives for Underground Load Haul Dump Vehicles. I.E.Aust. Trans.Mech Eng. 124-129.

Chiba, J. (1982). Electronic Control system for Construction Equipment. SAE Paper 820921. (Komatsu)

Day, E. & Frank, H.L. (1982). Electronic Diesel Fuel Controls. SAE Paper 820905. (Cummins)

Du Russell, E.N. & Muirhead, J.L. (1981). Development of Electrical Equipment at the Fox Mine. CIM Reporter. May 4.

Edwards, B.D. (1982). Energy Storage Systems - Worldwide. SAE Paper 820899. (Lucas - Chloride)

Evans, P.A. & Karlsson, A. (1981). The Volvo City Bus. I Mech E. C157/81 143-151.

Farber, A.S. (1982). Electronic Transmission Controls for Off-Highway Applications. SAE Paper 820920. (John Deere)

Flemming, W.J., & Wood, P.W. (1982). Noncontact Miniature Torque Sensor for Automotive Application. SAE Paper 820206.

Flick, J.F., & Salinger, J.A. (1982). Vehicle Mounted Management Information Systems. SAE Paper 820908. (Rockwell)

Geppert, S. (1980). AC Propulsion System for an Electric Vehicle. Intl. Conference on Transportation Electronics, IEEE 80CH1601-4 Paper C-2,2.

Komiyama, K., Okazaki, T., Hashimoto, H. & Takase, K., (1981). Electronically controlled high Pressure Injection System for Heavy Duty Diesel Engine - KOMPICS. SAE Paper 810997. (Komatsu)

Lawson, A. (1981). Progress in the Control of Underground Diesel Emissions. CIM Bulletin Nov 74,835 68-73.

Lebel, P. (1983). Methanol/Diesel Fuel Combination Can Increase Mine Productivity. Can. Mining J. May 38-39.

Magnusson, I. (1979). Energy Consumption, where does the money go? World Mining. April.

Reams, L.A., Wiemero, T.A., Levin, M.B., & Wade, W.R. (1982). Capabilities of Diesel Electronic Fuel Control. SAE Paper 820449. (Ford)

Regar, K.N. (1981). Stepless transmissions with hydrostatic power branch for energy saving traction drives. I Mech E. C153/81 113-118.

Salee,J. (1982). Engine Selection for Underground Use. World Mining. Nov. D34-D36.

Trenne, M.V., & Ives, A.P.(1982). Closed Loop Design for Electronic Diesel Injection Systems. SAE Paper 820447. (TRW, Lucas CAV)

Tsuka, W., McConnell, A.M., & Witt, P.A. (1982). Radar True Ground Speed Sensor for Agricultural and Off Road Equipment. SAE Paper 821059.

Tunley, J.D., & Preston, K.S. (1981). A Hydrostatic Transmission for a Railway Passenger Vehicle. I Mech E. C155/81 127-136.

Wallace, F.J., Tarabad, M., & Howard, D. (1983). The Differential Compound Engine - A New Integrated Engine Transmission System for Heavy Vehicles. Proc Instn Mech Engrs. 197A.

Watson, N. (1979). Turbochargers for the 1980's - Current Trends & Future Prospects. SAE Paper 790063.

Wearden, T. (1983). Automotive Electronics '83. Automotive Engr. 8,6 Dec. 68-70.

Wearden, T. (1984). Fibre optics for Automotive Applications. Automotive Engr. 9,3 June/July 64-67.

Wearden, T. (1984). UK Research in Automotive Electronics. Automotive Engr. 8,7 Feb/Mar 58-59.

Wheeler, W. (1978). Precision Position-Sensors in Automotive Application. SAE Paper 780209.

Zachary, A.T., & Ginsburg, B.R. (1982). Flywheel Power Module. 17th Intersociety Energy Conversion Engineering Conference, Los Angeles Aug.

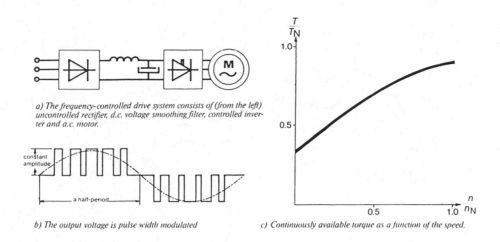

a) The frequency-controlled drive system consists of (from the left) uncontrolled rectifier, d.c. voltage smoothing filter, controlled inverter and a.c. motor.

b) The output voltage is pulse width modulated

c) Continuously available torque as a function of the speed.

FIGURE 1   Design, function and characteristics of a frequency converter (ASEA)

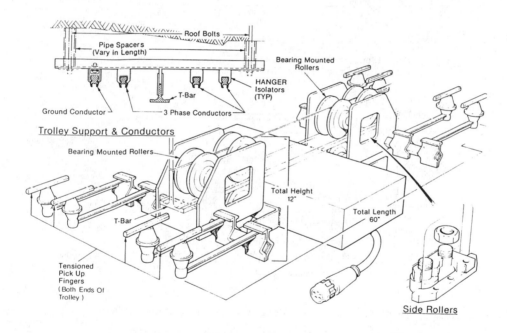

FIGURE 2   The Travelling Plug (Jarvis Clark)

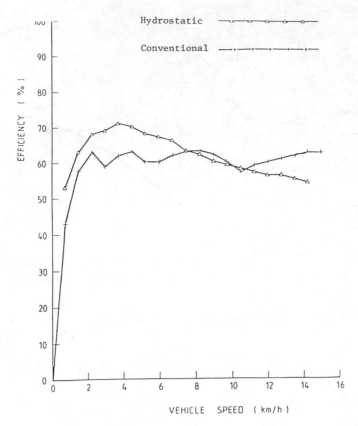

<u>FIGURE 3</u>   Comparisons of the efficiencies of hydrostatic and torque
             converter/4 speed power shift transmissions

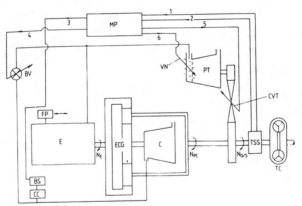

<u>FIGURE 4</u>   The Differential Compound Engine Concept (Wallace, 1983)

BV bypass valve; BS boost sensor; C compressor; CC charge cooler; E semi-
adiabatic engine; ECG epicyclic gear train; FP fuel pump; PT power turbine; TC
torque converter; VN variable turbine nozzles; TSS output torque and speed
sensor; N_E engine speed; $N_{o/s}$ output shaft speed; N_{pc} planet carrier speed; MP
micropressor;
Input signals: 1 torque transducer; 2 speed transducer; 3 boost transducer
Output signals: 4 bypass valve control; 5 CVT control; 6 nozzle control

Future Applications of Computers in the Design and Control of Mineral Beneficiation Circuits

D.J. McKEE

Julius Kruttschnitt Mineral Research Centre, University of Queensland, Brisbane, Queensland, Australia

Abstract. Twenty years of effort in developing the use of computers
in the design and control of mineral treatment circuits has resulted
in a sound base for future widespread application of the techniques.
The intent of this paper is to speculate to some extent about future
applications of computer related techniques in the mineral
extraction industries. The use of simulation methods for the
analysis and design of circuits is discussed. In the area of process
control, computers have been largely devoted to process monitoring
and reporting tasks. The application of computers for the total
control and optimization of complete plants is considered. Finally,
the role of computer based modelling studies for definitions of
optimum treatment flowsheets is suggested.

Keywords. Computer applications; mineral beneficiation circuits
design; control.

INTRODUCTION

After more than twenty years of effort, a range of
computer based techniques has been developed which
is applicable to the mineral beneficiation
industries. The techniques span the areas of
mathematical modelling and simulation, process
control, on line data acquisition and general
information systems for management. While the
degree of successful application of the techniques
in industry has varied widely, it is reasonable to
conclude that the potential for widespread
application of these methods is considerable.

The intention of this paper is to look to the near
future and consider the ways in which the computer
methods could and should be used in the area of
mineral beneficiation. To illustrate the ideas,
those stages of the development and operation of a
new treatment plant which can be associated with
computer based techniques are considered. Further,
a complex sulphide type ore has been selected, as
it is clear that this type of deposit, within
Australia at least, is now becoming very important
as a potential source of future mineral
production.

THE CHALLENGE IN DEVELOPING COMPLEX OREBODIES

There are increasing numbers of relatively small,
high grade sulphide deposits being identified in
Australia and overseas. The definition of small
is quite arbitrary, but can be considered to be
less than an ore reserve of 15–20 million tonnes.
The deposits are widely known as complex
sulphides, as they invariably possess two
qualities:

i. high grade, usually containing two or more
   base metals and precious metals (eg. Cu, Pb,
   Zn, Ag, Au).

ii. the mineralogy is complex, usually of a fine
    grained nature, and subsequent metallurgical
    treatment is difficult.

In such ores, the contained metal value is high,
and thus there is immediate economic incentive to
develop the deposit. However, conventional
treatment methods of comminution followed by
differential flotation frequently result in poor
grades and recoveries. The small ore reserve
dictates a relatively short life of the
operation, and the operators do not have years to
experiment and develop suitable treatment methods
on line. In short, while the apparent economic
attractiveness of small, complex sulphide deposits
is high, the barriers to successful development
and profitable operation are frequently immense.
It is not surprising that those charged with the
responsibility of developing such reserves are now
looking beyond conventional methods and circuits.

Computer methods alone will not solve all the
problems. However, many of the techniques
developed are relevant to the problem of complex
sulphide ore treatment, and the challenge is to
develop the application of the techniques.

A number of the steps involved in developing a
metallurgical operation for a new ore body are
detailed in Figure 1. Some of the possible
computer applications associated with each stage
of the development are also indicated, and it is
apparent that there is considerable potential for
computer techniques.

Four of the most important areas are listed below.

i. development/specification of optimum flowsheets,

ii. use of simulation for flowsheet design,

iii. on line data acquisition, process control and
     optimisation, and

iv. sophisticated methods of operator training.

Each of these topics is elaborated upon in the
ensuing sections of this paper. In some cases, the

basic work and methods exist, and it is really a case of applying existing technology. In other instances, much of the basic research must still be undertaken.

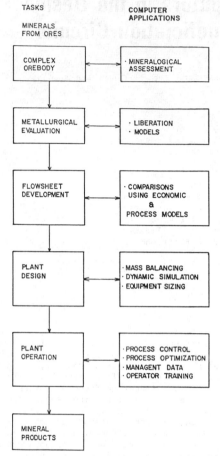

TASKS

MINERALS
FROM ORES

COMPUTER
APPLICATIONS

COMPLEX
OREBODY → · MINERALOGICAL
ASSESSMENT

METALLURGICAL
EVALUATION → · LIBERATION
· MODELS

FLOWSHEET
DEVELOPMENT → · COMPARISONS
USING ECONOMIC
&
PROCESS MODELS

PLANT
DESIGN → · MASS BALANCING
· DYNAMIC SIMULATION
· EQUIPMENT SIZING

PLANT
OPERATION → · PROCESS CONTROL
· PROCESS OPTIMIZATION
· MANAGENT DATA
· OPERATOR TRAINING

MINERAL
PRODUCTS

Figure 1   Computer applications involved
in the different stages
leading to mineral production from a complex orebody.

DEVELOPMENT OF OPTIMUM FLOWSHEETS

For the purpose of discussion, the task is to develop a flowsheet for a complex copper-lead-zinc orebody. The conventional beneficiation approach for the ore would use differential flotation to produce separate concentrates of copper, lead and zinc. If these concentrates were destined for conventional smelting, then the following grades would be necessary: Cu, 20%+; Pb, 50%+; Zn, 50%+. With many such complex ores, it is only possible to achieve these grades at quite unacceptably low mineral recoveries. As an added disincentive, because of the generally fine grained nature of such ores, it is necessary to grind very fine, with flotation feed sizings of 80% passing 40 microns not uncommon. The cost of such fine grinding becomes a very significant factor in total and operating costs.

Faced with these problems, there appear to be two possible approaches:

(i)   improve the efficiency and selectivity of the
flotation process, or

(ii) accept that flotation will not achieve conventional concentrate recoveries, and explore methods or integrated processes which use more than one beneficiation method.

Research on the problems of flotation efficiency for such ores is being actively pursued by many groups and it is not the intention to comment on that work.

There is now increasing interest in the integrated process approach, and there will be increasing research interest on the general topic. In the broad spectrum, an overall approach which utilizes comminution, flotation, hydro metallurgy and pyro metallurgy is feasible. The objective of such an approach is to maximise the overall treatment efficiency and economic return by best matching the characteristics of the unit processes to particular mineral components in the ore and the required final product specification.

The integrated treatment approach used could rapidly become much more complicated for two reasons. The abovementioned unit processes are certainly not exhaustive, and heavy media separation, gravity concentration and other processes may have a role in a particular integrated process. There is also the problem of splitting the original feed in the optimum manner so that each process is treating that component of the feed which it is best able to concentrate.

There are clearly major problems in turning such an integrated process concept into a reality. It is possible to identify two major areas for consideration. The first is concerned with the metallurgical performance of individual processes and matching the processes to the particular characteristics of the ore. The second consideration is the absolute necessity to develop a rigorous method for assessing the performance, both metallurgical and economic, of various treatment processes. There is scope for computer analysis and application in the metallurgical areas. The question of economic assessment of various treatment approaches is ideally suited to computer analysis.

The fundamental question can be simply asked. Is one flowsheet, involving one set of processes, capital and operating costs and one metallurgical performance level superior to another treatment approach with a different set of costs and metallurgical performance level? Further, it is necessary to assess the impact of different product grades from the integrated process on subsequent downstream processes. The answer to this question is likely to be very complex, yet it is quite critical in the final analysis.

The role of computer analysis here is to assist in answering such questions. The models required will contain both metallurgical performance and economic components. It is envisaged that such analysis would proceed in parallel with the traditional metallurgical testwork. Some of the components of the approach are available in other disciplines, but the task remains to develop the approach for use in the metallurgical flowsheet development for complex ores.

As an example of the type of processing alternatives available, the comminution stage of the operation may be considered. Two possible flowsheets are suggested in Figure 2. If it is assumed that the ore is amenable to both autogenous grinding and heavy media concentration, then the choice between the comminution flowsheets of options 1 and 2 will be complex and will involve an assessment of both the capital and operating costs of the two flowsheets, as well as considerations of the metallurgical performance of the different options. There is a clear role for combined economic and metallurgical models to assist in defining the optimum flowsheet. One of the important advantages of the computer analysis is the ease with which sensitivity studies, both economic and metallurgical, can be performed.

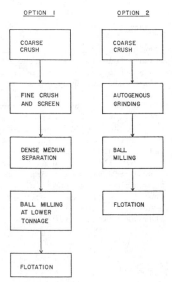

Figure 2 Two possible flowsheet
          approaches for comminution.

USE OF SIMULATION FOR FLOWSHEET DESIGN

Having developed a conceptual flowsheet for a new ore, the next task involves the detailed design of the circuit, including the selection and sizing of process equipment. Computer techniques are being used for this task, but it is considered that there remains wide scope for far more detailed application of process models and simulation in the design stage.

In the normal mineral processing areas, suitable mathematical models of most of the processes of interest are available. These include models of crushing, grinding, screening, classifying, flotation and to a lesser extent, dewatering. These models all provide predictive capability.

Currently, most computer related flowsheet work is involved with the generation of material balances for often very complicated flowsheets. A number of packages are available and a typical one is the Flexmet system (Richardson et al, 1981). While it is possible to incorporate detailed process models in such packages, generally the user specifies the split of product achieved in a particular unit. For example, the size fraction and solids and water splits from a cyclone are usually specified, rather than calculated by a process model.

Far more powerful simulation studies can be performed if true predictive models are applied. Such models take into account the feed characteristics of the material, as well as the size and operating parameters of the process unit. Typical examples of the use of steady state simulation models for flowsheet design problems have been outlined by Kavetsky and McKee (1984) where a grinding circuit study and a fine cyclone classification circuit were assessed using simulation.

A further use of mathematical models and simulation in design is in the prediction of

grinding mill sizes using information derived from laboratory breakage tests on the particular ore. Such tests have been described by Narayanan and Whiten (1983) and the method provides an alternative to the empirical design procedures based on the Bond Work Index approach. With comprehensive mathematical models available of most of the unit processes found in mineral beneficiation, the capability exists to greatly expand the reliability and accuracy of design methods.

Similar work on the use of models for the analysis and design of grinding circuits has been carried out by Herbst and Fuerstenau (1980).

A still more powerful application of computers in design studies is the use of dynamic simulation to analyse a proposed circuit flowsheet. When coupled with comprehensive process models, it is possible to assess potential flow constraints in a design and to test a variety of control strategies. The development of a general purpose dynamic simulation package for design purposes is currently being undertaken at the JKMRC.

Many of the process models have been available for many years. The widespread use of simulation in design has been hampered by the fact that the models and simulation packages have been difficult to use. In practice, only specialists highly familiar with the software were capable of applying the methods. Major effort is now being devoted to the task of developing simulators which are designed for use by engineers who are not necessarily modelling and simulation experts. One such system has been described by Hess and Wiseman (1984) and others are being developed. The use of such simulations for flowsheet design should become increasingly commonplace.

DATA ACQUISITION, PROCESS CONTROL
AND OPTIMISATION

The application of computers for on line process control was pioneered in grinding control applications over 15 years ago. Such installations are now commonplace. Automatic control of flotation circuits has proceeded much more slowly, mainly because the task of controlling a flotation operation has proved very difficult.

Automatic control of hydro and pyro metallurgical processes makes extensive use of process control computers. The task involves monitoring of large numbers of process variables, and implementing flow, temperature and pressure control loops. Optimizing control of these processes is an active area of research and development.

The control problem associated with an integrated process flowsheet for the complex ore referred to earlier will be challenging and critical. The control task will involve two separate functions:

i.   stabilizing the various unit processes, and

ii. optimizing total plant performance from an
     economic view.

It is in the latter area that computer based control systems will have a critical role. Just as the ability to select the most economically efficient flowsheet was a critical aspect of the original circuit specification, so will be an accurate knowledge of the economics associated with operating the plant to a particular set of targets. It will be extremely difficult to establish operating targets unless a model of the total plant is developed which can quantify the effect of a change in targets in one part of the plant on the total plant economics.

It is envisaged that a combined process and economic model of the entire circuit will run on-line in parallel with the plant. Data inputs will be available from plant instruments and less frequently, from shift and daily results. Cost and price factors will be incorporated and it will be necessary for the model to calculate the economic performance on a regular basis, as well as having the ability to predict the effect of a change in one part of the circuit on the total result. The model will provide targets for management, plant operators and the control system.

The sophistication of the control loops is likely to be less important than the ability to set the best operating targets for the various sections of the plant. A variety of modern process control techniques are now being developed and tested in mineral processing plants, and many of these will find application.

A block diagram of such an overall control system is sketched in Figure 3. The heart of the system is the on-line process and economic model which will provide data for management, allow management to test the effect of different plant operating targets and strategies, and finally will provide the necessary input information for the optimizing control calculations.

Development of such a system will be a major task, because it will also be necessary to specify constraints on many process variables, so that process alarms will occur automatically. Simulator based training will complement expensive on the job training and it will decrease the requirement for the latter. The standard of training should rise, and this will certainly be beneficial, particularly as the on-the-job approach is often poorly done.

CONCLUSIONS

Computer based technology is set to play an increasingly important role in the development and operation of new mineral beneficiation operations. Using a complex sulphide orebody as an example, four distinct computer applications have been identified. The first two of these are associated with the specification of flowsheets and actual flowsheet design. The other applications are those of optimization and control and operator training. The metallurgical problems associated with economic treatment of the new high grade, low tonnage complex sulphide orebodies are such that it is likely conventional methods of design, operation and control must be substantially improved, and a range of computer based methods offers very encouraging possibilities.

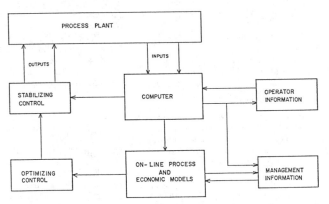

Figure 3   Block diagram of an integrated computer-based control system.

REFERENCES

In recent years more attention has been paid to providing operators with the best possible information in the most appropriate format. Although colour displays are now commonplace, there is certainly scope for improving the information presentation to make the operators task more simple and more efficient.

OPERATOR TRAINING

The use of process simulators for flowsheet design has been discussed. Simulators also possess potential for engineer and operator training. A number of groups are actively working on the development of dynamic simulators for training purposes, although this is largely related to the instruction of engineers where the emphasis is on technical aspects of the process.

A dynamic simulator based on a high resolution colour graphics terminal has the potential to become a powerful training aid. If this is coupled with some form of film or slide presentation of relevant physical equipment and incidents involving that equipment, then the basis for a highly realistic training simulator will exist.

Herbst, J.A. and Fuerstenau, D.W., (1980). Scale-up procedure for continuous grinding mill design using population balance models. Int. Jnl. Min. Proc., 7, 1-31.

Hess, F.W. and Wiseman, D.W., (1984). An interactive colour graphics process analyser and simulator for mineral concentrators. In Application of Computers and Mathematics in the Mineral Industries, Inst. Min. and Met., London. pp. 235-244.

Kavetsky, A. and McKee, D.J., (1984). Analysis and design of industrial grinding and classification circuit by use of computer simulation. In Application of Computers and Mathematics in the Mineral Industries, Inst. Min. and Met., London., pp. 57-68

Narayanan, S.S. and Whiten, W.J., (1983). Breakage characteristics of ores. Proc. Aust. Inst. Min. and Met., 286, 31-39.

Richardson, J.M., Coles, D.R., and White, J.W., (1981). Flexmet - a computer-aided and flexible metallurgical technique for steady-state flowsheet analysis. Eng. and Min. Jnl., Oct. pp. 88-97.

The Development of Instrumentation and its Impact on the Mineral Processing Industry

M.H. KERR

Senior Instrumentation Engineer, MIM Holdings Limited, Brisbane, Australia

Abstract. The origins of instrument application in the Mineral Industry are discussed. Development of measurement techniques pertinent to the industry is described as is the early application of mini-computers. The current application of microprocessor based equipment and the benefits obtained from this range of equipment are discussed. A speculative opinion as to what future instrument development may offer the industry is outlined.

Keywords. Metallurgical industries; computer control; density control; human factors; modelling; mineral processing.

INTRODUCTION

The ability to measure and control parameters accurately in processes has been the key to successful automation in industry for many years. The mineral processing industry, in the past, has not been renowned as a pioneer of automation for many and varied justifiable reasons.

Today, this reputation of conservatism in utilizing automation is no longer valid. Each application of automation has permitted small but acceptable increases in metal recovery which, due to the large tonnages of ore handled by the processes, has resulted in very attractive financial returns. In some instances, it has also permitted reduction in operating staff.

Consequently, investment in automation has been encouraged by management.

EARLY APPLICATION OF INSTRUMENTATION

In the early 1960's, instrumentation in the Mining Industry was very rudimentary. Most measurement techniques available at this time had been developed for other industries, predominantly the Petrochemical Industry, and many were not suitable or reliable for measurement and control in the harsh mining environment. Additionally, the industry was not renowned for pioneering automation, with the result that it could not attract what was, at this time, a very limited number of qualified and experienced Instrument Engineers. The consequence of this was a compounding problem of unsuitable instruments being poorly applied by inexperienced people; a guaranteed recipe for disaster.

Even today, there are still insufficient Instrument Engineers who have had the necessary experience in our industry. This early experience of poor results has to some extent had a lasting effect in the industry.

The parameters that were most commonly being measured and/or controlled at this time were:-

Water addition to mill feeds.

Ore mass to the grinding lines.

Level in the cyclone feed sumps.

Level in the flotation machine banks.

Reagent addition was generally being manually regulated by means of rotameters.

The actual controller instruments were large case instruments with pneumatic action housed in field located panels adjacent to the plant equipment they were monitoring or controlling.

Major plant disturbances were detected by means of laboratory analysis of manual samples, the results of which were available to the operator many hours later. In fact, it would be true to say that, up to the mid 1960's, control of both the grinding and flotation processes was more of an art than a science.

DEVELOPMENT OF MEASUREMENT TECHNIQUES

Measurement techniques that were developed in the mid to late 1960's and were significant to mineral processing were:-

Mineral content in slurries (On-stream Analysis).

The early models of on-stream analysers were the X-ray fluorescence type. they are the type of analyser used by Mount Isa Mines Limited in their Copper and Lead/Zinc Concentrators. They are relatively complex and require a high level of technical competence to maintain them. They work on the principle of multi element analysis of up to sixteen samples, which are presented to the analyser. This involves the pumping of representative samples to the centrally located analyser and the dispersal of the samples after they have been through the analyser. This means that each stream is analysed about once every 45 minutes. To be cost effective, these systems need to be applied in relatively large plants where the mineral stream is being analysed throughout the grinding and flotation process.

Today, analysers are available that permit single element in situ analysis. They permit the new user to gain the benefits of real time analysis for a much smaller investment. This type of system can be easily expanded to meet the needs of the process concerned. Due to the analysing probe being sited in the sample stream, they are referred to as In-Stream Analysers.

Density

The early method of measuring density involved measurement of the differential pressure between two vertically separated pressure sensing probes. However, this method required very careful application to ensure that only pulp density variation caused pressure changes to the probes, i.e. the pulp stream needed to be relatively slowly flowing around the probes, but not slow enough to permit saltation of heavy particles. Also, the stream velocity needed to be steady, fluctuations would not necessarily affect the probes evenly. This method of density measurement was open to many sources of errors and was a very high maintenance item.

The introduction of the nuclear radiation density gauge overcame many of the foregoing problems. The gauge consists of three components: a source head, a detector, and an electronics unit. The source head and detector are installed at a convenient point on a pipeline. Gamma rays from the source pass through the walls of the pipe and the slurry and are absorbed in proportion to the material density. The rays reaching the detector produce a current, which is inversely related to the slurry density. This current signal forms the input to the electronics unit, which contains the necessary power supplies, amplifiers and signal handling circuitry (Sigal 1974).

Today, the gauge still consists of the three components of source, detector and electronics unit. However, technological development has significantly changed the electronics unit used in these gauges. The modern density gauge uses a microprocessor together with a scintillation counter, thus permitting the use of much smaller radiation sources, in most applications, resulting in a very accurate, reliable and easily maintained gauge.

Magnetic Flowmeters

Magnetic flowmeters used for slurry flow applications differ from standard magnetic flowmeters only in the increased thickness of their linings and in the choice of material used for the electrodes. The magnetic flowmeter is used in conjunction with a field-mounted converter, which provides on-site range adjustment facilities and generates a current output signal (e.g. 4-20mA) proportional to the volumetric flowrate through the meter. The output of a magnetic flowmeter is independent of density variations in the metered slurry.

However, the operation and accuracy of magnetic flowmeters are seriously affected if the measured slurry contains significant or widely varying percentages of magnetic particles (Sigal 1974). Once again, development of the electronics has resulted in a more accurate, reliable and easier maintained instrument. Thus magnetic flowmeters used in conjunction with nuclear density gauges permit the calculation of slurry mass flow.

These measurements opened up dimensions of control that had hitherto been impossible in this industry. Concurrently, significant research took place in the development of mathematical models of mineral processing equipment. This was successfully applied in a number of operating plants. A local example of this research and development was carried out by the Department of

Mining and Metallurgical Engineering at the Queensland University, in The Julius Kruttschnitt Mineral Research Centre. Mathematical models were developed that resulted in automatic control strategies being successfully applied to both the crushing and grinding processes at Mount Isa.

During the past fifteen years, further refinement and additional research (Lynch and Elber 1980) has resulted in many more plants applying automation to their mineral processing.

EARLY APPLICATION OF COMPUTERS

It was coincident that the late 1960's and the early 1970's heralded the introduction of cost effective mini digital computers, which undoubtedly accelerated the development of automation in the industry.

Prior to this, there was a short period when analog computers were used (Dredge 1969). Although they were an improvement on conventional analog controllers, they were difficult and expensive to reprogramme when changes were necessary, consequently, their use was short lived.

Since the introduction of mini computers, there have been major improvements in both the hardware and software i.e.

The cost to performance ratio has been greatly improved.

The physical size has been considerably reduced.

The relative cost has dramatically reduced.

The choice of different computers increased.

The range of available software has greatly increased, and thus, the ease of use of the computers significantly improved.

These improvements have resulted in the mini computer becoming an indispensible tool to research and development.

CURRENT APPLICATION OF MICROPROCESSOR BASED EQUIPMENT

Today, the microprocessor has more computational power than the original mini computers, and many instrument manufacturers are marketing microprocessor-based instruments (Krigman 1984, Laduzinsky 1984, Ledgerwood 1984). These range from complete instrument control systems, capable of controlling large complex process plants, to individual instruments such as differential pressure transmitters, belt weighers, recorders, density gauges, single loop controllers, to mention a few. These instruments have many built-in features, which were either too difficult, thus too expensive, or even impossible to implement with analog techniques. Namely:-

Extensive and easy to configure algorithms in controllers.

Extensive built-in self diagnostics.

Transmitters with ability for remote calibration.

Transmitters with built in temperature compensation and linearisation resulting in better accuracy and repeatability, and many more features.

In the mining industry, where process plants are remote and labour costs high, the use of distributed instrument control systems, which use cathode ray tube (C.R.T.) displays, are particularly attractive. They require smaller control rooms, less signal and control cable due to the distributed input/output signal devices, are easier to maintain due to built-in diagnostics, and are more reliable. They also give the operator far more meaningful information and permit the implementation of more advanced control, where small efficiencies can be gained, but due to the large tonnages that are processed, result in significant cost benefits.

THE BENEFITS OBTAINED FROM INSTRUMENTATION

The increased automation of mineral processing plants has given the industry benefits, which fall into two categories:-

Tangible benefits.
Intangible benefits.

Tangible benefits.

: Increased recovery of finished product.

: Reduction in manpower requirements.

: Increased throughput of product.

: Deferment in capital equipment.

: Reduction in raw material or fuel etc.

the above are the major tangible benefits and, of course, are not necessarily applicable in all applications of automation but are more easily substantiated because they can be related back to a dollar value.

Intangible benefits.

: Provision of early warning of change in plant performance allowing fast response and thus limiting the amplitude and duration of change.

: Optimisation of plant performance.

: Increased visibility of operator and /or metallurgist actions and their results.

: Optimising maintenance scheduling due to improved records of equipment run times.

Once again, not all of the above benefits are applicable to all applications of automation, and the intangible benefits are not so easily substantiated and relatable to a dollar value. Where plants are fully automated, particularly with computers, there are many more benefits which can be gained.

WHAT DOES TOMORROW OFFER

The rapid development of microprocessors over recent years has caused an unprecedented proliferation of both instrument control systems and individual instruments and there is no sign of this development abating. Consequently, the future looks extremely favourable for inovative design of new systems and instruments with resultant benefits to instrument users.

The areas in which users can expect improvements to occur are:-

Greater intelligence in field located devices.

Faster and more secure communication data highways, with increased use of fibre optic techniques.

A common secure communication network which can handle many different requirements e.g. radio communication, closed circuit T.V., plant process information etc.

The ability to communicate economically between different control systems installed in different plants, permitting an economical company-wide management information system.

The ability to communicate with different plants from central locations or corporate headquarters for design or central maintenance purposes.

In fact, the possibilities are endless and are only bound by the imagination. Less than five years ago, the few items listed above would have been thought of as wishful thinking, but today they are very feasible. With the rate of development still increasing, who can confidently predict what the future offers?

CONCLUSION

Over the past two decades the development of automation in the industry commenced conservatively but accelerated to an acceptable level once credibility with management was established and technology permitted.

By inferential methods using simplified mathematical models (Lynch and Elber 1980), technological development has assisted in automating the industry where conventional methods are either too difficult, too costly or just impossible.

Development in instruments and allied equipment is continuing at a frenetic pace, which can only be to the benefit of our own industry. Our industry still has tremendous scope for automation. It is up to all of us to keep abreast of these developments, and to utilize them judiciously for the benefit of mankind.

REFERENCES

Sigal, P.M. (1974). Measuring slurries. Australian Process Engineering, pp 33-35.

Lynch, A.J., and Elber, L. (1980). Modelling and control of mineral processing plants. Paper presented to the 3rd IFAC Symposium, Montreal, Canada.

Dredge, K.H. (1969). Instrumentation and control in mineral dressing plant. Notes presented at the second residential school on mineral processing, University of Queensland, May 1968, Revised 1969. pp. 141-152.

Krigman, A. (1984). Distributed control pipe dreams to reality. Intech April. 1984. pp. 9-18.

Laduzinsky, A.J. (1984). Control business directions. Control Engineering February 1984. pp. 13.

Ledgerwood, B.K. (1984). Trends in control. Control Engineering February 1984. pp. 166.

Trends in Control System Equipment and Associated Data Networks

M.A. PEARCE

Foxboro Proprietary Limited, Maroondah Highway, Lilydale, Victoria 3140

Abstract. With the continuing move towards digital signal based control and
measurement, it is inevitable that such technology will be increasingly employed on
remote sites. The benefits of increased metallurgical performance, of reduced energy
and manpower requirements far outweigh the problems associated with high technology in
remote locations, provided that the installation is well thought out from both
technical and manpower viewpoints. The new accessibility of vast quantities of process
data will uncover previously unthought of savings and profit opportunities but brings
with it the responsibility to carefully design the data presentation methods. The
resulting changes in installed hardware will force changes in the maintenance and
engineering workforce and will affect even the stores and purchasing department
methods.

Keywords. Artificial intelligence; control system architechture; digital systems;
human factors; integrated plant control; management systems; technological forecasting.

1. INTRODUCTION

For the remote site control system user, the
issues to be faced are identical with those faced
everywhere:-

(1) Will the system increase the
profitability of the process

(2) Will the system be maintainable.

If the answer to both questions is 'yes', then the
system can be installed with confidence. The
factors to be considered when answering these
questions include the current performance of the
various types of equipment, their future
availability, and the availability of suitable
manpower to maintain the equipment. Will the
equipment be acceptable to the operators, plant
managers and engineers? How will the equipment
fit the organisational structure?

2. SYSTEM MAINTENANCE

The question of maintainability is perhaps the
easiest to answer, with consideration of both
hardware and people factors to be assessed.

Hardware maintainability is relatively simple to
answer. Look at the equipment together with the
manufacturers information and ask the question "Is
it reasonable to expect a maintenance technician
to handle this, and if so, how much training would
he need". By comparing each option against that
question, a good qualitative feel can be obtained
for the problem.

A more difficult question is "will the people be
available to maintain the equipment," and this
question is best answered by looking at the
current product types and the likely future
developments. These are the products for which
technical schools will be producing a skilled
workforce. For instance it is becoming unusual to
find a young instrument apprentice trained and
wanting to work with pneumatic equipment. Modern
control systems increasingly rely on
microprocessor and digital signal technology. It
is with this in mind that you should consider the
question of labour availability for system
maintenance and modification for the next 10 to 15
years.

In summary, microprocessor based controls and
computers are here to stay, and your future plans
will need to take this fact into account.

3. PROFITABILITY

The question of profitability is much harder to
answer. Many of the economic returns that accrue
are not obvious before installation, mainly
because the required process information is not
available. To develop an understanding of this
factor, we must first discuss the nature of the
control system.

3.1 Control System Characteristics

Control systems handle data. This data is of
two types; one being the hardwired field
signals, the other being visual data to and
from the operator. The interface between them
is formed by the control system itself.

The information flow indicated by the diagram above is common to all process control systems, small or large, with the hardwired field signals creating an essential difference between these and other data processing systems. In other systems, where all data is both input and used by people, the task is portable. Should the hardware fail, the problem can be overcome, if necessary, by transporting the task to, for instance, a bureau machine. By contrast, the process control system must work, and must work in 'real time'. This has lead to the development of robust hardware and software for the solution of process control problems.

Over the years, some especially effective concepts and methods of control have developed, these reflecting the demand for improved productivity and system reliability. The most common of these concepts is the traditional 2 or possibly 3 term controller [Proportional + Integral (P+I) or Proportional + Integral + Derivative (P+I+D)]. Despite many attempts to unseat it, this controller has proved to be the most versatile and robust of control methods, probably because it reflects, in a simplistic way, the basic characteristics of almost all controlled processes. Variations to this controller include non-linear gain and, more recently, expert self-tuning. It is confidently predicted that this controller, with its enhancements, will continue to be the mechanism by which the plant is ultimately manipulated. What is changing however, and what will continue to change, will be development of advanced control strategies that manipulate the setpoint of the P+I or P+I+D controller in order to achieve some higher level of performance. The concepts of advanced control are being actively pursued by the equipment manufacturers, by institutions such as the Julius Kruttschnitt Mineral Research Centre in Brisbane, and by end users of such equipment. These techniques typically draw together a number of inputs from a wide area of the process, and use this information to generate an 'intelligent' manipulation of the process control set points or control parameters.

This feature of wide data use is made possible by the nature of the microprocessor based control systems and is the source for much of the benefit to be obtained. Two examples of recently developed techniques of general applicability are described below.

3.2 Examples of Advanced Control Techniques

•   Hill Climbing

Simple hill climbing, or peak searching algorithms can now be easily implemented using perhaps 8% of the capacity of a multiloop controller, instead of 3 racks of hardwired equipment as used to be the case. This

technique can be used to search for peaks (or troughs) in one variable caused by variation of another. For instance, in an autogenous mill:

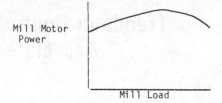

In this case, by changing the setpoint of the ore feed controller, the ore feed to the mill will vary, the peak motor load can be sought, and production then maintained at this peak. If the peak shifts due to an outside influence, then so will the production rate.

In seeking this peak, the controller is beginning to act in an 'intelligent' manner - it is seeking a performance peak in much the same manner as do operators, by reacting according to changes in process conditions.

•   Expert Self Tuning

It is now possible for a microprocessor based controller to review the performance of that control loop, to determine the quality of control achieved against a predetermined standard, and to calculate new tuning parameters as needed to maintain specified performance. In this way, without creating process upsets, the controller is maintaining loop performance in an 'Expert' manner - in other words, in the same way as would an instrument engineer when he 'tunes the loop'. This is another example of 'intelligent' control that depends upon the microprocessor's capacity; 13 years ago the use of a large computer was needed to perform the necessary calculations.

This type of selftuning control adapts not only to load dependent dynamic gain changes, but also to time related and/or random dynamic characteristic changes. Such applications include:

pH	-	variation caused by changing titration curves, electrode response etc.
Level	-	Dynamics change with plant throughput
Temperature	-	Heat exchanger fouling, variable heat source conditions
Flow	-	Pumps and valves wear, pipes foul up, etc.

Experience has shown that this type of controller can save many thousands of dollars per year in reagent costs and in energy expended.

3.3 Field Measurements

Of course, this new control capacity is wasted if the measurements are not accurate and reliable. Wherever possible, direct measurements of the variable to be controlled should be made. Inferential measurements are subject to too much outside interference (noise) in most 'real world' situations. The measurements should be 'on line' wherever possible in order to minimise time delays and to reduce the chances of error. By using the currently available measurement techniques for slurry density, concentrate and tailings assay, together with the magnetic flowmeter, many of the critical parameters can be measured directly for maximum economic benefit.

3.4 The Operator Interface

Having considered the field measurement and control problems, the remaining factor in the application of control systems is that of matching the information presentation method to the organisational needs. The collection and dissemination of data that is useless to the receiver is obviously counter productive, and is a real danger that is only now being effectively addressed.

Clearly, for a small system, up to, say, 10 loops of control, the quantity of data is small and is unlikely to have a large impact upon the organisation. The method of data presentation will be dictated by current operational practices and will range from local indications in the field at the most basic level, through discrete instruments in a control room panel, to a small Visual Display Unit (VDU) based operators console at the most sophisticated level. Which system is used will depend on many factors. There would be little point, for instance, in changing to the VDU system if the plant is designed for local indication and for operation by a roving operator, as potential savings would be unlikely to cover the installation costs.

For the larger plant , and for some smaller plants, advanced metallurgical calculations applied automatically to control the process conditions will create large gains. In these circumstances, the VDU based system is practical, and even necessary, in order to co-ordinate the plant operation. It does, however, bring with it some new issues - namely control room design and data availability. It is not the intention of this paper to discuss the control room design issues, other than to say that they are ergonomic, and are increasingly well understood. The issue of data availability is discussed in more detail below.

3.5 Communication Structures

With almost all VDU systems, a data network is implied. Typically this data network is designed to communicate data from the distributed control and data acquisition modules to and from the display processor. This network can be small or extensive, fast or slow. An example is shown schematically in Fig. 1. This network creates the potential for an information explosion which, if not well handled, will cause chaos in the user organisation.

For small networks, both the problems and the quantity of new data are correspondingly small. It is unlikely that a small network concerned with controlling, say, 30 loops, will have more than two consoles, and these are likely to be adjacent to one another for operational reliability reasons. This results in the screen(s) being under the direct

control of the operator, and they are likely to be optimised for his use. It is in the larger systems, where very large quantities of data become available, that problems can arise. These larger systems typically extend beyond the direct process control role, and data availability has not in the past been matched to the organisation's structure.

Figure 2 shows the typical data requirements of an operating plant, and the level of person in the company structure who is likely to use that data. Fig. 3 shows how the people and data typically interact in an organisation. If say, the plant manager now has data that was previously available only to the plant operator and his manager, two things are likely to occur:-

(1) The plant manager becomes involved in day to day 'nitty-gritty' details instead of plant management.

(2) The supervisors, engineers and operators feel threatened by this direct access to their performance.

Unless the well tried and tested organisational structure is to be changed (and there may be good reason for so doing) then the control system data structures must reflect the actual organisations needs both explicit and implicit.

The concept that is developing to accommodate this need is depicted in Fig. 4, where data is handed up the line to the next level. This data is 'owned' by the lower level system, and is conveyed up the line following manipulation to reflect the higher level's requirements. The computer system then reflects much more closely the command structure and traditional reporting roles of the user organisation, and therefore tends to be much more useful.

4. IMPLICATIONS FOR REMOTE SITES

How does all of this relate to remote sites, and what developments are foreseen?

Fig. 5 shows how a typical site system may look in the future, each computer network reflecting a definable level in the organisation and performing tasks relevant to that level. Information is displayed at VDU screens associated only with that network's computer (or control system) with data from adjacent systems being passed under control of the adjacent network's computers.

The positive aspect of this new data availability is the comparative ease with which energy and mass balances around the process can be accessed in real time. This leads to the possibility of implementing some very profitable co-ordination strategies between plant areas (say crushing, milling and flotation) and the likelihood of discovering opportunities for improved productivity and reduced losses.

From an implementation and maintenance viewpoint, it is unlikely that all hardware and software will be supplied from one source. It would be unusual for the supplier of the accounting computer to have any real knowledge of the control requirements of the mine, concentrator or smelter. The implications of this are:-

(1) To minimise training requirements,all hardware must be reliable and easy to repair, preferably by card replacement.

Equipment such as disc drives that do not respond well to this type of maintenance must be present, up and running on the installation, in sufficient numbers so that the loss of one unit will not kill the system. Skilled repair technicians must be available on short notice to fix the problem, so as to reduce the risk exposure due to operating with no available spare unit.

(2)   The hardware and software must be of a line that will be actively supported by the supplier for the foreseen life of the equipment, and for which you can confidently expect to employ knowledgeable technicians. This latter can be difficult at remote sites, especially if the total pool of suitable labour is small. In this case, a maintenance agreement with the supplier is of great benefit and can solve problems associated with the maintenance of skill levels of the site work force.

(3)   On-site technical support people will need to know computer and microprocessor based equipment sufficiently well to manage its 'front line' maintenance and modification. This includes the electrical control scheme, which will probably be PLC based, and the process control scheme, both of which will probably change with time. It is likely that the older barriers between trades will disappear, and that instead of having electrical and instrument

technicians, the maintenance work force will include people having an electrical background, some specialising in microprocessors, some in power distribution and electric motors, and some in field measurement devices.

(4)   The site workforce will be largely 'computer literate'.

(5)   On-site storage for high value, environmentally sensitive electronic components will be required, with appropriate security precautions.

These changes are imminent. Most organisations are using more and more computers, often at different levels in the organisation, largely reflecting the structure of Fig. 4, but without the communication networks. It is time to rationalise, to consider the data needs of the organisation, and to design for the inevitable.

References:

Beaverstock, M. C. (1984). Responsibility Oriented Systems for Plant-wide Control. The Foxboro Company Internal Paper.

Beaverstock, M. C. (1984). Management Style of Networks. ISA Conference, Houston.

Sawyer, R.D. (1984).Plant (Millwide) Automation. The Foxboro Company Internal Paper.

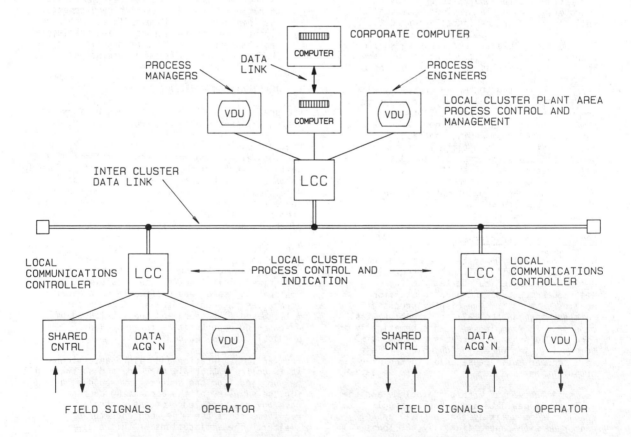

Fig. 1 Typical process control data network

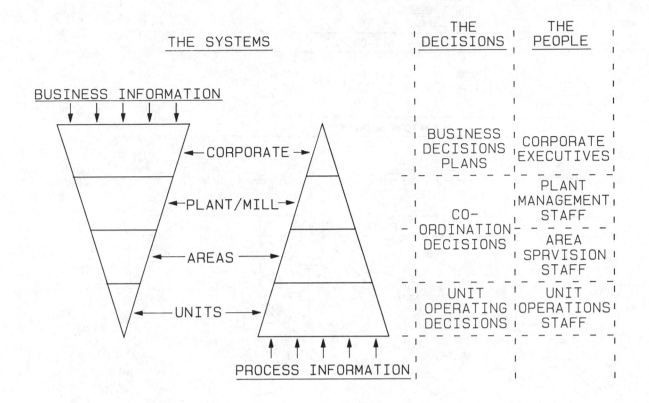

Fig. 2   Information flow requirements

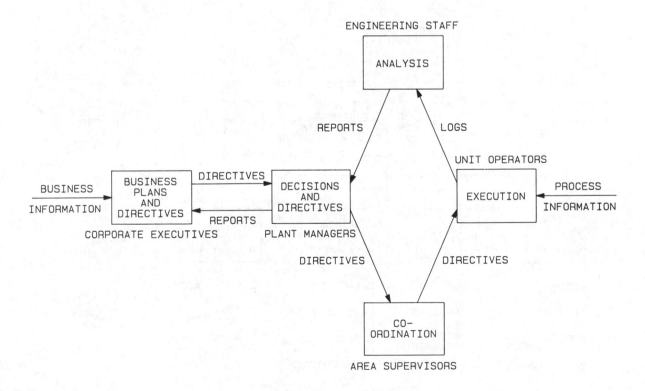

Fig. 3   Traditional plant roles

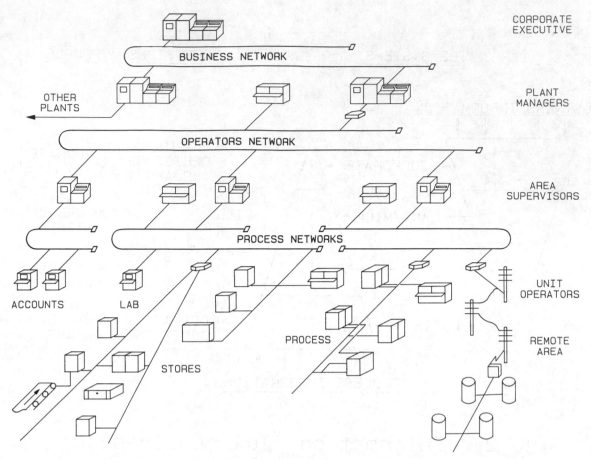

Fig. 4   Distributed management network

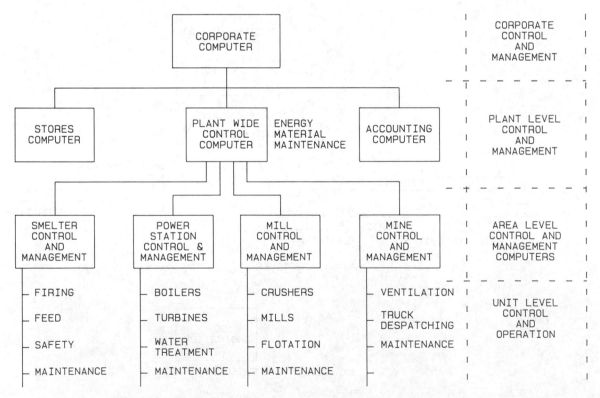

Fig. 5   Distributed data management
          applied to a mining complex

Failure Rate of Process Computers

H.J. STÜBLER

Hoesch AG Dortmund, West-Germany

W. WEBER

Lehrstuhl für Datenverarbeitung, Ruhr-Universität Bochum, West-Germany

Abstract. Methods of calculation of the reliability of process computers are getting more and more important. The paper presents long period observations of 21 process computers with on-line applications in the iron and steel industry. About 5000 failure events during 10 years have been systematically recorded and evaluated. On the basis of these long time observations it is possible to develop a probability model of the failure behaviour of process computers.
This failure model is constructed with four different states of the process computer:
Working phase, working phase with undetected reasons, repair phase, failure phase resulting from external influences.
The model is defined by differential equations for the probability of one of these four phases.
Long time observations show that all of these four phases are closely correlated with: the capacity of a process computer, restarting without detection of failure reason, changes of programs and software, costs and time of repair. So the result of this analysis is a model of failure for process computers, which allows to predict breakdowns of process computers on the basis of external parameters.

Keywords. Process control; data reduction and analysis; reliability theory.

INTRODUCTION

In this article we wish to present a statistical model that allows a failure rate estimation of process computers. Furthermore it can also be used to influence the reliability of process systems. This model abstraction is based on data material that was collected over 15 years at 21 process computers of different size. The storage capacity of these computers range from installations with 5 million words to small systems with 16 K-words only. These systems control the production machinery in a large industrial iron and steel factory.

We wanted to answer the following questions:
- Which are the affectable frame conditions to reduce breakdowns of a process computer?
- Is it possible to derive the interruption rate from the identification data of a process computer?
- How can a system designer profit from the informations of failure rates of process computers?
- What efforts must be made to guarantee the necessary working reliability?

DEFINITIONS OF TERMS

Within these investigations the terms breakdown, error and disturbance must be distinguished.
1. A breakdown means either the violation of at least one frame condition or the termination of operation capability of a previously intact modular processor unit.
2. An error is a prohibited deviation that may cause the breakdown of a process computer. That may be errors in software as well as those in construction (system design), operation or maintenance.
3. A disturbance means the interrupt or drawback of the defined process computer task from the user's sight.
4. A failure is the collective term for breakdowns and errors.

DATA MATERIAL

The disposable data allow a far ranging subdivision. On the one side the percentage partition in Fig. 1. relates to the number of disturbances - right scale - on the other side to the total number of all failures with or without processor outage - left scale. The scales themselves correspond to the relation disturbances:failures and so allow a comparison between each other.

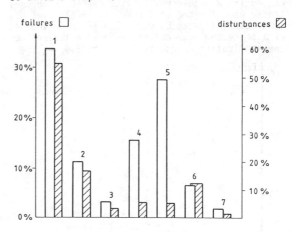

Fig. 1. Percentage Partition of Hardware Failures and Disturbances onto the Different System Parts.
Hardware failures are found in:
1. The central unit
2. The drum and the disk memory
3. The high-speed printer
4. The standard periphery
5. The process periphery
6. The power supply
7. Other resources.

Fig. 2. gives an effective look at the repair works and at expedients for the removal of production disturbances. Here, the necessary actions are presented in the same way as in Fig. 1.

In more than 36% of all cases the reoperation was performed by a restart, that means without knowing the exact reason of failure, or any actual repair.

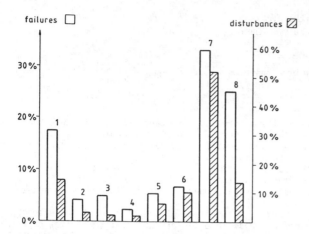

Fig. 2.   Percentage Partition of Executed Operations.
   1. Exchange of components
   2. Repair of components
   3. Elimination of contact error, definition of working point
   4. Elimination of program error
   5. Change of program
   6. Program newly loaded
   7. Restart performed
   8. Maintenance or other works.

Furthermore the investigations shown above indicate that a dependence upon integration degree, i.e. upon the extensive use of integrated circuits (ICs) is not detectable. Upto now no assured statements concerning a higher reliability caused by an extended use of integrated circuits in place of discrete elements, can be made. The introduction year of a processor type can be taken as a measure for the application degree of ICs (Fig. 3.).

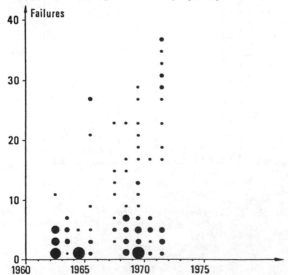

Fig. 3.   Amount of Failures per Quarter of a Year n Relating to the Introducing Year of the Processor Type.

The diameter of each point shows the third dimension. It reflects the number of failures, that means it reflects the number of times one special type of processor broke down during a quarter of a year. The number of failures per quarter of a year, relating to the introducing year of a process computer type, is rather of a rising than falling tendency.

This result gives the information, that hardware failures depend more on the system resources, i.e. on the memory and the input and output lines, than on the control units.

STATE MODEL

The user of a process computer is mainly interested in the total of it's reliability influences. Four different processor states can be defined:
1. In the effective working phase Z_p, the processor is fully disposable for the demanded process. There are no defects or interrupts. The processor executes all demanded and expected functions.
2. An "unstable" state Z_q, that means: failures and latent insufficiencies exist, but do not yet influence the present task or function. This can be caused by errors in programming, construction or handling. This state is unsafe but may be still tolerated.
3. During the state of repair Z_r, the system is analyzed, the cause is diagnosed and the repair will be done completely or just partially.   In case the cause can not be recognized, the processor is set into the "unstable" state Z_q by restart, change of fuses etc.. So, the failure has been bridged only.
4. Environmental conditions,   e.g. a breakdown of the power supply, of air conditioning or network data etc., influence the operation capability of the process computer. Therefore, the processor is in an "inactive" state Z_s, based on other external circumstances.

The transition between the states is illustrated in a state model to be seen in Fig. 4.

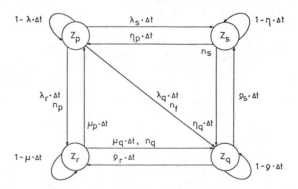

Fig. 4.   State Model of a Process Computer with
   - Effective working phase Z_p
   - "Unstable" state Z_q
   - Repair state Z_r
   - External conditioned, inactive state Z_s and the transition rates belonging to it as well as the number of
   - Hardware failures n_p
   - Software failures n_f
   - Restarts (unknown reasons) n_q
   - External conditioned failures n_s.

The corresponding differential equation system stems from the derivatives of the probabilities

$Y(t)$ of states Z_y. Using average time constant transition rates one can write:

$$\frac{P(A)}{dt} + (\lambda_r + \lambda_q + \lambda_s) \cdot P(t) \quad - \mu_p \cdot R(t) - \eta_p \cdot S(t) = 0$$

$$\frac{dQ}{dt} - \lambda_q \cdot P(t) + (\rho_r + \rho_s) \cdot Q(t) - \mu_q \cdot R(t) - \eta_q \cdot S(t) = 0$$

$$\frac{dR}{dt} - \lambda_r \cdot P(t) - \rho_r \cdot Q(t) + (\mu_p + \mu_q) \cdot R(t) \qquad = 0$$

$$\frac{dS}{dt} - \lambda_s \cdot P(t) - \rho_s \cdot Q(t) + (\eta_p + \eta_q) \cdot S(t) \qquad = 0$$

The asymptotical probability of rates can be derived as follows:

$$P = \frac{A_p}{A}, \; Q = \frac{A_q}{A}, \; R = \frac{A_r}{A}, \; S = \frac{A_s}{A}$$

The terms A_p, A_q, A_r and A_s as well as the denominator A are substituted by:

$$A_p = \rho_r \cdot \mu_p \cdot \eta + \rho_s \cdot \mu \cdot \eta_p$$

$$A_q = \lambda_r \cdot \mu_q \cdot \eta + \lambda_q \cdot \mu \cdot \eta + \lambda_s \cdot \mu \cdot \eta_q$$

$$A_r = \lambda_r \cdot (\rho_r \cdot \eta + \rho_s \cdot \eta_p) + \lambda_q \cdot \rho_r \cdot \eta + \lambda_s \cdot \rho_r \cdot \eta_q$$

$$A_s = \lambda_r \cdot \rho_s \cdot \mu_q + \lambda_q \cdot \rho_s \cdot \mu + \lambda_s \cdot (\rho_r \cdot \mu_p + \rho_s \cdot \mu)$$

$$A = A_p + A_q + A_r + A_s$$

The transition rates λ_r, λ_q, λ_s and ρ_r, ρ_s and the repair rates μ_p, μ_q and η_p, η_q may be determined in aid of failure count and repair times.

The calculation refers to the amount of different failure types. The repair times of every failure type must be taken in account.

So the transition rates λ, ρ, μ and η can be shown in dependence on
- the number of inputs and outputs, x_a
- the program size in K-words, x_b
- the capacity of annual programming works in man years, x_d
- the repair times, t_r.

The data material is quite heterogeneous and based on observation over 124 report years, 21 process computer systems and 14 different processor types. The following average measures of different failures and standard deviations have been registred.

Hardware failures:	6,5	$0 \leq$	n_p	$\leq 13,5$
Program errors :	4,1	$0 \leq$	n_q	$\leq 12,9$
Unknown reasons :	22,0	$0 \leq$	n_a	$\leq 70,0$
Other reasons :	3,1	$0 \leq$	n_s	$= 6,4$

FAILURE RATE

Depending on the program size the failure rate rises with the number of inputs and outputs, as well as with increase of memory capacity (size). The importance of memory capacity concerns smaller process computers, whereas in bigger systems the number of in- and outputs has more influence. Therefore, a greater number of inputs and outputs and a bigger program lead to a reduction of availability (Fig. 5.).

So, the failure probability is influenced reciprocally by the number of inputs and outputs x_a and by the storage capacity or program size. The result is the following relation:
A program amplification x_b of 100 K-words of a

smaller process computer corresponds with an enlargement of about 1000 inputs and outputs.

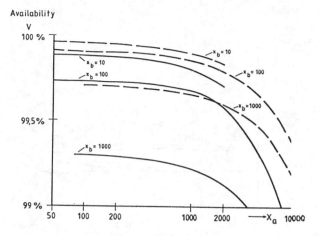

Fig. 5.   Availability of a Process Computer in Dependence on
- The number of inputs and outputs x_a
- The program size in K-words x_b and as a mean value and upper terminal value

In bigger process computers this relation is reduced. An amplification x_b of 100 K-words corresponds with an enlargement x_a of about 300 units.

Every programming work, expressed here in man months x_e, leads to a decrease of availability. Also the probability of the states Z_p and Z_q is independent from the size of a computer, but heavily influenced by the amount of x_e (Fig. 6.).

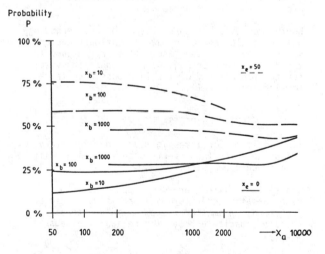

Fig. 6.   Influence of Programming Work for a Process Computer on the Probability of State Z_p Referring to the Values $x_e = 0$ and $x_e = 50$ Man Months (Process Computers of Different Size (x_a, x_b)).

From long to short repair times the probability of rate Z_p changes immensely, for example from 20% to 35%. The availability can be reduced by more than 3% by higher repair times. Especially in big and therefore complex process computers this influence can be registred (Fig. 7.).

There are only small differences of the probability P for small process computers based on the values $x_b = 10$, 100 and 1000, to be seen in Fig. 7..

The shorter the repair time, the more probable is state Z_p.

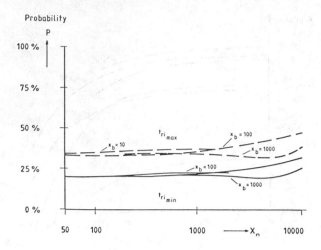

Fig. 7.   Influence of Repair Times on Probability of State Z_p Concerning Process Computers of Different Size (x_a, x_b), with Minimum Repair Times $t_{ri_{min}}$ and Greater Repair Times $t_{ri_{max}}$.

The effect of external influences upon the failure rate lies in the range of 10%. Corresponding to this, availability is reduced.

Although the investigated data material is inhomogeneous, dependencies can be derived clearly. So, the results may be applied for other users and other process computers, as long as there are no systematical differences between those computers and the ones investigated here.

Furthermore, even future systems of any technology will be influenced by hardware, software and environmental conditions and will so correspond to the model, shown in Fig. 4.. The equation system for the probability of the four possible states remains true as well.

FUTURE ASPECTS

The result of the investigations show for every influence value the effects upon the reliability of a process computer.

Big systems, with over 3000 inputs and outputs, range unsatisfactorily and a partition should be preferred. In smaller systems mainly external storage capacity effects the reliability negatively.

An important aspect for a smaller failure rate is the reduction of restarts. A number of suitable actions is possible:
- Facilities, aiding to recognize and localize hardware breakdowns and software errors, must

be used in a bigger frame.
- Education and experience exchange of the service staff must be increased.
- Hard- and especially software documentation must be clear and sufficient.

Further consequences refer to repair times. On the one hand the necessary actions for a restart must be taken, on the other hand repair times can be reduced by:
- Switching of a module to a redundant one in case of a failure.
- Use of reliability tested components or correspondingly constructed units easy to service.
- Faster access upon spare parts, so that the maintenance can be performed by exchange.
- External diagnosis, to avoid the coming time of the service personnel.

The negative influence of programming works upon process computer reliability is proved. Errors can be avoided by stronger testing of altered or increased software. Also the use of new programming methods can lead to more reliable programs.

An improvement of reliability of environmental conditions of a process computer, as there are power supply, air conditioning and the data supply in linked systems, is necessary.

The effect of a data transmission error or breakdown of a transmission line in linked systems, upon the reliability of a process computer, can be estimated as an external influence. In this case, the state Z_s especially refers to those external influences. The failure rates n_p and n_q then depend on the data transmission line, on the linked computer and on the transmitted data.

In future, the programs will define the reliability of a process computer much more than the higher integrated and error tolerating hardware. The elimination of programming errors will gain more importance and the repair of hardware units will lose influence.

The method for failure rate determination is especially devoted to the formulation of an "unstable" state Z_q, that will be of interest in future, where the equation terms have to be adapted to new technologies.

REFERENCES

European Purdue Workshop (1976). TC 7 System Reliability. Safety and Security Comittee. Technische Universität Braunschweig.
Military Standardization Handbook. Reliability Prediction of Electronic Equipment. Department of Defense, USA/Technische Verbände und Normung. (1974)
Stübler. H.J. (1984). Methode zur Bestimmung der Ausfallhäufigkeit von Prozeßrechnern (A method for prediction of failure rates of process computers). Doctor-Dissertation, Ruhr-Universität Bochum, West-Germany.
Weber, W., Schiefer P. (1976). Automatisierung von Anlagen der Stahlindustrie. Springer Verlag, Berlin.

Copyright © IFAC Automation for
Mineral Resource Development
Queensland, Australia, 1985

Process-Data Acquisition in Hydrometallurgy

D. ROYSTON

Hydrometallurgical Research Manager, MIM Holdings Limited, GPO Box 1433, Brisbane Queensland 4001, Australia

J.H. CANTERFORD

Principal Research Scientist, CSIRO Division of Mineral Chemistry, PO Box 124, Port Melbourne, Victoria 3207,

Australia

Abstract. Current and future methods of data acquisition in hydrometallurgy are
reviewed. Hydrometallurgical operations involve a wide range of chemical and
physical processes for the recovery of metals and other products from ores and
concentrates. Control of such operations requires information on the different
liquid/solid and liquid/liquid mixtures being processed. Continuous data acquisition for the liquid phases uses procedures already commercially exploited in
other types of chemical plant. The major problem is the lack of robustness of
measuring probes. Continuous analysis of the solid phases is much more difficult
to achieve because of the sampling and analytical techniques involved.

Keywords. Hydrometallurgy; leaching; data acquisition; mineralogy; on-stream
analysis.

INTRODUCTION

Hydrometallurgy involves a wide range of liquid/
solid and liquid/liquid chemical processes
directed toward metal recovery. The general process scheme involves the comminution of a solid
phase (ore, concentrate, or metallurgical by-
product), leaching of the resulting material in an
agitated vessel, purification of the leach liquor,
and recovery of the components of interest. Recovery and regeneration of the leachant, as well
as disposal of solid and liquid wastes, are
important related processes. In-situ leaching
(solution mining) is a relatively new hydrometallurgical development. A leach liquor is injected
into the orebody, and metals are recovered from
the resulting solution. In heap leaching, the
leachant is sprayed over heaps of crushed but not
ground ore, and metals are again recovered from
the solution. Apart from the difference in the
particle size of the feed, heap and in-situ
leaching processes differ from conventional
leaching systems in that the reaction system is
not agitated.

The long-term potential of process-data acquisition in hydrometallurgy has been discussed by
Brown and Bhappu (1982), Warren (1984), and, in a
more practical context, Behrend (1977) and Jansen
(1984). Specific examples have been described by
Redd and Kongas (1982), Van Zyl (1983), and Sarkar
(1984). Anthony (1984) and Andrews and others
(1984) provide excellent overviews of current
research and developments in the mineral industry,
briefly discussing various aspects of process
control.

Process data are acquired on solution chemistry,
solid-phase composition (mineralogy), and pulp
chemistry, especially pulp rheology. The methods
used to acquire these data are common to other
chemical and metallurgical processes, and some are
subject to routine process control. This paper
covers the determination and interpretation of
process data in hydrometallurgical operations.

Particular attention is given to bench-scale
research techniques and to the emerging automatic
data-acquisition methods that will lead to an
improved precision and understanding of hydrometallurgical process control.

SOLUTION CHEMISTRY

Data Required - Solutions

Hydrometallurgical process solutions include both
aqueous and organic media that vary in composition, pH, and oxidation state, according to a
wide range of reaction environments. Reaction
systems for very refractory ores are required to
be either very acidic (silicate uranium ores) or
very alkaline (caustic zirconia frit). Highly
oxidizing solutions are required for the dissolution of uranium and copper sulphide ores, while a
reducing environment is necessary for the hydrogen
reduction of nickel. Aggressive chemical corrosion is expected in chloride systems, while
other systems are remarkably benign even though
the materials involved appear to be refractory
(e.g., cyanidation of gold and silver). The
pressure of the system can be well above atmos-
pheric pressure (e.g., zinc sulphide pressure
leaching).

Several process conditions may occur within one
flowsheet, especially in processes for complex
(multi-element) ores (Barbery, Fletcher, and
Sirois, 1980; Davies, 1981). The newer hydrometallurgical processes often incorporate both
aqueous and organic phases that have separate and
peculiar control requirements. Murray (1982) has
noted that chemical interactions in copper solvent
extraction/electrowinning plants result in critical process-control parameters originating in any
one of several places.

Solution-composition parameters that are measured
include:

193

pH -
 by glass electrode, titration with indicators,
 conductometric titration, pH paper. Note the
 crucial properties of pH probes: low-sodium
 error glass for high pH levels, appropriate
 glass for the desired response time and temperature, a reference system compatible with
 the reaction chemistry.

Eh -
 by platinum electrode and reference electrode,
 or gold if S^{2-} is present. A "real" number can
 be obtained only for a system containing an
 oxidized/reduced pair of species in significant
 concentration.

Temperature -
 by thermometer, thermocouple.

Dissolved oxygen -
 by dissolved-oxygen probe, iodometric
 titration.

Colour -
 by UV/visible spectrophotometer, tristimulus
 filter.

Density -
 by hydrometer, pycnometer, weight/volume
 measurement.

Viscosity -
 by off-line viscometer, on-line vibrating
 probe, mixer power.

Corrosion potential -
 by corrosimeter, corrator, pH probe, calcium
 carbonate concentration (Langelier Saturation
 Index/Ryznar Stability Index).

Conductivity -
 by conductivity meter.

Ion mobility -
 by moving-boundary method.

Dissolved gases -
 by CO_2 analyser, dissolved-oxygen probe, SO_2 or
 HCl or pH probe.

Heat capacity/heat of fusion and reaction -
 by calorimeter.

Flow-rate -
 by flowmeter, rotameter.

Species characterization -
 by nuclear magnetic resonance, infrared and
 UV/visible spectroscopy, mass spectrometry, gas
 chromatography, polarography, voltametry,
 coulometry, specific-ion probe.

Metal concentration -
 by atomic absorption spectrophotometry (AAS),
 inductively coupled plasma-emmision spectrophotometry (ICPES), ion exchange, wet
 chemistry.

There are numerous excellent reviews that discuss
the measurement of the above parameters. Two
recent and extremely useful articles are by Dealy
(1984) and Hoeppner (1984), covering viscosity and
density, respectively.

Development of the Measurement - Solutions

Development of the above techniques has generally
taken three paths:
● Mechanisation of off-line methods to improve
  assay rates.
● Development of methods capable of measuring
  several components simultaneously.
● Development of in-situ methods giving direct,
  rapid measurements.

Mechanisation of off-line methods involves well
known techniques for automating wet methods or
methods such as AAS. The aim is to increase
productivity by increasing the number of samples
analyzed and/or the number of components measured
in each sample as a function of time.

With respect to the second path, ICPES has been
the method of greatest interest and potential. It
has provided a substantial step forward in
determining the detailed elemental composition of
samples.

In-situ methods have the most promise for both
research and commercial applications in that they
allow rapid feedback for experimental and plant
control. Well tried methods include:
● Specific-ion electrodes.
● Isotopic measurements (for radioactive solutions).
● Isotope-driven in-stream X-ray analysers.

More novel methods that have potential for
increasing the speed of detection are:
● Chemical-sensitive surface-implanted
  semiconductors.
● Fibre-optic devices that use colour-sensitive
  sensing tips on fibre-optic probes to measure
  pH and composition (Covington, 1982).
● Flow-through analysis using porous electrodes
  (Anderson, 1983).
● Ion chromotography (Andrew, 1984) and polarography (Bond, 1981).

Limitations on Use - Solutions

Wet analytical techniques and instrumental methods
allow almost all element-composition parameters to
be measured off-line. However, analytical difficulties can arise, such as in cases where:
● The species of interest is dilute within a
  concentrated matrix.
● The oxidation states of the anion/cations
  require measurement (e.g., aqueous sulphur
  species, or the ferrous/ferric ratio).
● The composition of the mixture of species in a
  buffered solution is required (e.g.,
  carbonate/bicarbonate).

The turn-around of samples is largely limited by
the mechanical systems used to transport the
samples to the analytical instruments, which are
constantly increasing in sophistication. Data can
be logged directly from many analytical instruments and relayed to a control centre. Jacobs,
Badian, and Proudfoot (1984) describe the
computer-aided design of systems for regulating
pH. Direct on-line instrumental techniques, while
increasing in range and application, are still
limited in the range of variables that they can
assay. Probes can be mechanically fragile and are
often sensitive to poisoning, subject to blockage,
or limited in the range of environments in which
they can operate (e.g., temperature, total ionic
strength, corrosive solutions). For example, pH
probes suffer these limitations, but they can be
improved by the development of special glass
bulbs, porous plugs, etc., to suit particular
applications.
Lambert (1982) has decribed the use of ultrasonics
to clean pH probes.

Future Developments - Solutions

Future developments will involve improvements in
both analytical techniques and instrumentation.
Areas of particular interest to both research and
process hydrometallurgists are:
● Direct measurement of minor species (especially
  Au and Ag) or significant impurities (Sb, Se,
  Te).
● Measurement of single species within a complex,
  concentrated or strongly acidic (or alkaline)
  background.
● Rapid simultaneous analysis of a range of
  species in complex mixtures.

For on-line analytical instrumentation, the emergence of element-sensitive semiconductors and
fibre-optic devices will complement the existing
range of specific-ion, X-ray, and isotopic
devices, among others. Conductive plastics may
also be applicable to instruments in this area
(Mascone, 1984).

Development of on-line probes is needed in order
to:

- Improve selectivity and operating range in terms of both concentration and environment.
- Increase the robustness of probes to withstand mechanical shock, drying, and coating with residues.

SOLID CHEMISTRY

Data Required - Solids

Leaching metals from an ore or concentrate is the essence of extractive hydrometallurgy. The mineralogical and physical characteristics of the solid phase determine the success of leaching, and so data are necessary on:
- Bulk chemical analysis.
- Basic comminution characteristics.
- Physical size, shape, and surface properties.
- Mineralogical composition, including the size, nature, distribution, and juxtaposition of mineral types within an assemblage.

Detailed chemical and mineralogical data are required for process design, whereas only semi-quantitative data are necessary for process control. Nuclear techniques (X-ray fluorescence, neutron capture, neutron activation, gamma-ray scattering, etc.) represent the most powerful tool for on-line analysis, and Sowerby (1983) provides a valuable discussion of the use of such techniques in the Australian mineral industry. The techniques have obvious applications in pulp chemistry as well as in in-situ leaching.

General milling data are common to all aspects of extractive metallurgy. Additional aspects of concern to hydrometallurgists include the need for fine particle sizes (for adequate leaching rates) and the distribution of deleterious gangue minerals (e.g., reagent consumers, clays, and silica) that can affect downstream processing costs and operation.

The physical characteristics of particles can affect leaching kinetics and the degree of reaction completion. For example, particles may be coated with an inert species before and/or during leaching.

Mineralogical composition is an obvious primary determinant of leaching performance. The refractory character of target minerals determines the basic opportunity for leaching, the particle size determines the exposure to leaching, and the juxtaposition of different minerals with different electrochemical potentials affects the rate and selectivity of leaching reactions. These data become particularly important where valuable trace minerals such as gold are of interest, where the ore is particularly fine-grained, where mineral types (e.g., chalcocite) change during the initial stage of leaching and so limit the rate and degree of extraction, or where the mineral assemblage is particularly complex. Ideally, the liquid/solid interface should be monitored continuously as leaching proceeds.

Data Acquisition - Solids

The derivation of comminution data and the related process control of milling and flotation circuits is now widely practised (Lyman, 1981; Lynch, 1977; Toop, Wilkinson and Wenk, 1984). Process simulation and control procedures at the Mt Isa operations of MIM Holdings Limited have been recently reviewed by Bernard and others (1984). The most pressing areas of concern are the applied mineralogical issues. The traditional methods are point counting and microscopic examination of individual grains and thin sections. Use of the electron microscope to examine the composition of individual grains is now well established. Unfortunately, all of these techniques are indirect and time-consuming. Automatic or partially automatic techniques using X-ray diffraction (Whitehead, O'Hara, and Frost, 1973) are now commonplace, and together with recent developments in electron microscopy (Reid and others, 1984), they promise sufficient detail in a short enough time to permit feedback control of production circuits on a macro scale.

Future Developments - Solids

Mineralogical assessment of materials as they undergo reaction is a long-established tool of hydrometallurgical research. It provides very useful data on, for example, the dissolution of gangue minerals, the selectivity of leaching, changes in porosity, and the formation of insoluble reaction products. If the latter form a coherent layer on the reacting particle, then the diffusion of soluble species to and from the reaction face may be affected. The use of mineralogical assessment as a control device is limited owing to the time-consuming nature of obtaining results, from sample preparation onwards, and the need for highly trained applied mineralogists, etc. However, the continued development of instrument-based techniques such as QEM*SEM (Reid and others, 1984) promises great rewards for both research and process control. There is probably no greater single area of endeavour that could provide a breakthrough in the development and commercialization of present and future hydrometallurgical operations.

PULP CHEMISTRY

Data Required - Pulps

The data required on pulps are common to the metallurgical and process industries and cover:
- Physical behaviour (e.g., settling, thickening, dewatering, filterability).
- Rheology (e.g., concentration, mixing power, foaming/aeration potential, particle behaviour in dilute pulps).
- Solid/liquid reaction effects (e.g., effects of solids concentration on mixing, heating, reaction rates, and reagent requirements).
- Liquid composition (e.g., pH, Eh, chemical composition).

Data Acquisition - Pulps

Data acquisition on the physical behaviour and chemical composition of pulps is common to many extractive-metallurgy and chemical processes (Klimpel, 1982; Nguyen and Boger, 1984). Rheological properties emerge during bench testwork as the degree of mixing, aeration, etc., is determined for a particular reaction. In plants, rheological parameters can be measured by feedback on the power consumption, the solids density, and the foam height (Anonymous, 1983; Asher, 1982; Langdon, 1983). Developments using ultrasonics and vibrating probes are of particular interest in this area. For dilute pulps, data on particle behaviour and shape, etc., are valuable; acquisition techniques include direct sampling and light scattering (Ahmed and others, 1983).

Data on the solid/liquid interaction in pulps also emerge from bench testwork as the effect on reaction rates of pulp density, temperature, reagent concentrations, and metal recovery is determined. Here, data acquisition is concerned with the liquid phase, where pH, Eh, and chemical composition are typically required. There is often a close electrochemical interaction between the solid and liquid phases that requires monitoring

(Hiskey and Wadsworth, 1981). The practical difficulties of data acquisition on the bench (as in
plants) relate to the mechanical robustness of the
instruments (especially pH probes) and to gaining
realistic information on the liquid composition in
the presence of solids (Ormrod and Vida, 1980).

Future Developments - Pulps

The most important area for development is the
assessment of in-pulp liquid pH, Eh, and chemical
composition and of pulp density and viscosity.
The improvement and development of in-pulp methods
for measuring these parameters is necessary to
advance understanding and control of the reaction
processes. Critical to this development are more
robust probes that can detect the properties of
liquids separate from those of solids. Interesting developments in fibre-optic devices hold some
promise in this area, as do vibrating probes for
measuring rheological properties.

HEAP AND IN-SITU LEACHING

As Davidson, Huff, and Sonsteile (1979) have
pointed out, instrumentation developed by the oil-
production industry provides very useful data on
the physical properties of injection and recovery
boreholes used in in-situ leaching. Table 1 lists
the types of wireline instruments now in use.

Table 1 Wireline Instruments Used in In-Situ
         Leaching

Instrument	Information
Thermal neutron	Porosity
Spontaneous potential	Zones of porosity
Gamma density	Porosity, bulk density
Sonic velocity	Porosity, degree of fracturing
Electrical resistivity	Porosity, fluid saturation
Natural gamma	Intensity of natural radio-activity
Neutron capture	Element compos-ition
Induced polarization	Sulphide content
Temperature	Thermal gradient, zones of permea-bility
Spinner	Fluid velocity, zones of permea-bility
Radioisotope	Fluid velocity, zones of permea-bility

The productivity of heap and in-situ leaching
operations is largely controlled by the concentration of the desired value(s) in the production
liquor (Canterford, 1983). If this falls below a
preset value, then the production system must be
moved to a new area. However, if the production
system consists of a number of separate units, as
is normally the case, then each unit must be
evaluated. The chemistry of the process liquids
in each unit can be determined by downhole probes,
but these are generally fragile, difficult to
calibrate, and prone to a loss of sensitivity. A

number of devices have been developed for recovering liquid samples from various locations within
the leaching zone; one such device is SIROSAMPLER
(Canterford, Kuester, and Miles, 1982).

Bacteria play an important role in some heap and
in-situ leaching operations (Canterford, 1983).
Monitoring of these operations requires specialized microbiological techniques, but a discussion
of these is beyond the scope of this paper.

Examination of the solid phase by mineralogical
methods involves extensive coring, which is both
expensive and time-consuming. In addition,
because of the likely horizontal and vertical
variations in gangue and ore mineralogy, the
mineralogical methods are unlikely to be useful
for control purposes. However, preproduction core
analysis is essential for process design, particularly with regard to leachant type and concentration. Postleach coring is rarely carried out,
but the wealth of information that can be acquired
is substantial.

In the short term, there does not seem to be a
significant scope for rapid development of suitable downhole control procedures in in-situ
leaching, even though the technology is likely to
become more important. This is because each discrete solid/liquid interface is extremely variable
in comparison with the "average" interface in
agitated leaching systems. However, as in conventional leaching systems, more robust probes for
liquid analysis will certainly help process control downstream from the leaching circuit.

CONCLUSIONS

Process control in hydrometallurgical operations
is becoming more sophisticated and is placing
increasing reliance on accurate and "continuous"
data acquisition. Hydrometallurgical operations
involve both liquid/solid and liquid/liquid
systems. Data acquisition for the latter is
relatively straightforward and, in many cases,
uses already developed chemical plant systems.
However, many probes suitable for liquid-phase
data acquisition suffer from their inability to
withstand the chemically and physically abrasive
nature of process streams. For this reason, off-
line methods are often used. Nevertheless, recent
developments in probes and instrumentation are
increasing the practicality of automatic in-line
systems.

Rapid analysis of solid phases, particularly in
the leaching stage, is much more difficult but no
less important than control of the liquid phase.
Although solid sample preparation is still time-
consuming, the rapidity and depth of knowledge
that can be obtained by modern mineralogical
techniques, particularly electron microscopy, is
significant in process design and control.
Instrument developments now taking place can only
further the effectiveness of control procedures.

REFERENCES

Ahmed, N., D.E. Landberg, J.A. Raper, and G.J.
    Jameson (1983). Comparison of liquid-borne
    particle sizing techniques. In Proc. Chemeca
    '83. IE Aust., Canberra. pp. 723-727.
Anderson, J.L. (1983). Flowthrough porous elec-
    trodes as analytical detectors. Chemical
    Engineering, 90(4), 65-67.
Andrew, B. (1984). Potential applications of ion
    chromatography in process control. In Proc.
    Chemeca '84. IE Aust., Canberra.
    pp. 751-759.

Andrews, S.J., D.J. Barrett, A.D. Fernie, and R. Pendreigh (1984). Recent developments in the mineral processing industry. In M.J. Jones and P. Gill (Eds.), Mineral Processing and Extractive Metallurgy. Institution Mining Metallurgy, London. pp. 87-98.

Anonymous (1983). Equipment survey: flowmeters. The Chemical Engineer, No. 396, 37.

Anthony, M. (1984). Review of current research and development activities in the mineral industry. In M.J. Jones and P. Gill (Eds.), Mineral Processing and Extractive Metllurgy. Institution Mining Metallurgy, London. pp. 1-13.

Asher, R.C. (1982). Ultrasonic techniques for non-invasive instrumentation on chemical and process plants. The Chemical Engineer, No. 383, 317.

Barbery, G., A.W. Fletcher, and L.L. Sirois (1980). Exploitation of complex sulphide deposits: a review of process options from ore to metals. In M.J. Jones (Ed.), Complex Sulphide Ores. Institution Mining Metallurgy, London. pp. 135-150.

Behrend, G.M. (1977). The installation of instruments in mineral processing plants. CIM Bulletin, 70(777), 63-69.

Bernard, N.G., J.H. Fewings, I.S. Schache, and R.M.S. Watsford (1984). Review of process simulation and control developments at Mount Isa Mines, Ltd., Australia. In M.J. Jones and P. Gill (Eds.), Mineral Processing and Extractive Metallurgy. Institution Mining Metallurgy, London. pp. 99-112.

Bond, A.M. (1981). Developments in polarographic (voltametric) analysis in the 1980's. In D.A.J. Rand, G.P. Power, and I.M. Ritchie (Eds.), Progress in Electrochemistry. Elsevier, Amsterdam. pp. 381-394.

Brown, M.C., and R.B. Bhappu (1982). Recent trends in instrumentation of hydrometallurgical plants. In K. Osseo-Asare, and J.D. Miller (Eds.), Hydrometallurgy - Research, Development and Plant Practice. TMS-AIME, New York. pp. 739-757.

Canterford, J.H. (1983). Solution mining: general principles and Australian practice. In Jobson's Mining Year Book. Dun and Bradstreet, Melbourne. pp. 215-226.

Canterford, J.H., A. Kuester, and J.G. Miles (1982). Borewater sampling device. Australian Mining, 74(3), 24-30.

Covington, A.K. (1982). The measurement of ions in solution. Laboratory Practice, 31, 239-251.

Davidson, D.H., R.V. Huff, and W.E. Sonsteile (1979). Measurement and control in solution mining of copper and uranium. In E.F. Kursinskii (Ed.), Instrumentation in the Mining and Metallurgy Industries, Vol. 6. ISA, North Carolina. pp. 149-156.

Davies, G.A. (1981). Unit operations in hydrometallurgy. Chemistry Industry, No. 13, 420-427.

Dealy, J.M. (1984). Viscometers for online measurement and control. Chemical Engineering, 91(20), 62-70.

Hiskey, J.B., and M.E. Wadsworth (1981). Electrochemical processes in the leaching of metal sulfides and oxides. In M.C. Kuhn (Ed.), Process and Fundamental Considerations of Selected Hydrometallurgical Systems. TMS-AIME, New York. pp. 304-325.

Hoeppner, C.H. (1984). Online measurement of liquid density. Chemical Engineering, 91(20), 71-78.

Jacobs, O.L.R., W.A. Badian, and C.G. Proudfoot (1984). Computer-aided design of systems for regulating pH. The Chemical Engineer, No. 401, 19-21.

Jansen, J.P. (1984). The control room of the future. The Chemical Engineer, No. 401, 22.

Klimpel, R.R. (1982). The influence of slurry rheology on the performance of grinding circuits. In Proc. Mill Operators Conference, North West Queensland. Australasian Inst. Min. Metall., Melbourne. pp. 1-14.

Lambert, W. (1982). Ultransonic cleaning - its role in the process industries. The Chemical Engineer, No. 383, 320.

Lyman, G.J. (1981). On-stream analysis in mineral processing. In A.J. Lynch, N.W. Johnson, E.V. Manlapig, and C.G. Thorne (Eds.), Mineral and Coal Flotation Circuits, Their Simulation and Control. Elsevier, Amsterdam. pp. 235-272.

Lynch, A.J. (1977). Mineral Crushing and Grinding Circuits. Elsevier, Amsterdam.

Mascone, C.F. (1984). Progress sparks interest in conductive plastics. Chemical Engineering, 91(2), 25-29.

Murray, D.J. (1982). Process control in solvent extraction-electrowinning. In D.J. Spottiswood and D.H. Davis (Eds.), Instrumentation in the Mining and Metallurgy Industries, Vol. 9. ISA, North Carolina. pp. 53-57.

Nguyen, Q.D., and D.V. Boger (1984). Rheology - fundamentals put into practice in Australian industry. In Proc. Chemeca '84. IE Aust., Canberra. pp. 791-803.

Ormrod, G.T.W., and J. Vida (1980). An automatic sampling probe for use in the carbon-in-pulp process. NIM Report 2086.

Redd, R.W., and M. Kongas (1982). Automation systems improves refinery operation. In D.J. Spottiswood and D.H. Davis (Eds.), Instrumentation in the Mining and Metallurgy Industries, Vol. 9. ISA, North Carolina. pp. 193-202.

Reid, A.F., P. Gottlieb, K.J. McDonald, and P.R. Miller (1984). QEM*SEM image analysis of ore minerals: volume fraction, liberation and observational variances. ICAM 84, Los Angeles.

Sarkar, K.M. (1984). Temperature control of acid pressure leaching of uranium ores. Transactions Institution Mining Metallurgy, 93, C23-C29.

Sowerby, B.D. (1983). Nuclear techniques of analysis in Australian mineral industry. In Proc. Third Australian Conference on Nuclear Techniques of Analysis. AINSE, Lucas Heights (NSW). pp. 129-132.

Toop, A., L.R. Wilkinson, and G.J. Wenk (1984). Advances in in-stream analysis. In M.J. Jones and P. Gill (Eds.), Mineral Processing and Extractive Metallurgy, Institution Mining Metallurgy, London. pp. 187-194.

Van Zyl, R.M. (1983). Computer controlled Merrill-Crowe processing. Engineering Mining J., 184(4), 58-59.

Warren, G.W. (1984). Hydrometallurgy - a review and preview. J. Metals, 36(4), 61-66.

Whitehead, D.G., K. O'Hara, and M.T. Frost (1973). Improved data collection for an X-ray diffractometer. IEEE Transactions - Instruments Measurement, IM-22(1), 47-52.

A Review of Automation at Kalgoorlie Nickel Smelter

C.W. HASTIE

Resident Manager

J.M. LIMERICK

Technical Services Superintendent

and

R.A. CAMPAIN

Chief Engineer

N.J. HORNER

Electrical Engineer

Kalgoorlie Nickel Smelter, P.O. Box 448, Kalgoorlie, Western Australia 6430

Abstract. The review covers the different concepts of automation which are appropriate to non-ferrous smelting and the degree to which the smelter, and the furnace in particular, is amenable to total automation. Due to the problems peculiar to a smelter, it is unlikely that a fully closed loop system could ever be achieved and that the smelter will continue to depend upon extended manual control supported by computer based information sorting and optimisation.

Generally, loop control is performed using Taylor's Mod III process control system. For distinct plant systems involving a combination of process loops, sequencing, and data acquisition, the company has developed its own microprocessor based systems. Process optimisation using computer based analysis of operating data is well established. Utilising this data, computer modelling provides manual set point supervision for the furnace process control system.

The impact of automation on the workforce and on operating costs is also examined. Predictions on future developments in the field of automation, optimisation and sequencing are propounded and related to the non-ferrous industry at large.

Keywords. Non-ferrous smelting; closed loop system; process control; data acquisition; microprocessor; optimisation; computer modelling.

INTRODUCTION

The word automation was coined by one Delmar S. Harder as late as 1948 to describe the automatic control of the manufacture of a product through a number of successive stages. Its definition, however, does seem to have become rather nebulous since. More recent definitions suggest that automation is the use of machinery to save mental or manual labour or, alternatively, to reduce human intervention to a minimum. Clearly the ideal for automation in a process would be to have a plant which runs completely by itself without human intervention. Currently certain inherent difficulties associated with smelting effectively preclude full automation. However there are many opportunities for automation of parts of the overall process. The degree to which these have been taken up at Kalgoorlie Nickel Smelter (KNS), their subsequent effect and the direction for the future form the basis of this paper.

LOCATION AND PROCESS

Western Mining Corporation Limited (WMC) operates a smelter to treat 500 000 tonnes per year of nickel concentrate from mines at Kambalda, Laverton and Leinster located in the sparsely populated, semi arid Goldfields region of Western Australia (Fig. 1). The smelter is located 15 km south of Kalgoorlie, the main commercial centre of the region. KNS produces a 72% nickel matte product which is railed 600 km to the west coast for export. A proportion of the matte produced is further refined at the WMC Kwinana Refinery south of the capital, Perth.

Central to the complex is an Outokumpu style flash smelting furnace. In the flash smelting process, combustion of nickel sulphide concentrate in preheated air provides the primary energy source for the high temperatures (~1500°C) required.

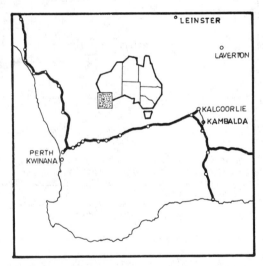

Fig. 1. Location of flash smelter

Combustion is enhanced by the use of oxygen-enriched air, provided by a 3 MW, 200 tpd oxygen plant. Off-gasses from the furnace enter a waste heat boiler (WHB) at ~1450°C generating up to 100 tph steam which is then superheated and fed to two steam driven turbo-alternators of 21 MW combined capacity. Optimisation of the complex mass and heat balances within the flash furnace is of paramount importance for achieving an energy efficient operation. Although connected to the state grid, KNS is generally self-sufficient for electrical power.

Silica flux is added to the furnace to remove iron from the melt as an iron silicate waste slag. The

199

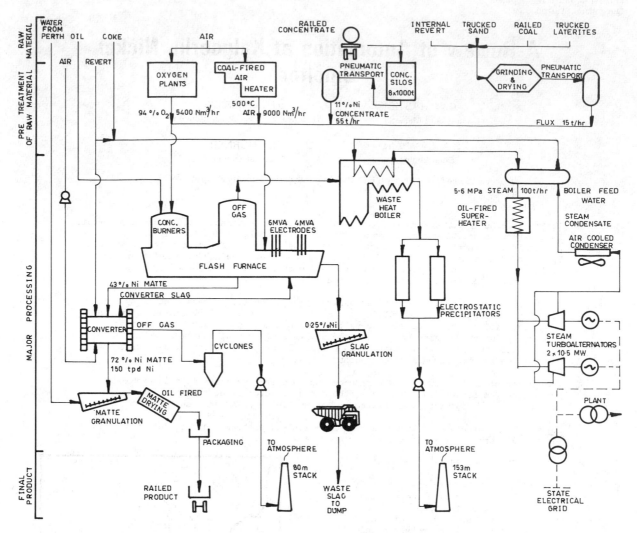

Fig. 2. Process flow diagram of Kalgoorlie Nickel Smelter

flux is prepared by mixing sand, coal and recycled material ("revert") through a crushing and grinding plant and an oil fired drying kiln. The dried flux is pneumatically conveyed to the furnace, as is the concentrate feed. A schematic process flow diagram is shown in Fig. 2.

SYSTEMS OVERVIEW

When KNS was originally commissioned in 1972 instrumentation consisted mostly of discrete controllers, indicators and recorders, panel mounted in the power house and flash furnace control rooms. Sequencing control was performed using relay logic.

In 1977 construction commenced on a new larger flash furnace at KNS. Taylor Mod III control systems were purchased for the new flash furnace control room and also to upgrade the instrumentation in the power house control room.

In early 1978 the first in-house designed microprocessor system was installed to control reverts charging of the converters.

Based on the success of this system a similar design was installed to control cold charge addition to the new flash furnace. Before this furnace became operational in November 1978,

another system designed around a new generation microprocessor was installed to control pneumatic transport of concentrate to the furnace. Previously this had been achieved by relay logic.

During 1979 the laterites crushing and grinding plant was commissioned. The latest microprocessor type was utilised to control the sequential operation of this plant. Instrumentation in the converter control cabins was changed from pneumatic to electronic also during this year.

A remotely run computer model of the flash furnace was introduced for operational control in 1980. Initially the model was used for comparison with daily manual set-points of the Mod III. By early 1981 sufficient confidence in the model had been gained for its full utilisation for set-point control and new terminals were added in the flash furnace control room, and later in the analytical laboratories, to facilitate usage.

In the last three years two proprietary programmable logic controllers (PLC's) have been installed on small pnuematic transport systems.

During the same period four in-house designed microprocessor systems for data acquisition have been installed, the most complex being that for power supervisory control.

PROCESS CONTROL

The Taylor Mod III process control equipment, installed during the building of the new furnace in 1978, was chosen because it provides for secure dedicated analog control yet incorporates centralised process management, utilising visual display unit (VDU) based command consoles. From the command consoles the operator can select manual control, automatic control with set point adjustment, or turn the system over to computer control.

While a computer has been connected to the Mod IIIs bus by an in-house designed microprocessor based interface, output data from the furnace computer model has to date been entered as Mod III set-points by hand.

The furnace process control system supervises all furnace fuel and air requirements including process air preheating, together with the furnace feed and off gas systems. The process control system presently comprises 44 control loops, 10 manual outputs and 100 inputs.

Supporting the furnace system is another Mod III installation which supervises the steam and feed water controls for the furnace WHB, and the remainder of the steam based power plant.

The process control systems provide effective closed loop control for many discrete loops. For example, the WHB drum level is effectively achieved with the usual cascade and feed forward control developing from drum level, steam flow and water flow.

However, closed loop control of the integrated flash smelting and slag cleaning furnace presents serious problems. For example, continuous temperature measurement of a furnace gas at about 1450°C, containing a high level of SO_2 and entrained dust is an almost insoluble problem. Likewise the percentage nickel in the molten matte bath along the length of the furnace cannot be directly measured on a continuous basis. Hence, with two of the main control parameters denied direct continuous measurement, optimisation by inference through an indirect control model is the only possible solution. This topic is returned to in the section on Process Modelling.

LOGIC CONTROL

In 1977 WMC opted to use a microprocessor as the basis of control for a new converter flux and reverts feed system. The microprocessor was chosen as it offered lower cost, increased flexibility, hardware standardisation and a lower number of components. As no complete microprocessor systems were then available it was decided that, although proprietary boards should be used wherever possible, in-house design would be required to complete the task. National's 16 bit PACE microprocessor was selected as it was well supported in Australia and offered flexible, minicomputer-like, addressing modes. The PACE system also was expandable with a reasonable selection of peripheral boards being available.

A few months after the converter project was completed a similar PACE system was used to control the coke and reverts charging of the new flash furnace. However at this time it was becoming evident that the PACE system was most unlikely to receive further international support. As Intel microprocessors were now more popular an 8085 board was utilised for the next project. Although this 8085 single board computer was not expandable this was not a problem for this project involving simple pneumatic transport.

When six months later a decision was required on a control system for the laterites crushing and grinding project, the extent of the proposed plant ruled out this 8085 single board computer. In its proposed form the control system was required to have approximately 180 digital inputs and 110 digital outputs. At the time of this project in 1979 there were still no suitable microprocessor based control systems available commercially at a reasonable cost.

The S100 computer bus, though, did seem sufficiently flexible to allow expansion to cater for the large number of inputs and outputs required. A Z80 microprocessor was the obvious choice at this time as it was the most advanced microprocessor commercially available in Australia and Z80 based S100 boards were being offered by a number of different manufacturers. There was also a reasonable selection of peripheral S100 boards from which to choose. Finally, programs written for the 8085 could still be run on a Z80 based system.

All S100 boards selected were proprietary. In-house hardware design was only necessary for input and output (I/O) interface boards. One feature of the control system design was the replacement of mimic and alarm panels, and stop start buttons by VDUs and keyboards. This reduced panel size and enabled more information to be presented to the operator.

The configuration of the laterite control system when installed consisted of a processor, memory, I/O, VDU and keyboard in each of three control rooms. There was an additional processor, memory and I/O installed in the laterites switch room. These four processors all communicated serially over distances of up to 165m.

Monitor and compiler programs for this system were produced in-house. These programs have made this control system design extremely flexible. The actual logic program is similar in principle to those now used for PLC's but has the added advantage of being able to accept keyboard commands and write messages to the VDU on a real-time basis. The monitor, compiler and logic programs are all stored in a memory type (EPROM's) which is not disrupted by power failures. With the system off-line using the monitor and compiler, it is possible to either temporarily or permanently change the logic program. This enables changes to plant control logic to be easily performed and facilitates fault finding.

This laterites control system has proved to be reliable, the only weak point being the optocouplers. Designs to date have always used some form of optical isolation for inputs and outputs in the microprocessor control systems to isolate the electronics from the electrically harsh plant environment. The isolation has been successful but at the expense of the occasional failure of an optocoupler.

After the success of the laterites control system a number of very similar systems have been installed at the company's Kambalda operation.

Recently PLC's have successfully been used for controlling small systems at KNS. This practice will continue in future. Where a proposed control system requires more than 50 I/O or a greater degree of sophistication, however, the laterites type would be used. To further increase the versatility of this type of control system, it is intended to make improvements in the area of analog signal processing.

DATA ACQUISTION

Subsequent to the laterites control project a
number of projects of a data acquisition nature
have arisen. As a Z80 S100 development system
had already been established, the same hardware
base was utilised for all of these projects.

The first of such projects involved accessing,
processing and printing information for an aisle
crane weigh system. The next undertaking was a
gas turbine temperature monitoring, alarm and trip
system. The multiplexing of low level signals
made this a more complex project. A S100 board
was designed in-house to handle the multiplexing
and analog to digital conversion.

The most complex data acquisition type project has
been the power supervisory and control system
installed mid last year. The State Energy
Commission (SEC) required that a supervisory
system be installed before interconnection with
their grid was completed. Although proprietary
systems were available it was decided to produce
an in-house design on the basis of cost and
previous success with microprocessor systems. The
result was certainly the most powerful control
system yet produced within the company.

A data acquisition computer was installed in the
power house substation at KNS to accept
approximately 80 analog inputs and 70 digital
inputs. The analog inputs were real power loads
for feeders on site and real and reactive loading
for both generation at KNS and lines to external
substations. The status of the KNS grid and
interconnections to external points defined the
digital inputs.

The data computer serially communicates digital
inputs and processed analog inputs in both
directions with an identical data computer at
Kambalda. This communication is performed over a
30 km radio link. Some digital data from Kambalda
and KNS is relayed to the SEC in parallel form
along a 300m cable. As well the data computer
accepts some analog and digital information from
the SEC.

Visual outputs from the data computer consist of a
VDU monitoring processed analog data and a mimic
panel, both of which are located in the power house
control room.

As power generation at KNS is governed by the
furnace offgas and hence by the WHB steam flow
rather than by the plant electrical load, the
consequences of being islanded from the state grid
are severe. Therefore, one of the important
duties of the data computer is to monitor internal
generation and load, and to match these by load
shedding should an island situation occur. The
operation of this is critical in the first few
minutes before standby generation comes on line.

A graphics computer was installed in the power
house control room at both Kambalda and KNS. It
requests analog and digital information from the
data computer and is responsible for short and
long term data storage, reporting, accounting and
driving the colour graphics display. This colour
monitor presents detailed mimics, load flows,
trends and peak demand management advice. A hard
disk is used for short and long term storage.
Permanent records are maintained on floppy disks.

Previous data acquisition type systems at KNS have
been programmed in assembly language. The
complexity of this supervisory system though, made
it more expedient to use a combination of the
monitor and compiler previously developed for the

laterites logic system and assembly, Basic and
Fortran languages.

After the power supervisory system was installed a
computer interface was developed for the flash
furnace Mod III. This interface was designed to
allow computer access to the loop trend data that
is available on the Mod III's system bus. The
computer running the flash furnace model is now
able to immediately update the model parameters
available from the Mod III bus. The interface
allows communications in both directions and it is
intended to use the computer to directly update
setpoints.

One of the critical parameters for the operation
of the flash furnace is the preheated air flow.
To obtain the best possible accuracy commensurate
with reasonable cost a Z80 S100 system was
installed to calculate standardised air flow.
This data acquisition system presently inputs via
field transmitters, ambient temperature and
pressure and the differential pressure signal from
a venturi. From these it calculates air flow
correcting for the variable expansibility factor.
A humidity transmitter may be added in the future
and the program modified to take this factor into
account.

The low signal level multiplexing and analog to
digital conversion board, developed for the gas
turbine temperature monitoring project appears to
have potential. It is intended to further develop
this board, to produce a data acquisition input
board capable of accepting a very wide range of
signals. This board would have immediate application in data logging projects.

PROCESS MODELLING

A computer model of the furnace has been developed
with the principal objective of minimising energy
costs and enhancing operational flexibility.
Figure 3 shows a flow sheet of what is essentially
a thermodynamically based model supported by
empirical correlations. As KNS is unique in the
world, in that flash smelting and electric slag
cleaning operations are incorporated within the
one furnace, the model has a complexity not
encountered with other flash furnace models.

The first use of the model for furnace control
purposes was to produce a table of target values
for operating parameters at various feedrates.
This table was generated on a Cyber mainframe
computer in Perth and proved to be of such value
to operating personnel that the model was put on
a dedicated minicomputer located at the smelter.
By solving the complex heat and mass balances
within the furnace in seconds, the computer model
provides advice to the control room operator on
how to fine tune furnace conditions.

In the KNS model a distinction is drawn between
the actual grades of matte and slag tapped from
the furnace and the grades produced in the flash
smelting reaction shaft. This is necessary to
take account of revert and converter slag
additions to the slag cleaning section. The
various intermediate matte and slag weights and
compositions along the furnace length are
unknown so there is no strict mathematical
solution to yield reaction shaft grades. However,
an empirical correlation based on regression
analysis of operating data yields a satisfactory
estimate.

From these estimates of reaction shaft grades the
stoichiometric oxygen requirement for concentrate
combustion is calculated. The necessary preheated
air flow is then either increased or decreased

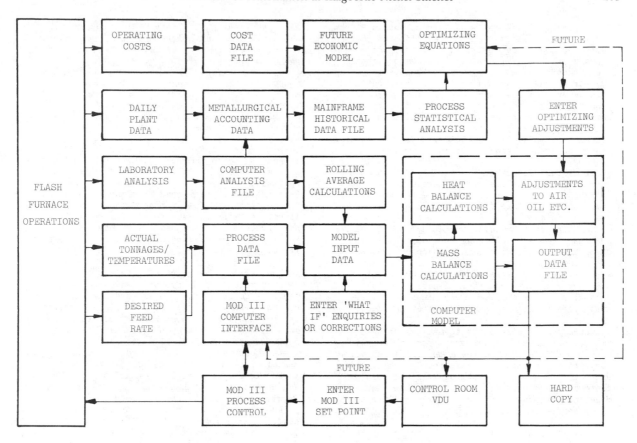

Fig. 3. Flow sheet of flash furnace management systems

according to an empirical correlation between reaction shaft matte grade and the efficiency of oxygen utilisation. This feature is used to control matte product grade. The net heat of combustion from concentrate and coal in flux is also computed, along with the additional heat required to account for heat losses. Any heat deficit is made up by oil injection, for which the requisite combustion air is also calculated.

In practice, the reaction shaft is treated as a separate entity from the slag cleaning section. Power input to the electrodes in the latter section is calculated to provide not only the heat necessary to maintain the temperature of slag coming forward from the reaction shaft, but also the energy for melting the revert additions and for bringing the returned converter slag up to furnace temperature.

Data input to the model is of two types, process operating data and analytical data, both of which are held on regularly-updated computer files. Some process operating data is accessed from the Mod III. However, some of the parameters which go into the model are not available on the Mod III. One problem is measurement of the high temperatures involved. Temperature measurement is complicated by the aggressive nature of matte and slag and by the high dust loads in the furnace. Consequently it is not possible to continuously measure either liquid or gas temperatures accurately. Fortunately the furnace is a large heat sink and intermittent measurements of matte and slag temperatures suffice as they are not subject to rapid fluctuation. These values are entered manually into the computer. The model is quite sensitive to waste gas temperature and, until

an accurate and reliable method is found for measuring this parameter, a constant in the model will continue to be used, as it gives steadier control than to input measured temperatures.

Analytical data is entered into the computer from a terminal in the laboratory and an eight-hour rolling average analysis of the individual two-hourly samples of matte, slag and furnace feed materials is computer-calculated when the model is run. The volume of matte and slag held in the furnace bath is such that the residence time, with normal tapping frequency, is of the order of 24-36 hours. Consequently there is no need to analyse matte and slag on a continuous basis, even if it were technologically feasible.

Likewise there is no need to have the model running continuously and updating setpoints when the response time for the furnace is so long. However, individual control loops for feeds to the furnace have much shorter time constants, so it is important to use time averaged values for such model data. The Mod III computer interface provides the facility for doing this automatically.

PROCESS OPTIMISATION

There are three main considerations in non-ferrous smelter process optimisation: compliance with environmental standards, metallurgical production control and cost control. Having ensured that the first of these is met, there is a need to set objectives for which an optimal mode of operation can be established.

KNS is one processing unit in an integrated nickel production organisation. There are three distinctly different nickel concentrate suppliers upstream of the smelter and several matte refinery customers downstream. Thus simply from the aspects of feed availability (and composition) and market requirements, the constraints on smelter operation are numerous. What is optimal furnace operation as such may conflict with process optima at other upstream or downstream units in the organisation. Flexibility in furnace operation is therefore crucial and what constitutes optimality will vary from time to time as the objectives and constraints vary.

To allow for the number of possible permutations which occur in the real world, a computer-based optimisation procedure for the smelter would be unavoidably complex. For this reason it is the practice at KNS simply to use the computer model of the furnace to predict operating conditions for specific hypothetical cases in order to assist management to select the case which most nearly meets the objective at the time. In the current world climate of depressed non-ferrous metal prices and excess production capacity, cost control will normally be the dominant consideration in establishing optimum operating conditions.

Computer generated reports to management on smelter performance incorporate comparisons of actual performance with the model-predicted values for various parameters, thus greatly facilitating management control of metallurgical and cost factors. For instance, use of the flash furnace model has helped to reduce energy costs per tonne of concentrate feed by 50% between 1981 and 1984.

Control of matte and slag composition has also improved significantly. For example, the model calculates the silica flux addition rate needed to achieve a preset $Fe:SiO_2$ ratio in slag and it is now possible to control the variance in the $Fe:SiO_2$ ratio in slag to within less than 1% (recent plant data taken over 252 days). Matte grade is controlled by varying the furnace oxygen to concentrate weight ratio either side of the stoichiometric requirement; the variance in matte grade tapped from the furnace is now close to the limit of routine analytical accuracy.

Variance in matte grades has been reduced to a level where extraneous influences on optimal furnace performance are now readily identifiable. As an example, a recent shift in matte grade was traced to variations in concentrate feed particle size, and as a consequence a minimum fineness in concentrate receipts is now specified.

In addition to using the furnace model for short term optimisation, it is regular practice at KNS to apply statistical analysis to historical operating data to establish and refine longer term empirical correlations. The existence of a computer data base facilitates this work. The outcome of such analysis is normally applied to improving the metallurgical performance of the smelter, and correlations used within the furnace model are updated by this means.

IMPACT OF AUTOMATION

Non-ferrous smelters are typical of industries in which the concept of production line automation is difficult to apply. Automation of discrete process units within a smelter can be achieved, however, greatly improving reliability and flexibility which leads to increased productivity. Automation also facilitates information gathering which allows optimisation techniques to be fully utilised. The benefits of automation have more recently been achieved through the agency of integrated micro electronics.

Although integrated electronics are susceptible to the dust, heat and moisture typically found in non-ferrous smelters they can survive when suitably packaged and protected. The very nature of integrated electronics allows a multitude of components to be packaged down to a small size less accessible to the environment and prying fingers. This reduction in size, together with a minimisation of moving components and inter-connections all assist to enhance reliability.

In particular, since microprocessor based equipment is readily reprogrammable without the need for hard wiring changes, it is very adaptable to changes in operating philosophy. Within a single decade technology has progressed from the point where it was necessary to fully specify the logic of an operation before it was possible to specify the hardware, to a point where it is now possible to specify the hardware before determining the plant logic.

Although it may be surprising to some, automation has had ready acceptance by a wide cross-section of the work force at KNS. Operators have accepted the new equipment primarily because of the additional information which can so readily be provided, enabling them to better understand what is happening to the plant in their care. Digital systems can provide not only numeric displays but graphic representations and analog bar displays which provide an excellent interface between man and machine. Maintenance workers have just as readily accepted the new equipment; firstly because the inherent reliability means that it seldom breaks down and secondly, because should it fail it is capable of providing extensive diagnostic information. Management effectiveness has been considerably enhanced by computer based data collection and reporting, together with modelling and optimisation techniques.

Naturally enough with operators, maintenance personnel and management all better informed the plant has run more smoothly with consequential improvements in manpower productivity and plant availability.

Computer modelling allows tighter control of operating parameters by avoiding the fluctuations and wasteful practices frequently associated with 'seat of the pants' type operation. It is only through computing that the complex heat and mass balances of a modern day smelter can be accomplished quickly enough to be of operational value. Computer modelling also allows theories to be tested and the sensitivity of major process variables to be determined without using a large operating plant as an expensive pilot plant.

In seeking optimisation, the sensitivity of the process becomes such that, to be self correcting, a plant needs to have a high level of automation. In other words optimisation tends to a condition where the process becomes increasingly sensitive to small changes, just as a sailing craft pointing high on a strong wind is sensitive to wind and rudder. High level automation will allow larger and more complex plants to be built with improved energy efficiency and increased output per man.

Automation has always raised the spectre of lost jobs and will continue to do so until the problems and consequences are properly addressed. It is suggested that the spectre need not exist in the non-ferrous smelting industry. When one considers the nature of the work and the varied tasks required of the operators in a smelter there is no reason foreseen why automation in particular should significantly reduce the number of jobs available, although some redistribution will be necessary as new techniques evolve and old skills disappear. Employers must be cognisant of this need for change and provide retraining for these new jobs whilst employees must be tolerant and adaptive towards necessary change. At KNS automation is not seen as a means to reduce manpower but rather as a means to improve operating efficiency and cost effectiveness thereby increasing market competitiveness and, if the market allows, increased production from an existing plant and work force without additional capital expenditure. Gains in energy efficiency are likely to be more cost effective and long lasting than any paring of manpower.

Community demands for improved living standards, including increased leisure time and environmental awareness, create a dichotomy: community demands place greater costs on industry but, for the community to prosper, industry must remain competitive. Industry must automate to achieve this viability, with some possible redistributions of labour. The only alternative is plant closure with resultant massive community disruption.

THE FUTURE

Until the technical problems associated with reliable, continuous furnace off gas temperature measurement and hot metal analysis can be overcome, total closed loop control of flash smelting will not be possible. Furthermore, although it can be foreseen that the temperature measurement problem could be resolved in the next few years, it is most unlikely that the hot metal analysis problem will ever be resolved.

As new technologies become commercially available they will be considered for application in the inevitably increasing role of automation. For example, at KNS the use of robots for tapping and stopping matte and slag holes will be investigated. One new technology which has been recently implemented with the power house data computer is fibre optics. Fibre optics have great potential for data transmission in heavy industry as they are immune to electrical noise. This, together with their high transmission rate capability, should see them become increasingly utilised.

Although initial in-house microprocessor development was primarily devoted to logic control, more recently the emphasis has swung to data acquisition. Such development will continue in the near future with projects being implemented to enhance the centralised reporting of all significant plant data, which will aid management and facilitate optimisation in other areas such as converter operations. This in-house development has allowed investment to be minimised whilst confirming that plant control and operator performance can be improved and management reporting enhanced. In the process, both designers and users have gained valuable experience in how to make the best use of these technologies. However, whilst customised software and interface developments will continue to remain the province of the end user, the present trend for proprietary equipment manufacturers to develop distributed networks with an integration of microcomputing and process control, could signal an end to in-house hardware development later in the 1980s. Cost and flexibility will be the final determinants.

At KNS direct control of the process set points by the furnace model will be introduced, although data relating to various inputs, such as analytical data, are likely to be entered manually, at one point or another, for some time to come. Data may eventually be entered directly from the analysing instruments but this will depend upon the replacement of some existing analytical methods and instrumentation.

Although models for copper converting have been developed elsewhere they are not translatable to nickel converting. As a consequence a thermodynamically based model of the converting operation is being developed. This will enable the blow interval to be determined from known charge weights and analyses, together with the measured process gas flow and oxygen percentage.

Future non-ferrous smelter models will need to include an economic component, especially when the consequences of further environmental awareness confound the complexities of furnace feeds with secondary recycle materials. Environmental regulations are already having a profound effect on the operations of smelters in developed countries where operating conditions must be controlled not merely to meet production schedules, product specifications and budget constraints but stringent waste specifications as well.

Future smelter automation will ideally incorporate a self learning, fully adaptive, process control model responsive to meteorological conditions, the economic climate, the vagaries of numerous feed materials and (hopefully) the requirements of a product or two.

ACKNOWLEDGEMENT

The authors wish to thank the board of directors of Western Mining Corporation Limited for permission to publish this paper. We wish to acknowledge the efforts of WMC personnel, in particular Mr H. McKay, who have contributed so much to the success of automation at Kalgoorlie Nickel Smelter.

The Current State of Automation and Areas for Future Development in the Electrolytic Zinc Industry

B.R. CHAMPION

Senior Instrument Engineer, Electrolytic Zinc Company of Australasia Limited, Risdon, Tasmania

Abstract. The paper discusses how automation is steadily proceeding throughout all areas of the electrolytic zinc industry. Fluid bed roasters with integral sulphuric acid plants are now on computer control. Highly efficient, continuous leaching and purification stages, with feed-forward predictive and optimisation computer control, precede electrolytic cells with on-stream solution analysers and automatic stripping of deposited zinc. Controlled electric induction furnaces feed continuous casting lines, equipped with automatic stacking and strapping machines, after computer based metal analysis has confirmed metal purity or alloy composition. The paper also discusses the use of computer models tied to process control systems, and the equipment and solution measurements that could be automated with advantage, using modern micro-based controllers and analysers.

Keywords. Computer control; electrolytic zinc; feed-forward; predictive control; modelling; on-stream analysis.

INTRODUCTION

Automation in the electrolytic zinc industry started many years ago when manually raked hearth roasters were replaced, first by mechanically raked roasters, and then by flash roasters. Flash roasters have now been replaced by fluid bed roasters. Continuous leaching superseded batch leaching, and continuous casting has replaced ladle casting. The relatively recent introduction of electronic and pneumatic logic, computers, microprocessors and P.L.C.'s (programmable logic controllers) has enabled the electrolytic zinc industry to rapidly automate, and to reduce the number of arduous, hazardous and manpower intensive jobs in the modern electrolytic zinc plant.

The following operations are common to most modern zinc plants:

(1) blending of the zinc concentrates (mainly zinc sulphide but including lead sulphide and many impurities) from several sources;

(2) fluid-bed roasting the concentrates to produce calcine (mainly zinc oxide) and, sulphur dioxide, which is then converted into sulphuric acid;

(3) leaching the calcine in spent electrolyte (100-200 gram sulphuric acid/litre) and purifying the solution to provide a zinc sulphate solution (100-150 gram zinc/litre); followed by

(4) electrolytically depositing high purity zinc metal onto aluminium cathodes, stripping the zinc deposit, and induction furnace melting the metal. Then casting the zinc metal and a variety of alloys into suitable ingots for uses such as galvanising and die casting.

CONCENTRATE HANDLING

Concentrate handling varies according to the means of delivery. Cominco Ltd. (Cominco), at their Trail plant in Canada, receive concentrates by rail from their two major mines (Broster, 1976).

They have installed automatic rail car unloading and conveyor belt feeding with closed circuit television monitoring of the tramp material screening station and of the automatically positioned tripper above six fluid bed roaster day-bins. The extractor belts for the day-bins are variable speed driven and nuclear weightometers provide reliable measurement for computer control of the feed rate, and the ratio of two types of concentrate for satisfactory operation of the fluid bed roasters.

At Budelco B.V.'s Budel zinc plant in Belgium, up to 34 different concentrates are received by rail wagon, and computer logged and blended before being fed to fluid bed roasters.

The blending of concentrates can be critical to fluid bed roaster operation and Rhur Zinc, at their Datteln plant in Germany, computer program their feed of 5 to 7 concentrates to control the total lead content to $\cong 2\%$, as $>3\%$ can result in setting of the bed.

The Timmins plant in Texas Gulf in Canada is adjacent to their ore concentrator. The concentrator is computer controlled with two hourly manual input of metal prices to optimize the return on their marketable concentrates by adjusting the concentrator performance. This has been in operation since 1972.

AMR–H

At the Risdon, Tasmania, plant of the Electrolytic Zinc Co. of Australasia Ltd (E.Z. Risdon), two new dual purpose gantry wharf cranes were recently commissioned to handle incoming concentrates from at least four sources, phosphate rock and other raw materials, and export materials in bulk or container, such as slab zinc, die casting alloys, lead/silver residues and cobalt and copper oxides. These unique cranes are P.L.C. supervised for position and speed (Jones, 1984).

CONCENTRATE ROASTING AND ACID PLANTS

Most modern zinc works have installed Dorr-Oliver or Vieille Montagne (V.M.) - Lurgi type fluid bed roasters, a notable exception being Hoboken -Overpelt in Belgium, who have developed a pelletized concentrate fed fluid bed furnace, which is capable of handling concentrates containing high concentrations of lead. The fluid bed roasters operate at about 900°C and are followed by waste heat boilers, cyclones, electrostatic precipitators, and gas cleaning equipment prior to conversion of the sulphur dioxide gas into sulphuric acid. Most roaster/acid plant complexes are heavily instrumented, especially in the furnace and gas conversion stages, where temperatures are critical.

Norzink AS, at their Odda plant in Norway (Norzink AS), have, for many years, data logged their roaster/acid plant processes and are probably now on computer control.

Cominco have had an IBM 1800 Direct Digital Control computer system controlling three roasters and three acid plants since the mid 1970's and are believed to be upgrading the system.

Painter (1980) reports that at Jersey-Miniere's Clarkesville plant in Tennessee, USA, the roaster/acid plant is controlled by a Foxboro Videospec system.

This type of system, backed up by two Fox 300 computers and working with three Modicon P.L.C.'s was commissioned by Canadian Electrolytic Zinc (C.E.Z.) at their Valleyfield plant in Canada in January, 1983, to replace the analogue/control equipment on three roasters and two acid plants, and to provide measurement and control of a fourth roaster and a third acid plant (Gibb, 1984). The control system is fully backed up, a desirable feature for control of this part of a zinc plant, where there are fast and critical loops such as boiler feed and drum levels.

The Akita Zinc Co. Ltd (Akita) at the Akita plant in Japan has a central plant computer operating a distributed control system on their roaster acid plant system (Mealey, 1973).

Other zinc plants appear to have individual loop analogue control systems. Many have turbo alternators to recover power from the waste heat boiler systems before the system is used for process and building heating.,

LEACHING AND PURIFICATION

Most modern electrolytic zinc plants operate continuous leaching and purification sections, and some form of residue treatment plant to recover zinc from ferritic residue not dissolved in the conventional neutral leaching stage. Neutral leaching involves dissolving zinc oxide from the calcine in spent cell room electrolyte under pH control, precipitation of some ferric iron to remove certain impurities, followed by thickening to separate undissolved residue and precipitated solids. Residue treatment plants usually involve a hot acid leaching step to dissolve zinc ferrite, separation of undissolved lead/silver residue if warranted, then precipitating dissolved iron as jarosite, goethite or haematite. Impure solution from the neutral leaching stage is purified of major impurities such as copper, cadmium, nickel and cobalt by cementation with zinc dust in one or more stages.

All of these processes involve chemical reactions, and can require the measurement of pH, flow, density and conductivity, and the use of colorimetry, selective ion and polarography and other on-stream or sample analysis techniques. Many of these measurements are used for feedforward and/or feed-back control.

Some of these processes are fairly slow, but require a high degree of accuracy on process measurements, many of which can be affected by the settlement of solids, precipitation of basic zinc sulphate and, worst by far, gypsum precipitation. The latter has such a detrimental effect on pipes and cooling towers that most zinc plants are adding or incorporating separate gypsum removal stages.

Many zinc plants are incorporating computer or computer backed distributed control in their modernizing of existing leach/purification plants. As mentioned earlier, Akita have central computer control on their whole plant, and, Jersey-Miniere use Foxboro Videospec to cover their whole plant.

Maughan (1977) reported that Norzink AS had installed a Noratam computer system to data log and remote set-point control the leach/purification plant. Budelco B.V. have also installed a computer for this purpose, as have Mitsui at their Hikoshima plant in Japan (Mukae, 1977).

E.Z. Risdon, in 1981, as a prelude to its modernization program, set up a A$1.5M computer controlled pilot leach/purification plant (Matthew, 1983). This, 1/1000th scale plant incorporated strong-acid, high-acid, neutral leaching, preneutralization and low contaminant jarosite precipitation stages, and sections for gypsum removal and activated carbon treatment, two stage hot zinc dust purification and selective zinc precipitation. It also controlled two of the cascade cells in the main plant cell room, which were used to electrolyse the pilot plant solution. Full computer control by a Foxboro FOX3 computer enabled time proportioning electric temperature control of many reaction tanks, frequency proportioning control of double acting plug type "poppet" valves for accurate slurry metering and pH and conductivity control, mass-loss control of screw and plunger ("horizontal poppet" valve) feeders and complex feed-forward control loops for predictive neutralization and flocculant flow control. Data logging (one disc/week) of the 90 controlled and 80 monitored-only process variables greatly simplified subsequent computer interpretation of plant results in E.Z.'s Research Department. The Risdon Pilot Plant continued to operate through September 1984 to compare hot zinc dust purification processes, evaluate heat exchanger performance and optimize electrolytic cell performance.

E.Z. Risdon also put on line in 1983 a Honeywell TDC 2000 system to monitor and control a two stage iron purification process. The ready acceptance, by plant operators and supervisors, of the VDU (Visual Display Unit) plus keyboard, for process monitoring and control has led to many other peripheral measurements being tied into the TDC 2000 system.

ELECTROLYSIS AND CASTING

The electrolysis process in most zinc plants is at a controlled acidity (inferred from electrodeless conductivity measurement or manual titration) and/or a controlled specific gravity.

These parameters are controlled by the addition of pure zinc sulphate solution to recirculating or cascading spent electrolyte. Temperature control may be effected by cooling coils, by varying the speed of cooling tower fans for recirculating solution or by solution bypass around the cooling towers. Power control to meet demand limits or deposition times is also practised.

However, the very high labour content, and the reluctance of many workers to endure hard physical labour in short bursts during a shift, has led to one of the major automation phases in the electrolytic zinc industry. Mechanical stripping machines, either electronic or pneumatic logic controlled, have been developed to strip the zinc deposits from the aluminium cathodes while minimising damage to the cathodes and with as few rejections as possible. Operating cycles as short as 6 seconds have been obtained with 1.1m cathodes, and 15 seconds is typical for "Jumbo" (2.6 or 3.2sq.m.) cathodes. Three basic types of stripping machines are used:

(1) V.M. type, which lifts the cathode up between two starting and two ploughing knives to initiate and then peel off the zinc deposits;

(2) Mitsui type, which hammers the zinc to loosen the deposit, then inserts, with the aid of air jets, stripping wedges to complete the separation; and

(3) Montedison type, which relies on a hinged flap on the plastic edge strips. The hinged flap is swung aside to allow horizontal penetrators to detach the upper edge of the zinc from the aluminium and to open a gap for shears to slide vertically down the whole length of the cathode to strip the zinc deposit (De Michelis, 1974).

Akita developed a multi cathode, multi knife stripping machine for its highly automated cell room and Asturiana de Zinc have also tested their own stripping machine at their Aviles plant in Spain. Toho Zinc Co Ltd (Toho) at Annaka in Japan have also developed a stripping machine and an automatic monorail system for transporting cathodes to and from the stripping machine.

Some of the more recently built zinc plants have P.L.C.'s on their stripping machines.

Cathode handling involves lifting up to half of the cathodes in a cell at a time (50 Jumbo cathodes) with electric or air motor hoists or cranes, transporting them to washing and/or fluxing baths, then to the stripping machine and returning the stripped cathodes to the cells.

Vieille Montagne have fully computerized this process (Freeman, 1980), as have Budelco V.B., Cominco and Hoboken - Overpelt. Toho use operators to control the cathode lifting and replacement, as do National Zinc at their Bartlesville -US plant, and Rhur Zinc. Others use operators to control the transportation to the stripping machine.

The new Cominco electrolytic and melting plant (on stream in October 1983) has extended automation from the cell house to the cathode fabrication shop where a computer controlled mechanical robot fabricates 200 cathodes per shift (10 workers per shift manually produced 100 new cathodes per day). Cominco's totally automated cell house features four computer controlled cranes for lifting, transporting and replacing cathodes, four Cominco designed (based on the Montedison type) stripping machines and a microprocessor based electrolyte analyser. This on-line Continuous Electrolyte Quality Monitor (CEQM) instrument has been patented by Cominco. It operates by measuring the overpotential of the nucleation of zinc from the cell feed solution on a moving aluminium wire. The analysed measurement is then used to control the addition of reagents to optimize the electrolysis current efficiency.

The Cominco melting plant uses three electric induction furnaces to feed three automatic casting machines with automatic skimmers, and then automatic stacking and strapping machines bundle the finished zinc slabs (Honey, 1983; Lewis, 1984).

E.Z. Risdon commissioned a new casting plant in 1971. It features three melting and three alloying furnaces, which feed three automatic casting machines with their electropneumatic logic controlled automatic stacking and strapping machines. Casting is also tied to metal analyses, which are performed on a mini-computer-controlled, direct reading, emission spectrometer.

C.E.Z. have automated their casting plant, and were commissioning an automated Jumbo casting line (250 kg) as well as an automated slab (25 kg) casting, stacking and strapping machine (Tanner, 1983).

Most other zinc plants have also automated all or part of their slab casting lines and especially the stacking and strapping processes, which are manually arduous, dangerous and labour intensive.

The electrolysis and melting processes consume 80-90% of the electrical power used in electrolytic zinc plants. At E.Z. Risdon, manual control of transformer tap changing to match maximum demand limits (as various parts of the plant switch on and off) is being replaced by a G.E.C./D.E.C. PDP11E computer control system. This will reduce power to the cell room as the total power demand approaches the contract limit. Its use will increase the maintenance on the tap changing gear and a microprocessor control system is being investigated to smooth and synchronize the step changes due to the temperature control on the three melting furnaces.

Jersey-Miniere at their Clarkesville plant, have been using a load-optimizing computer since start up in 1979.

STORES, ACCOUNTS AND PAYROLL

Most zinc plants have installed large mainframe Electronic Data Processing (EDP) systems and the E.Z. Risdon system is probably typical. Stores inventory includes reagents, consumeables and plant spares. Ordering and accounting is mainly handled by the E.D.P. system. Works payroll is prepared by the computer, at this time from data entered from time cards, but, in the near future, conceivably from works-wide time clocks. Metallurgical records are now stored in the E.D.P. system, providing a data base for Research Department and plant metallurgists on plant operations. The mechanical, electrical and instrument equipment registers are being entered into the E.D.P. data base to provide 24 hour on-line maintenance data.

COMPUTER MODELLING

At E.Z. Risdon the Research Department have been using computer modelling for the leaching and purification stage modernization program and solution circuit evaluation. A joint exercise with C.S.I.R.O. and Sydney University to accurately model an electrolytic cell has just commenced. A computer-backed process-control system for a zinc plant can effectively optimize zinc extraction by using the plant computer models to determine setpoints for strategic control loops. These models are also invaluable in determining the effect of, and the control scenario for, different concentrate blends with different levels of impurities. The chemical industry, including Monsanto, Shell and Dupont, are using computer model process control (Haggin, 1984).

ON-STREAM ANALYSIS

The complex chemical nature of zinc plant solutions have led to the development of a wide variety of special monitoring systems. Norzink -AS have developed a pH and conductivity sampling system to extend the life of the electrodes. Used in many zinc plants, this system operates by drawing up a pulp sample into a measurement chamber via a cooler, stabilizing the electrodes, reading and holding the pH or conductivity value, back flushing with water and/or dilute acid, blowing the liquid from the sample tube and then sucking up a new sample. E.Z. Risdon developed a variation on this system with a long insulated water-cooled tube to enable a 25°C sample to be obtained from reaction vessels operating at over 90°C and in which the level varies over $1\frac{1}{2}$ metres. The pH electrode life has been extended from 2 hours to several months. E.Z. Risdon have also developed deep-tank electrodeless conductivity sensors and analysers to measure and control the acidity of residue leaching pulps with up to 300g H_2SO_4/L, at temperatures exceeding 90°C.

The E.Z. conductivity systems are also used to monitor and control cell-room electrolyte and jarosite discard pulp. They are used with modified sensors to measure and control variables in the Risdon pilot plant.

V.M. have developed a solution purity meter, which, unlike Cominco's CEQM system, is a stripping voltammetry type instrument with a rotating graphite cathode. It provides a plot each 30 minutes, which shows the effect of Cu, Cd, Fe, Pb and solution clarity.

Montedison and Norzink AS and others have also developed Purity Index Meters, which give a reading that can be related to the expected electrolysis current efficiency.

Mitsui have developed and patented a variety of onstream pH, Fe++, electrolyte impurity, conductivity and weighfeeder systems for their computer controlled Hikoshima zinc plant (Mukae, 1977).

Square-wave polarographs using anodic stripping voltammetry are used in many zinc plants, together with atomic absorption mass spectrometers and colorimeters to manually analyse zinc solutions during purification. Only limited continuous onsteam use is made of these instruments due to complex interpretation and the common problem in electrolytic zinc plants of fouled sampling lines, due to basic zinc sulphate and gypsum precipitation.

FUTURE DEVELOPMENTS

As with most complex hydrometallurgical processes, instruments for measuring chemical or physical parameters must be ultra-rugged and ultra-reliable. Process operators quickly tire of erratic or unreliable instruments and find alternative ways to run their plants. The instruments are often suspected first and frequently sensors are requested to be changed before an evaluation of the plant operation is made to determine whether the fault is a plant malfunction or a sensor failure. Thus, the use of computer backed control systems with their ability to diagnose plant performance and reject suspect measurements should lead to improved plant operating performances.

Computers are only as good as the data that is fed to them, and therefore time must be spent by the zinc industry and its suppliers to develop reliable specific-ion on-stream analysers for zinc, cadmium, copper, iron, lead and antimony and a host of less-important elements. Simple specific ion electrodes for most of these ele ments are not suitable due to interference from other elements in the solution. Recent test work with a copper specific ion electrode at E.Z. has shown that it is usable but requires hard polishing at hourly intervals to retain its sensitivity. The solution presented to the electrode requires vigorous agitation about the electrode, and temperature control is also essential. Unfortunately, its sensitivity to copper below 1 mg/L is poor and its use for detecting filter cloth breakthrough does not look promising. It was successfully used for feed forward control of zinc dust addition in E.Z.'s Risdon pilot plant.

Filters and heat exchangers can be high maintenance manpower consumers. Automatic filters can be used by cycling solution and flushing operations under time and/or pressure drop control.

The heat transfer co-efficient of heat exchangers can be computed and used to detect a drop in performance. Automatic change over to a flushing cycle can be made, reducing operator intervention and maintaining continuous plant throughput.

The petrochemical industry with its "clean" products can in most instances be laboratory simulated. Not so the hydrometallurgical zinc extraction process. Pilot plants provide scaled down operating conditions but present problems with control devices and sensors. The controlled feeding of zinc ferrite residue and zinc dust

into the E.Z. Risdon pilot plant were two such
exercises. The former was found to compact as
the scraper in a screw feeder attempted to keep
it free flowing. The speed control on the feeder
could only be operated down to 80% due to the
constant torque required by the scraper. A
horizontal "poppet" valve was developed to
provide 60:1 turn down and computer controlled
mass loss maintained better than 1% feed rate
accuracy. The smallest available screw feeders
were purchased to control the zinc dust additions
to the two stages of purification, however, these
were found to provide 10 to 20 times too much
zinc dust at their minimum speed. Diverter
chutes to split the feed 10:1 and super slow
gearboxes with tiny hoppers for zinc dust in the
feeder enable computer controlled mass loss feed
rates as low as 2 grams per minute to be
obtained.

Scaling up to full plant size for the zinc dust
feed control presents additional problems as wet
zinc dust dries like cement and excellent dust
control and freedom from steam clouds from tanks
is required to maintain consistent feed rates
while the high bulk density of zinc dust presents
the same problems for weigh feeding whether pilot
or full scale.

Instrument manufacturers are busy converting
their equipment to micro-processors. Beckman and
Leeds and Northrup have microprocessor based pH
analysers but neither have developed pH sampling
systems. Honeywell has released its
microprocessor-based differential-pressure
transmitter, which can be control-room calibrated
and checked via the signal wiring, but what use
is that if the sensing ports to the orifice plate
or "Annubar" are blocked with scale or dust ? It
can only be hoped that when the instrument
manufacturers have finished playing with
microprocessors in their laboratories, that they
come out into the field in the hydrometallurgical
industries and help the industry to solve the
solution interface problem so that it may
continue to reliably automate its zinc plants.

The zinc industry would be delighted to have a
reliable microprocessor-based solution analyser
that doesn't block up due to basic zinc sulphate
or gypsum precipitation, provides better than 2%
accuracy for zinc, iron, lead, copper and cadmium
etc. in solution and does not require a PhD in
Chemistry to drive it.

REFERENCES

Argall, G.O. Jr. (1984). Cominco's new zinc
    refinery. International Mining, 1(1), 58-61.
Broster, J.D. and Delong, O.J. (1976) Zinc conc
    entrate roasting at Trail. A.I.M.E.
    Conference Paper.
De Michelis, T (1974). The stripping of zinc pro
    duced by electrolysis.
    La Metallurgia Italiana, 3, 154-158.
Freeman, G. and others (1980). Comparisons of
    cell house concepts in electrolytic zinc
    plants.Proceedings   Lead-Zinc-Tin   '80
    Symposium.
Gibb, R.S. and Mezl, Z. (1984). Zinc roasters
    analog/digital process control conversion -
    A case study. The Metallugical Society of
    A.I.M.E. Conference Log Angles, Feb/March,
    Paper No. A84-11.
Haggin, J. (1984) Process control no longer
    separate   from   simulation   -   design.
    C & E N, 62(14), 7-16.

Honey, R.N. (1983). Evolution of electrowinning at
    Trail. C.I.M. Metallurgical Society, 13th
    Annual Hydrometallurgical meeting - Zinc '83.
Jones, R.M. (1984). Features of the new grabbing/
    container handling cranes at the E.Z. Risdon
    wharf. Paper presented to the Institute of
    Engineers - Australia (Tasmanian Division),
    Sept.
Lewis, A. (1984). Cominco's new electrolytic zinc
    plant. E & M.J. 185, 4, 38-43.
Matthew, I.G., Haigh, C.J. and Pammeter, R.V.
    (1983) Initial pilot plant evaluation of the
    low contaminant jarosite process. Presented
    at the 3rd International Symposium on
    Hydrometallurgy, 112th A.I.M.E. Meeting,
    Atlanta U.S.A. March.
Maughan, R. (1977). Private report to E.Z. Risdon.
Mealey, M. (1973). Hydrometallugy plays big role
    in Japan's new zinc smelter. E/MJ, 174(1),
    82-84.
Mukae, S. and Iseda, M. (1977). Implementation of
    a computer controlled system for an electro
    lytic zinc plant. ISA Annual Conference,
    137-148.
Painter, L.A. and others (1980) Jersey - Miniere
    zinc   plant   design   and   startup.
    E & M.J, 181(7), 65-88.
Tanner, C.D. (1983). Casting modernisation pro-
    ject. C.I.M. Met. Soc. 13th Annual Hydromet.
    Meeting Zinc '83.

Automatic Control of Agglomeration and Smelting Processes

R.J. BATTERHAM and W.T. DENHOLM

CSIRO Division of Mineral Engineering, PO Box 312, Clayton 3168, Australia

Abstract. Current control practice is reviewed for agglomeration processes, including pelletising, sintering and induration; and for smelting processes, including blast and reverberatory furnaces, flash smelting, the intensive bath smelting processes together with converting. The importance of sensors for on-line control and the status of mathematical modelling is discussed.

Future progress in automatic control will require improved component specific sensors and the use of validated mathematical models running in parallel with the process. The possibilities in both these areas are reviewed in light of the trends towards more intensive smelting processes.

Keywords. Adaptive control; computer control; modelling; metallurgical industries; metals production; iron and steel industry.

INTRODUCTION

The development of mineral resources involves all processes from the definition of an unmined mineralization to the final production of metal. This review of automatic control covers that part of the mineral to metal spectrum regarded as extractive metallurgy, and in particular discusses smelting and agglomeration, the main high temperature processes.

For completeness, we have included control of pelletising and granulation, given the integral part such feed preparation processes play in subsequent high temperature stages. Further, the modern trend towards intensive smelting processes makes the distinction between metal production and metal refining less clear. Control of intensive smelting processes is therefore discussed.

This short paper does not attempt to review the entire field of automatic control of the pyrometallurgical process as several conferences and reviews have been published in recent years. Control in steelmaking is well covered by Lu (1981), while advances in smelting have also received attention (Sohn, George and Zunkel, 1983). Process control and modelling have been addressed in several conferences: Weiss (1979), O'Shea and Polis (1980), Molerus and Hufnagel (1981), Kuhn and others (1982), Anon (1983a) and of particular relevance, the international conference on automatic control in mineral processing and process metallurgy (Herbst, George and Sastry, 1984) which serves as a state of the art summary of the rapidly developing field covered by this paper.

Sensors

Automatic control in extractive metallurgy is made difficult by the general problems of time varying, non-linear systems of high dimensionality. The problems are compounded by the lack of adequate measurements of process variables. The availability of suitable sensors for the harsh environments encountered has long dominated the efficacy of process control schemes. Much ingenuity and a certain degree of opportunism has been shown in using indirect measurements and local computing power to provide on-line estimates of the real states of a process. The trend has been that off-line laboratory measurements become the on-line measurement 5-10 years hence.

In 1980 a task force of the American Iron and Steel Institute identified 18 sensor needs for serious consideration. Development has now been organized in four areas (Anon, 1983b); direct in process analysis of steel using Laser Induced Breakdown Spectroscopy (also described in Anon, 1984); estimation of temperature in slabs, castings and other sections in heating furnaces using a combination of temperature and flux measurements and on-line heat transfer models; porosity measurement in steel; and on-line detection of surface defects. The laser measurement is a clear example of a laboratory technique becoming an on-line measurement while the temperature estimation example illustrates a recurring theme in this review, viz. the use of existing measurements together with on-line, phenomenological models to estimate otherwise unmeasurable states. The sophistication of some of these measurement methods under development should not blind the control engineer to the elementary consideration that the effectiveness of any measurement is governed by how representative and timely is the sample used for the measurement.

In non-ferrous smelting the most important control parameter is oxygen potential which largely determines the distribution of elements between gas, slags and mattes or metals. This parameter can be directly measured using disposable tip probes which employ electrolytic cells with a solid electrolyte based on zirconia stabilized with lime or magnesia (Floyd and others, 1984). Unfortunately the electrolytes degrade rapidly in slag environments and cannot be used for continuous measurements. Some progress towards a

213

continuous sensor of oxygen potential has been reported by Etsell and Alcock (1983). They use a non-isothermal probe, using a zirconia rod inserted through the refractory wall into a metal bath, with the reference electrode located at the cooler outer end of the rod. Fully automatic control of non-ferrous smelters will continue to be limited until better sensors are developed.

Models and Automatic Control

The non-stationary, multi-variable nature of most metallurgical processes has meant that, regardless of the on-going development of better sensors, the design and implementation of control systems has tended to be fairly simple in concept. Control has been limited to the few key variables that are thought to dominate the whole process. In most cases, fundamental understanding of the processes is limited. As a consequence, the most visible effort in control continues to be in model development and validation, with a strong emphasis on deterministic models, see e.g. the contents of Control '84, (Herbst and others, 1984). This paper will therefore highlight the state of modelling for individual processes.

As the level of understanding improves, so the sophistication of control schemes should also improve. Regulatory, or stabilizing control moves on to advisory control which in turn moves on to closed loop supervisory control. An interesting trend is the increasing use of comprehensive, deterministic models with key parameters identified in real time for advisory or supervisory control. The approach has been applied in heating furnaces (Cook and others, 1982), for the heat hardening of iron oxide pellets (Thornton and Batterham, 1982a) and has been proposed for blast furnaces (Christiansen and others, 1984).

Despite the lack of process understanding, significant control achievements are available. The conventional approach of using a complex, deterministic, dynamic model to develop simpler models for on-line control has been applied to heating furnaces (e.g. Geskin, 1980 and Vesloeki and Smith, 1982). Self tuning controllers and adaptive control have been successfully applied to a glass furnace (Haber and others, 1981), heating furnaces (Lumelsky, 1983) and to blast furnace control. In particular they have been applied to the single variable control of fuel/ore ratio based on the silicon level in the product (De Keyser and Cauwenberghe, 1981) and the multi-variable control of Rajbman (1980).

The question can well be asked why modern control theory is still finding so few applications in the metallurgical industries. It is certainly not due to any lack of installation of computers; distributed, central, hierarchial, mini, micro or otherwise! The answer, unfortunately, is that many of the processes remain poorly understood although current progress is encouraging.

CONTROL OF AGGLOMERATION PROCESSES

Smelting processes, such as blast furnaces require agglomeration of the feed material to withstand the mechanical load conditions in the furnace and to maintain an open structure for the flow of gas through the burden as it moves downward. The agglomeration itself is achieved by blending or mixing of the feed materials, followed by granulation or pelletisation then heat hardening of the granules by induration or sintering into lumps of adequate size for smelting.

These individual steps in the agglomeration chain are each well established with the automatic control trends reflecting the general requirements: to eliminate disturbances in feed materials, to use the most advanced sensors and finally, to incorporate the best understanding of the process into the control system. In recent years, as the cost of manpower has increased, as environmental constraints have been better defined and as processing units have become larger, control per se has increased in importance. Computer control is now common place, see e.g. Corson (1980), Corson and Armstrong (1981), Anon (1983c), Eisenhut, Gratias and Worberg (1981).

Feed blending and mixing

Both in iron ore processing and in non-ferrous smelting, the last decade has seen increasing emphasis on blending and mixing (Dittrich 1981). Blending piles for dry ores and fluxes are a particularly attractive way of minimizing variations in composition to subsequent agglomeration and smelting steps. In iron ore processing, and to some extent in cokemaking (Flowers, DeVanney and Ostrowski, 1982), it is common to see up to 2 weeks supply of feed material blended in a single pile. The pile is then sampled and laboratory tests used to determine the optimal process conditions for sintering and smelting. Chemical changes, e.g. the level of lime or other fluxes, are then minimized.

When handling wet concentrates, large bedding piles may be inappropriate and control of flux levels for smelting is then best handled by feed forward control based on on-line measurement of grade of the concentrate (Droste, 1980).

Granulation and pelletising

When agglomeration is achieved by sintering, some of that portion of the product which is still too fine for use as product sinter is generally returned to be mixed with the fresh feed. The resulting granulation of fresh, fine feed material onto larger pieces of return sinter has been well studied, Yoshinaga and others (1980), Rankin and others (1983, 1984) although the process dynamics are not well understood. Process control is limited by the lack of understanding of the dynamics and the lack of cost effective sensors for moisture level measurement. The current practice is to rely on minimizing disturbances by sound engineering plant design.

The moisture level in granulation critically affects the subsequent permeability to gas flow of the material in the sintering process (e.g. Barton and others, 1980), and hence the productivity of the sinter machine. Most control schemes have therefore concentrated on measurement of permeability of the granulated feed as a basis for manipulating moisture levels in the granulation stage, as described by Lückers, Voll and Boelens (1980).

When the agglomeration is achieved by pelletising and subsequent heat hardening by induration, the pelletising does not involve the recycling of such large quantities of hardened, undersize product. Recycling is still involved however, but is of unfired, undersize "green pellets". On balling discs, the recycle is within the disc. In the more common balling drums, undersize pellets are screened separately and returned to the feed end of the drum. The cycling performance of drum/screen circuits is very well documented and has been the subject of several controller synthesis papers, e.g. the definitive work of Wellstead, Munro and Cross (1977, 1978).

The causes of cycling have been attributed to the inherent dynamics of pellet growth and have been extensively studied and modelled e.g. Capes and others (1975), Cross (1977), Sastry (1981), Kapur, Sastry and Furstenau (1981). Despite this body of work, plant control is still fairly elementary and fluctuations are hard to eliminate. Indeed, an ingenious attempt was once made by LKAB to utilize a measurement of the fluctuations as the input variable in a control scheme (Sjoberg and others 1977). Other instrumental techniques have included measures of the surface reflectance of the pellets (by the British Steel Corporation), and on-line measurement of pellet size and moisture (Batterham, Hall and Barton, 1981). The latter work showed that the interactions between the tumbling pellets and the wall of the drum was a major cause of the fluctuations.

Sinter plant control

The grate sintering process is utilized extensively in both iron ore processing and in non-ferrous smelting. A great deal of expertise has been accumulated and most modern plants utilize some form of computer control as described by Corson and Armstrong (1981) and the Kawasaki announcement (Anon, 1983d).

Sinter plant control has been reviewed by Sastry and Cross (1979) and Luckers, Voll and Boelens (1980). Present control is fairly basic with, of order 3 measured and manipulated variables. The emphasis in modern plants is still on the elimination of input disturbances through blending of feed materials rather than on line and in situ measurement of such key variables as the peak sinter temperature. Such a process engineering solution, as opposed to a control engineering solution, emphasises that control problems can often be eliminated, rather than solved, by appropriate process design and engineering.

Better control awaits the availability of comprehensive, validated mathematical models of the whole sintering process. Such models are now being published, e.g. the work of Young (1977), Toda and Kato (1984), and Cumming and co-workers (1980, 1981, 1985).

Heat hardening of pellets

The induration or heat hardening of pellets, also referred to as pelletising, involves much more complex plants than sintering. The inherent, high energy consumption has prompted much research into process development, optimization and control. Induration plants tend to involve fairly complicated arrangements of gas streams passing sequentially through the bed of pellets in order that the maximum heat can be recovered from the pellets, the potentially damaging initial drying of the pellets can be controlled and the main firing can be manipulated so that fuel consumption and over-firing are minimized. Comprehensive computer control schemes are commonplace, as reviewed by Sastry and Cross (1979) and described in detail by Corson (1980) and Turner and co-workers (1981).

The major control problem is easily defined: pellets must be maintained at a certain temperature for a few minutes. The temperature and time required for particular feed materials has been extensively studied, e.g. the work of

Wynnyckyj and Fahidy (1974), Thomas, Carter and Gannon (1981), Dalal and Heerma (1982) and Lobanov (1983). In addition, complete and validated models have been developed to describe how pellet time/temperature profiles are determined by process conditions, e.g. Lebelle, Kooy and Hasenack (1973), Voskamp and Brasz (1975), Drugge (1975), Pape, Frans and Geiger (1976), Cross and Young (1976), Young, Cross and Gibson (1979), Thurlby, Batterham and Turner (1979), Breitholtz and Hillberg (1980), Thurlby and Batterham (1980), Thurlby, Batterham and Turner (1981), Cross and co-workers (1982), Batterham and co-workers (1982), Thornton and Batterham (1982b). While some of this extensive effort has been directed at solving control problems (Voskamp and co-workers, 1972), the major activity has been for energy minimization (Thurlby, Batterham and Turner, 1982).

The problem remains that there is still no on-line measurement available for pellet temperatures in the firing zone. The nearest measurable equivalent, the gas temperatures leaving the firing zone, has been shown to be a poor measure of the actual pellet temperatures due to significant interactions between process variables.

One solution is to use, on-line, a detailed, dynamic, deterministic model of the heat transfer to predict the pellet temperatures. The more easily measured gas temperatures and flows become the initial and boundary conditions for the model. The boundary conditions are satisfied by on-line identification of key parameters, in this case the pellet diameter and void fraction of the bed of pellets, (Thornton and Batterham, 1982b). The model then predicts the pellet temperature profiles which are used for on-line control.

Such combinations of large, deterministic models and on-line adaption of key parameters are a powerful methodology. In the induration case, the method solved the combined problems of lack of suitable measurement, significant process time delays and severe interactions between process variables. Over the period of this particular project, fuel consumption per tonne decreased 25%.

Given the emergence of detailed process models in other areas of pyrometallurgy, we can expect to see more examples of the combination of deterministic models with on-line identification of key parameters.

CONTROL OF SMELTING PROCESSES

The traditional methods of smelting in blast furnaces and reverberatory furnaces have relied for control purposes on the establishment of a consistent quality in feed materials and a standardized operating procedure. Owing to the long residence times in furnaces and the slowness of response of such systems corrections to the operating conditions are generally made by operator intervention on the basis of chemical analysis of the products.

Blast furnaces

Blast furnaces are employed extensively in the steel industry for the production of large tonnages of pig iron, and in the non-ferrous industry for the production of lead, zinc and, to a lesser extent copper, and tin.

Dissection of iron blast furnaces by Japanese workers (Kanbara and others, 1976; Shimomura and others, 1976; Sasaki and others, 1976; Kojima and others, 1976) revealed six recognizable zones of burden structure, namely from top to bottom:

1. Granular zone, in which the charge material remains unchanged in shape.
2. Cohesive zone, in which the ore particles reach an advanced stage of reduction, to yield a semi-molten mass with relatively high flow resistance.
3. Active coke zone, in which molten iron and slag flow through a matrix of loosely packed coke.
4. Raceways, void spaces in front of the tuyeres, generated by displacement of coke due to the momentum of the hot blast.
5. "Deadman", a stagnant coke layer between the raceways.
6. Hearth, the zone below the tuyeres packed with coke, slag and liquid iron.

According to Wakayama and others (1979), optimum furnace operation depends largely on the formation of a desirable cohesive zone shape, which is in turn determined by the mechanical properties of the burden and the burden charge distribution. In seeking an external furnace measurement which will reliably predict the shape of the cohesive zone, measurements have been made of profiles of gas velocity, temperature, CO/CO_2 ratio and pressure distribution. However, in the production situation, only a limited number of measurable quantities are available, e.g., production rates, top gas composition and hot metal temperature. None of these provides sufficient information for full dynamic control of the furnace.

Given the impetus of the findings of the Japanese dissection work, mathematical modelling of the blast furnace is proceeding rapidly, e.g. the work of Burgess and co-workers (1982), Taguchi and others (1982) and Kubo and others (1982), Christiansen and co-workers (1984) and Cross and co-workers (1984). There is now an attempt to run comprehensive dynamic models on-line, initially as an operator's guide to the internal state of the furnace. The ultimate aim of the work is to incorporate the models into a comprehensive control strategy to maintain hot metal quality. This is a similar trend to that described for iron-ore indurators. The proposed scheme (Christiansen and others, 1984) uses the available measurable quantities as inputs to an operating dynamic model. As with the indurator work, the available measurements would be regularly compared with simulated values from the model for on-line identification of key model parameters.

In seeking to achieve a degree of automatic control of a blast furnace, current strategy is, to maintain a consistent quality in burden charge materials, to establish a set of operating conditions and to make a change only when the operation departs from the target. Control is then fairly elementary. Other attempts at more sophisticated control, e.g. adaptive control (Rajbman, 1980 and De Keyser and Cauwenberghe, 1981) have not gained wide acceptance.

Reverberatory furnaces

In view of the high capital cost of approximately U.S.$2,500 per annual tonne (Traulson, Taylor and George, 1982) for the adoption of new technology on a greenfield site, many existing smelters are closely reviewing the operation of their reverberatory furnaces. Keran (1983) quotes total copper smelter energy consumption of 20.2 MBtu/ton (21GJ/tonne) at Mount Isa, of which 65% is used in the reverberatory furnace. This energy

consumption is better than for most reverberatories, but it does not compare favourably with that reported for the newer processes. However, the cost of the energy compares favourably owing to the use of pulverized low grade coal as fuel.

Measures which can improve the performance and production rate of reverberatory furnaces are:
.   Computer aided scheduling of operations
.   Oxygen enrichment of furnace combustion air
.   Conversion to oxygen sprinkle smelting
        (Johnson and Jackson, 1983)

The main opportunity for automatic control in a reverberatory furnace is in the control of combustion.

Keran (1983) and McCain and Rana (1983) describe variants of conventional boiler control techniques using sensors for residual oxygen and combustibles in the furnace uptake.

One interesting control strategy employed at Mount Isa has resulted in a 5% increase in smelting rate. The Mount Isa copper concentrate is generally self-fluxing, producing a slag of 34 to 38% silica with a matte grade of 42 to 44% copper. The concentrate is first roasted in a fluidized bed to eliminate about 47% of the sulfur and the calcine is melted in the reverberatory with removal of approximately a further 5% sulfur. The Fe/Si ratio in the slag is governed mainly by the extent of oxidation of iron in the roaster. It was found that slag problems caused by abnormal Fe/Si ratios were causing reduced throughput when smelting in the conventional way to produce constant slag and matte grades. The problems were overcome by smelting to produce a desirable slag composition, while allowing the matte grade to vary between 40% and 50%. Control was achieved by using a flux control model based on a computer program which balances major elements around the fluid bed roaster. This allows the iron distribution between matte and slag to be defined as a function of fluidizing air rate for any given concentrate feed rate and composition. This balance is made on-line via a direct link with the assay laboratory which updates assays on a four-hourly basis.

Electric furnace smelting

Where cheap electric power is available, the thermal energy for either ferrous or non-ferrous smelting may be supplied from this source. Computer control applications are widespread, e.g. Evans and Ravenscraft (1983), Sannia (1983) and Sommer and co-workers (1981). Aune and Strom (1983) describe recent developments in furnace design which make electric smelting attractive for a variety of sulfide smelting operations.

As in reverberatory smelting the major control strategy is to schedule the flux and feed additions and to optimize the electric power input (Winter et al, 1983). Furnace electrical parameters are readily controlled to arbitrary set values. The most important operating parameter is arc length, which governs both the mode and efficiency of heat transfer to the bath. Most control systems alter the electrode position to maintain the desired power settings.

Szekely and McKelliget (1984) have suggested a control strategy based on a mathematical model which permits optimization of the cost effectiveness of the furnace by a trade-off between power efficiency and arc consumption. They acknowledge that the mathematical complexity of their heat transfer model might preclude its use for on-line control. However, they suggest that, for each furnace, a data base be set up with correlation of the results of the heat transfer model in terms of furnace operating parameters. Since arc length is also the major factor affecting cost, the set values from the furnace control computer would be continuously modified by a decision loop addressing both a cost optimization routine and the model's data base.

Flash smelting

The distinguishing feature of flash smelting is that the fuel and the powdered concentrate are contacted with oxidizing gases in the flame zone of some form of concentrate burner delivering into a hot enclosure. Melting and partial reaction of the feed material occurs in the flame, sometimes in very short times, and with the attainment of high particle temperatures, which encourages volatilization of impurities. Gases are separated from the liquid phases, i.e. slag and matte or metal, which pass into a settler zone. The hardware configurations for the overall flash smelting process may take many forms. Concentrate burners may be of the linear or cyclone type, and may be fired with oil, gas or pulverized coal, using air, oxygen or oxygen enriched air as combustant. In each case the gases pass over a settling zone containing the liquid phases, and are discharged through an uptake shaft.

As examples: the INCO oxygen smelter dispenses with the reaction shaft and uses technical grade oxygen to oxidize the concentrates which are injected horizontally through burners placed in the end walls (Antonioni and others, 1982). In the Contop (Weigel and Melcher, 1982) and Kivcet (Melcher, Muller and Weigel, 1976) copper processes flash smelting is done in a cyclone furnace where some of the liquid phase interactions also occur as the liquids flow down the cyclone walls. In some applications of the top blown rotary converter (TBRC), the concentrate can be delivered through the water cooled lance, and be partially flash smelted before it is injected into the mechanically mixed bath of the TBRC. "Oxygen sprinkle smelting" (Johnson and Jackson, 1983) has recently been introduced as a method of retrofitting existing reverberatories by placing oxygen-fired concentrate burners of special design in the roof.

Additional details for copper smelting are reviewed by Mackey and Tarassoff (1983).

Control of flash smelting was covered in the review by Richards and George (1979). Control requires the maintenance of both furnace temperature and oxidation conditions. The sulfide feed will, in most cases, have a considerable fuel value. This will be closely related to the matte grade, i.e., to the extent to which iron is oxidized from the sulfide phases into the slag. Carbonaceous or hydrocarbon fuel is also required to supply process heat in air blown systems. The heat from its combustion will depend on the degree of combustion which controls the oxygen partial pressure (i.e., the CO_2/CO ratio) in the off-gases. Thus, for each operation, there will be closely defined ratios between the mass flows of concentrate, fuel and oxygen fed to the furnace. These flows must be accurately measured and regulated.

The calculation of these flow rates is by no means straight forward, since the concentrate tends to partially over-oxidize in the reaction shaft and re-equilibrate in the settler. The dependency of furnace performance on operating strategy for the Outokumpu nickel flash furnace at BCL Limited, Botswana is described by Elliot, Robinson and Stewart (1983).

In the BCL operation, pulverized coal was blown into the settler gas space in an attempt to control the formation of magnetite accretions in the settler. However, this led to the reduction and resulfidization of iron oxides in the settler, causing a low-grade matte to be tapped even though a higher-grade matte was produced in the reaction shaft. When lump coal was charged in place of the injected pulverized coal, the operation improved considerably. The accretions disappeared, and reliable formation of a protective settler lining was established at a slightly lower matte grade. Most importantly, the oxygen ratio (i.e., oxygen usage per tonne of concentrate) was lowered by 16% and the proven capacity of the furnace improved from 100 to 115 tonnes per hour.

In a flash smelting operation the most precise indicator of the state of the process is the oxygen potential of the matte and slag baths. This parameter is estimated with reasonable precision at BCL by measuring the distribution of cobalt between matte and slag and calculating the oxygen potential from this distribution. Using XRF analysis of the slag and matte, the oxygen potential derived in this manner may be used as a control parameter.

Control of the flash furnace operation at the Tomano smelter is now achieved by running a process model in parallel with the process (Okada and others, 1983). The parameters of the model are modified on-line on receipt of plant data and assay results through a distributed computer system until a good fit with current operating results is achieved. New set-values required for the control of the furnace are then calculated by the model and the corrections implemented.

Intensive bath smelting: ferrous

On a tonnage basis, the most important intensive bath smelting process is oxygen steelmaking. This is a batch process in which hot metal from the blast furnace, containing 4% carbon and about 2% each of silicon and manganese, is blown with oxygen to oxidize the silicon, manganese and carbon and produce low carbon steel. The most important parameters requiring control are temperature, final carbon content and phosphorus content. In current practice, the targets are achieved by strict adherence to a standardized operating procedure and close control of total additions of scrap, oxygen and lime respectively. It is customary to blow down to a very low carbon content and to recarburize the steel in the ladle to the required carbon content.

Measurement of the temperature and carbon content at turn down of the vessel is achieved by using disposable thermocouples and taking a sample for rapid carbon analysis. More recently, these measurements have been made even more rapidly by the use of a "sublance" assembly, which incorporates the necessary components for temperature measurement and estimation of carbon by thermal analysis. By the use of sophisticated mechanical handling arrangements the sublance may be inserted into the vessel in its upright blowing position, return its measurements to the control room and returned to its cradle on the operating floor to deliver samples for final confirmatory analysis in about two minutes.

Given the economic importance of the process,

computer control is widespread, e.g. Kawawa
(1981), Buchanan and others (1983), Anon
(1983e). On-line control schemes using dynamic
models have been developed for estimating such
major variables as, the carbon content and
temperature of the molten steel, the degree of
oxidation of the slag, and many minor variables,
e.g. the detailed composition of the melt and the
slag.

For accurate mass and composition control, and for
economic and safety reasons, it is essential for
the molten metal and slag to stay in the
converter. Prevention of foaming or slopping is a
major control problem. Iida and others (1984),
have described a method for fully automating basic
oxygen steelmaking. Early attempts were based on
a variety of indirect measurements such as
acoustic output, furnace vibration and off-gas
analysis, none of which were entirely
satisfactory. The system finally adopted involves
measurement of the acceleration, in two directions
at right angles, of a suspended lance in a vessel
which is operated under programmed and precisely
controlled conditions of feeding and blowing. The
averaged output signal from the accelerometers
gives a reliable measurement of the height of the
foaming slag, which is used to regulate oxygen
flow and lance height. This mode of control
permits 95% success in achieving carbon content,
bath temperature and extent of dephosphorization
at turndown. Control is achieved at high smelting
rates and without foaming or slopping of the
converter contents. The Iida system is the first
to be published that covers all the current types
of oxygen blown converters.

Oxygen steelmaking processes have developed along
two paths. The top-blown LD process which was
widely adopted from about 1950, employs a water
cooled lance, located above the liquid metal
surface. One or more supersonic oxygen jets are
directed downwards through this lance into the
bath. This causes the formation of a slag-metal
emulsion in which the bath reactions take place.

Following the demonstration of the shrouded oxygen
injector, the Q-BOP process was developed in
Germany, initially by retrofitting existing Thomas
basic converters. The injectors, placed in the
bottom of the vessel, employ a central oxygen pipe
protected by an annular space through which a
hydrocarbon or other suitable cooling gas is
passed. Powdered coal or limestone may also be
fed through the injectors. This reactor geometry
promotes more efficient mixing of the metal bath
than occurs in the top-blown process and the metal
acts as the oxygen carrier. As a result the metal
bath operates closer to equilibrium with both the
gas phase and the slag. Metallurgical control is
therefore easier and the injection of fine lime
through the bottom prevents the slag from foaming,
making process control easier.

Modern practice tends towards combining the best
features of both top and bottom blown systems by
mixed blowing (Chatterjee and others, 1984). The
chemistry of the reactions between oxygen, carbon
and the numerous species in the metal, slag and
gas phases has been extensively studied. For full
scale converters, however, modelling the partition
of components between metal and slag involves
empiricism, see e.g Ho and co-workers (1982), and
Nilles (1982). This limits development of more
advanced control schemes.

The uncertainties in bath smelting systems,
ferrous or non-ferrous, stem largely from a lack
of understanding of the major physical processes,
viz:

- behaviour of the gas envelope structure
  in moving through the liquid metal
- global movement of the liquid metal
- momentum interaction between the liquid
  metal and injected gas
- heat transfer and mass exchange via the
  chemical reactions.

In less intense bath smelting processes, more
progress has been made. The blowing of gases into
steel held in ladles is used for refining and
adjustment of the composition of the final
steel. Gas and steel flow patterns have been
measured and modelled. There has been some
success, e.g. the work of El Kaddah and Szekely
(1981), in modelling the desulfurization of steel
in argon-stirred ladles. Recent work by Cross and
others (1984) describes a computational framework
that takes account of the multi-phase nature of
the flow in bath smelting systems and such
computations may well be extended to most
intensive smelting processes.

Intensive bath smelting: non-ferrous

The need to tighten pollution controls and
conserve primary energy, has encouraged the
development of intense smelting processes. These
processes use smaller sealed reactors with high
specific smelting rates which make use of the fuel
value of the mineral concentrates.

In smelting of non-ferrous sulfide concentrates,
the reactors are generally blown with oxygen, to
achieve autogenous operation, to limit the gas
volume and to produce a concentrated SO_2 stream
for recovery of sulfur. In such cases the
response times of the system are short.
Successful operation depends on rapid detection of
deviations from the target operating conditions
followed by prompt corrective action. Intensive
bath smelting thus requires far more stringent and
effective control than blast furnaces or
reverberatory furnaces.

In most cases of intense bath smelting of non-
ferrous metals, there is a direct counterpart with
flash smelting. The distinguishing feature is
that oxidizing gas (and fuel if required) is
injected into a molten bath containing liquid
matte and slag, and the reactions all take place
in the bath. Reactors may assume many shapes.
They may be top-blown from above the bath, as in
the top-blown rotary converter (TBRC), the
Mitsubishi copper process (Sweetin and others
1983) and the Contop process (Weigel and Melcher
1982), side blown as in the Noranda process
reactor (Bailey and others, 1983), bottom blown
using shrouded tuyeres, as in the QSL process
(Fischer, 1983), or top blown with a submerged
lance as in the Sirosmelt process (Floyd and
Conochie 1979,1984).

As in the case of oxygen steelmaking, the
performance of an intensive non-ferrous bath
smelting process is determined in large measure by
the manner in which the reactants are introduced
into the reactor and mixed. A comprehensive
mathematical model has not yet been developed for
any of the emerging processes, to the extent that
it can be used on-line for automatic control.
However, equilibrium-based models such as that
described by McClelland and Lightfoot (1984) and
McClelland and Denholm (1984) for Sirosmelt
processes have been run in parallel with the
corresponding process.

At Associated Tin Smelters, Sydney, Australia, the
Sirosmelt tin smelting process is being developed
to the commercial scale. When correctly operated

without interruption the final slag-cleaning stage
of this process shows a first-order response with
a half-time of 30 minutes. The end point can be
predicted with reasonable precision by rapid tin
analysis of the slag during the course of the
reduction, followed by extrapolation on the basis
of the first order response. The reliability of
this method depends to a large extent on the
maintenance of steady blowing conditions with an
excess of reductant to drive the reaction. The
predictions of the model follow the furnace
performance quite closely and the model is of
value in setting the operating parameters.
However, a knowledge of the half-time of reduction
is adequate for control purposes in the absence of
a rapid sensor to measure the actual tin content
of the slag.

In general, as with flash smelting processes, the
control of intensive bath smelting processes
requires accurate metering of the mass flows of
concentrate, fuel and oxygen to achieve constant
temperature at the required oxygen potential. The
most urgent requirement before fully automatic
control can be realized is a reliable sensor of
oxygen potential.

CONCLUSIONS

This paper has ranged across the extensive field
of control of pyrometallurgical processes. The
general conclusions are that, as in other fields,
improvements in control are dictated by
developments of better and component specific
sensors, and by the availability of comprehensive,
validated process models.

Pyrometallurgical processes are becoming more
intensive with smaller, primary smelting zones and
this is forcing the development of more effective
sensors and control schemes.

An interesting trend is the use of large,
deterministic models run on-line, with key
parameters adapted so that model predictions of
secondary variables match plant measurements.
Primary variables available from the model are
then used in controllers as timely and accurate
estimates of otherwise unmeasurable variables.

REFERENCES

Abreviations used

"Advances in sulphide smelting": Sohn, H.Y., D.B.
     George, , and A.D. Zunkel (Eds), (1983).
     Advances in sulfide smelting. Vol. 2.
     Technology and practice. Met. Soc. of AIME,
     Warrendale.
"Agglomeration 81": Molerus, O. and W. Hufnagel
     (Eds), (1981). 3rd International Conf. on
     Agglomeration. 2 vols., Schuster, Nurnberg.
"Automation in mining, mineral and metal
     processing". O'Shea, J. and M. Polis (Eds),
     (1980). Automation in mining, mineral and
     metal processing. 3rd IFAC Symposium,
     Montreal, August 1980, Pergamon, Oxford.
"Computer methods for the 80's": Weiss, A. (Ed),
     (1979). Computer methods for the 80's in the
     mineral industry. Soc. Mining Engrs. of
     AIME, New York.
"Control 84": Herbst, J.A., D.B. George, and
     K.V.S. Sastry (Eds), (1984). Control '84.
     Mineral/Metallurgical Processing, Proc. 1st
     Int. Symp. on Automatic Control in Mineral
     Processing and Process Metallurgy, March
     1984, Los Angeles, Soc. Min. Engrs. of AIME,
     New York.

"3rd Process Technology Conference": Kuhn, L.G.,
     C.M. Hiles, J.W. Hlinka, and J.K. Brimacombe,
     (Eds), (1982) Application of mathematical and
     physical models in the iron and steel
     industries. 3rd Process Technology
     Conference, Pittsburgh, March 1982, Iron and
     Steel Soc. of AIME, Warrendale, PA.

References

Anon. (1983a). Improved productivity and
     automation of iron and steel industry. South
     East Asian Iron and Steel Soc., Kaohsuing
     Conference, Sept. 1983, SEAISI, Manilla.
Anon. (1983b). Process control and sensor
     development. Iron and Steelmaker, 10(6), 14.
Anon (1983c). Unique use for computers at
     Bethlehem. Iron and Steelmaker, 10(6), 12.
Anon (1983d). KSC computerizes its No.4 sintering
     plant. Iron and Steelmaker, 10(6), 14.
Anon (1983e). Bethlehem licences its sensor lance
     control system. Iron and Steelmaker, 10(6),
     13.
Anon (1984). Molten steel analysis thrust of new
     research program. Iron and Steelmaker,
     11(1), 10.
Antonioni, T.N., C.M. Diaz, H.C. Garven, and C.A.
     Landolt (1982). Control of the Inco Oxygen
     Flash Smelting Process: In D.B. George and
     J.C. Taylor (Eds) Copper Smelting: an
     Update. Proceedings of a Symposium at the
     111th AIME Annual Meeting, Dallas, Feb.
     1982. The Metallurgical Society of AIME,
     Warrendale, PA, 17-31.
Aune, J.A. and K.H. Strom (1983). Electric
     sulfide smelting technology - Elkem
     Engineering Division's new process/design
     concept for the '80s. "Advances in Sulfide
     Smelting", 635-658.
Bailey, J.B.W., G.D. Hallett, and L.A. Mills
     (1983). The Noranda Horne smelter -1965 to
     1983. "Advances in Sulfide Smelting", 691-
     707.
Barton, G.W., J.R. Siemon, R.J. Batterham and B.W.
     McLoughlin (1980). Improved control of the
     conditioning stage in lead and zinc
     sintering. Chemeca 80, Aust. Annual Conf. on
     Chemical Eng., I.E. Aust., 30-34.
Batterham, R.J., J.S. Hall and G. Barton (1981).
     Pelletizing kinetics and simulation of full
     scale balling circuits. "Agglomeration 81",
     A136-A140.
Batterham, R.J., J.A. Thurlby, G.J. Thornton, T.E.
     Norgate and B. Povey (1982). The use of
     mathematical models in the processing of iron
     ore. "3rd Process Technology Conference",
     110-116.
Breitholtz, C. and C. Hillberg (1980).
     Development of a reference model for the
     drying zone of a travelling grate pelletizing
     plant. "Automation in mining, mineral and
     metal processing", 415-424.
Buchanan, D.J., B.K. Fulcher and M.D. Harshaw
     (1983). Installation of a new computer at
     Dofasco's #1 BOF melt shop. Iron Steelmaker,
     10(6), 29-36.
Burgess, J.M., M.J. McCarthy, D.R. Jenkins, W.V.
     Pinczewski and W.B.U. Tauzil (1982). Gas
     liquid distribution and hearth liquid
     drainage models for the analysis of the
     ironmaking blast furnace. "3rd Process
     Technology Conference", 14-24.
Capes, C.E., A.E. McIlhinney and R.D. Coleman
     (1975). Some considerations on the dynamics
     of balling circuits. Trans. SME-AIME. 258,
     204-208.
Chattergee, A., C. Marique and P. Nilles (1984).
     Overview of present status of oxygen
     steelmaking and its expected future trends.
     Ironmaking and Steelmaking, 11(3), 117-131.

Christiansen, E.L., C.M. Ufret, R.H. Spitzer, and T.J. Williams (1984). A Dynamic simulation model of the iron blast furnace. "Control '84", 337-346.

Cook, J.R., D.F. Ellerbrock, T.R. Dishun and J.T. Nenni (1982). Minimization of fuel consumption in soaking pits using a cylindrical equivalent model for on-line estimation of ingot thermal profiles. "3rd Process Technology Conference", 122-133.

Corson, R.C. (1980). On-line computer control of straight grate indurators. "Automation in mining, mineral and metal processing", 425-434.

Corson, R.C. and R.K. Armstrong (1981). Sinter plant control at China Steel through distributed microcomputers. "Agglomeration 81", I96-I111.

Cross, M. (1977). Mathematical model of balling drum circuit of a pelletizing plant. Ironmaking and Steelmaking, 4, 159-169.

Cross, M. and R.A. Young (1976). Mathematical model of rotary kiln used in the production of iron ore pellets. Ironmaking and Steelmaking, 3(3), 129-137.

Cross, M., E.C. Bogren, J.S. Wakeman, and R.D. Frans (1982). Mathematical models of iron-ore pellet induration-validation and application. "3rd Process Technology Conference", 101-109.

Cross, M., R.D. Gibson, M.G. Tonks and F.B. Traice (1984). Towards a comprehensive model of the blast furnace. "Control '84", 347-354

Cross, M., N.C. Markatos and C. Aldham (1984). Gas injection in ladle processing. "Control '84", 291-298.

Cumming, M.J., G. Barton and R.J. Batterham (1980). Simulation of sintering - a complex physico-chemical process used in the smelting of sulphide concentrates. Fourth biennial Conf. Simulation Soc. Aust., Brisbane, 108-119.

Cumming, M.J., G.W. Barton and R.J. Batterham (1981). Dynamic simulation of a moving grate combustion process. Advances in computer methods for partial differential equations, IV. Vichnevetsky, R. and Stepleman, R.S. (Eds). I.M.A.C.S. 286-292.

Cumming, M.J., W.J. Rankin, J.R. Siemon, J.A. Thurlby. G.J. Thornton, E. Kowalcyzk and R.J. Batterham (1985). Modelling and simulation of iron ore sintering. Fourth International Symp. on Agglomeration, Toronto, 1985. (To be published)

Dalal, J.G. and R.H. Heerema (1982). Optimization of iron ore pellet production variables. "3rd Process Technology Conference", 92-100.

De Keyser, R.M.C. and A.R. Van Cauwenberghe (1981). A self tuning multi-step predictor application. Automatica, 17(1), 167-174.

Dittrich, W. (1981). The homogenization of aggregates and ores for pig iron production in blending bed installations. World Steel Metalwork, Export Man., 1980-81, 42. 44. 46-50.

Droste, W. (1980). On line analysis of iron ore for pellet production. "Automation in mining, mineral and metal processing", 435-442.

Drugge, R.B. (1975). Modelling of heat exchange phenomena during the indurating of magnetite pellets on a travelling grate. Paper presented to 104th AIME annual meeting, New York, Feb. 1975.

El-Kaddah, N. and J. Szekely (1981). Mathematical model for desulfurization kinetics in argon-stirred ladle. Ironmaking and Steelmaking, 8, 269-278.

Elliott, B.J., K. Robinson and B.V. Stewart (1983). Developments in flash smelting at BCL Ltd. "Advances in Sulfide Smelting",

875-900.

Eisenhut. W., A. Gratias and R. Worberg (1981). Automation of coking plant operations. World Steel Metalwork. Export Man., 1981-82, 15-18, 20-21.

Etsell, T.H. and C.B. Alcock (1983). Continuous oxygen determination in a concast tundish. Can. Metall. Quarterly, 22, 421-427.

Evans, P.D. and J. Ravenscroft (1983). Automation of the arc furnace steelmaking process. Paper presented to SEAISI conference: Productivity and Automation of Iron and Steel Industry. Kaohsiung, Sept., 1983. SEAISI, pp22.

Fischer, P. (1983). The QSL Process : A new lead smelting route ready for commercialization. "Advances in Sulfide Smelting", 513-528.

Flowers, W.E., K.F. De Vanney and E.J. Ostrowski (1982). Mathematical optimisation of coal mixes in cokemaking. "3rd Process Technology Conference", 57-68.

Floyd, J.M. and D.S. Conochie (1979). Reduction of liquid tin smelting slags. 2 - Development of submerged combustion process. Trans. I.M.M., Section C, C123-C128.

Floyd, J.M. and D.S. Conochie (1984). Sirosmelt - The first ten years. Proceedings of Symposium on Extractive Metallurgy, Melbourne, November 1984 (The Aus. I.M.M.) (In press).

Floyd, J.M., W.T. Denholm, R.N. Taylor and L.D. Mitchell (1984). Application of oxygen potential probes for process control in non-ferrous smelting. "Control '84", 39-46.

Geskin, E.S. (1980). The application of automatic control for energy saving in heating furnaces. "Automation in mining, mineral and metal processing", 623-632.

Haber, R., J. Hetthessy, L. Keviczky, I. Vajk, A. Feher, N. Czeiner, Z. Csaszar, and A. Turi (1981). Identification and adaptive control of a glass furnace. Automatica, 17(1), 175-185.

Herbst,J.A., D.B. George and K.V.S. Sastry (Eds), (1984). "Control '84".

Iida, Y., K. Emolo, M. Ogaiva, Y. Masuda, M. Orisina and H. Yamada (1984). Fully automatic blowing technique for basic oxygen steelmaking furnace. Trans. Iron Steel Inst. Japan, 24, 540-546.

Ito, S., M. Kitamura, S. Koyama, H. Matsui and H. Fujimoto (1982). On the new refining process by the top and bottom blowing converter. 65th Steelmaking Conf. Proc., Pittsburgh, March, 1982, Iron and Steel Soc. AIME, Warrendale, 123-131.

Johnson, R.E. and T.D. Jackson (1983). Oxygen sprinkle smelting at the Morenci Smelter. "Advances in Sulfide Smelting", 473-488.

Kanbara, K., T. Hagiwara, A. Shigemi, S. Kondo, Y. Panayama, K. Wakabayashi and N. Hiramoto (1976). Report on the dismantling of blast furnaces. 1. Dismantling of blast furnaces and their inside state. Tetsu To Hagane, 62, 535-546.

Kapur, P.C., K.V.S. Sastry and D.W. Furstenau (1981). Mathematical models of open circuit balling or granulating devices. Ind. Eng. Chem. Process Des. Dev., 20(3), 519-524.

Kawawa, T. (1981). Historical review and present status of BOF computer control system of N.K.K. BOF end point determination. McMaster Symp. No.9, Hamilton, Ontario, May, 1981, 5.1-5.18.

Keran, V.P. (1983). Current commercial sulfide smelter operations. "Advances in Sulfide Smelting", 667-690.

Kojima, K., T. Nisi, T. Yamaguchi, H. Nakama and S. Ida (1976). Report on the dismantling of blast furnaces. 4. Change in the properties

of coke in blast furnaces. Tetsu To Hagane,
62, 570-579.

Kubo, H., T. Nishiyama, G. Kyoguchi, M. Yasuno, S.
Taguchi and J. Kurihara (1982). A dynamic
one dimensional simulation model of the blast
furnace process. "3rd Process Technology
Conference", 39-48.

Kuhn, L.G., J.W. Hlinka, C.M. Hiles and J.K.
Brimacombe (Eds) (1982). "3rd Process
Technology Conference".

Lebelle, P.A.M., J.J. Kooy and N.A. Hasenack
(1973). The induration process of pellets on
a moving strand. Paper presented to ISI
Conf. on mathematical process models in iron
and steelmaking, Amsterdam.

Lobanov, V.I. (1983). Arranging efficient firing
regimes of iron-ore pellets. BISI English
translation 21281 of Izv. V.U.Z. Chernaya
Metall., 4, 12-15, Apr. 1982.

Lu, W.K. (Ed) 1981. BOF end point
determination. McMaster Symp. on Iron and
Steelmaking, No.9 McMaster Univ. Press,
Hamilton.

Luckers, J., H. Voll, and J. Boelens (1980).
Automation in the iron and steel industry.
Main problems and trends. "Automation in
mining, mineral and metal processing".
Separate paper, 14pp.

Lumelsky, V.J. (1983). Estimation and prediction
of unmeasurable variables in the steel mill
soaking pit control system. IEEE Trans on
Automatic Control, AC28(3), 388-400.

McCain, J.D. and I.A. Rana (1983). New techniques
and efficiencies at the San Manuel Smelter,
"Advances in Sulfide Smelting", 709-726.

Mackey, P.J. and P. Tarassoff (1983). New and
emerging technologies in sulphide smelting.
"Advances in sulfide smelting", 399-426.

McClelland, R.A. and W.T. Denholm (1984). A
mathematical model of the Sirosmelt
process. Proceedings of Symposium on
extractive Metallurgy, Melbourne November
1984, The Aus. I.M.M. (In press).

McClelland, R.A. and B.W. Lightfoot (1984).
Modelling of the Sirosmelt Tin Process.
"Control '84", 350-360.

Melcher, G., E. Muller and H. Weigel (1976). The
KIVCET Cyclone Smelting process for impure
copper concentrates. Journal of Metals,
28(7), 4-6.

Molerus, O. and W. Hufnagel (Eds) (1981).
"Agglomeration 81".

Nilles, P.E. (1982). New techniques in basic
oxygen steelmaking. Iron Steelmaker 9(6),
27-36.

Okada, S., M. Miyake, A. Hara and M. Vekawa
(1983). Recent improvement at Tamano
Smelter. "Advances in Sulfide Smelting",
855-874.

O'Shea, J. and M. Polis (Eds) (1981). "Automation
in mining, mineral and metal processing".

Pape, P.O., Frans, R.D. and K.H. Geiger (1976).
Magnetite oxidation kinetics and thermal
profiles in a magnetite pellet plant
cooler. Ironmaking and Steelmaking, 3(3),
138-145.

Rajbman, N.S. (1980). Adaptive control of
metallurgical processes: principles and
applications. "Automation in mining, mineral
and metal processing". 1-14.

Rankin, W.J., P.W. Roller and R.J. Batterham
(1983). Quasi-particle formation and the
granulation of iron ore sinter feeds.
Aus.I.M.M. - I.S.I.J. joint symposium,
Tokoyo, Oct. 1983, 13-28.

Rankin, W.J., R.J. Batterham, and P.W. Roller
(1984). A laboratory study of the
permeability of granulated iron ore sinter
feeds. Paper 84-82, presented to SME-AIME
Annual Meeting, Los Angeles, Feb. 1984, AIME,
12pp.

Richards, K.W. and D.B. George (1979). Control
Strategies. Pyrometallurgical process
control. "Computer methods for the 80's",
806-810.

Sannia, A. (1980). Applications of a computerized
control in an aluminium smelting plant.
Alluminio, 7-8, 358-364.

Sasaki, M., K. Ono, A. Suzuki, Y. Okuno, K.
Yoshizawa, and T. Nakamura (1976). Report on
the dismantling of blast furnaces. 3.
Formation and melt-down of the softening-
melting zone in blast furnaces. Tetsu To
Hagane. 62, 559-569.

Sastry, K.V.S. and M. Cross (1979). Control
Strategies. Pelletization and Sintering.
"Computer methods for the 80's", 799-805.

Sastry, K.V.S. (1981). Mathematical modelling of
pellet growth processes and computer
simulation of pelletizing circuits.
"Agglomeration 81", A122-A135.

Shimomura, Y., K. Nishikawa, S. Arino. T.
Katayama, Y. Hida and T. Isoyama (1976).
Report on the dismantling of blast
furnaces. 2. On the inside state of the
lumpy zone of blast furnaces. Tetsu To
Hagane, 62, 547-558.

Sjoberg, B., B. Fageberg, Q. Lotgren and H.
Ornstein (1977). Some continuous measurement
and control methods for application in iron
ore processing. XII Int. Min. Process.
Cong., Iron., Vol.1, Sao Paulo, 66-78.

Sohn, H.Y.. D.B. George and A.D. Zunkel (Eds)
(1983). "Advances in Sulfide Smelting.

Sommer, G., N.A. Barcza, I.J. Barker, M.S. Renni
and A.B. Stewart (1981). Computer aided
control for the production of ferrochromium
in a submerged arc furnace. INFACON 80, 2nd
Int. Ferro Alloys Congress, Lausanne, Oct.
1980. Inst. des Prod. de Ferro-Alliages,
Lausanne, 53-70.

Sweetin, R.M., C.J. Newman and H.G. Storey,
(1983). The Kidd Smelter: Start-up and early
operation. "Advances in Sulfide Smelting,"
789-816.

Szekely, J. and J. McKelliget (1984). A
mathematical model of electric furnace
operations and its use for control and
process optimization. "Control '84", 329-
336.

Taguchi, S., N. Tsuchiya, K. Okabe, H. Kubo and K.
Ichifufi (1982). Mathematical simulation of
silicon transfer in the blast furnace. "3rd
Process Technology Conference", 25-38.

Thomas, C.G., N.B. Carter and J.F. Gannon
(1981). Optimising iron oxide pellet
properties. "Agglomeration 81", B104-B119.

Thornton, G.J. and R.J. Batterham (1982a).
Adaptive mathematical models for real time
control of pellet induration processes. "3rd
Process Technology Conference", 69-75.

Thornton, G.J. and R.J. Batterham (1982b). The
transfer of heat in kilns. Chemeca 82, Aust.
Annual Conf. on Chemical Eng., I.E. Aust.
260-266.

Thurlby, J.A., R.J. Batterham and R.E. Turner
(1979). Development and validation of a
mathematical model for the moving grate
induration of iron ore pellets. Int. J.
Miner. Proc., 6, 43-64.

Thurlby, J.A. and R.J. Batterham, R.J. (1980).
The measurement and prediction of drying
rates and spalling behaviour of hematite
pellets. Trans. Inst. Min. Metal. C89, C125-
C131.

Thurlby, J.A., R.J. Batterham and R.E. Turner
(1981). The role of mathematical models and
pot grate simulators in studies of new and
existing pelletization plants.
"Agglomeration 81", B23-B35.

Thurlby, J.A., R.J. Batterham and R.E. Turner
(1982). Minimizing energy costs in straight

grate iron ore pelletization. _Ironmaking Steelmaking_, 9,(5), 200-206.

Toda, H. and K. Kato (1984). Theoretical investigation of sintering process. _Trans Iron Steel Inst., Japan_, 24, 178-186.

Travisen, H.R., J.C. Taylor and D.B. George (1982). Copper Smelting - an overview. In D.B. George and J.C. Taylor (Eds) _Copper Smelting: an Update : Proceedings of a Symposium at the 111th AIME Annual Meeting, Dallas, Feb. 1982._ The Metallurgical Society of AIME, Warrendale, PA, pp1-15.

Turner, R.E., R.J. Batterham, G.J. Thornton and J.A. Thurlby (1981). The use of digital computing aids for control of pellet processing. _Conf. on Productivity and Maintenance_, Tokyo, March 1981, South East Asia Iron and Steel Institute, Singapore, 191-210.

Vesloeki, T.A. and C.C. Smith (1982). Application of a dynamic mathematical model of a slab reheating furnace. _"3rd Process Technology Conference"_, 134-141.

Voskamp, J.H., J. Brasz, and W. Krijger (1972). Control analysis of a pellet indurating machine. _Iron and Steel_, 45, 635-642.

Voskamp, J.H. and J. Brasz (1975). Digital simulation of the steady state behaviour of moving bed processes. _Measurement and control_, 8, 23-32.

Wakayama, S., Y. Kanayama and Y. Okino (1979). Characteristics and control of burden distribution in the blast furnace. _Ironmaking and Steelmaking_, 6, 261-267.

Weigel, H. and G. Melcher (1982). Method for the smelting of materials such as ore concentrates. U.S. Pat. 4,362,561, December 7 1982.

Weiss, A., (1979). _"Computer methods for the 80's"_.

Wellstead, P.E., N. Munro and M. Cross (1977). Modelling and control of the balling drum circuit of an iron ore pelletizing plant, _Automation in Mining, Mineral and Metal Processing_, Proc. 2nd IFAC Symp., Johannesburg, Sept. 1976, 639-648.

Wellstead, P.E., M. Cross, N. Munro and D. Ibrahim (1978). On the design and assessment of control schemes for balling drum circuits used in pelletizing. _Int. J. Min. Proc._ 5, 45-67.

Winter, J.B., J. Draffen, L. Lee and D. Timis (1983). Process control modernization at Inspiration Smelter. _"Advances in Sulfide Smelting"_, 727-748.

Wynnyckyj, J.R. and T.Z. Fahidy (1974). Solid state sintering in the induration of Fe ore pellets. _Metall. Trans._, 5(5), 991-1000.

Yoshinaga, M., S. Sato and T., Kawaguchi (1980). Fundamental study on granulation of sinter raw material mixture. _Australia/Japan Extractive Metallurgy Symp._, Sydney, July 1980, Aus. I.M.M., 145-155.

Young, R.W. (1977). Dynamic mathematical model of sintering process. _Ironmaking and Steelmaking_, 4, 321-327

Young, R.W. (1977). Dynamic mathematical model of sintering process. _Ironmaking and Steelmaking_, 4, 321-327.

Young, R.W., M. Cross, and R.D. Gibson (1979). Mathematical model of grate-kiln-cooler process used for induration of iron ore pellets. _Ironmaking and Steelmaking_, 6, 1-13.

Development of a Process Simulation Package for Mineral Processing Plants

L.G. LISSIMAN

Mechanical Engineer, ACE-T Pty. Ltd., 58 Jersey Street, Jolimont, WA 6014

and

T.W. RILEY

Mechanical Project Engineer, ACE-T Pty. Ltd., 58 Jersey Street, Jolimont, WA 6016

Abstract. This paper describes a Process Simulation Package for modelling the steady-state operation of a mineral processing plant. There are two levels of input allowed into the package. The first allows simple manipulation of the process models within a defined modular format which enables a wide range of process systems to be simulated. The second level of input is via a Command file which allows plant models to be defined and changed without reference to the code or the requirement of any recompiling of the programme sources. The Command file interface was designed for use by metallurgists and assumes a sound knowledge of the system being modelled. As such it is an effective tool for metallurgists rather than a user friendly generalised package for the layman.

Keywords. Metallurgical industries; modelling; interactive methods; hierarchical systems.

INTRODUCTION

The development of a processing circuit to concentrate or extract efficiently the economic mineral content from a complex ore is itself a complex and time consuming process involving many activities and people with widely varying skills. The circuit resulting from the design process of sampling, laboratory scale testing and finally pilot plant testing must be capable of inherent, balanced steady-state operation.

This process simulation computer programme, which enables the steady-state condition in a mineral processing plant to be studied, was developed in co-operation with an Australian mining company. The company was embarking on an investigation into the treatment of a complex ore from a potential mine and their metallurgists were piloting many of the required processes. To aid the rapid development of the overall process, they requested assistance with the computation of the steady-state conditions.

Their initial laboratory test work and small scale pilot plant studies had shown that the circuit had to be a complex multi-stage process with a number of recycle loops, which made manual computation tedious and very time consuming. This software package enabled the process metallurgists to make a change in one section of the plant and quickly study the effect of that change on the other sections of the process. This allowed them to study a much wider range of options and test only the potentially optimum solution.

A major requirement of the package was for a system which could be developed as more information about the processes being modelled became available through the pilot plant and testing work which was proceeding at the time. This was a fundamental reason for not using an existing process simulation package. The metallurgists needed process models which they could define themselves and change easily as required. The result is a package which is flexible, open and can cope with a wide range of mineral processing scenarios.

The software package was developed on a Hewlett-Packard HP-1000 series mini computer, and is now also running on a Cyber mainframe. The package was written in FORTRAN 77 (Pollack, 1983).

STRUCTURE OF THE PACKAGE

The structure of the package is designed around two levels of input. The lower level is the systems level which allows changes and additions to the actual process models being used. New process models can be developed around a given format and existing process models can be altered to reflect the current knowledge of the system. The top level is the metallurgist or general user level which allows plant models to be built from the existing process models via a Command file. Only the lower level of input requires knowledge of programming as the Command files are simple to construct.

The Unit Model

The package is based on a modular structure where each identifiable process is called a Unit (Wells, 1973). The treatment plant to be examined is constructed by assembling the appropriate Units. These Units have the general form shown in Fig. 1.

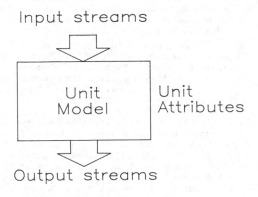

Fig. 1. General form of a unit

223

In the programme the Unit Model is a subroutine consisting of a set of equations that, together with the Unit Attributes (a small number of parameters which define the unit settings), define the relationship between the various mass flows into the Unit to the mass flows out of the Unit.

In this package each flow is called a Stream and consists of 11 separate components. These Stream components represent the materials of interest in the processing system. When setting up a system for a new processing environment, the flow variables that need to be tracked are identified and assigned to separate components in the Stream. An example would be:

TABLE 1   Example of Stream Component Assignments

Stream Component Number	Description of Component	Units Used
1	Water	m /hr
2	Cu as solid	kg/hr
3	Ag as solid	kg/hr
4	Au as solid	kg/hr
5	S as solid	kg/hr
6	Fe as solid	kg/hr
7	Solid gangue	kg/hr
8	Cu in solution	kg/hr
9	Ag in solution	kg/hr
10	Acid	kg/hr
11	Gangue in solution	kg/hr

Every separate Stream in a processing system is given a number called the Stream number, and a unique position in the programme's memory where the component values are kept. The input and output Streams of a Unit are then identified by their Stream numbers and the component values can be easily found or updated.

The Block Structure

What is required to enable the steady-state solution of a system consisting of recycle loops is an ordered system of evaluating each of the Units in turn. The calculation sequence used in this package was developed on a hierarchial structure. Essentially each recycle loop represents a step up in the level of the hierarchy. A processing section with no recycle loops would remain on level 1. A single recycle loop would achieve level 2 and a recycle loop within another recycle loop would achieve level 3. This can be continued up to level 5. Each step up the hierarchy groups together a number of elements of lower hierarchies into a set called a Block (see Fig. 2).

The Blocks are the means by which the calculation sequence is defined. Each of the elements of a Block, whether they are lower level Blocks or Units, are given a rank within that Block which assigns the order of calculation.

At the lowest level only Units are elements of a Block and to evaluate a Block consists of evaluating each of the Units in the order given. The evaluation of a Unit is achieved by calling the relevant subroutine which finds the value of output Streams of the Unit given the current value of the input Streams to the Unit. A pigeon-hole type of system operates with the Streams where the values from the various input Streams are read

from the relevant pigeon-hole and the values of the output Streams are written over the current values in the corresponding pigeon-holes.

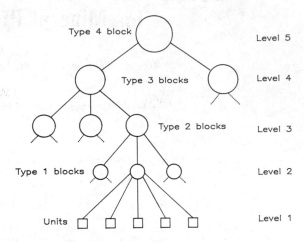

Fig. 2. The hierarchial block structure

This system of calculation assumes that initial values are given for all Streams in the system. For all except the main input feeds to the system, the values for each Stream can only be estimated since the purpose of the package is to solve for the actual values of each Stream in the steady-state condition. Naturally the better the initial estimate the fewer the iterations or calculation cycles necessary before convergence to steady-state occurs, but the package has proven to be quite well conditioned to poor estimates of the initial values.

In the instance where the elements of a Block are lower level Blocks then each of the Blocks are evaluated in the order given. This essentially passes the calculation sequence down to a lower level. This continues until the lowest level is reached and the Units are evaluated. The evaluation of a Block is considered complete when convergence is reached. Convergence is the condition where all the input Streams in the Block have altered less than a given percentage between two calculation cycles. Until this condition is reached the evaluation of each of the elements of the Block is continued in turn.

The convergence is based on successive changes in the input Streams, as a steady state condition has been reached when successive estimates of the input Streams to that block converge to constant values. Consequently the calculation sequence requires local steady-state conditions to be evaluated on the way to the overall steady-state solution.

The User Interface

A Command File is used as the major form of setting up a particular process system. The information in the Command File includes all the Units in the system, their individual attributes and all the Stream connections for the various Units.

The Command File is a text file created and altered using the standard text file Editor on the particular computer system. Each file is based on a standard format which is shown in Fig. 3.

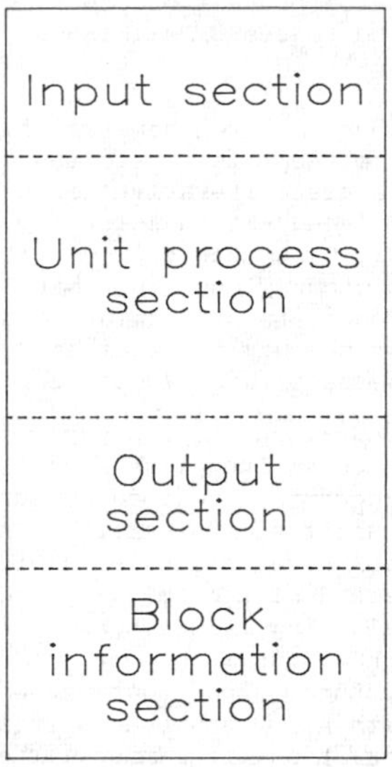

```
┌─────────────────────────┐
│                         │
│     Input section       │
│                         │
│- - - - - - - - - - - - -│
│                         │
│    Unit process         │
│      section            │
│                         │
│- - - - - - - - - - - - -│
│                         │
│     Output              │
│     section             │
│- - - - - - - - - - - - -│
│     Block               │
│   information           │
│     section             │
│                         │
└─────────────────────────┘
```

Fig. 3. Overall format of a command file

The input section. The input section is used to specify all the overall system variables such as a title for the particular simulation run, the current date and the run identification number. Also included is the value of the components of the feed to the system and initial estimates for any variable flows into the system such as makeup water or acid additions.

The unit process section. The unit process section is actually made up of a number of subsections, one for each Unit included in the system. All the information relevant to a particular Unit is included in its corresponding subsection.

Each unit process subroutine is the set of equations required to describe fully the mass transfer occurring in the particular process. It is fundamental that the total mass entering the process by all the input Streams must be equalled by the mass leaving the process. The processes therefore manipulate the component Streams between the input and output to achieve the desired results.

Many of the unit processes, involving solid/liquid separation are very simple using an attribute such as percentage solids in a particular output Stream to control the process. Other processes, such as leaching, or Counter Current Decantation (CCD), may require many more attributes to define the mass transfer required resulting in a much more complex set of equations in the subroutine.

The Unit processes were developed in close conjunction with the process metallurgists involved. In several processes they supplied or altered the basic empirical equations proposed to more precisely represent the experimental results which had been achieved in testwork. This ability to easily modify the unit process subroutines was one of the principal reasons for writing the package.

Each identifiable process in the circuit is a Unit and is given a unique number called a Unit number. This number is then used for any subsequent identification of that Unit and also makes up part of the Stream numbers of all Streams which are outputs from that Unit. The process types may be

used any number of times.

The information in each subsection includes:

a) a process identifier which defines the type of process (and thus the subroutine referenced) that the Unit represents;

b) the Unit attributes which together with the type of process, completely defines the Unit Model;

c) the inputs to the Unit, i.e. the Stream numbers of the flows into the Unit;

d) the initial values of all the components of the output Streams of the Unit;

e) and a set of parameters called Flow Attributes which define particular control conditions which are placed on some of the Streams into the Unit. For example, the amount of acid added in a leach stage could be dependent on the flow rates of other Streams entering that Unit. These Flow Attributes can be used in four different modes:

1. To control the amount of water added to a Unit to maintain a particular percentage of solids in the output Stream of the Unit.

2. To control one of two flows in order to keep a particular ratio of given components in the combined Stream.

3. To enable bleed Streams to be set to a percentage of the total Stream flow.

4. To enable the flow in a Stream to be set to maintain a particular percentage of a component in that Stream relative to the similar component in another Stream.

The output section. The output section merely specifies those Streams which are the outputs of the system.

The block information section. The Block information section details all the Blocks in the system, their position in the hierarchial structure and the rank of each element in each Block. This information is decided upon with reference to the recycle loops present in the system and as described in the Block Structure section above, defines the calculation sequence for the system.

Package Output

The essential output of the package is the values of each component of each Stream in the system when the whole system reaches steady-state. At present this information is printed out in two forms, one as the flow rates of each of the components of the Stream in the units as given in Table 1 and the other as the assay values for each Stream with regard to the particular components that were tracked.

Another form of output of the package consists of a plotted flow diagram of the section of the plant being modelled over which relevant information about the flow values at steady-state can be added. This flow diagram is drawn directly from information contained in the Command File and so also forms a ready means of checking the flow connections specified. An example of a typical output of this section of the package is shown in Fig. 4 where the numbers in the bottom right corner of the Unit boxes are the Unit numbers and the four digit numbers by the flow lines are the Stream numbers. This example illustrates a part of the overall plant that processes the slag from another section of the plant to recover as much payable material left in the slag as possible. It illustrates how particular sections of an overall plant can initially be tested independently. The whole plant can then be tested together by merging and editing the separate command files.

detecting the occurrence of high calculating loads, for testing the effect of a bleed stream, and examining the results of contaminant build up. Variations to the process or the flow rates could then be made to reduce or eliminate unwanted characteristics.

In order to keep the programme as simple as possible and to keep its development time to a minimum only a basic iterative technique was used to obtain the steady-state solution. The hierarchial Block structure promotes the convergence of the system by ensuring that local convergence occurs in particular sections of the plant model on the way towards the overall plant model convergence to steady-state.

The package described in this paper is similar to the PRIMER (PRocess Iterative Mass-flow EvaluatoR) simulation programme developed by Burgess, Robson and Wells at Sheffield University (Wells, 1973). The application here though, to a mineral processing plant instead of a chemical processing plant where heat balancing is necessary, required a different emphasis on some sections of the programme's structure. The programme was also written to conform with ACE-T's standard structure with regard to simulation programmes. The use of the Command file for the input of information for example, has proven over many years experience with simulation programmes to be a flexible and simple method for setting a large number of variables before a simulation run.

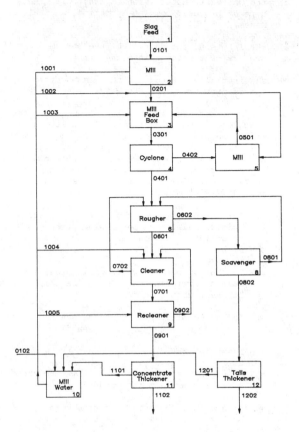

Fig. 4.   Typical section flow diagram

DISCUSSION

In the present application of the package 19 separate process subroutines have been written in order to model the required plant processes. Its principal application has been for establishing the capacity of the various sections of a plant being designed. By testing various feed rate and control scenarios the flow rates through the various sections of the plant can be quickly assessed for steady-state operation.

The package was also useful for testing the sensitivity of particular Unit processes, for

CONCLUSION

This Process Simulation Package represents an example of a computer programme produced to replace a complex manual calculation which had to be repeated many times. By reducing this process to the simple running of a computer programme a much wider range of scenarios can be tested in the available time giving more confidence that the optimum solution has been reached.

Further development of the package could be directed towards more detailed modelling of the individual processes or increasing the range of process types available when the need arises. Also the flow diagram form of output could be expanded since it provides an excellent form of output for reporting purposes.

REFERENCES

Pollack, S. V.   (1983) Structured Fortran 77 programming of Hewlett Packard computers Boyd & Fraser, San Francisco.

Wells, G. L. and Robson, P. M. (1973) Computation for Process Engineers. Leonard Hill, London.

A Preliminary Model of a Water-Only Cyclone and its Application to Product Quality Control

K.J. PILLAI, D.J. SPOTTISWOOD and W.R. BULL

Colorado School of Mines, Golden, Colorado, USA

and

E.G. KELLY

University of Auckland, Auckland, New Zealand

Abstract. An empirical model of a 254mm water-only cyclone treating minus 1700µm coal has been developed. The model includes the prediction of by-passing, SG_{50} corrected, water split and the use of a single equation for the reduced performance curve of particles in the range 1700-106µm. Particles finer than 106µm did not separate according to specific gravity. Simulations using the model showed the water-only cyclone performance to be dependent more upon feed size distribution than washability when the feed contains high-ash fines, but that, with a deslimed feed, the yield and ash content of the overflow should be controllable by altering feed rate or feed pulp density.

Keywords. Coal preparation; models; modeling; quality control.

INTRODUCTION

The use of hydrocyclones for cleaning coal in the finer sizes has become widespread in the past 20 years (Draeger and Collins, 1980; O'Brien and Sharpeta, 1976; and Visman, 1966, 1968). Such hydrocyclones, widely known as "water-only" cyclones, differ from classification hydrocyclones and dense medium cyclones primarily in their larger cone angle and longer vortex finder. Their major use is in cleaning coal in the 850-150µm range, although they are used for coarser sizes and have been proposed for cleaning to "zero" (O'Brien, 1980).

While water-only cyclones in coal cleaning have the advantage of low capital and operating cost, the separation possible is not as efficient as that in alternative equipment (Gottfried and Jacobsen, 1977; Gottfried, 1978; and Kelly and Spottiswood, 1982). So, it is particularly important for them to be operated at maximum efficiency and within their operating limitations. The work described in this paper was aimed at developing a mathematical model to describe water-only cyclone performance under a wide range of operating conditions. Although no such model has previously been available, the value of such models has been demonstrated and widely accepted in many other areas of the mineral processing industry. Mathematical models to describe size reduction processes, classification, concentration by flotation, etc., are now available.

The use of these models has provided an improved understanding of the processes, leading to greater operating efficiency. Their use has permitted better control, and they have been used in automatic control, both in the development of control strategies and as part of the control method itself. Also, when available, mathematical models have been used for equipment selection and flow-

sheet design. The ability to test alternative equipment selections and circuit configurations by computer simulation using mathematical models has allowed many options to be considered without extensive plant testing.

In gravity separation processes, such as those widely used in coal cleaning, the models have been limited (Karantzavelos and Frangiscos, 1984; Kelly and others, 1984; Kelly, Pillai, and Spottiswood, 1985; Kelly and Spottiswood, 1982, 1983; Pillai, 1985; and Subasinghe and Kelly, 1984). It was however, in the coal preparation industry that one concept, now widely used, was developed. It is the use of a "washability" or "separability" curve to describe mineral characteristics and hence separation potential, and "performance" or "Tromp" curves to describe separator performance (Kelly and Spottiswood, 1982, 1983; and Leonard 1979). Extensive studies at the U. S. Bureau of Mines (Gottfried and Jacobsen, 1977; Gottfried, 1978) established the concept of the reduced performance curve for predicting the products obtainable from a given gravity separator. The method was based on the premise that each type of separator has its own characteristic reduced performance curve (Fig. 1). This approach has also been used in classification (Kelly and Spottiswood, 1982; Lynch and Rao, 1965, 1968; Plitt, 1971, 1976; and

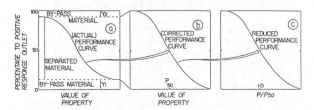

Fig. 1. Actual, corrected and reduced performance curves.

Reid, 1971) where size is the property of prime
importance rather than specific gravity (density).

The performance curve as obtained directly from
experimental data is shown in Fig. 1(a). This
curve represents the actual separation occurring,
which includes both material that undergoes
separation and material that passes through the
separator without separation (by-pass material).
The corrected performance curve [Fig. 1(b)] is
obtained from the actual performance curve by
elimination of the by-pass material. The reduced
performance curve [Fig. 1(c)] is obtained by
dividing the property scale by the property at
the separation point, that is, SG/SG_{50}.

As the mechanism resulting in particle movement
must be the same, or at least similar, in both
classifying hydrocyclones and water-only cyclones,
the form of the models successfully applied for
classifying hydrocyclones was considered to offer
a good basis for the present work.

In the present work, a Krebs D10 LB water-only
cyclone was modeled. While it might be expected
that the form of the model developed would be
adaptable to other sizes and configurations, and
to water-only cyclones from other manufacturers,
this has not as yet been determined.

EXPERIMENTAL

The tests were carried out using a conventional
closed circuit test rig capable of handling
400-500m³/hour of feed slurry. The slurry was
pumped from a sump to the water-only cyclone and
the overflow and underflow products were returned
to the sump. The details of the rig are shown
in Fig. 2. The total mass of feed added to the
sump was sufficiently large so that even if high-
ash material were held up in the "bed" in the
water-only cyclone, this would not significantly
change the quality of the remainder. The sample
points were set up so that timed representative
samples from the overflow and underflow streams

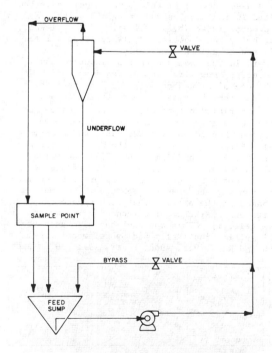

Fig. 2. Schematic diagram of experimental test
rig.

TABLE 1   Experimental Conditions

Test	Solids Flow Rate t/h	Pulp Density mass % solids
1	18.93	19.68
2	11.97	19.10
3	13.76	16.70
4	8.67	14.62
5	9.29	9.23
6	5.94	9.01
7	4.26	4.71
8	2.69	5.24
9	23.92	25.35
10	15.11	24.69

could be cut simultaneously. The water-only
cyclone used for all tests was a Krebs Model D10
LB (nominal diameter 10 inch, 254mm). The feed
sample was taken from a stockpile of roughly
3 tonnes, then crushed to pass 1700μm.

A total of ten tests were carried out. Five feed
pulp density conditions at two nominal levels of
volumetric flow rate were used during the tests.
The details of the experimental conditions are
given in Table 1.

Clean coal (overflow) and refuse (underflow)
samples were collected simultaneously for each
set of conditions. After collection, all samples
were weighed wet and dry, wet screened at 38μm,
and then dry screened at 850μm, 300μm, and 106μm.
Each fraction was weighed and subjected to float-
sink separations at specific gravities of 1.3,
1.4, 1.5, 1.6, 1.7, 1.8, 1.9, 2.0, 2.2, and 2.55.
The dense liquids used were made up from suitable
mixtures of tetrabromoethane, tetrachlorethylene,
and carbontetrachloride. Detailed experimental
results and description of experimental procedures
are available elsewhere (Pillai, 1985).

RESULTS

From float-sink analyses, the performance curves
were determined for each size fraction in each
test. The experimental values from all tests are
shown in Table 2 and the performance curves from
Test 3 are shown in Fig. 3.

Regression equations, as discussed in subsequent
sections, were generated using the Statpack Interactive Statistical package developed by Western
Michigan University. This versatile statistical
package enables stepwise regression analysis with
user-flexibility in selection of the form of
equation to be tested.

TABLE 2   Experimental Results

| Size, μm | Mass % in fraction | |
	Not deslimed	Deslimed
1700-850	25.7	31.2
850-300	38.3	46.6
300-106	14.2	17.3
106- 38	6.2	2.5
- 38	15.6	2.4
	100.0	100.0

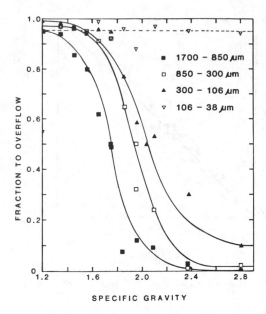

Fig. 3. Performance curves for Test 3.

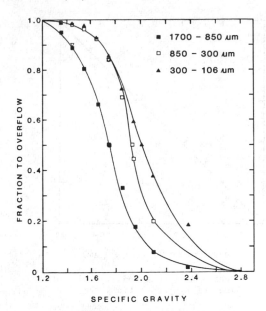

Fig. 4. Corrected performance curves for Test 3.

MODEL DEVELOPMENT

Performance Curves

Figure 3, taken from the results of Test 3, shows that the three coarsest size fractions displayed typical performance curves, with values of SG_{50} that could be determined with reasonable accuracy. The curve for the 106-38μm fraction shows very limited separation. These results are typical of those from all tests performed. The separation of the -38μm fractions were not analyzed by float-sink tests.

Each of the three performance curves from each of the 10 tests was analyzed as follows. It was assumed (Fig. 1) that a representative fraction y_1 of the feed by-passed the separation process and reported directly to the overflow product, and a representative fraction y_2 of the feed reported directly to the underflow product, leaving a fraction $1 - (y_1 + y_2)$ of the feed that was separated by the water-only cyclone. The actual products of the separation process were calculated by subtracting fraction y_1 of the feed from the total overflow product and fraction y_2 of the feed from the total underflow product. The corrected performance curves, and the interpolated values of SG_{50} were then determined for each size fraction for each test. Typical corrected performance curves (from Test 3 only) are shown in Fig. 4, and all values of SG_{50c} appear in Table 4. The reduced corrected performance values for all tests were then calculated, and are shown in Fig. 5. These values appeared to fit the expected pattern well enough to justify searching for a common equation for all points. The Plitt-Reid (Plitt, 1971, 1976; Reid, 1971) equation was linearized, and the equation:

$$Y_c = \exp\left[-0.693\ (SG/SG_{50c})^{7.02}\right] \qquad (1)$$

was found to fit the data with a coefficient of determination of 0.94.

By-Pass Ratios

Empirical relationships between the by-pass ratios (y_1 and y_2) and the operating conditions of the

water-only cyclone were then sought. Although many equations were found with coefficients of determination of over 0.95, it was decided to seek relatively simple equations with a common form and reasonable coefficients of determination. Additionally, equations containing variables in forms that could be justified rationally would be preferred, although our success in achieving this may be questioned. The results were as follows, with P the feed pulp density and SF and WF the solids and water mass flows, respectively.

$$y_1 = K_1 + K_2 SF^2 + K_3 P^2 + K_4 SF/P + K_5\ SF/WF \qquad (2)$$

SIZE μm	K_1	K_2	K_3	K_4	K_5	CD
1700-850	0	0	0	0	0	
850-300	0.11029	0.00979	0.007512	-0.01531	-2.584	0.94
300-106	0.07099	0.01750	0.010780	-0.02739	0	0.91

$$y_2 = K_1 + K_2 SF + K_3 P + K_4 P^2 + K_5 WF + K_6\ SF/WF \qquad (3)$$

SIZE μm	K_1	K_2	K_3	K_4	K_5	K_6	CD
1700-850	0.4543	0	0	0.00204	-3.906	-4.637	0.79
850-300	-0.0925	0.0244	0	0.00119	18.320	-4.140	0.94
300-106	-0.0048	0	0.000836	0	2.671	0	0.74

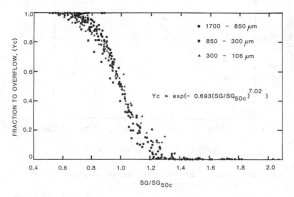

Fig. 5. Reduced performance data for all tests.

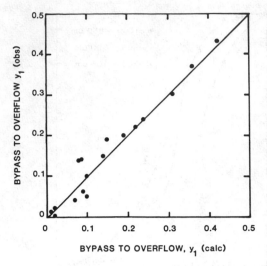

Fig. 6. Comparison of observed and calculated
values of by-pass ratio y_1.

Figures 6 and 7 compare calculated values for y_1
and y_2 respectively with the corresponding
observed values.

Corrected SG$_{50}$

From the values of SG$_{50c}$ obtained from the
corrected performance curves, empirical equations
were sought correlating the SG$_{50c}$ value for each
size fraction with the value of the operating
variables. The results were as follows:

$$SG_{50c} = K_1 + K_2 P^2 + K_3 WF \qquad (4)$$

SIZE μm	K_1	K_2	K_3	CD
1700-850	1.2011	0.000975	0.002918	0.89
850-300	1.3990	0.001139	0.004014	0.88
300-106	1.8166	0.001022	0	0.84

Figure 8 compares the values calculated using the
model with the corresponding observed values.

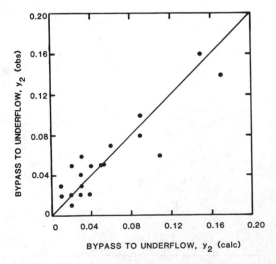

Fig. 7. Comparison of observed and calculated
values of by-pass ratio y_2.

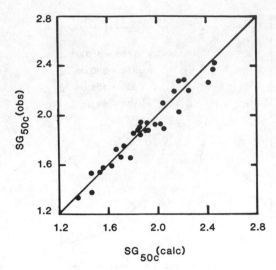

Fig. 8. Comparison of observed and calculated
values of SG$_{50c}$.

Separation of Finer Sizes

Figure 3 indicates that there was no significant
separation of particles in the 106-38μm size
fraction based on specific gravity. The model,
therefore, includes provision for a constant
value (95%) of this sized material reporting to
the overflow, the remainder to the underflow.
This value remained sensibly constant in all tests.

The -38μm particles were assumed to follow the
water. The water split was therefore correlated
with the operating variables, producing the following empirical relationship:

$$R_f = 0.1568 + 0.000322 P^2 - 0.3445 \, SF/WF \qquad (5)$$

where R_f is the fraction of feed water reporting
with the underflow product. This equation had
a coefficient of determination of 0.97. Calculated
values are compared with the corresponding observed
values in Fig. 9.

The model, therefore, calculates that a proportion
R_f of the -38μm solids reports to the underflow,
with the remaining $(1 - R_f)$ to the overflow.

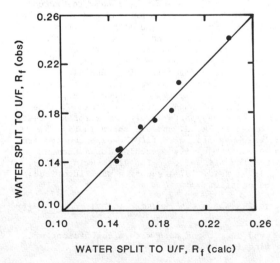

Fig. 9. Comparison of observed and calculated
values of water-split ratio R_f.

SIMULATION

A computer program was written to simulate the operation of the water-only cyclone using the model described above. The steps in the program are outlined below.

1. The overall feed conditions are defined by the solids flow rate (SF, t/h) and pulp density (P, mass % solids) from which the water flow rate (WF, t/h) is calculated.

2. The feed is characterized by the size distribution (mass % in each size fraction) and by the specific gravity distribution of particles in each size fraction. The feed is then defined by calculating the solids flow rate of material fed to the water-only cyclone in each SG fraction of each size fraction, referred to hereafter as the tonnage of each "element". The material in the top three size fractions is defined in this way—a total of 3 x 11 = 33 elements. Since the particles in the two finer fractions are not separated according to specific gravity, no such complicated definition is necessary.

3. The specific gravity of the SG fractions are defined as follows: lighter than 1.30 as 1.20, heavier than 2.55 as 2.8; other fractions are defined as the arithmetic average of the two extremes (e.g., for 1.3-1.4, the value 1.35 is used).

4. Using Equations 2 and 3 with their appropriate coefficients, the mass of material in each element in the three top size fractions that by-pass to the overflow and underflow respectively is calculated. By difference, the mass of material in each element that goes to the separation process is obtained.

5. The SG_{50c} for each size fraction is calculated from Equation 4.

6. The ratio SG_{50c} is calculated for each element in each size fraction.

7. Using Equation 1, the fraction Y_c of each element in the three top size fractions reporting to the overflow as a result of separation is calculated, and then the mass of each such element. The mass of each element reporting to the underflow by separation is found by difference.

8. Again, only for the top three size fractions, the total mass of each element in the overflow is found by adding the mass of each element reporting as a result of separation to the mass of each element reporting as a result of by-passing. Similarly, the total mass of each element in the underflow is found.

9. The gross mass of solids in the overflow is found by adding together all of the elements in the overflow from the top three size fractions, plus the mass of the fourth and fifth size fractions reporting according to the 95:5 split of the 106-38μm fraction and the water split ratio $(1 - R_f):R_f$ for the −38μm fraction, respectively.

10. Steps 1-9 above deal only with the masses of material reporting to the overflow and underflow according to the by-passing, separation, and "splitting" mechanisms proposed. The simulation can be modified to include the quality of the products by additional input data, namely (i) the ash content of the material in each specific gravity fraction

TABLE 3 Feed Size Distributions

Specific gravity fraction	Mass % in fraction
1.20	19.5
1.35	15.9
1.45	11.7
1.55	8.5
1.65	5.9
1.75	3.9
1.85	4.1
1.95	5.1
2.10	5.9
2.38	6.8
2.80	12.7

in the three top size fractions of the feed, and (ii) the ash contents of the two finest fractions, and using these data to calculate total ash in the products. The simulation can now predict the yield, ash content of the products, and flow rates (solids and water) of these products for any feed condition (feed rate, pulp density) within the limits of the model, for any feed size distribution, and any specific gravity distribution within each size fraction.

RESULTS OF SIMULATION

The simulation program was used to test the "reasonableness" of the model and to investigate some of the operating characteristics of the water-only cyclone. Four cases of different feed "types" were tested, each over a range of solids feed rate and pulp density.

The feed "types" were the four possible combinations of two size distributions and two washabilities, as shown in Tables 3 and 4.

For the coal with poor washability, it was assumed that each of the eleven specific gravity fractions contained 9.09% of the total material (except that the 2.80 fraction contained 10.10%).

TABLE 4 Specific Gravity Distribution—Good Washability Coal

TEST	SIZE μm	SG_{50c}	y_1	y_2	R_f
1	1700-850	1.86	0.00	0.00	
	850-300	2.20	0.20	0.05	0.205
	300-106	2.32	0.30	0.02	
2	1700-850	1.66	0.00	0.06	
	850-300	2.10	0.15	0.03	0.183
	300-106	2.31	0.24	0.04	
3	1700-850	1.74	0.00	0.02	
	850-300	1.93	0.00	0.02	0.176
	300-106	2.01	0.10	0.01	
4	1700-850	1.57	0.00	0.05	
	850-300	1.95	0.04	0.03	0.170
	300-106	1.89	0.22	0.03	
5	1700-850	1.55	0.00	0.06	
	850-300	1.95	0.01	0.02	0.152
	300-106	1.90	0.19	0.02	
6	1700-850	1.54	0.00	0.10	
	850-300	1.75	0.06	0.02	0.146
	300-106	1.94	0.06	0.03	
7	1700-850	1.37	0.00	0.28	
	850-300	1.65	0.02	0.05	0.141
	300-106	1.85	0.14	0.02	
8	1700-850	1.35	0.00	0.14	
	850-300	1.60	0.00	0.16	0.153
	300-106	1.83	0.05	0.05	
9	1700-850	1.95	0.34	0.05	
	850-300	2.25	0.69	0.04	0.282
	300-106	2.42	0.84	0.02	
10	1700-850	1.88	0.00	0.08	
	850-300	2.20	0.37	0.04	0.241
	300-106	2.37	0.43	0.03	

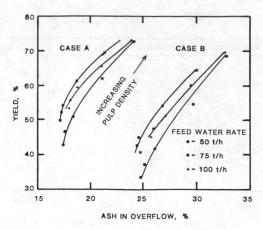

Fig. 10. Relationship of yield and ash content of
overflow product: Case A - deslimed
feed, good washability, Case B - deslimed
feed, poor washability.

The treatment of each of these four combinations
over a range of feed rates (covering the manufacturer's recommended range) and pulp densities was
simulated; the results are shown in Figs. 10 and
11.

DISCUSSION AND CONCLUSIONS

The test work carried out has been limited to the
investigation of the performance of a single water-
only cyclone, with a single physical configuration
and a single feed material. The model derived is
by no means general for this type of equipment,
but several qualitative conclusions may be drawn.

1. By-passing occurs in the particle sizes from
   1700-106µm except for the heavier fractions
   of the coarser particles.

2. Particles in this range that undergo separation
   do so in the expected manner, the SG_{50c} values
   being dependent upon feed conditions and
   particle size.

3. The reduced performance curves of all particles
   in this size range appear to be common, and
   are quite accurately described by a single
   equation of the Plitt-Reid type.

4. The SG_{50c} values can be correlated with the
   operating conditions of the water-only cyclone.

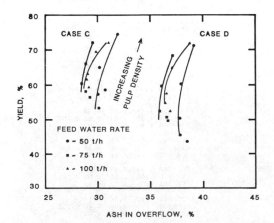

Fig. 11. Relationship of yield and ash content of
overflow product: Case C - feed not
deslimed, good washability, Case D - feed
not deslimed, poor washability.

5. The by-pass ratios of the operation are dependent on particle size, and are less accurately
   correlated with operating conditions.
   Generally, the proportion y_2 by-passing to
   the underflow product is small; the proportion
   y_1 by-passing to the overflow is greater for
   the finer particles and can be very significant.

6. Particles finer than 106µm appear not to be
   separated according to particle specific
   gravity, the finest separating according to
   the water split, and the 106-38µm separating
   in a constant proportion somewhat different
   from that of the water split.

7. The water-slit ratio is accurately correlated
   with feed conditions.

From the results of the simulations shown in
Figs. 10 and 11, it is clear that when the fines
(-106µm) have a high ash content, the quality of
overflow product is more sensitive to the size
distribution of the feed than to its washability.
In Cases A and B, in which most of the feed consisted of particles in the three coarser size
fractions, a wide range of overflow products would
be possible from a range of feed rates and feed
pulp densities, and control of product quality
should be possible by adjusting either of these
feed variables.

On the other hand, Cases C and D show that, when
the feed contains a high proportion of high-ash
fines (-106µm), the yield-ash curves are almost
vertical. This indicates that the separation of
the coarse fractions is overwhelmed by the non-
separation of the fines, which go predominantly
to the overflow product. This, in turn, means
that with this type of feed, the water-only
cyclone is capable of producing an overflow
product of almost constant ash content regardless
of the yield, and that little can be done to
reduce this ash content in a single stage operation. The model, therefore, confirms the wisdom
of either desliming the feed to, or the overflow
product from, a water-only cyclone.

The work reported above involved a very large
number of sample analysis procedures, yet the
model developed still applies only to a single
water-only cyclone with fixed dimensions, and a
feed with fixed characteristics. No data have
been obtained regarding the effects of changing
apex or overflow dimensions, or regarding scale-
up.

Much work, therefore, remains to be done in this
area, but it is thought that the limited model
presented gives indications of the characteristics
of the separation possible in a water-only
cyclone, and of the limitations of its application. Only when a more comprehensive model is
developed will it be possible to apply it to the
design of control systems and strategies.

ACKNOWLEDGMENTS

The authors would like to express their thanks
to Krebs Engineers, Inc. for the provision of the
water-only cyclone, to the Colorado School of Mines
Research Institute for use of their gravity
separator/classification test rig, and to the
Desarado Mine of Western Fuels Association for
providing the coal sample.

REFERENCES

Collins, J. W. (1970). The hydrocyclone: a most
   useful preparation tool. Coal Min. and
   Process., 6, 53-55.

Draeger, E. A., and J. W. Collins (1980). Efficient use of water only cyclones. Min. Eng., 32, 1215-1217.

Gottfried, B. S. (1978). A generalization of distribution data for characterizing the performance of float-sink coal cleaning devices. Int. J. Miner. Process., 5, 1-20.

Gottfried, B. S. and P. S. Jacobsen (1977). Generalized distribution curves for characterizing the performance of coal cleaning equipment. U.S. Bureau of Mines Report of Investigations, RI 8238.

Karantzavelos, G. E., and A. Z. Frangiscos (1984). Contribution to the modeling of the jigging process. In J.A. Herbst (Ed.), Control 84, Chap. 12, AIME/SME, 97-105.

Kelly, E. G., and D. J. Spottiswood (1982). Introduction to Mineral Processing, Wiley Interscience, New York.

Kelly, E. G., and D. J. Spottiswood (1983). Coal preparation: prediction and analysis of plant performance. Miner. and Energy Resources, 26, 3.

Kelly, E. G., K. J. Pillai, and D. J. Spottiswood (1985). Performance assessment of modern gravity concentration equipment. Proc. 15th Inter. Miner. Process. Congr., Cannes, France (in press).

Kelly, E. G., D. J. Spottiswood, D. E. Spiller, and C. N. Robinson (1984). Performance of compound trough profile spiral cleaning fine coal. AIME/SME Preprint, 84-48; Trans AIME (in press).

Leonard, J. W. (1979). Coal Preparation, 4th. ed., AIME.

Lynch, A. J., and T. C. Rao (1965). Digital computer simulation of cominution systems. Proc. 8th Comm. Min. Metall. Congr., 597-606.

Lynch, A. J., and T. C. Rao (1975). Modelling and scale-up of hydrocyclone classifiers. Proc. 11th Inter. Miner. Process Congr., Paper 9, Cagliari, Italy.

O'Brien, E. (1980). Fine coal cleaning in the 28 x 0 fraction - a technical overview. Min. Eng., 32, 1213-1214.

O'Brien, E. J., and K. J. Sharpeta (1976). Water only cyclones : their functions and performance, Coal Age, 81, 110-114.

Pillai, K. J. (1985). Mathematical modeling of a water-only cyclone. Ph.D. Thesis, Colorado School of Mines.

Plitt, L. R. (1971). The analysis of solid-solid separation in classifiers. CIM Bull., 64, 42-47.

Plitt, L. R. (1976). A mathematical model for the hydrocyclone classifier. CIM Bull., 69, 114-123.

Reid, K. J. (1971). Derivation of an equation for classifier reduced performance curves. Can. Met. Quarterly, 10, 253-254.

Subasinghe, G. K. N. S., and E. G. Kelly (1984). Modelling pinched sluice type concentrators. In J.A. Herbst (Ed.), Control 84, Chap. 11, AIME/SME, 87-96.

Visman, J. (1966). Bulk processing of fine materials by compound water cyclones. CIM Bull., 15, 333-346.

Visman, J. (1968). Integrated water cyclone plants for coal preparation CIM Bull., 61, 74-79.

Ore Sorting and Artificial Vision

R. MANANA, J.J. ARTIEDA and J.C. CATALINA

Department of Electrical Engineering, Escuela de Minas, Rios Rosas, 21 28003-Madrid, Spain

Abstract.- The aim of this paper is to analyse the possibilities of applying signal processing and artificial vision techniques to the ore sorting problem. In order to overcome the limitations imposed by the classical approach to the problem, which essentialy consists of using only punctual information, neglecting the relationships among different points, we are studying these other methods.

This paper also describes the preliminary works of design and the corresponding experimentation of an artificial vision system, dealing with coal sorting problems.

Keywords.- Mining; sorting; pattern recognition; robots; vision.

INTRODUCTION

A major problem in automation is instrumentation due to the frequent difficulties, in practice, of directly measuring the desired attributes. This results in the need for instrumentation with enough intelligence to decode the signal supplied by the transducer. This problem is greater and also more important in robotic.

It is possible to use sensor instruments that at first might not seem useful, as is the case of artificial vision. The possibility of using an image (either optical or gathered by non-conventional methods, e.g. ultrasounds) as an information source for a control system has many advantages, including physical contact becoming unnecessary and therefore there is no need to normalize the work environment, which is extremely useful in mining applications.

The fact that there are already commercially available cameras (image sensor) sturdy enough to work in the mining process environment and that processors able to meet real time processing needs for the image sensor supplied information are about to appear, means that this idea might soon become a reality.

As a practical application of artificial vision in mining, a system able to distinguish different rocks and ores placed on a conveyor belt is being developed.

At present, the coal-shale sorting process is the most promising field of its application in Spain, incorporating the capability of detecting extraneous bodies.

Neither the ore sorting process nor vision system are new concepts. They have been described in the literature extensively (Schapper, 1976; Schapper,1979; Mclaughlin 1979; Cunin, 1982; Maenpaa, 1982; Eerola, 1983; Matikainen, 1983; Pugh, 1983; Basañez, 1984; Mañana,1984).

But in this case the problem is not solved, because :

1.- Ore sorters :

- We do not know of an ore sorter system for coal. Experiments with South African Coal showed that they are not sortable with low power HeNe Laser, (Shapper, 1976).

- Traditional ore sorters require that :

  . Valuable fractions must be liberated.
  . Fractions must be identifiable by single parameters.
  . Special material feed presentation system is needed because stones must not be in contact with each other.

2.- Vision Systems

- Vision systems have not either speed or adequate algorithms to process this kind of image.

- This system has to work in the mine.

HARDWARE DESCRIPTION

The sensing device is a TH 7802 circuit of 1024 pixels arranged perpendicularly from above the conveyor belt, therefore, while the camera performs the transversal sweep, the conveyor belt executes the longitudinal one.

The frequency (per pixel) of the transverse sweep is 1MHz, but the longitudinal one is a function of the sampling needs, the conveyor belt speed, illumination and computational needs.

The camera output already sampled is con<u></u>verted into digital by two flash conver-<u></u>ters with 8 bits resolution. The camera and A/D converter control is carried out by Z-80 microprocessor, which also controls the conveyor belt speed, the illumination conditions and the data transfer to the main computer.

The microprocessor also has the following assigned tasks :

- Correction of camera imperfections ari<u></u>sing firstly, from photodiode irregula<u></u>rities and secondly, from the losses in the retard line.

- The camera samples 1024 pixels. It has been found that 256 are enough to carry out the survey, when the conveyor belt is 20 cm wide as in our case.

The main computer is a PDP 11/23 plus, with all the necesary peripherals for graphical I/O and fast processing.

VISION SYSTEM DESCRIPTION

The vision system developed consists of three stages Fig. 1 :

Preprocessing

Processing at line level. It includes signal processing methods implemented in hardware as well as in software, in the followings steps :

1.- Analogue processing

A low pass filter eliminates the switching transients and transmission noise.

2.- Sampling

The conversion is perfomed in 8 bits over the 1024 pixels.

3.- Discretization

In order to reduce the amount of informa<u></u>tion to be subsequently processed, one out of each   N consecutive pixels is transmitted. We can choose :

- 1 or 2 or Nth of each set of N consecu<u></u>tive pixels.

- The pixel that has the minimal value.

- The pixel that has the maximal value.

- The average of the N pixels.

- Randomly.

4.- Correction of camera imperfections

The original linear image is multiplied by a vector whose elements are obtained previously for each sensing circuit (Artieda, 1984)

$$\{Sr\} = \{Sc\} \cdot \{C\}$$

Sr = Corrected image
Sc = Original image
C   = Coefficients vector

5.- Transmision

The image already converted into numbers and corrected is transmitted to the PDP 11, used for numerical processing of the image.

Model generator

At this stage, processing is perfomed on a 2D image and can be divided into two stages.

Image processing

This aims at separating the rocks into individual specimens. In this case this process is a numerical one and not a mechanical one as with ore sorters.

To achieve that, a series of classical processing stages are used, among them (Rosenfeld, 1982; Ballard, 1982).

. Filtering. Up to now, non - recursive filters are being applied according to

$$f'(x,y)= \sum_{-\nu}^{\nu} \sum_{-m}^{m} g(x+i,\ y+j)S\ (ij)$$

where

f(x,y)   = original image
f'(x,y) = filtered image
S(i,j)   = matrix of filter coefficients
u, m   = dimensions of S(i,j)

. Conversion to binary images
. Opening and closing. These two mathema<u></u>tical morphology algoritms (Serra, 1982) have been implemented. The grid used is squared and the basic element of these operations is

$$B = \begin{vmatrix} 0 & 0 & 0 \\ 0 & 0 & 0 \\ 0 & 0 & 0 \end{vmatrix}$$

. Homogenization. As a complement to the previous stage, the system can per<u></u>form the following operation :

100 - 01 ===== 111-11

. Equalization of the image.
. Logic operation and mask between images.
. Determination of connected sets.
. Determination of borders.

These processes change images so that the next stage can extract features of every stone.

Model generator

The second part of a model generator is to make a semantic model of the stone. So some algorithms have been implemented.

Among these are

- Statistics of gray levels (1st, 2nd 3rd & 4th moments).
- Rate base/height.
- Minimal and maximal values
- Histogram.

The research up to now shows that none
of the measured characteristics are de-
terminative, in general, so more elabora
te and complex models are being studied.

Matching

The next stage has to match the model of
the stone with a set of patterns of stones to determinate the most similar.

EXAMPLE

An experiment with coal and shale from
the Puertollano coal field has been done,
to demostrate the efficiency of this sys
tem. A set of 50 stones has been ran -
domly placed on a laboratory conveyor
belt. The size of the stones ranges from
60 - 140 mm. The stones are not on top
of each other, nor are they separated.
The stones had previously been air-dryed
20 pictures have heen taken of this belt.

It is 256 x 128 pixels x 6 bits in size.
The Fig. 2 is a diagram of the installations. The ilumination has not any special features.

Algorithm

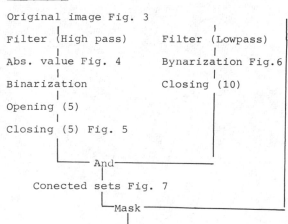

Original image Fig. 3

Filter (High pass)   Filter (Lowpass)

Abs. value Fig. 4   Bynarization Fig.6

Binarization   Closing (10)

Opening (5)

Closing (5) Fig. 5

And

Conected sets Fig. 7

Mask

Image of separated stones Fig. 8

The model for this experiment had only
one parameter. This parameter was the
average of the gray levels of the pixels
of a stone.

Results

In the first stage the perfomance was
95% and the second one 87%. The total
perfomance was 82%. This value is considered good enough for a man, doing this
work.

IN THE FUTURE

The system presented here is laboratory
system, and there are still two problems :

- Discrimination parameters have to be
  better, so we are studing :

  . Acquisition information about 3D.
  . Algoritms, to see typical elements
    like textures, brightness , interca-
    lations, etc.

. More intelligent systems

- To make the system work faster.

A new set of cards based upon specialized. I.C. are being developed. With
these cards the system will be able to
process several images per second.

CONCLUSIONS

. Laboratory equipment has been developed to carry out studies in the recog
nition of rocks and ores through application of image processing techniques.

. An algorithm for separating the coal
fragments from waste has been presented. Now, a special material feed pre
sentation system is not needed.

. A specific case has been presented
with good perfomance.

ACKNOWLEGEMENTS

This project is supported by the Comi--
sion Asesora de Investigación Cientifi-
ca y Técnica, Ministerio de Educación y
Universidad Politecnica de Madrid.

The author wishes to thank Paul Landes-
man for revising this paper.

REFERENCES

Artieda, J.J., and J.C. Catalina (1984).
Aplicacion de la visión por computador
al escogido óptico de rocas y minerales.
In M. Silva (Ed.), II Simposio Nacional
IFAC sobre Automática en la Industria,
Zaragoza 14-16 Noviembre. pp 433-447.

Ballard, D.H. and C.M. Brown (1982).
Early processing. Computer Vision, Chap
3, Prentice-Hall, Englewood Cliff, New
Jersey, p.p. 63- 118.

Basañez, L. (1983). La percepción visual
en los robots industriales : realidades
y tendencias. Revista de Robótica, 5,
17-23.

Cunin, P. (1982). Essai d'une unité pilo
te de tri optique a la Societé des Talcs
de Luzenac. L'Industrie Minerale, Febre-
re, pp 63-66.

Eerola, P.I. (1983). Improving the ove-
rrall economics of mining and ore processing with preconcentration. The Second
International Symposium on Small Mine
Economics & Expansion. Helsinki, June,

12-16.

Mañana, R. and others (1984). Nuevos de-
sarrollos en la monitorización de maqui-
nas y sistemas en la minería Española.
12 Th world Mining Congress, New Delhi,
November 19-23.

Matinkainer, R. and others (1983). The
technical and economical aspects which
make underground and open pit mining
of limestone competitive in Filand,
The second International Symposium on
Small Mine Economics & Expansion. Helsin
ki, June 12-16.

Maenpaa, I. G P. Malinen, and R. Soders-
trom (1983). A computer system for photo
metric mineral sorting. In T. Westerlung
(Ed.) 4TH Symposium on automation in mining, mineral and metal processing. Finland, August 22-25, pp 471-481.

Mclaughlin, D. (1979). Operation of an
optical ore-sorting system. American Mining Congress. Los Angeles, September
23-26.

Pugh, A (1983). Commercial robot vision
systems. In A. Pugh (ed.), Robot Vision,
Springer Verlag, Berlin. Chap. 7. pp
305-354.

Rosenfeld, A, and A. C. Kak (1982). En-

hancement and Restoration. Digital Picture Processing, Vol 1, 2 nd ed. Academic Press, Orland, pp 209-352.

Schapper, M. A. (1976). Beneficiation at
large particle size using photometric
sorting techniques. In F. H. Lancaster
(Ed.), II IFAC Symposium of Automation
in Mining, Mineral and Metal Processing.
Johannesburg 13-17 September. pp 277-287.

Schapper, M.A. (1979). The gamma sort.
Reprint from Nuclear Active, July.

Serra, J. (1982). The hit or miss trans-
formation, erosion and opening, Image
Analysis and Mathematical Morphology,
Chp. 2, Academic Press, London. pp 34-62.

Fig. 1. Vision system

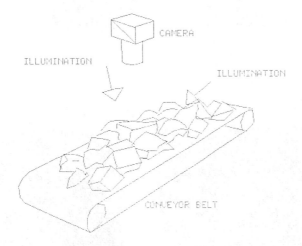

Fig. 2. Diagram of the installation

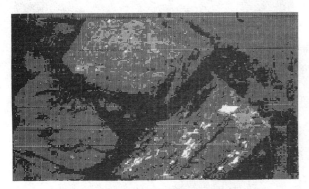

Fig. 3. Example. Original image

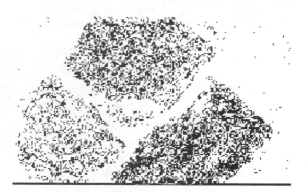

Fig. 4. Example. Absolute value

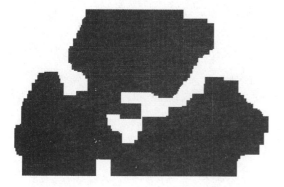

Fig. 5. Example. Stone area

Fig. 6. Example. Shadow area

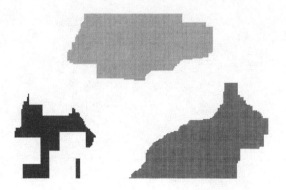

Fig. 7. Example. Connected sets

 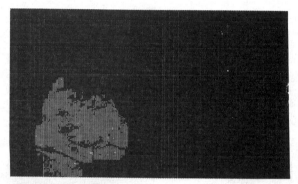

Fig. 8. Example. Separated stones

Control Strategies for Coal Flotation Circuits

R.A. SEITZ and S.K. KAWATRA

Department of Metallurgical Engineering, Michigan Technological University, Houghton, MI 49931, USA

Abstract. The incentive to control coal flotation circuits arises
because of disturbances, from plant operating practices and variations
in feed coal characteristics, which result in large amounts of fine
coal being lost in circuit tailings streams and lower grade concentrates
being produced. Additionally, the flowrates and compositions of concentrates and tailings streams can have significant effects on dewatering processes. A universal control strategy for these circuits cannot
be developed. Rather, three major groups of variables must be considered during strategy development: coal characteristics; circuit
design and operating practice; and the variables selected to control
the process. The second and third groups are discussed in this paper
and examples from studies in operating plants are presented.

Keywords: Coal flotation; flotation control; stabilizing control;
optimizing control; flotation; mineral processing.

INTRODUCTION

For many reasons, the design, operation,
and control practices for coal flotation
circuits differ drastically from those
used in the processing of metallic mineral
ores. In particular, the feed is not
ground to liberation size; rather only a
fraction of the raw plant feed is processed by flotation, which generally
results in larger and more frequent disturbances in the circuit feed. Additionally, coal is a light, readily flotable
mineral and the bulk of the feed is recovered in the concentrate, hence coal
flotation circuits are normally operated
with overloaded froths.

The incentive to control coal flotation
circuits arises because of disturbances,
from plant operating practices and variations in feed coal characteristics, which
result in large amounts of fine coal being
lost in circuit tailings streams and lower
grade concentrates being produced. Additionally, the flowrates and compositions
of concentrates and tailings streams can
have significant effects on dewatering
processes.

The objective of any control system for
coal flotation circuits is twofold: to
minimize the effects of feed disturbances
on circuit grade-recovery performance; and
to operate on the optimal grade-recovery
curve for a particular coal. However, a
universally applicable control strategy
for coal flotation circuits cannot be
developed; due to the fact that raw coals
are extremely heterogeneous materials with
widely varying response to flotation and
because specific circuit performance objectives vary. In addition, the effects
on strategy development of the various
types of flotation circuit configurations
in common use has not been considered

previously. Rather, three major groups of
variables must be considered during strategy development: coal characteristics;
circuit design and operating practice; and
the variables selected to control the process. The second and third groups are
discussed in this paper and examples from
studies in operating plants are presented.

When developing strategies, two levels of
control must be considered: stabilizing
control, to minimize the influence of disturbances; and optimizing control, to
achieve the optimal grade-recovery behavior
for the circuit. The former must be implemented initially in order to provide a
stable circuit for optimizing purposes.
The effects of circuit design and operating practice are very important factors in
maintaining stable and optimal circuit
operation. This should be considered as a
part of strategy development. The use of
circuit design and operating practice to
obtain some stability and optimization can
drastically reduce the burden on any control strategy.

EXPERIMENTAL WORK

The experimental work, performed by personnel from the Department of Metallurgical Engineering, Michigan Technological
University, consisted of in-plant and
laboratory flotation tests. Concentrates
and tailings from these tests were screened into coarse, intermediate, and fine
size fractions. The choice of sizes for
these fractions was arbitrary, but designed to permit analysis of the three
significant particle size ranges.

Tests were conducted in plants treating
coals from several different seams. In
each plant, a series of tests at frother
and collector addition levels, both above

and below the standard operating conditions were run. Fuel oil was the collector used in each of the plants, while the frother used in two of the plants was MIBC, and in the third plant, a polypropylene glycol frother was used. For each test, the reagent feeders were set at the appropriate rate and a suitable time was allowed for the circuit to reach steady state. The following samples were then collected: (1) flotation feed, (2) froth from each cell, and (3) pulp from each cell. Each sample was then analyzed to determine the following: (1) percent solids, (2) size distribution, and (3) ash and sulfur (as required) for the entire sample and individual size fractions. Froth mass flow rates from each cell were also measured in some tests. Subsequently, laboratory tests were performed on samples from each plant to permit further examination of the flotation behavior of each of the coals. Additional discussion of the tests on bituminous coal is presented in Kawatra and Waters (1982); Kawatra and Seitz (1984); and Kawatra, Seitz, and Suardini (1984) and details of the tests on anthracite coal will soon be published (Seitz, 1985).

Plant 81-1. The feedrate to this plant is 800 TPH; with the minus 28 mesh (600 μm) coal being the feed to flotation. The flotation circuit consisted of two banks of four 300 ft³ WEMCO cells each treating approximately 75 TPH of raw coal. During the test period the flotation feed varied from 20 to 26% ash, with 80 to 90% of the ash in the minus 150 micron size fraction. The coal in this seam typically contains a large amount of ultrafine clay. Table 1 shows the typical size and ash distribution for Plant 81-1 flotation circuit feed.

Table 1.   Typical Flotation Circuit Feed
           Analysis for Plant 81-1.

Size (microns)	Weight %	% Ash
-600 + 300	13.97	10.76
-300 + 150	20.42	11.00
-150 + 106	16.15	14.48
-106 + 74	10.09	17.99
-74 + 53	10.09	20.78
-53 + 44	6.06	21.02
-44	23.22	37.34

Reagents used in the flotation circuit were fuel oil as a collector, MIBC as a frother, and a cationic flocculent to depress the clay slimes. The fuel oil and MIBC were both added to the feed box of each circuit and the flocculent was added directly to each cell. In this study, only the effects of variations in fuel oil and MIBC addition rate were investigated while the flocculent addition rate was held constant for the following reasons: the large amount of ultrafine clay in the circuit feed which it was required to depress; and the potential problem associated with filtering a concentrate containing ultrafine clay.

Plant 82-1. At the time of these tests, the feedrate to Plant 82-1 was 600 TPH; with the minus 28 mesh coal being the feed to flotation. The flotation circuit consisted of a bank of four 300 ft³ WEMCO cells treating approximately 60 TPH of raw coal. Table 2 shows the typical size

and ash distribution for Plant 82-1 flotation circuit feed.

Table 2.   Typical Flotation Circuit Feed
           Analysis for Plant 82-1.

Size (microns)	Weight %	% Ash	% Sulfur
-600 + 300	5.01	9.39	1.67
-300 + 150	16.83	11.89	1.85
-150 + 74	23.02	15.19	2.34
-74 + 44	8.51	17.67	2.80
-44	45.83	29.71	3.67
Total	100.00	21.16	2.88

Reagents used in this flotation circuit were fuel oil as a collector and MIBC as a frother. Both reagents were added to the feedbox of each bank.

Plant 84-1. At the time of these tests the feedrate to this plant was 600 TPH; with the -28 + 200 (600 x 75 μm) mesh coal being the feed to flotation. The -28 mesh coal was deslimed at 200 mesh and the slime product from this operation was rejected as tailings. The flotation circuit consisted of three banks of 3-150 ft³ WEMCO flotation cells, treating approximately 25-35 TPH of feed. Table 3 shows the typical size and ash distribution for Plant 84-1 flotation circuit feed.

Table 3.   Typical Flotation Circuit Feed
           Analysis for Plant 84-1.

Size (microns)	Weight %	% Ash
+1180	0.62	9.54
-1180 + 600	7.50	9.98
-600 + 300	22.60	15.85
-300 + 212	13.16	18.36
-212 + 150	14.80	18.91
-150 + 74	28.16	19.86
-74	13.16	44.87

The reagents used in this flotation circuit were fuel oil as a collector and polypropylene glycol as a frother. The fuel oil was added to a conditioning tank with approximately 20 minutes residence time and the polypropylene glycol was added to the line carrying the feed to flotation.

RESULTS AND DISCUSSION

When developing a control strategy, circuit stabilization must be considered first, with optimization as a secondary goal. Both circuit design and operating practice have a strong influence on stability and optimization. Hence, close attention should be paid to this influence prior to attempting to develop and implement any control strategy.

The variables that are available for monitoring and controlling circuit performance are listed in Table 4. After considering all of the various combinations of these variables that can be used to control a circuit, a decision was made to examine the use of reagent control, as this alternative is conceptually simple and requires the lowest capital expenditure to implement. The effects of circuit design, operating practice, and reagent

Table 4. Variables for Control of Coal
          Flotation (based on Sutherland
          and Wark, 1955, 13-23).

Control Variables
  Chemical
    Frother Type and Addition Level
    Collector Type and Addition Level
    Reagent Addition Points
  Mechanical
    Aeration
    Conditioning Time
    Impeller Speed
    Pulp Level
Measured Variables
  Composition (% Ash, % Fe)
  Flowrate
  Pulp Percent Solids
  Pulp Level
  Froth Level
  Power Input
Disturbance Variables
  Coal Rank and Type
  Coal Oxidation
  Feed Size Distribution
  Presence of Clay Slimes
  Water Chemistry (pH, water hardness, etc)
  Feedrate
  Feed Percent Solids

control depend on the following: (1) feed
composition (size distribution, liberation, etc); (2) the behavior of feed
components in flotation (coal, ash, and
sulfur minerals); and (3) product quality
requirements.

Circuit Design and Operating Practice

Many circuit configurations have found use
in the U.S. coal industry, yet no logical
analysis of this factor has previously
been attempted. However, based on a
detailed survey by the authors, U.S. coal
flotation circuits can be divided into
three groups: single-stage rougher
flotation, rougher-cleaner flotation, and
rougher-scavenger flotation. In Australia,
coal flotation circuits can also be
divided into three groups (Mishra, 1982):
single-stage rougher flotation, rougher-
scavenger flotation, and separate flotation of size classified feed. The choice
between various circuits depends on: the
particle size distribution; liberation
characteristics; the surface chemical
properties of coal particles and the
associated gangue minerals; and product
quality requirements. Feed slurry
generally comes to the circuit from a DSM
screen, hydrocyclone, or hydroclassifier.
There is usually no serious attempt to
control pulp solids content and it typically varies from 4 to 12 percent
(Zimmerman, 1968). The feed size
distribution is widely variable in some
plants, depending on the coal being
treated, and the top size treated is
generally 600 microns (28 mesh).
The froth concentrate generally flows by
gravity to a vacuum filter and the
tailings flow directly to a settling
thickener, possibly followed by filtration,
or to a tailings pond.

In many plants, operation of the flotation
circuit is restricted by subsequent
dewatering circuits (Olson, 1983). For
example, the froth filter is often under-
sized and therefore unable to handle the

volume of froth produced when the circuit
is operated to maximize coal recovery.
This may result from changes in feed size
distribution or coal flotability from the
original design specifications. Whatever
the reason, the result is that the flotation circuit must be operated according to
the dewatering circuit limitations rather
than according to flotation circuit limits.

Consider Table 4, five of the seven disturbance variables listed are somewhat
controllable by circuit design, both of
the entire preparation plant and of the
circuit directly feeding the flotation
circuit, and by operating practice. A
considerable degree of improvement can be
achieved by stabilizing these variables as
much as possible by design and operation.
Additionally, some degree of circuit optimization can be achieved by altering these
variables to provide the best feed for
flotation. As examples, consider the
following. Because of the critical importance of particle size, the removal of
+600 micron (28 mesh) particles
prior to flotation is necessary. The
difficulty of rejecting ultrafine clay
particles by flotation necessitates the
use of desliming circuits prior to flotation when such particles are present in
such an amount as to affect flotation
results. Water pH and hardness should be
such that coal recovery is not depressed.
One final consideration is that automatic
pulp level control is necessary to achieve
any degree of circuit stability.

By achieving some degree of stability and
optimization using the factors discussed
above, it is possible to considerably
reduce the burden of stabilization on the
control strategy. This in turn may permit
a greater degree of optimization from the
control strategy.

Reagent Control of Circuit Performance

The variables selected for control should
be used, as much as possible, to reduce
the effects of disturbances while at the
same time yielding optimal circuit behavior. Pulp percent solids and flowrate
are the easiest of the circuit operating
variables to monitor, and reagent addition
levels are conceptually simple and relatively economical variables to use for
control. These variables will certainly
provide an adequate basis for circuit
control, and the following discussion is
based on their use.

Particle size distribution and liberation
play a major role in controlling the
performance of flotation circuits. The
different size and liberation fractions
respond differently to reagent addition
levels. Free particles of coal are
largely recovered by bubble attachment;
free ash particles are recovered by
entrainment; and composite particles of
coal and ash are recovered by both
mechanisms, to an extent dependent on size
and composition.

Recovery-size curves for the recovery of
bubble attached and entrained particles
both have characteristic shapes, which
may be divided into three size regions:
coarse, intermediate, and fine particles.
Because the boundaries between these

regions are frequently ill-defined and their locations variable, it is more often convenient to refer to coarse, intermediate, and fine particle behavior (Trahar, 1981). The characteristic curve for bubble attached coal particles is shown in Fig. 1. The characteristic shape for entrained particles consists of a concave curve, increasing in magnitude as particle size decreases. Composite particles are recovered by bubble attachment and entrainment and the shape of the size-recovery curve depends on the relative contribution of each mechanism. By this argument the data in Fig. 1 reveals that the bulk of the ash particles, in these circuits, are composites recovered by bubble attachment.

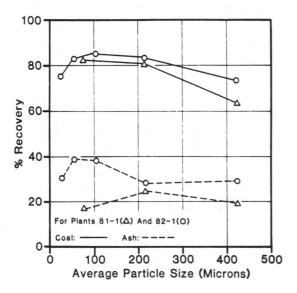

Fig. 1. The variation in percent recovery as a function of size for coal and ash minerals in Plants 81-1 and 82-1 operating with typical reagent addition levels.

When an attempt is made to recover the coarser or finer coal particles through increased reagent addition or by altering other control variables, an increased quantity of gangue particles will be recovered through bubble attachment and entrainment. Hence, some acceptable compromise between coal recovery and loss must be found in order to achieve an acceptable concentrate grade. The obvious goal is to achieve maximum recovery at a specified grade.

The important concerns at this point are: (1) what should the steady state frother and collector levels be; and (2) how should the frother and collector addition levels be used to control circuit behavior. Consider Figs. 2 and 3; where the frother addition level in Fig. 3 is 50% greater than in Fig. 2. It is apparent that equal coal recoveries can be achieved through higher frother/lower collector or lower frother/higher collector addition levels. However, the use of increased frother addition levels to improve recovery of the fine particles leads to higher ash contents at the same recovery. For coarser particles, this effect is

either considerably reduced or not noticeable at all. These conclusions hold for all of the in-plant and laboratory test work that we have conducted (Kawatra and Waters, 1982; Kawatra and Seitz, 1984; Kawatra, Seitz, and Suardini, 1984). The significance of this finding lies in the following. Consider the three flotation circuits reported on in this paper. The feed to Plant 81-1 contains large amounts of ultrafine clay, while the feed to Plant 82-1 contains considerably less ultrafine clay, and the feed to Plant 84-1 contains almost no ultrafine clay. As a consequence, the control system for each of these circuits must be different. For Plant 82-1, the frother addition level should be minimized, to reduce water recovery and entrainment of fine ash, and collector addition should be used to maximize recovery. The minimum achievable frother level in any case must be high enough to prevent froth overloading, a condition of reduced froth mobility and reduced rate of particle transfer from the pulp to concentrate launder, which occurs due to the high rate of flotation of coal and the recovery of the bulk of the circuit feed in the concentrate. The collector addition level should be based on the solids feed rate to the circuit. The frother addition level should be based on the volumetric flowrate of water fed to the circuit, so that the frother concentration in the pulp is constant, and thus, pulp aeration processes are relatively constant. For Plant 82-1, frother can be used to some additional extent to improve recovery without harming grade (Kawatra et al, 1984). For Plant 84-1, it is necessary to use frother to achieve additional recovery, because the lack of -200 mesh fines drastically reduces froth mobility and the frother is required to improve mobility and hence achieve good recovery (Seitz, 1985).

CONCLUSIONS

Circuit design is an important factor contributing to the ease of developing a process control strategy. Good circuit design will reduce the burden placed on the control strategy; by presenting to the circuit a relatively stable feed devoid of particles that are either too coarse or too fine and hence are easier to beneficiate by processes other than flotation.

A control strategy cannot be developed that is universally applicable to all coal flotation circuits. Rather, site-particular factors, such as coal characteristics, circuit design and operating practice, and selected control variables, must be considered during strategy development. This control should include elements of both stabilization and optimization. Reagent addition levels can be used to stabilize circuit performance, often in an optimal manner. Consider the mode of recovery of different particles during flotation. Large composite particles of coal and ash and liberated particles of coal are recovered by bubble attachment, while fine liberated particles of ash are recovered by entrainment. Consequently, the manner in which reagents are used for stabilization is critically dependent on the size and liberation characteristics of the coal being treated.

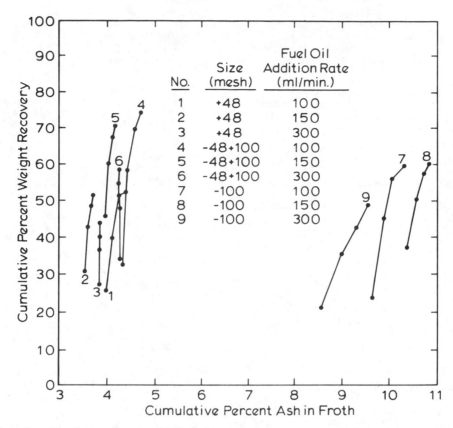

Fig. 2.   Grade-recovery performance of the flotation circuit in Plant 81-1
for individual size fractions at a MIBC addition rate of 0.04 lb/
ton and fuel oil addition rates of 100, 150 and 300 ml/min (0.14,
0.21 and 0.42 lb/ton, respectively).

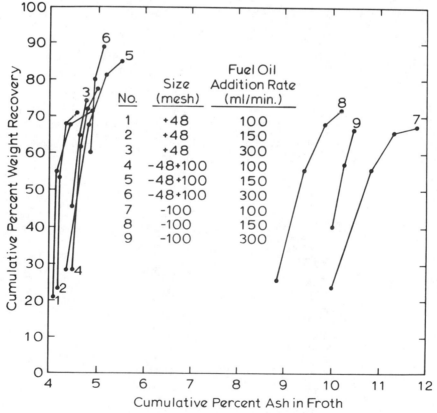

Fig. 3.   Grade-recovery performance of the flotation circuit in Plant 81-1
for individual size fractions at a MIBC addition rate of 0.06 lb/
ton and various fuel oil addition levels.

As examples, consider the plants discussed
above. In Plant 81-1 large amounts of
ultrafine clays are present in the raw
feed to flotation and the frother addition
level should be minimized in order to
minimize the entrainment of this gangue
into the froth. However, in Plant 82-1,
little ultrafine clay was present in the
feed and use of the frother and collector
to maximize yield while minimizing froth
percent solids is desirable.

ACKNOWLEDGEMENTS

The authors would like to thank the
following people and organizations:
F. M. Lyon of Kitt Energy Corp., for
extensive discussions concerning this
work; the managements of Plants 81-1,
82-1, and 84-1 for financial assistance
and facilities; the U. S. Department of
Energy for Research Grant No. DE-FG01-
81FE00102; and the U. S. Department of
Interior, Bureau of Mines, State Minerals
Institute Program, for funding portions
of this investigation.

REFERENCES

Kawatra, S. K., and Seitz, R. A. (1984).
The effect of froth structure on coal
flotation and its use in process
control. In Proc. of 18th APCOM, IMM,
London, 139-147.

Kawatra, S. K., Seitz, R. A., and Suardini,
P. J. (1984). The control of coal
flotation circuits. In Herbst, J. A.
(Ed.), Control '84 Mineral/Metallurgi-
cal Processing, AIME, 225-233.

Kawatra, S. K., and Waters, J. L. (1984).
An investigation of the effect of
reagent addition on the response of a
fine coal flotation circuit. In Proc.
of 14th IMPC, Toronto, CIM, preprints.

Mishra, S. K. (1982). Fine coal flotation: a mineral processing problem.
In Proc. of Aust. IMM - North West
Queensland Branch Mill Operators Conf.,
383-393.

Olson, T. J. (1983). How to get higher
recoveries via improved flotation.
Coal Min. & Process., 20(6), 60-65.

Seitz, R. A. (1985). An analysis of the
theory and industrial practice of
coal flotation, Ph.D. Dissertation
(unpublished), Michigan Technological
University.

Sutherland, K. L., and Wark, I. W. (1955).
Principles of Flotation, Aust. IMM,
Melbourne.

Trahar, W. J. (1981). A rational interpretation of the role of particle size
in flotation. Int. J. Mineral Process,
8, 289-327.

Zimmerman, R. E. (1968). Froth Flotation.
In Leonard, J. W. and Mitchell, D. R.
(Eds.), Coal Preparation, 3rd Ed.,
AIME, 10.73-10.88.

The Potential for Automation and Process Control in Coal Preparation

C.J. CLARKSON

Superintendent, Utah Development Company Limited Research and Development Laboratory, Mackay, Queensland

Abstract. Several coal preparation plants are presently "automated" to the extent of automatic sequencing and interlocking via digital means, together with closed loop control on various level and density circuits. This paper will explore the potential for second generation automation utilizing on-stream ash analysers and other measuring elements so as to optimize the total metallurgical process. This would enable maximization of recovery, improved product homogeneity and minimum reagent consumption.

Keywords. Coal preparation; process control; optimization; froth flotation; dense medium process; on-stream ash analysis.

INTRODUCTION

The basic aim of a preparation plant is to produce the required tonnage of product coal at specified quality, maximum recovery, and minimum cost. These objectives are best achieved by considering the following factors.

Reliability. Availability of plant is essential to produce the necessary tonnage - any overdesign required to compensate for downtime represents excess capital outlay.

Efficiency. Mining costs range from approximately $10/tonne in low cost open cut operations to greater than $60/tonne for some of the older, underground European mines. Thus it is essential to recover to product every possible tonne of coal that is mined. Up to the above amount ($10-$60) may therefore be spent in the plant on each additional incremental tonne and still produce a net benefit. In addition, as low cost reserves become depleted, mining costs inevitably rise and so present inefficiencies in the plant represent increased opportunity costs for the future.

Control. The quality of product, whether coking or steaming coal, is becoming of greater importance. Consistency of ash, sulphur and coking properties, as well as satisfying contract specifications, is of great importance to coking coal users. Product quality is of equal importance to steaming coal users for a variety of reasons. Improvements in the quality of coal can maximize the availability of existing generating capacity (Harrison, 1983), and so delay the need for expensive capital outlay as power demand increases. Emission control laws are becoming more stringent and studies in the U.S. have illustrated that coal cleaning is the most "cost-effective" solution (Green, 1984). Finally, world trade in steaming coal is expanding and so unnecessary shipping costs are incurred on any incombustible material in the coal.

Whilst proper pre-blending of the coal is essential to ensure long term consistency of product, tight control of the plant process is still required.

This enables the product to be produced as close as possible to, but not above the target specification (hence maximizing yield), as well as allowing the plant to respond to changes in feed characteristics (e.g. washability) that may result in changes in product quality.

Flexibility. Coal preparation plants are usually designed for at least 15 years operation, and in many cases operate for far greater periods of time. They are normally specifically designed to treat a carefully evaluated feed to produce a very specific product. However, with the extremely volatile markets of the past decade, exacerbated by technological changes affecting both mining and end user, the specifications for a particular plant can be expected to change substantially during the course of its life. In addition, the greater expected growth in export steaming coal compared to coking coal can be expected to create substantial changes in product requirements for marginal coking coals.

Thus it is essential to design plants that are capable of responding to these changes with minimal capital expenditure and disruption to production.

Low Cost. The ever increasing mining costs associated with depletion of open-cut reserves for older established mines or increasing infrastructure costs and government charges for new mines makes it essential that coal preparation costs are minimal to maintain a competitively priced product on the world market.

EVOLUTION OF CONTROL TECHNIQUES

The purpose of this paper is to explore the benefits of automation and process control in the preparation plant, and so the evolution of process control in the plant will be evaluated in relation to the above criteria.

Hard-wired interlocking and discrete loop control. The earliest form of automation involved hard-wire interlocking of related units (e.g. conveying systems) and use of local, dedicated analogue controllers for closed loop control of discreet process variables such as flotation level, medium density, etc. These are usually reflected in a central control room by a full array of contactors, alarm lights, controller outputs and ammeters. The appropriate displays are at times incorporated in hard-wired mimic panels.

AMR-I*

The advantage of this system is that it frees the operator to concentrate on the overall process and to respond to non-routine problems such as blocked chutes, motor failures, etc. By and large this approach is very reliable, especially during commissioning where complete local control allows maximum flexibility and non-interference from software bugs and sensor failures.

The major limitations include:-

(i)   Significant initial capital cost with discreet PID analogue controllers for each loop and hard-wiring of contactors for all interlocks.

(ii)   Lack of flexibility if interlocks or control procedures require changing.

(iii)   In some situations, this approach has been found to lead to excessive physical and mental strain on the operator if the plant is large (Brown and Elsworth, 1984). However, experience with this style of plant with up to 1500 tph capacity in the various UDCL[1] Central Queensland Mines has not shown the above shortcoming.

Central digital control. The next step involved the use of a centralized digital computer to undertake all control facets and co-ordination of sequencing of the preparation plant. Tasks required of this unit included:-

(i)   Sequencing operations (usually via PLC's).

(ii)   Operation of all control functions such as level control, density control, etc.

(iii)   Provision of a central data management system, used as the basis for automatic shift reports, alarms, etc.

Such units are usually based on fairly large mini-computers with several VDU's being used for mimic displays, numeric status displays and for accessing control parameters. Alarms are usually also logged on a printer. These units may or may not incorporate manual backup, although if not available commissioning has generally been found to be extensive, tedious and costly until sensors, data highways, and software were all debugged.

The compact arrangement allowed easier operation by the controller who, by calling up any specific section of the plant that he was interested in, had all the necessary information (and more) at his fingertips. The unit also provided the management with an extensive array of data not previously available (Manackerman, 1982). Perhaps the major long-term benefit was in maximizing the flexibility of the system. Rapid changes of interlocking or control strategies were possible by software changes together with perhaps some relocation of sensors.

However, the hardware cost savings in having a single unit are increasingly being overwhelmed by the software development costs (Green, 1981). A large centralized system depends on a complex, multi-tasking programme which is usually only comprehensible to the specialist. Moreover, it has been found that data highways can be overwhelmed with unnecessary information, with priority ranking on control and information becoming extremely difficult e.g. several minutes have been known to elapse before the operator was able to shut a critical valve (Brown and Elsworth, 1984).

[1] UDCL - Utah Development Company Limited

Distributed control. The present evolutionary stage, assisted by the availability of lower-cost microcomputers, has been to decentralize all control and sequencing functions wherever possible. Thus small discreet digital units are used to automatically control local loops, and are merely networked to a central unit to convey the status of the individual operation. Local control then continues if the central system fails and facilities are always available for local manual override (Hoffman, 1983).

Thus all the benefits of central control are available: that of overall supervision from a central region, flexibility of control and sequencing procedures, and accumulation of data for management purposes. In addition though, software is simpler and more accessible by being distributed to discreet units with minimal requirements for priority ranking and data transfer. Hence overall costs are usually similar due to reduced software costs. Of most importance, reliability of the total system is enhanced.

Overall process control. However, such systems still do not address some of the basic requirements of a preparation plant - that of maximizing the yield of coal of specified quality from a potentially variable feed. To achieve such an aim, instantaneous knowledge of the coal quality being processed must be known so that immediate corrective action may be taken. Until such time as reliable sensors that continuously and accurately monitor the coal quality are available, true automation of the coal preparation process is not feasible.

The product specifications for a plant are written such that the product ash is usually the limiting quality parameter. Where possible, variation in sulphur and other properties (either coking or steaming) are blended out. Thus the control philosophy should be based on producing the maximum yield at the limiting product ash.

In the next section, available primary elements for monitoring coal ash will be discussed.

ASH MONITORS

Coarse Coal

A variety of ash monitors for coarse coal have been available for well over a decade, the most common units being the Gunson-Sortex and the Humboldt-Wedag. Both have been extensively described in the literature (Boyce and others, 1977; Clarkson and others, 1983; Hardt, 1962). Neither of these units have enjoyed universal application due to a variety of shortcomings.

The Gunson Sortex, developed by the NCB, requires a complex sample preparation system involving crushing a sample to -5mm and carefully smoothing it onto a rotating table. Only a few mm of coal are effectively analysed by a low energy gamma backscatter technique. The Humboldt-Wedag unit offers a superior sample preparation system, taking up to 50mm topsize coal through a screw feeder. Ash is measured by a medium energy (60 keV) backscatter technique. Almost three years operating experience at the UDC Blackwater Mine suggests that the unit has in excess of 99% mechanical availability.

Both units suffer from inaccuracy due to variability in coal ash composition. Changes in Fe are of greatest concern, although the Gunson-Sortex will compensate to some extent by use of an Al filter. Despite the above, and especially if suitable

modifications and calibration procedures are employed, the above units can offer approximately
5% RMS relative error, which is certainly useful
for process control.

Within the past few years, several new devices have
become available. The Coalscan unit (Smith and
others, 1982), developed in Australia, is based on
the pair production technique, and so is far less
vulnerable to ash composition changes. It can also analyse up to 50mm topsize coal, so only sub-
sampling and not pre-crushing is required. The
reliability and accuracy of this unit (better than
0.5% RMS ash error for up to 20% ash coal over
prolonged periods without recalibration) is only
made possible by the exceptionally stable pulse
analysis circuitry which was recently developed.
The same company also supplies an"LET" gauge
utilizing twin beam transmission through the belt,
and so obviates the need for sampling. However,
this unit shares the same vulnerability to ash
composition as does the Humboldt-Wedag monitor. A
similar gauge has been developed in the U.S.
(Bematowicz and others, 1984) and is marketed as
the rapid ash component of the Nucoalyser series.
This company also produces a neutron activation
gauge capable of a complete multi-element analysis
of the coal either on a belt or through a chute.
This unit is presently undergoing field trials.

Thus there certainly appears to be suitable
monitors for measuring ash of coal on the belt.

Fine Coal Slurries

There are several systems currently under develop-
ment, but no commercial units presently available.
The ASHSCAN system, developed jointly between the
JKRMC[2] and UDCL[1] has had several thousand hours
continuous operation on-line at the Peak Downs
Preparation Plant in Central Queensland.

ASHSCAN is based on twin beam (660 keV Cs137 and
60 keV Am241) transmission of γ rays through a
slurry that has been pressurized to remove the
influence of entrained air. The present system is
based on the same concepts as outlined by Lyman
and Chesher (1980), but has been redesigned by the
Utah R&D Laboratory to provide a reliable, accurate
industrial prototype. Up to four streams (feed,
concentrate, tails and spare) can be selected,
their condition being automatically monitored by
computer control. They are passed sequentially
through a ceramic lined piston pump to present the
slurry at >1800 kpa to the ABS measurement section.
A vezin sampler on the discharge is co-ordinated
by the controller computer for accurate sub-
sampling. "Coalscan" type MCA units and scintillation detectors are incorporated in the rig together with two HP9915 computers to process the
count-rates and to output stream type, ash and %
solids, and operational alarms. An extensive set
of level, flow and pressure sensors together with
the appropriate logic allows fully automatic
operation of the rig in co-ordination with plant
operation.

The signals are relayed to the preparation plant
office where trending and alarms are displayed on
an HP86 computer. An automatic calibration procedure is incorporated on the basis of entering
shift composite samples into the HP86. The HP86
also interacts with the flotation circuit where it
interfaces with the level control and reagent
addition for use in developing automatic flotation
control.

---

[2] JKRMC - Julius Kruttschnitt Mineral Research
      Centre, University of Queensland

Overall electrical and mechanical availability has
been shown to be 95%, and accuracy for the three
streams is approximately 0.5% for concentrates,
1-2% for feed and 3-5% for tails (RMS absolute %
ash). The gauge is expected to be commercially
available in the near future.

OVERALL PLANT CONTROL

Overall optimization of several parallel loops or
different feed types has already been described in
detail based on equal incremental ash of each process or feed type (Clarkson and Leach, 1981;
Clarkson and others, 1982). The setpoint can
therefore be determined for each process, contributing towards a global optimum. The following
example relates to a simple two process preparation
plant - dense medium cyclone for "coarse" +0.5mm
coal and flotation for the -0.5mm feed.

Coarse Circuit

In practice, the coarse circuit usually accounts
for 70-80% of the product, has a very low residence
time (say 1-2 minutes), and very precise operating
characteristics (single control parameter - density). Therefore a feed-back loop to the dense
medium circuit can be used to ensure overall quality according to a coarse coal ash monitor on the
product (see Fig. 1).

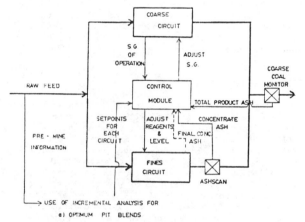

Fig. 1. Schematic of an Overall Control Strategy
         for the Plant

In practice, there are two incompatible optimizing
criteria:-

(i)   Maximize recovery by operating at constant
        incremental ash. Based on a constant ash/
        density relationship, this in turn means
        operating at a constant density.

(ii)   Maintain constant ash of product
        (specification).

If the feed washability varies though, either the
density will have to be changed to maintain constant product ash, or else if operated at constant
density to maintain maximum recovery, the product
quality will vary.

With the delays in conventional ash sampling and
analysis, historically a compromise was achieved
by varying the density in the longer term (several
hours) so as to minimize the operating density

TABLE 1   Factors Affecting Rate Determining Steps in Froth Flotation

Step \\ Component	Particle/Bubble Attachment	Froth Transfer
Combustibles (determines yield)	Rank of coal Reagent addition (Aeration)	Level control Frother/diesel ratio (determines "mobility" of froth)
Incombustibles (determines ash)	Reagent selectivity	Level control (pulp entrainment) Reagents (affects froth drainage)

band (hence loss of recovery) but still maintaining specification ash in lots of several thousand tonnes. Subsequent blending and mixing from materials handling on product and shipping stockpiles certainly provided additional homogenization.

However, with the use of on-stream analysers, it may be possible for feed-back to the density much sooner (say 20-30 minutes), thus requiring much smaller changes so that overall recovery gains may be achieved by operating in a narrower S.G. band. In addition, by producing a product ash with less variation, the target ash may be set higher with the same probability (say 98%) of being within penalty levels for each unit of production (say 4000 tonnes). Hence an appropriate incremental recovery may be gained.

At present, the Utah R&D Laboratory is undertaking a dynamic plant simulation of the coarse circuit utilizing plant data to quantify the benefits of the above control philosophy.

Fine Circuit

Unlike the coarse gravity circuits, flotation is a far more complex process with a wider range of variables available to control the process. A number of control strategies (Baseur and Herbst, 1982; Laurila, 1983; Leach and others, 1981; Lyman and others, 1979) have been proposed for flotation but virtually all require on-stream analysis (OSA) to really be of practical use. In addition, depending upon the rank and other properties of coal, there can be several different rate limiting steps to the process, but in virtually all cases, these are confined to four steps, illustrated in Table 1, to determine overall product quality and recovery.

The following strategy is the proposed skeleton on which the Utah flotation control programme is based, using both feed-back (concentrate and tails) and feed-forward (feed) control. ASHSCAN is utilized to provide on-stream analysis of % ash and % solids for each stream.

Due to the high rank (20% volatiles) and high vitrinite content (68%) of the coal, the rate-determining step, especially in the early cells, for both combustibles and incombustibles is usually within the froth. A strategy incorporating two tiers of control is being evolved as follows, based on the assumption that the major controllable ash contamination is due to fine gangue entrainment.

First tier. The immediate control involves two concurrent loops:-

Level Control. The basic aim is to maintain the concentrate % solids >30% solids. This has two effects. Firstly, extensive testwork indicates that normal froth always contains 32-35% solids,

and so any reduction of concentrate density below 30% suggests too high a level and pulp overflowing to the concentrate. This results in excessive ash in the concentrate (Fig. 2 illustrates on-line data showing a distinct inverse correlation between concentrate % ash and % solids). Secondly it maximizes the filter capacity and so minimizes operational restraints on the flotation circuit.

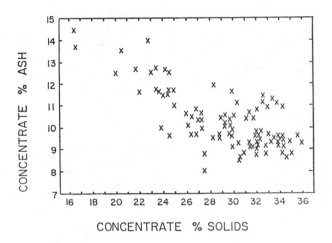

Fig. 2. Influence of Concentrate % Solids on Concentrate % Ash (feed ash range 20-44%)

Tailings Ash. The aim is to maintain the tailings ash >70% ash. The tailings ash is probably the most sensitive indicator of the efficiency of the operation, irrespective of feed variations. There are three possible causes of low tailings ash:-

(i)   Low pulp level, so that insufficient froth is able to be recovered. Hence the level should be raised as high as possible so that 30% solids in concentrate is just maintained.

(ii)   Reagent starvation. Either insufficient diesel with resultant inadequate attachment of coal particle to the air bubble or frother starvation (excess diesel/frother ratio) leading to froth overloading. For a given feed, previous testwork and modelling should indicate the broad range of reagent addition required - changes in sizing will affect the reagent requirements and so the latter can be adjusted on-line, according to feed ash and % solids and tailings ash.

(iii) Oxidized Coal. If the product is to be used as coking coal, coal which is too severely oxidized (or surface affected) to float should be rejected as the coking properties may be expected to be impaired accordingly.

Second tier. A longer term control strategy may
then be superimposed, depending on the concentrate
ash. The target ash for flotation may be determined by assessing equal incremental ash between
the coarse circuit (density of separation) and fine
circuit (concentrate ash of final cell). Thus,
with flotation being immediately controlled on-line
by the first tier, a secondary optimization may be
made utilizing both level and reagents, based on
models of the flotation process to trim the flotation circuit according to the target ash.

In future studies, incorporating a set of fully
instrumented 10 cu.ft. pilot cells in open circuit,
the influence of aeration rate and impeller type
and speed, especially in relation to cell position,
will be investigated.

Secondary Circuits

With the installation of a distributed control
system maintaining appropriate sequencing, loop
control of levels, densities, etc. and overall
process control using OSA, secondary control circuits within the preparation plant should also be
included.

By and large, reliable sensors for flow rate,
density and level are now commercially available
and extensively utilized. Experience has suggested that non-contact types such as nucleonic,
ultrasonic and electromagnetic are preferable to
reduce the effects of blockage and abrasion
(Jenkinson and others, 1982). Even so, the reliability of associated electronics and transmission
systems within the harsh environment of a preparation plant is still of paramount importance.

Filters. Depending on feed, operational and
maintenance conditions, the filters may at times be
the rate determining step affecting the fine circuit. The basic criteria for filter control
should be:-

Minimize moisture: Operational procedures such as
good maintenance, flocculation and maximizing pulp
density of feed all contribute to reducing moisture. Nevertheless, with the availability of a
variety of on-stream moisture meters, appropriate
level, density and pressure (vacuum) sensors, together with considerable modelling of the filtration process (Brown and others, 1981; Wakeman,
1978), it should be possible to utilize microprocessor technology to optimize the filtration process on-line. This constitutes another arm of the
Utah R&D research programme using both a twin disc
and a drum pilot filter rig.

Maintain filter level without overflow: Any overflow recycled to flotation must impair the flotation process and complicates any control strategies. Simple feedback loops for maintaining
level by adjusting filter speed have proven effective at plants such as Norwich Park.

Flocculation. A number of models, such as by
Macdonald and others (1983), have been published
describing automatic control procedures for flocculation. These can be incorporated within the
overall distributed control system by utilizing
tailings information from the ASHSCAN unit to
allow an element of feed forward control.

Dense Medium Recycling. Both magnetite consumption
and the dynamics of the dense medium control circuitry are vitally dependent on the operation of
the magnetite recovery circuit. The basic aim of

producing clean, high density make-up medium together with 100% recovery of magnetite can in all
likelihood be assisted by suitable modelling and
process control, especially of variables such as
feed density, flow rate and level.

Note that there is also considerable potential for
implementation of new technology in this area such
as utilizing HGMS[3] for secondary scavenging, thus
ensuring very high recoveries of finer magnetite
while allowing the conventional drum separator as
a primary unit to be tuned to maximize the density
and minimize the % non-magnetics of the make-up
medium.

There are a variety of other areas within the preparation plant suitable for control, according to
the local needs. Again, however, the need for
decentralizing such control loops and relaying only
status or adjustment of setpoint to the central
operator cannot be overemphasized.

GENERAL FACTORS

Whatever the system of control that is installed,
the plant still has to be managed by an operator.
Therefore, if the system is to be fully utilized,
the manner in which the system interacts with the
operator is crucial. In general, the trend in this
area has evolved to the use of the following
tools:-

VDU's. These can mimic requested areas of operation, output alarms, be used for adjusting control
parameters and sequencing options, and provide
trending of process variables.

Closed Circuit T.V. Monitors. To give the operator
a display of critical operating areas, especially
in remote areas such as loadouts.

Printers. Especially for hardcopy of alarms.

Plants are often configured in a transition stage
and still retain elements of dedicated loop controllers, contactors, etc.

By and large, the keyboards are evolving towards
dedicated process control layouts rather than the
conventional typewriter layout relying upon codes
to call up various segments.

Experience with implementing process control in
the metalliferous industry a decade or more ago
suffered from operator unfamilarity with process
control equipment. Experience within Utah where
small microprocessors are being used suggests that
with the advent of the personal computer and the
general increasing public awareness of computers,
the only novelty for many operators is the application and not the computer itself. This factor
alone, together with the improvements in software
and hardware compared to a decade ago, must allow
the adoption of total computer control of preparation plants to be easier than for the metalliferous predecessors of the 70's.

Similarly, advances in hardware now dictate that
careful management of software development and
implementation will have most effect on the overall cost/benefit of introducing process control.

Finally, the massive increase in process data
available to management from such systems needs to
be carefully considered. Increased use of EDP
almost inevitably leads to increased bureaucrat-
ization, and so one of the basic aims of implementing process control, that of relieving the
operators and management from tedious, repetitive

manual operations to concentrate on the real task at hand, can become submerged with the overwhelming mass of data available.

ACKNOWLEDGEMENT

The author acknowledges and thanks the staff of the R&D Laboratory and of all Utah operations for their useful discussions and input into this overall topic. The paper represents the views of the author and not necessarily those of UDC Limited.

REFERENCES

Basuer, O.A., and J.A. Herbst (1982). Dynamic modelling of a flotation cell with a view toward automatic control. *Proceedings XIV International Mineral Processing Congress, Toronto Canada*.

Bematowicz, H., D.R. Brown, T. Gozani, and C.M. Spencer (1984). On-line coal analysis for control of coal preparation plants. *Presented at 1st Annual Coal Preparation Conference, Lexington USA*.

Boyce, I.S., C.G. Clayton and D. Page (1977). Some considerations relating to the accuracy of measuring the ash content of coal by x-ray backscattering. *Proceedings Symposium on Nuclear Techniques and Mineral Resources, Vienna Austria*. pp. 135-164.

Brown, C.H., G.O. Allgood, G.S. Canright, and W.R. Hamel (1981). Transient modelling of froth flotation and vacuum filtration processes. *Presented at Symposium on Instrumentation and Control for Fossil Energy Process, San Francisco USA*.

Brown, T.A., and A. Elsworth (1984). Automation and control of large-scale coal preparation plants. *Proceedings 18th APCOM Conference, London England*. pp. 149-157.

Clarkson, C.J., and K.R. Leach (1981). A control strategy for automatic optimization and control of Central Queensland preparation plants. *Proceedings 1st Australian Coal Preparation Conference, Newcastle*. Paper F1, pp. 267-279.

Clarkson, C.J., K.R. Leach and D.J. Walker (1982). Optimization and control in coal preparation. *Proceedings IX International Coal Preparation Congress, Delhi India*. Paper G2.

Clarkson, C.J., V.A. Fiedler and D.A. Tonkin (1983). Use of on-stream ash analysis for coal. *Proceedings 2nd Australian Coal Preparation Conference, Rockhampton*. Paper 6C, pp. 310-334.

Green, P. (1981). Computer controls flow to cyclones. *Coal Age*. June, pp. 90-96.

Green, P. (1984). A rational approach to acid rain. *Coal Mining*. June, pp. 9.

Hardt, L. (1962). A rapid method for determining the ash content of coal by means of low-energy radiation. *Proceedings IV International Coal Preparation Congress, Harrogate UK*. pp. 101-118.

Harrison, C.D. (1983). Historical deterioration of U.S. coal quality - effects on the power industry. *Presented at 2nd Annual Seminar on Electric Utility Research and Development, Raleigh USA*.

Hoffman, W. (1983). Problems in the application of digital automation systems in the process industry. *Process Automation*. pp. 69-76.

Jenkinson, D.E., P. Cammack, and D. Vaillie (1982). Transducers for coal plant control system. *Proceedings IX International Coal Preparation Congress, Delhi India*. Paper D4.

Laurila, M.J. (1983). Coal flotation: The evaluation of controllable variable and the development of a distribution model. *Proceedings 4th Symposium on Automation in Mining, Minerals and Metal Processing, Helsinki Finland*. pp. 447-453.

Leach, K.R., H.I. Lief, and C.J. Clarkson (1981). A two parameter steady state model for froth flotation of coal. *Proceedings 1st Australian Coal Preparation Conference, Newcastle*. Paper D2, pp. 171-182.

Lyman, G.J., C.K. Mackenzie, G.V. Walter, K.R. Leach, and K.W. Bateman (1979). Automatic monitoring of the coal flotation process. *Proceedings VIII International Coal Preparation Congress, Donetsk USSR*, Paper E-1.

Lyman, G.J., and R.J. Chesher (1980). On-stream analysis for ash in coal slurries. *Proceedings 5th International Conference on Coal Research, Düsseldorf, FRG*. Paper C-1, pp. 145-163.

Macdonald, J.R., S.A. Mays, E. Gallagher, and J.E. Lewis (1983). A mathematical description of factors affecting settling rates of slurries. *Proceedings of 2nd Australian Coal Preparation Conference, Rockhampton*. Paper 8A, pp. 397-420.

Manackerman, M. (1982). The computer in coal preparation at Renishaw Park. *Colliery Guardian*. August, pp. 373-376.

Smith, K.G., H. Crowden, B.D. Sowerby, G.J. Lyman, and W.J. Howarth (1983). Performance of a Coalscan on-line ash measurement system at Ulan Coal Mines Ltd, Australia. *Seminar on Applications of Continuous Coal Analysis, Nashville, USA*.

Wakeman, R.J. (1979). Low-pressure dewatering kinetics of incompressible filter cake, I. Variable total pressure loss or low-capacity systems. *International Journal of Mineral Processing*. 5, pp. 379-393.

Audiometric Study to Detect Overload on Rod Mills

FERNANDO DE MAYO, HUMBERTO SOTO, RUBÉN HERNÁNDEZ, GUILLERMO GONZÁLEZ

Department of Electrical Engineering, University of Chile, Casilla 5037 Santiago Chile

CRISTIAN GONZÁLEZ, CLAUDIO ZAMORA

Department of Engineering and Development, CODELCO-CHILE

ABSTRACT: The performance of a system for detecting the onset of overloads on rod mills is described. The system has been tested at CODELCO El Salvador Concentrator where, due to variations on size distribution of the copper ore feed, the rod mill is frequently subject to overloads. The main input signal to the system is the rod mill sound, after being measured and analog conditioned. This signal and other signals measuring the fresh ore feed and the mill water addition are fed to a computer containing a simple mathematical model. The model parameters are empirically estimated and an analog output signal is generated, which gives an early prediction of the overload condition. The experimental results described in this paper show that the overload condition may be predicted with an anticipation of about three minutes.

Keywords. Grinding; rod mill; overload prediction; instrumentation; microcomputer application; sound analysis.

INTRODUCTION

An important problem in the operation of grinding plants is the overload that may occur in the rod mills. This problem, which may even reach a rod tangling condition, is apt to produce a decrease in overall plant production. Therefore, a reliable method for providing sufficient early warning of the overload is highly desirable.

There are commercially available devices to detect the rod mill overload condition, and in the published literature some developments to solve this problem may be found, based on the rod mill sound. In Bradburn (1977), a device that detects rod mill sound intensity was developed, and a plot of relative decibel reading versus frequency for the normal and the overload condition is shown, but the method followed to implement this device is not further discussed. Another paper (Uronen, 1976) presents an audiometric control system for a wet semiautogenous mill. In this case, the power spectrum of the sound from the mill was studied under different conditions to find a suitable frequency range to detect the overload.

The development programme of El Salvador Concentrator to improve grinding control includes a research and development for the implementation of a system for detecting the onset of rod mill overload.

This paper describes the development of a computer based system which has been found to permit a reliable and clear detection of the onset of overload conditions. The main input signal to the system is the rod mill sound, after being measured and analog conditined. The equipment for sound measurements, the sound monitoring device, and the results of the performance obtained are described.

PLANT DESCRIPTION

El Salvador is an underground mine and concentrator facility located in the northern part of Chile. It is one of the four plants owned by the national copper corporation of Chile (CODELCO). The El Salvador ore is described as a porphyry type containing copper and molybdenum.

The plant, which has been operating since 1959, is at present treating 33.000 ton/day.

In 1978, an expansion project was started and new sections of crushing, grinding and flotation were installed. A distributed control system for grinding control was incorporated so that the process control in El Salvador has grown from manual to a fully instrumented one in the last four years (Zamora,1983).

The grinding section of El Salvador concentrator consists of four identical sections, each one with a 10`x14` rod mill operating in open circuit. Each rod mill feeds a five cyclone battery operating in closed circuit with two 10`x14` ball mills.

In the expansion project a new section was installed, having a 13.6`x18` rod mill feeding a circuit formed by a twelve cyclone battery in closed circuit with one 16.6`x19` ball mill.

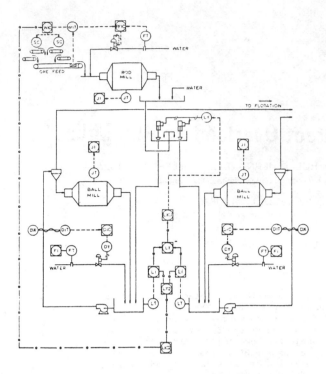

Fig. 1. Diagram of one section of the El Salvador Grinding plant containing control and instrumentation scheme.

Fig. 1 shows a schematic process and control diagram of the plant, where the control system was determined after testing and evaluating different control strategies (González,1982).

With this control system the plant operates in good conditions, and despite obtaning an increase in throughput of about 6%, the maximum tonnage has not been reached because of rod mill overloads. These overloads are due to variations in the size distribution of the copper ore feed. When the field operator detects this situation, he normally stops the feed conveyor and advises the central control room. The feed ore set point is then decreased, thus reducing the plant throughput.

PRELIMINARY TEST STAGE

To research the overload condition on rod mills based on the sound, the first step was to specify equipment for field data acquisition.

Next the optimal positions of microphones were investigated, to obtain the best posible quality of the signal. Finally, with all the information acquired, a complete sound analysis was performed to find a relationship between rod mill overload and sound. These stages allow the design and implementation of an online overload analyzer device.

Instrumentation and data acquisition system

The data acquisition system employed in the preliminary phase consisted of analog equipment for acquiring and recording the sound signal, together with microcomputer based equipment for data acquisition and digital signal processing. The analog equipment used was a set of highly directional microphones, very low distortion amplifiers, an oscilloscope and a tape recorder. These devices were used to analyze and store the mill sound with a frequency response up to 20 kHz. On the other hand, the microcomputer was specially

conditioned to receive analog signals from field transmitters, through an opto-isolated interface. For the analysis of data obtained in this preliminary test, a software package was developed containing programs for spectrum analysis using FFT algorithm, mathematical processing, graphic display and output generation (Soto H. 1984).

In addition, a commercial real-time frequency spectrum analyzer was used to observe dynamic sound spectrum variations.

Location of the microphones.

To select the optimum location of the microphones, a mechanical system that allows angular and axial microphone displacements was mounted close to the rod mill, (Fig. 2). Cylindrical coordinates were used.

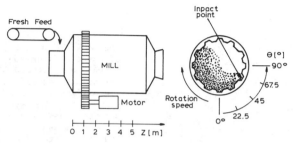

Fig. 2. Angular and axial coordinates for location of microphones at the rod mill.

Longitudinal Z values were measured using the mill feed side as origin and angular θ values sweep a 90 degrees arc, starting from the bottom of the mill. Sound data was acquired and processed to obtain information of sound intensity. Based on two parameters, signal to environment noise ratio, (Su/n), and sensor saturation time percentage, (SAT), it was found that the best location in angular position for the microphones was aproximately 30 degrees (Fig. 3).

This maximum value is theoretically corroborated, because the bars fall and strike the mill lining at about this angle by mill rotation effect (Fig. 2).

In the axial direction, the sound is stronger near the mill discharge end, but in this case the microphones reached saturation (Fig. 4).

Therefore the microphones were placed near the mill feed inlet, in spite of the lower sound intensity. Also, in this place large differences have been found between the signal in normal operation and in overload conditions.

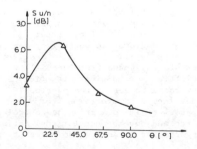

Fig. 3. Sound signal to noise ratio vs. microphones angular position.

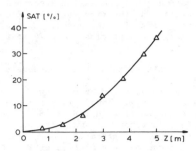

Fig. 4. Percentage of time in which the microphones
signal saturate vs. mill axial coordinates.

Sound analysis.

After a long period of data recording, the
differences in the sound wave form between normal
and overload condition were studied on the time and
frequency domain by means of the developed software
package.

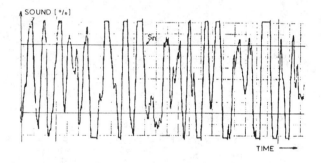

Fig. 5a. Rod mill sound intensity under overload
condition.

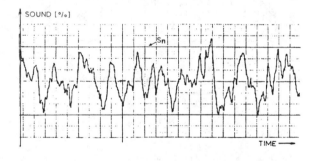

Fig. 5b. Rod mill sound intensity under normal
operation.

Figures 5a and 5b show the intensity vs time
behaviour of the sound signal under normal and
overload condition. It can be seen that when the
rod mill is overloaded that the sound signal
exceeds a larger number of times an experimental
determined value S_n. This result was useful on the
design of the overload prototype.
On the other hand the frequency domain study shows
that during mill overloads there are dynamic

variations in the frequency intensity of the mill
sound spectrum. Fig. 6 shows that the main
component of the frequency spectrum are in the
range of 50-2000 Hz. This result was also used in
the design of the prototype by filtering the
frequency above 2000 Hz.
The characteristic sound of bars hitting the inner
lining of the mill was simultaneously listened and
visually correlated with the FFT display and the
scope signal. This correlation allows to conclude
that the bars produce a different sound pattern in
spectrum and in time, under normal and overload
condition.

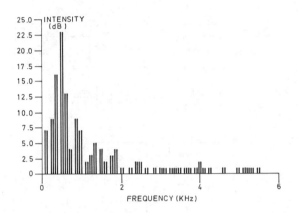

Fig. 6. Sound Spectrum as observed in the video
display terminal.

Disturbances from surrounding plant noise (horns,
ball charge, envirommental noise and neighboring
sound) were identified on the frequency spectrum,
since they also modify the sound intensity. It
follows that a device based only on the intensity
of the sound signal cannot be used for overload
detection.
Other results show that sound intensity increments
may be also produced by an underload condition, as
show in Fig. 7. Also, clearly, the fresh ore feed
rate that produces an overload is a function of
the size distribution and the hardness of the ore.

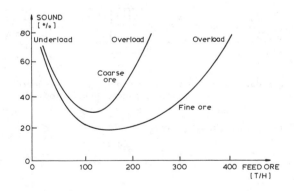

Fig. 7. Sound intensity vs. feed flow for different
types of ore.

THE ON LINE OVERLOAD ANALYZER PROTOTYPE

The results obtained during the preliminary test
of this study indicate that the structure of the
system should be implemented based in two stages.
The first stage has electronic circuits for sound
analog signal processing and the second stage is
based on a microcomputer for digital signal
processing and mathematical calculations.
In the analog processing stage, the sound signal is
amplified and filtered, and this signal is
conditioned in order to generate three new signals
S_1, S_2 and S_3 of slow evolution related to the rod
mill load condition (Fig 8a).

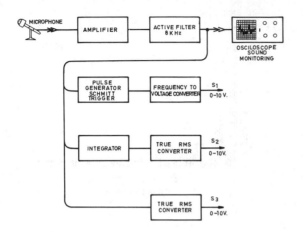

Fig. 8a   Stage 1: Analog   sound   signal
                conditioning.

S_1 is a signal measuring the average frequency of
the sound signal exceeding a determined threshold
S_n value. This signal represents the striking of
the bars against the lining.
S_2 is the real-time rms value of the low frequency
sound signal component (less than 2000 Hz) obtained
trough integration.
S_3 is the sound signal intensity (real time rms
value).
These three analog signals, together with S_4,
measuring the fresh ore feed, and S_5 measuring the
water to ore feed ratio, are fed to the second
stage where the digital and logical processing
takes place.(Fig 8b).

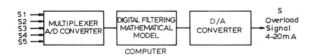

Fig 8b. Stage 2: Digital processing and standard
                overload signal generation.

The mathematical model used is:

$$S = \sum_{i=1}^{5} a_i \cdot S_i \qquad (1)$$

The weighting coefficients a_i are empirically
adjusted based on the gain of the microphone,
filtering coefficients and a compromise between
the phenomena represented by S_1 S_2 and S_3.
The situation shown in Fig. 7 has also been
considered and the coefficients of the equation (1)
are changed according to the relation between
sound and feed ore flow.

The signal S (normalized 4-20 mA) is the result of
the analog and digital processing described above
and represents a function of the overload
condition of the mill. In normal operation, S
signal remains constant close to a 20 percentage
value (Fig. 9a).

If an overload situation starts to develop, the
average value of S begins to increase. In Fig. 9b,
it is shown the behaviour of S when two succeeding
overloads take place. These last for about six
minutes. The value of signal S was 90 percentage
when the feed conveyor belt was stopped by the
field operator in an attempt to restore normal
operation. It can be seen in figure 9b, that the
overload condition could be accurately predicted
with about three minutes of anticipation, when the
S value exceeds the 60 percentage. In this case,
an override control strategy reducing the feed ore
when signal S is greater than 60 percentage can be
implemented avoiding conveyor stop.

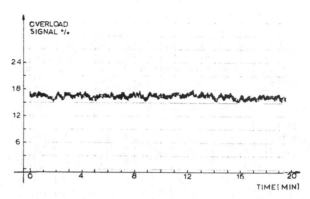

Fig. 9a. S signal recording during normal
                operation.

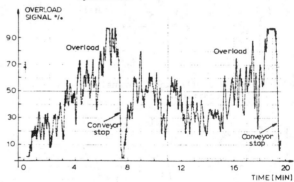

Fig. 9b. S signal recording during two mill
                overloads.

CONCLUSIONS

A system has been developed that allows prediction
of overload of the rod mills with about two or
three minutes in advance.
The signal obtained after analog and digital
processing for detecting overloads permits a clear
distinction between normal operation and the onset
of an overload situation. This signal is largely
insensitive to surrounding plant noise.
These favorable characteristics make this signal
suitable for automatic prevention of overload when
used, for example, with an override control ore
feed. This system will be implemented into the
existing El Salvador computer distributed control
system.

REFERENCES

Bradburn R.G, Flintoff C. and Walker R.A. Practical approach to digital control of a grinding circuit at Brenda Mines Ltd. Transactions Society of Mining Engineers, vol 262. AIME June 1977.

Uronen P., Taivainen M. and Aurasmaa H. Audiometric control system of wet semiautogenous and autogenous mill. Automation in Mining Mineral and Metal Processing. Proceeding of the second IFAC Symposium, 13-17 Sept 1976 Page 215-224

Zamora C. González C. and Cortez L. Distributed Control System Implementation in Chile, El Salvador Concentrator. CODELCO CHILE Internal report 1983

González C. De Mayo F. Zamora C. Automatic control of the grinding circuit of El Salvador. Proceeding Fifth ACCA (Chilean Automatic Control Asociation) Symposium. Valparaiso 22-27 Nov 1982 Page 504-525

Soto H. De Mayo F. Hernández R. Time and Frequency Domain Analizer System. Proceeding sixth ACCA Symposium. Santiago 15-19 October 1984. Page 181-186

Acknowledgements. The work described in this paper has received support from the following projects: CODELCO El Salvador N 782, Universidad de Chile DIB-I-1951 and DIB-I-1740, and Chilean National Research Fund N 587/82.

Column Flotation: Some Plant Experience and Model Development

G.S. DOBBY

Department of Metallurgy and Material Science, University of Toronto, Toronto, Ontario M5S 1A4

R. AMELUNXEN

Gibraltar Mines, PO Box 130, McLeese Lake, British Columbia VOL 1PO

J.A. FINCH

Department of Mining and Metallurgical Engineering, McGill University, Montreal, Quebec H3A 2A7

Abstract   Flotation columns represent an important addition to flotation practice. Columns have largely replaced conventional machines in molybdenum cleaning in Canada. Columns have been successfully tested on bulk copper/molybdenum cleaning at Gibraltar Mines Ltd. A rougher cleaner/scavenger cleaner combination was shown to be capable of giving higher grades and recoveries compared with three stages of conventional cleaning. A control strategy for the columns has been developed. An interaction between gas holdup and wash water rate has been exploited to control performance.

Keywords   Flotation columns; copper, molybdenum cleaning; operation-gas holdup; washwater; control.

INTRODUCTION

A significant development in flotation over the past few years has been the increasing industrial use of flotation columns, primarily in Canada. The column differs dramatically from conventional mechanical flotation machines, both in design and operating philosophy, which has been a principal reason for its slow acceptance by the mineral industry. The concept was developed in the 1960s (Boutin, Wheeler, 1967; Wheeler, 1966) and since then plant and laboratory columns have been marketed commercially by Column Flotation Company of Canada Ltd. In 1980-81 at Mines Gaspé, Canada, three columns replaced 13 conventional cleaners in a molybdenum upgrading circuit, with superior results (Coffin, Miszczak, 1982). Since then many of the copper/molybdenum producers in British Columbia have installed columns for molybdenum cleaning, including Gibraltar, Lornex, Highmont and Island Copper. There now appears to be widespread interest in flotation columns.

The column is particularly attractive for applications involving multiple cleaning stages and can upgrade in a single stage compared with several stages of mechanical cells. This results in simpler, more controllable circuits. Importantly, the column itself is well suited to computer control.

Extensive model and industrial-scale investigations have been conducted. This paper presents some of the findings, in particular the experience at Gibraltar Mines Ltd.

BACKGROUND

A flotation column resembles a counter-current bubble column. It is typically 10-13 m high and 0.5-2 m in diameter (either circular or square in cross section). A schematic diagram is shown in Fig. 1 with the flow of solids and water illustrated in Fig. 2. Reagent conditioned feed enters the column 2-3 m from the top and flows downward against a rising swarm of gas bubbles. Tailings are withdrawn at the column bottom. The feed and tailings flowrates are measured and the tailings flow is controlled at a rate slightly greater (1 to 15%) than that of the feed; this is called a positive bias. The bias is provided by washwater added from a distribution of pipes located a few centimeters below the lip of the column, beneath the top of the froth. Figure 2 shows that part of the washwater travels down the column and part exits with the concentrate. No feed water should exit with the concentrate. Gas enters the column at the bottom through a distribution of rubber or cloth spargers. There is no mechanical agitation.

Three zones of differing rheology can be identified (see Fig. 1). Zone 1 is the particle collection zone, a three phase counter-current bubble column, with gas bubbles 0.1-0.4 cm in diameter, superficial gas velocity v_g up to $\sim$5 cm/s, and superficial slurry velocity v_ℓ of 0.5-2.0 cm/s. Zone 2 is the cleaning zone, a deep (1-2 m) packed bubble bed generated by the downward flow of washwater; washwater superficial velocity v_w is 0.5-1.0 cm/s. The interface between zones 1 and 2 is very distinct. Zone 3 is a conventional froth, extending a few centimeters above the wash water pipes. Its sole purpose is to act as a medium for the transport of collected solids over the lip of the column.

The objective of wash water addition and operation with a positive bias is to prevent feed water, and therefore gangue particles, from reaching the concentrate. Liquid tracer tests at Mines Gaspé (Dobby, Finch, 1984) showed that less than 1% of the feed water reports to the concentrate. The column virtually eliminates the recovery of hydrophilic gangue particles, in contrast to conventional flotation machines where recovery by physical entrainment is inevitable (Trahar, 1981). The unique cleaning action of a flotation column is largely responsible for its ability to upgrade a fine sized concentrate in a single step.

COMPARISON WITH CONVENTIONAL CIRCUIT AT GIBRALTAR MINES LTD.

Gibraltar Mines is in British Columbia near the town of Williams Lake. It is an open pit operation with an average milling rate of 38000 mt/day grading about 0.3% Cu (as chalcopyrite) and variable Mo (< 0.016% MoS_2). In the conventional circuit a bulk Cu/Mo rougher concentrate, grading about 12-16% Cu, is produced which is reground and cleaned three times. Cleaner tails, grading <0.4% Cu, is discarded to final tails. When Mo head grades warrant a Cu/Mo separation is performed by

259

depressing Cu. Ten cleaning stages were used. Overall recovery of Cu is 85% at 30% grade. Average Mo recovery is about 40% at > 90% MoS_2.

The Cu/Mo separation circuit has now been replaced by four columns (1 10 in, 1 20 in and two 36 in diameter). This new circuit gave similar metallurgy but at significantly reduced reagent costs, principally by permitting an economic use of the nitrogen/hydrosulphide combination for copper depression.

Testwork was then started on the bulk Cu/Mo cleaners. Initially this involved replacing the third (final) cleaning stage to try to achieve higher grades. This was successful and research was redirected to investigate replacing the entire cleaning circuit with a single column stage. Final grade (30% Cu) was readily achieved by a single column. However, to ensure the high (> 98%) Cu recovery across the cleaners, (demanded because cleaner tails are discarded) a rougher/scavenger arrangement was eventually adopted, as shown in Fig. 3.

Figure 3 shows the column test circuit, treating about 1/6th of the primary rougher concentrate (i.e. about 6 mt/hr at 20% solids by weight), in parallel with the conventional circuit. Scavenger concentrate is typically 10-15% Cu and circulating load of solids is about 8%. Total retention time in the circuit is about 18-20 min.

The column test circuit was able to outperform the conventional circuit, as evidenced in Table 1. (Mo is not considered as Mo heads were frequently too low to operate the Mo circuit during this period).

TABLE 1   Comparison of Conventional and Column Circuit

Test Date	Copper Assays						Recovery	
	Feed	Conv. Conc	Tail	Feed	Column Conc	Tail	Conv	Col
06-6-84	14.8	29.3	0.62	14.2	29.6	0.34	97.9	98.7
18-6-84	18.0	27.3	0.70	19.9	31.4	0.50	98.6	99.1
28-6-84	6.5	26.2	0.23	6.5	26.9	0.22	97.3	97.4

It is now planned to replace the conventional circuit by columns.

COLUMN DESIGN

At Gibraltar, columns are made of readily available pipe; the launders of the smaller diameter columns are also made of pipe. It is important that the columns are vertical to avoid channeling.

Wash water is added through an array of perforated steel pipes. The water should be added at minimum pressure, to give a sprinkle rather than a spray. Launder spray water is also used; a nozzle giving a fan-spray was necessary to break the froth.

The air spargers are a series of 1 to 2 in diameter pipes, perforated and covered with filter cloth. These cloths last from about 2 months (in high hardness water) to 10 months. There is some evidence that the cloth gives large diameter bubbles (∿3 mm) and alternative materials are being examined. A back-up set of spargers can be included, to be switched on when a rupture in the running set is detected. Replacement of spargers will mean a column shut-down; a running spare column may, therefore, be worth considering.

The only troublesome area was the tail port, which is prone to high wear at the control valve. The control valve is critical, it is the only moving part of the column and its performance is crucial in controlling smoothly the small changes in flowrate demanded to meet the bias set-point. At times the flowrate can be low enough to sand the tails line, consequently pipe sizes should be designed for flow velocities well above the critical.

The original tail port design, a butterfly valve in the line to a pump box, was abandoned in favour of connecting the pump directly to the column and using an in-line dart valve (see Fig. 4). The dart and dart seat are made of silicon carbide ceramic. At the time of writing, this has not been replaced for seven months; as a comparison, neoprene lasted ten days.

This final design has been quite successful. A dart with little taper gives a smooth response to instrument control. The taper is, however, critical and must be determined for each application. Advantages of the system are that the pump uses some of the column's static head, no spillages occur and the flowmeter always operates under flooded conditions.

COLUMN OPERATION

Axial Mixing in Collection Zone

The collection zone is where particle recovery occurs. Mixing has a profound influence on recovery. A knowledge of the mixing characteristics should enable more reliable column scale-up.

Residence time distribution studies, both solid and liquid, were conducted at Mines Gaspé on 18 in. and 36 in. columns. The data were modelled by a plug flow dispersion model (Dobby, Finch, 1984) and two mixing parameters were extracted: (1) the mean residence time; and (2) the vessel dispersion number, $N = D/uL$, where u is the particle velocity (liquid velocity plus relative particle settling velocity), L is the length of the collection zone, and D the axial dispersion coefficient. The mean residence time for the particles could be computed from a knowledge of particle relative settling velocity; D for solids and water was found to be the same. D was related to d, the column diameter in metres, as follows:

$$D = 0.063 d \qquad (1a)$$

The exact dependence on gas rate is not known. At present the relationship is considered adequate up to superficial gas velocities (v_g) of about 3 cm/s. At v_g > 3 cm/s, a more suitable relationship may be

$$D = 0.083 d \qquad (1b)$$

The vessel dispersion number and mean residence time can, therefore, be estimated for a given column size. When combined with a suitable flotation model (e.g. first-order with fast and slow floating fractions) the capacity/recovery relationship of the column can be predicted. Such effects as increasing column diameter, or baffling an existing column, can be studied (Dobby, Finch, 1985).

Wash Water

With respect to column control, the wash water has three functions: (1) to supply the make-up bias (which provides the cleaning action); (2) to maintain pulp level (i.e. zone 1, zone 2 interface position); and (3) to achieve the required water content in zone 3 to permit unhindered overflow of concentrate.

The bias must be positive ($Q_T/Q_F > 1$ in Fig. 2). The bias is controlled at a set point, commonly Q_T/Q_F is 1.01 to 1.15. The minimum bias is desirable to minimize wash water demand and maximize residence time and, therefore, recovery. Concentrate grade does not seem to deteriorate at low bias, but close control is required.

The pulp level is sensed and wash water is adjusted to maintain the level. For example, an increase in gas rate will cause a decrease in pulp level as more liquid is held in zone 2, which eventually exits with the overflow. To compensate, wash water rate is increased.

The actual position of the pulp level (or depth of cleaning zone) does not appear to be important. The level has been brought within 1 m of the overflow without a decrease in grade. This suggests that the cleaning action may be localised at the interface.

Lastly, it has been found that the overflow must have a minimum water content for the solids to freely overflow, otherwise recovery drops. Overflow (i.e. concentrate) solids flowrate can be correlated with concentrate percent solids. Since concentrate grade does not vary greatly, concentrate solids flowrate can be calculated for a given combination of feed solids rate, feed grade and desired recovery (e.g. 98%). From the correlation, the quantity of wash water to the overflow (Q_C in Fig. 2) can then be estimated. This is exploited in controlling column recovery.

Gas Rate

An attempt has been made to distinguish between the role of gas rate in the collection zone from that in the cleaning zone. A laboratory 2 in. diameter column was modified to study the collection zone only. Mass recovery of a sample of final copper concentrate at Gibraltar was measured as a function of gas rate, v_g. Figure 5 shows that recovery (and, therefore, the collection rate constant, k) reaches a plateau around 1.7 to 2.5 cm/s. This effect can be understood from the dependence of k on v_g, bubble diameter, d_b, and bubble collection efficiency, E_k, namely:

$$k \; \alpha \; \frac{v_g \, E_k}{d_b} \qquad (2)$$

Increasing v_g increases k, but at the same time increases d_b which also causes a decrease in E_k; both these effects cause k to decrease. Therefore, at a given v_g, k can be expected to pass through a maximum. This is also the case in mechanically agitated flotation cells (Laplante, Smith, Toguri, 1983).

Figure 5 shows that, in contrast, the plant operates typically in the range 2.5 to 4 cm/s. This may in part reflect coarser bubbles (with lower E_k) in the plant, but the suggestion is that extra air is required to carry material through the cleaning zone. This has resulted in plant trials with additional spargers set just below the cleaning zone. The role of air in the cleaning zone is

difficult to quantify. It is known that increasing air will increase liquid retention in the cleaning zone, which activates additional wash water. The wash water in turn is related to the solids removal rate. Consequently the apparent requirement of air to carry material through the cleaning zone may reflect an interaction between air and wash water.

Gas Holdup

Mines Gaspé reported that recovery could be correlated with gas rate (Coffin, Miszckak, 1982); this was not found at Gibraltar. This may be because bubble size is also a factor; a change in frother would change bubble size and alter the dependence of recovery on gas rate.

A measure, which contains both gas rate and bubble size, is the gas holdup, the fraction of air in the air/slurry mixture. Holdup varies directly with gas rate and inversely with bubble size (smaller bubbles rise more slowly giving longer retention time and increased holdup). Average gas holdup across the collection zone is measured by two manometers, one tapped near the pulp level, the other near the bubbler (see Fig. 6). The difference in water level between the two manometers (ΔH, sensed by using Metri-tape) divided by the distance between the tapping points (L) is the measured gas holdup.

Figure 7 shows a relationship between concentrate solids flowrate and gas holdup for individual columns with similar recoveries (> 95%). The apparent maximum of 16% gas holdup can be further increased by frother addition. This relationship between gas holdup and concentrate rate is being exploited to control column performance.

The maximum holdup is about 20-24%. Above this, solids start to accumulate in the column and eventually exit to tailings. This condition is identified by rapid changes in holdup (2-4%/min). This phenomenon is probably related to the transition from a homogeneous bubbling regime to a churning regime (Shah and co-workers, 1982). Laboratory tests on a two phase (no solids) system tended to confirm this. Transition occurred around 30% holdup and $v_g > 5$-6 cm/s and was characterized by an ascending chanel of large bubbles and a descending chanel of fine bubbles and loss of the zone 1/zone 2 interface.

Apart from its potential in control, gas holdup also has diagnostic applications. The rapidly changing gas holdup referred to above is one; another is detection of burst bubbler when the increased bubble size will cause the holdup to drop suddenly.

INSTRUMENTATION AND CONTROL

Manual control of a column is not feasible, instrumentation and some degree of automatic control is a necessity.

The control configuration under test at Gibraltar is shown in Fig. 6. There are three control loops, bias control, level control and the wash water gas holdup interaction loop.

Bias control. In this loop, the feed and tailings flowrates are transmitted to a controller. The controller regulates the tail control valve to maintain the ratio of tailings to feed flowrate at the set bias, usually in the range 1.01 to 1.15.

Level control. The level of the pulp is sensed (in the upper manometer, Fig. 6, initially by a bubble tube, now by Metri-tape). The signal is transmitted to a controller, which maintains the level by means of a control valve on the wash water line.

Wash water-gas holdup interaction loop. This loop is designed to control the recovery. The evolving strategy is as follows. For a known feed solids rate, grade and fixed recovery, the concentrate solids rate is computed. This is related to gas holdup (e.g. Fig. 7) and, via the algorithm, controllers vary the gas rate to achieve the desired holdup. Because the concentrate rate demands a minimum water content to overflow, the wash water rate is checked.

At present, a variation in this strategy is employed. The concentrate solids rate is computed which is related to a wash water requirement. The operator is supplied with tables to calculate this requirement and he sets the controller algorithm, F(x), accordingly. The controller will vary the set point of the holdup which in turn causes an adjustment in the flowrate of air to the column.

Experience showed that the bias and level control loops were the minimum requirement for stable operation of the column. The last loop is designed to control column performance. In comparison to conventional cells, the goal of performance control seems more attainable.

CONCLUSIONS

Flotation columns outperformed conventional cells on bulk Cu/Mo cleaning at Gibraltar Mines.

The units have proved stable to operate and offer the potential, through the interaction of gas holdup and wash water rate, for control of performance.

Scale-up of column appears to be feasible.

An expanding application of columns is foreseen.

ACKNOWLEDGEMENTS

The authors wish to thank: Gibraltar Mines Ltd. for permission to publish results; Noranda Mines Ltd. for a scholarship to G. Dobby; CANMET for funding column flotation research; and J. Yianatos (McGill University), and D. Wheeler (Column Company of Canada) for helpful discussions.

REFERENCES

Boutin, P. and Wheeler, D.A. (1967). Column flotation development using an 18 in pilot unit. Can. Min. J., 88(3), 94-101.

Coffin, V.L. (1982). Column flotation at Mines Gaspé. Proc. 14th Int. Min. Proc. Congr., Toronto, Canada, Oct. 1982, paper 4-21.

Dobby, G.S. and Finch, J.A. (1984). Mixing characteristics of industrial flotation columns. Chem. Eng. Sci. in press.

Dobby, G.S. and Finch, J.A. (1985). Column flotation scale-up. Proc. 17th Operators Conference, Can. Min. Proc. Jan. 1985, to be published.

Laplante, A.R., Toguri, J.M. and Smith, H.W. (1983). The effect of air flowrate on the kinetics of flotation (part I). Int. J. Min. Proc. 11, 203-219.

Shah, Y.T., Kelkar, B.G., Godbole, S.P. and Deckwer, W.-D. (1982). Design parameters estimations for bubble column reactors, AIChE Journal 28(3), 353-327.

Wheeler, D.A. (1966). Big flotation column mill texted, E/MJ 167(11), 98-99.

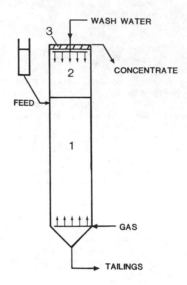

Fig. 1. Schematic illustration of Flotation Column (1) collection zone; (2) cleaning zone; (3) thin froth layer.

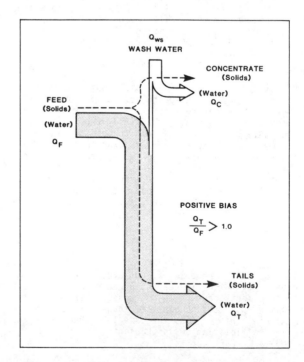

Fig. 2. Flow of solids and water through a column with positive bias. Note no feed water enters the concentrate stream.

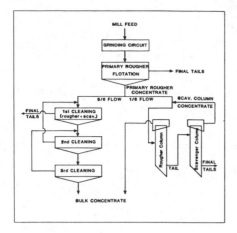

Fig. 3. Test circuit for flotation columns on
bulk Cu/Mo cleaning at Gibraltar Mines
(regrind on primary rougher concentrate
is not shown).

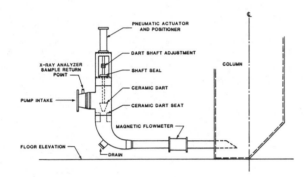

Fig. 4. Current design for tails port.

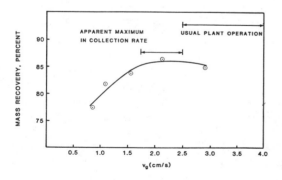

Fig. 5. Mass recovery of copper concentrate vs
superficial gas velocity. (Note, using
Cu concentrate, mass recovery is approximately Cu recovery).

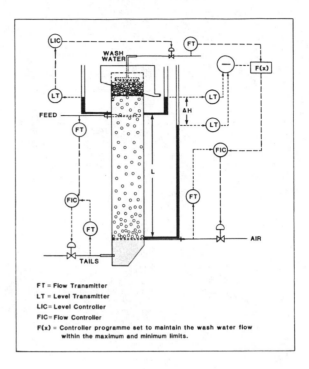

FT = Flow Transmitter

LT = Level Transmitter

LIC = Level Controller

FIC = Flow Controller

F(x) = Controller programme set to maintain the wash water flow
within the maximum and minimum limits.

Fig. 6. Control circuit for flotation columns
showing the three loops, Bias, Level
and Holdup-Wash Water Interaction.

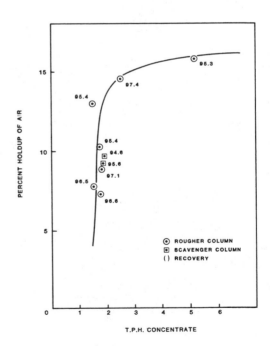

Fig. 7. Holdup vs concentrate solids mass flowrate for individual column recovery
greater than 95%.

Optimisation of Grinding Circuits at the Lead/Zinc Concentrator, Mount Isa Mines Limited

E.V. MANLAPIG, D.J. DRINKWATER

Research and Development Division, Mount Isa Mines Limited, Mouth Isa, Qld. Australia

P.D. MUNRO, N.W. JOHNSON, R.M.S. WATSFORD

Metallurgical Divison, Mount Isa Mines Limited, Mount Isa, Qld, Australia

Abstract. This paper describes the use of modelling and simulation techniques in the analysis and optimisation of grinding circuits at the Lead/Zinc Concentrator of Mount Isa Mines Limited.

The objectives of the studies were to: -

1. improve primary grinding circuit performance by reducing the production of -0.006 mm galena fines.

2. maximise the fineness of grind in the secondary grinding circuit using the existing grinding power and pumping capacities.

The study has shown that quantitative models can be useful in analysing plant operations. The models described plant operations, planned pilot scale tests and simulated the effect of process changes on plant performance.

The use of automated scanning electron microscope technology to understand the liberation characteristics of the major minerals in the Mount Isa Lead/Zinc ore is also described.

Keywords. Simulation; grinding; classification; flotation; optimisation; electron microscopy; liberation.

INTRODUCTION

Lead-Zinc-Silver ore geology and concentration practices at Mount Isa have been discussed previously (Mathias and Clark, 1975; Bush, 1984; Watsford and others, 1983). The Concentrator flowsheet consists of crushing, heavy medium separation, grinding and flotation. In the 1983/84 fiscal year just over four million tonnes of ore were treated.

After 30% of the run-of-mine ore has been rejected in the heavy medium plant the preconcentrate moves through the grinding and flotation processes shown in Fig. 1. Known as the Two Stage Grind-Flotation circuit, its apparent rationale has been to recover the more easily liberated galena into the cleaners at a relatively coarse primary grind of around 65% -0.074 mm without liberating too much of the sphalerite and spheroidal, carbonaceous pyrite. These minerals are difficult to depress and interfere with the selectivity of lead flotation.

The grinding section is operationally divided into two discrete lines, each one consisting of:

    2 x rod mills

    2 x primary ball mills

    2 x secondary ball mills

It has been important, and will continue to be so, that the best use is made of the installed grinding power to get acceptable liberation of the valuable sulphides from a refractory ore.

This must be set against the trend of having to handle ever increasing quantities of new feed.

PRIMARY GRINDING

Background

An 800 t/h heavy medium plant was commissioned in late 1982 as part of a programme to increase concentrator throughput from 2.5 million to 4 million tonnes of ore per year. The heavy medium process rejects at least 30% of the run-of-mine ore, mostly the low specific gravity non-sulphide gangue minerals, before the softer preconcentrate is treated by grinding and flotation.

Table 1 shows the increase in the fineness of the primary grind since the commissioning of the heavy medium plant. This has gone from 47% to 56% -0.037 mm with the -0.01 mm nominal quartz size increasing from 27% to 33% with the same rod mill feed rate.

TABLE 1 Lead Primary Rougher Feed Yearly Composite Sizing

Year	% -0.037 mm	% -0.01 mm Nominal Quartz
81/82	46.89	27.53
82/83	49.12	28.51
83/84	56.65	33.06

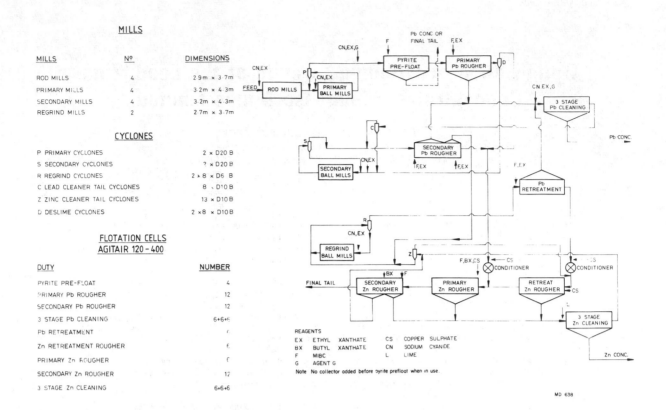

**Fig. 1 Grinding and Flotation Flowsheet for Mount Isa Mines
Lead-Zinc-Silver Ore Treatment**

The Problem

The increasing proportion of very fine galena
(Infrasizer F7 fraction, -0.01 mm nominal quartz
or -0.006 mm for galena) caused concern because
galena in this size fraction has a lower
recovery during lead primary rougher flotation.
Fig. 2 shows the recovery versus size
relationship for galena determined by surveys of
the lead primary rougher flotation section.

Objectives of the primary grinding study were
to: -

1. calculate the breakage and classification
    characteristics of the primary grinding
    section;

2. determine if it was possible to minimise the
    production of -0.006 mm galena;

3. achieve objective 2 above without decreasing
    the fineness of the overall size
    distribution. This is essential as it will
    be seen in subsequent sections that the
    final grinding product sizing must be as
    fine as possible to maximise the liberation
    of sphalerite.

Selective Classification

Figure 1 shows that the Lead/Zinc Concentrator
primary grinding section is conventional with
rod mills followed by primary ball mills in
closed circuit with a single stage of
hydrocyclone classification. This study
considered only the hydrocyclones and ball mills.

The primary ball mill hydrocyclone performance
is presented in Fig. 3. which shows the actual

and corrected efficiency curves for galena and
total solids. Galena cuts finer than the rest
of the solids primarily due to its higher
specific gravity. This has been observed by
other workers including Lynch (1977) and Finch
and Matwijenko (1977). It may also be seen that
a substantial proportion of the solids
(including galena) are short-circuited to the
hydrocyclone underflow without being subjected
to true classification.

From this it was suggested that galena particles
that should have reported to the overflow were
sent to the underflow because of poor
classification. These particles were subjected

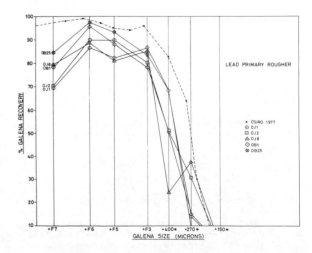

**Fig. 2 Flotation Frequency - Size
Relationship for Galena**

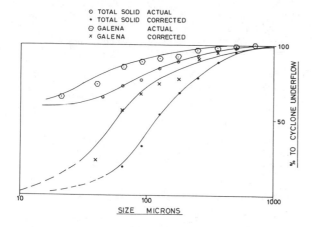

Fig. 3   Primary Ball Mill Hydrocyclone
         Efficiency Curves

to further grinding which may have produced more
-0.006 mm galena fines. This has led to a
number of authors including Ramirez-Castro and
Finch (1980), Williams (1984), Rawling and
Goyman (1984), Kallioinen and Tarvainen (1984)
and Gowans and Simkus (1984) suggesting that the
production of galena fines can be minimised and
recovery increased by flotation of grinding mill
discharge before classification. Such a
flowsheet is shown in Fig. 4.

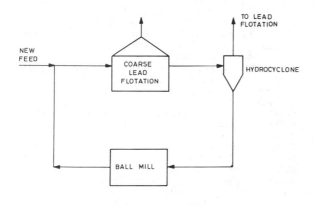

Fig. 4   Flowsheet for Lead Flotation within
         the Grinding Circuit

The configuration shown in Fig. 5 was used to
test the concept in the 1 t/h Pilot Plant at
Mount Isa. As presented in Fig. 1 the Lead/Zinc
Concentrator primary circuit consists of a rod
mill, a primary ball mill in closed-circuit with
hydrocyclone classifiers, followed by lead
primary flotation. For the Pilot Plant test
circuit, the flotation cells were installed
within the second stage of ball milling as they
were extremely difficult to operate on the first
stage mill discharge. Operational parameters
were adjusted to give a second stage ball mill
discharge sizing equivalent to that of the
Lead/Zinc Concentrator's primary ball mills.

Figure 6 shows the effect of <u>selective</u>
classification in a closed grinding circuit.
This effect is presented using classification
efficiency curves for galena where efficiency is

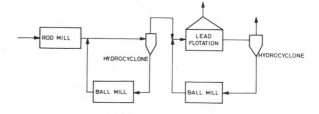

Fig. 5   Pilot Plant Flowsheet for Lead Flotation
         within the Grinding Circuit

calculated as the percentage of galena in the
classifier feed that reports to the classifier
underflow. The lower line shows the efficiency
curve for classification of galena for the
circuit with flotation cells on the second ball
mill discharge whereas the upper line is the
efficiency curve for galena using only the
hydrocyclone.

These results gave the following conclusions: -

1. the classification efficiency of galena was
    sharpened with flotation of the mill
    discharge as the cells acted as a selective
    classifier which prevented liberated galena
    from returning back to the mill;

2. liberated floatable galena particles between
    +0.006 mm and 0.053 mm were most affected.

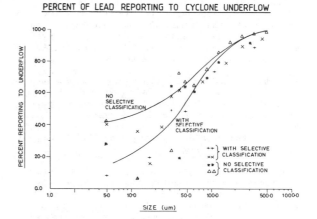

Fig. 6   Actual Classification Efficiency Curves
         for Galena with and without Selective
         Classification by Flotation

<u>Modelling and Simulation to Predict Plant
Performance</u>

It has been shown that it was possible to
prevent galena-containing particles between
0.006 mm and 0.053 mm from reporting to the
hydrocyclone underflow and thus being subjected
to further grinding. The next task was to
determine whether this would reduce the
production of -0.006 mm galena. This was simply
answered by finding out whether particles in the
(nominal quartz) 0.010-0.053 mm size range
produced significant amounts of F7 material.

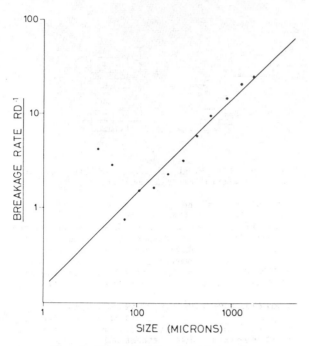

Fig. 7 Calculated Breakage Rates for
Primary Ball Mill

The Lead/Zinc Concentrator primary grinding section was sampled to provide data for a simulation.

The performance of the primary ball mills was modelled using the perfect mixing model of Whiten (1976). Fitted values for the breakage rates (RD^{-1}) are shown in Fig. 7. It may be noted that the form of the breakage rates closely follows that discussed by Whiten. Fig. 8 shows the predicted and observed values for the mill feed and discharge.

The primary ball mill performance was simulated with the breakage rate of particles in the 0.010–0.074 mm size fraction reduced to zero. This determined whether the primary ball mill breakage of particles in this size range affected the rate of production of the -0.010 mm size fraction.

Table 2 presents the comparison of the mill discharge sizings showing the two cases where the 0.01-0.074 mm particles are ground, and not ground. The results indicate that this size range of particles had relatively low breakage rates and that it did not significantly contribute to the production of the F7 material.

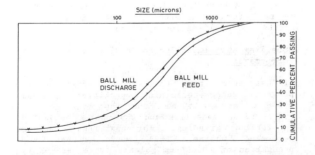

Fig. 8 Predicted and Observed Cumulative Size
Distribution for the Primary Ball Mill
Discharge

TABLE 2

Size (mm)	Mill Discharge -0.074 mm ground	Mill Discharge -0.074 mm not ground
0.2380	0.18	0.18
0.1680	1.14	1.14
0.1190	2.45	2.45
0.841	4.05	4.05
0.595	6.17	6.17
0.420	9.29	9.29
0.297	14.78	14.78
0.210	13.28	13.28
0.149	12.11	12.11
0.105	8.47	8.47
0.074	6.35	6.62
0.053	3.77	3.82
0.038	3.55	3.56
0.026	1.71	1.67
0.018	1.45	1.41
0.013	1.19	1.15
0.010	1.54	1.50
-0.010	8.53	8.35

Effect on Plant Operation

This study shows that a flotation bank or separation device installed on the mill discharge stream inside a ball mill-hydrocyclone circuit can improve the efficiency of classification. Liberated floatable particles can be removed instead of reporting to the hydrocyclone underflow.

However, this concept cannot reduce the amount of -0.006 mm galena produced in the primary grinding section at Mount Isa where liberation occurs in the 0.006-0.037 mm size range. The breakage rates of particles in this size range are so low that even if they were sent through the mill a number of times more than necessary, very little grinding would take place. Successes claimed by other workers must be based on galena that is liberated at a very coarse grind (Williams, 1984; Rawling and Goyman, 1984; Kallioinen and Tarvainen, 1984).

Thus the concept is not justified at Mount Isa's Lead/Zinc Concentrator based solely on the objective of minimising the amount of -0.006 mm galena in the lead primary rougher feed.

SECONDARY GRINDING

Background

Originally the two secondary mills in each grinding line were operated in parallel, each mill in closed circuit with 2 x operating Krebs D20B hydrocyclones fed by a pump with 110 kW variable speed drive. The main advantages of running the secondary mills of each line in parallel were:-

1. simpler to operate as the performance of each mill is independent of the other which ensures that process disturbances ocurring in one mill are not reflected in the other;

2. low pumping costs and unused capacity which allowed surges to be easily handled.

After the heavy medium plant was commissioned in 1982, parallel operation of the secondary mills on the softer preconcentrate had the following parameters: -

D20B hydrocyclone   102 mm dia. spigot
                           168 mm dia. vortex
                             finder

total feed to each set
of 2 x operating   130-150 t/h of ore
hydrocyclones on
each mill.

hydrocyclone overflow   50% solids w/w

hydrocyclone underflow   65% solids w/w

secondary grind   1.5% +0.21 mm
                           78%   -0.074 mm
                           64%   -0.037 mm

The Problem

Previous efforts to improve the secondary
grinding performance with the parallel mill
configuration focussed on better
classification. For example, using 76 mm
hydrocyclone spigots to reduce the amount of
fines short-circuiting to the underflow gave a
coarser size split with an overall coarser lead
secondary rougher feed. After it had been
confirmed that the hydrocyclone spigot and
vortex finder sizes were the best choices
available, there was the possibility of adding
water to the feed to improve the fineness of
split. This would have given a lower
hydrocyclone overflow density. Volume
restrictions in downstream flotation operations
ruled this out, especially in the light of
mounting evidence that much more residence time
was needed for zinc roughing and scavenging.

Series operation of the secondary mills on each
grinding line seemed to be the only way of
improving the product sizing within the
constraint of no extra volume to flotation. The
anticipated benefits of series milling were:-

1. improved classification from series
    hydrocycloning. Operating the first mill's
    hydrocyclones in a near "roping" underflow
    condition would reduce the amount of fines
    being bypassed to the mill whereas the
    second mill's hydrocyclones would place any
    tramp oversize in the first stage overflow
    into their underflow for grinding.
    Considering the secondary grinding circuit
    as a whole, the first stage would lower the
    intercept of the actual efficiency curve on
    the abscissa while the second would steepen
    up the top section of the curve.

2. improved ball mill performance from denser
    feed.

Series Secondary Grinding Test

The operating parameters for the trial of series
secondary grinding were:-

secondary No. 1 mill
(No. 27)

2 x D20B hydrocyclone   76 mm spigot
                           168 mm vortex finder

total feed to   240 t/h of ore
hydrocyclones

hydrocyclone overflow   50% solids w/w

hydrocyclone underflow   78% solids w/w

secondary No. 2 mill
(No. 26)

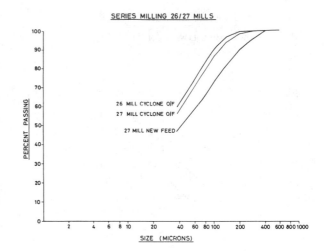

Fig. 9 Measured Cumulative % Passing Curves
        for Numbers 26 and 27 Mills in Series

2 x D20B hydrocyclone   102 mm spigot
                           168 mm vortex finder

total feed to   300 t/h of ore
hydrocyclones

hydrocyclone overflow   50% solids w/w

hydrocyclone underflow   70% solids w/w

Sizing results are shown in Fig. 9.

It should be noted that the change in fineness
of grind was greater in the first stage (No. 27
mill) than in the second.

Modelling Series Secondary Grinding

The grinding behaviour of No's 27 and 26 mills
was modelled using the perfect mixing model
described by Whiten (1976). Breakage rates for
the different size fractions for No's 26 and 27
mills are shown in Fig. 10. It is clear that
the breakage rates (RD^{-1}) for No. 27 mill were
higher than those for No. 26 mill. This
difference was attributed to the fact that the
former was being run at a higher density of 78%
solids compared with 70% solids.

Such a conclusion was hardly startling as it
followed grinding operating practice. Klimpel
(1982) observed similar changes in breakage
rates with increased pulp density using
laboratory data. Kanau (1982) found likewise in
large diameter ball mills at Bougainville Copper
Limited.

The performance of No's 27 and 26 ball mills
operated in series with No. 27 as the first mill
was simulated <u>assuming efficient classification</u>
under the following conditions:

Mill	Breakage Rate	Number of Operating Hydrocyclones	Spigot Dia mm
No. 27	high	2	76
No. 26	low	2	76
No. 26	high	2	76
No. 26	high	1	102

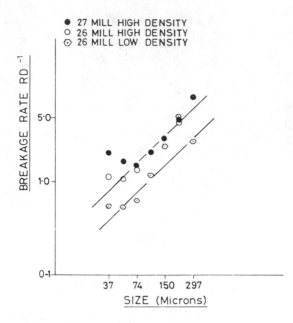

Fig. 10 Calculated Breakage Rates (RD^{-1}) for
Numbers 26 and 27 Mills Operating
at Different Densities

Simulated and actual results (the latter are
those of Fig. 9) are shown in Fig. 11. These
highlight the importance of operating ball mills
at sufficient pulp densities to get high
breakage rates. It was difficult to get a high
density underflow on the No. 26 mill
hydrocyclones even with 76 mm diameter spigots.
The simplest way to achieve a high density in
No. 26 mill was to run with only one
hydrocyclone.

Figure 12 shows the results of further
simulations which assumed that the high breakage
rates observed in No. 27 mill at a high pulp
density could somehow be reproduced, assuming
efficient classification, in both mills running
in parallel. Product sizing would approach that
observed with the mills in series when No. 27
had a high breakage rate and No. 26 mill a low

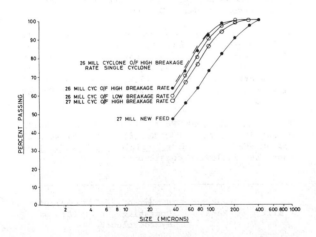

Fig. 11 Measured and Simulated Cumulative
percent passing for Numbers 27 and 26
Mills in Series at Differing
Conditions of Mill Densities and
Breakage Rates

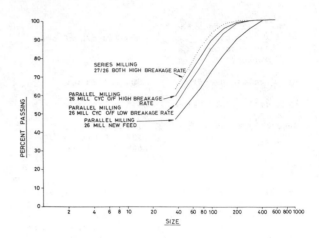

Fig. 12 Simulated Cumulative Percent Passing
of Cyclone Overflows with the Mills
using High and Low Breakage Rates

breakage rate as in Fig. 10. The broken line in
Fig. 12 shows the expected result if No's 27 and
26 mills were operated in series with both mills
having high breakage rates.

Effect on Plant Operation

The results of the simulations shown in Figs. 11
and 12 could not be reproduced in practice
because it was not possible to get both a high
underflow density (i.e. a near "roping"
discharge) and efficient classification in a
single stage of hydrocycloning. The
implications of this are discussed below.

With the available equipment, the high breakage
rate conditions of No. 27 ball mill could not be
re-created in No. 26 mill. More testwork was
therefore done to see if the performance of
No. 27 mill could be improved further.
Figure 13 shows that a reduction in No. 27 mill
hydrocyclone spigot diameter from 76 mm to 50 mm
gave a further 40% increase in breakage rates.
Plant operation has now standardised on 65 mm
diameter spigots for the hydrocyclones on the
first stage of secondary grinding as the 50 mm
diameter ones caused too many underflow sand-ups.

The secondary grinding circuit product sizing is
now: —

0%	+0.21	mm
84%	−0.074	mm
70%	−0.037	mm

The two regrind mills have also been converted
to series operation on the same principle i.e.
"rope" the first stage hydrocyclone underflow to
get a high mill pulp density.

Operation of some of the regrind and secondary
mills at high pulp densities has highlighted
questions such as whether smaller grinding media
sizes should not be used. Past testwork at the
lower mill pulp densities was always
inconclusive.

Another point that the work on series milling
has highlighted is the possible use of smaller
size hydrocyclones for the second mill in the
series to handle the finer material.

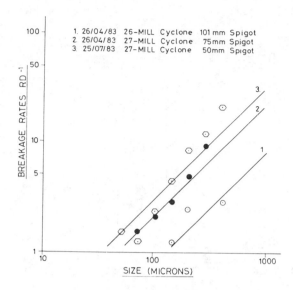

1. 26/04/83   26-MILL Cyclone   101 mm Spigot
2. 26/04/83   27-MILL Cyclone   75mm Spigot
3. 25/07/83   27-MILL Cyclone   50mm Spigot

Fig. 13 Comparison on Breakage Rates RD^{-1}
in Secondary Grinding Mills at
Different Mill Feed Density

Implications for Grinding Circuit Design

The work has led to an understanding of the
simple fact that for a ball mill in closed
circuit with a single stage of classification in
hydrocyclones: -

it is extremely difficult to have efficient
classification and the best condition for
grinding i.e. a high underflow density.

The single stage hydrocyclone in a grinding
circuit is usually operated at a compromise.
Moving to a higher underflow density to reduce
the amount of short-circuiting fines and giving
higher ball mill breakage rates inevitably
forces oversize particles into the hydrocyclone
overflow.

The conventional answer to improving
classification in the grinding circuit is to use
a two stage system with most of the literature
favouring re-treatment of the first stage
hydrocyclone underflow to remove the
short-circuiting fines.

The disadvantages of this approach rather than
re-cycloning the overflow are: -

1. operational problems from handling a large
   flow rate of coarse solids;

2. higher capital and operating costs from
   pumping a large volume of feed to the second
   stage hydrocyclone;

3. possible need to add water to the second
   stage hydrocyclone feed giving volume
   problems in downstream processing;

4. too efficient fines removal could give low
   viscosity mill feed and poor breakage rates.

The only advantage of re-cycloning the first
stage underflow is that finished product is
removed. Simulations of re-cycloning the
overflow with the second stage underflow
returning to the mill sometimes show a continual
build-up in the circulating load.

The experiences from this work show that two
stage hydrocycloning, the first stage operated
with a "roping" underflow and the second
scavenging the overflow of the first, should be
carefully evaluated for ball mills on fine
grinding duties. Existing hydrocyclone and ball
mill simulation models are capable of showing
the possible benefits from such an arrangement.
This approach seems especially fruitful when
considering the low grinding efficiencies at
very fine particle sizes that some concentrators
have to achieve for acceptable beneficiation
performance.

LIBERATION ANALYSIS

Background

As part of the project on grinding at the
Lead/Zinc Concentrator, samples from the primary
and secondary circuits were sent to the
Commonwealth Scientific and Industrial Research
Organisation (CSIRO) for mineralogical
analysis. The study was done using the new
instrument system QEM*SEM (for "Quantitative
Evaluation of Minerals by Scanning Electron
Microscopy") developed by the Division of
Mineral Engineering of the CSIRO (Miller and
others, 1982).

The objectives of the study were to:

1. Gain some understanding of the liberation
   process of the major minerals - galena,
   sphalerite, iron sulfides, and the
   non-sulfide gangues - as the ore is ground.

2. Assemble "snap-shot" liberation pictures of
   the grinding products.

Liberation Process

The QEM*SEM results provide some clues to the
process of liberation of the major minerals in
the ore as it was being subjected to grinding.
Figure 14 shows the intergrowth and liberation
characteristics of galena in the 0.014 to
0.027 mm size range in samples from the rod mill
discharge, primary hydrocyclone underflow,
primary ball mill discharge, primary
hydrocyclone overflow. The bar height in the
intergrowth histogram represents the number (or
volume) of particles falling into a given volume
fraction class. The dark shading represents the
amount of galena (the mineral under
consideration) contained in each class. For
example, in the case of the hydrocyclone
underflow sample, in the size range 0.014 to
0.027 mm, about 5.5% of all particles have
average composition of 90%-100% galena.

The cumulative liberation yield defines the
amount of a given mineral carried in those
particles for which the volume fraction of the
mineral is greater than or equal to a given
value. Starting from particles in the 90%-100%
volume class, the cumulative liberation yield is
formed of the amounts in each successively lower
volume fraction class.

The results indicate that the level of
liberation of the major minerals in each size
fraction is constant at different grind size
distributions. In other words, if the ore is
ground to 40, 50, 60 or 70% minus 0.074 mm, the
level of liberation of galena or sphalerite at
the 0.014 to 0.027 mm size fraction (or any
other size fraction) is constant. As the ore is
ground and becomes finer, the amount of material

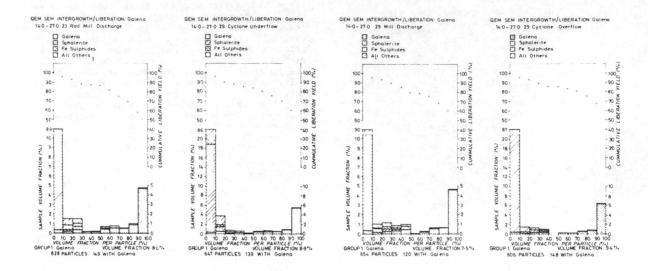

Fig. 14   QEM*SEM Intergrowth and Liberation Curves at Different Stages of Grind

in the 0.014 to 0.27 size fraction may increase but the fraction of liberated galena in that size fraction remains constant.

Similar liberation characteristics have been noted for sphalerite with samples taken from the rod mill discharge and samples taken from the secondary ball mill hydrocyclone overflow. This is significant because, although there was a wide difference in the fineness of grind between the two streams, the level of liberation was similar.

Liberation Properties of Secondary Grinding Products

Samples from the secondary grinding products were also sent for QEM*SEM analysis. The object was to gain information on the liberation characteristics of the major minerals - galena, sphalerite and non-sulfide gangue - after secondary grinding.

Figure 15 shows the intergrowth and liberation characteristics of galena and sphalerite in the +0.212 -0.106 mm,   -0.027 +0.014 mm   and -0.014 +0.010 mm size fractions of the secondary grinding product. It may be noted that the secondary grinding process produced liberated sphalerite in the fine size ranges between +0.027 -0.010 mm suitable for flotation in the zinc flotation circuit.

The secondary grinding process, however, did not appear to produce liberated galena. The galena remained in composite form and so concentrates from the flotation of these particles would probably have to be reground to produce suitable material for cleaning into a lead final concentrate. Data from microscopic grain counting of samples from flotation streams confirm these observations.

A different grinding and flotation concept compared with the current circuit configuration may be suggested based on the above

observations. As the galena tended to remain in composite form with other sulfides, and the non-sulfide gange appeared to be substantially liberated even at the coarse size of 0.106-0.212 mm, the concept of discarding the non-sulfide gangue after a minimum of grinding and redirecting the grinding power to comminuting sulfide minerals is worth considering.

CONCLUSION

This work has shown how useful quantitative models can be in analysing plant operations. They were employed to describe plant operations, plan pilot scale tests, and simulate the effect of process changes on plant performance.

Specific conclusions were:

1. Selective classification by putting a separation stage such as a flotation cell inside a ball mill-hydrocyclone primary grinding circuit, will only reduce overgrinding of a dense mineral if there is liberation at a coarse size.

2. Series hydrocycloning with the first stage having a "roping" underflow and overflow treated by the second stage hydrocyclone should be considered for fine grinding mills. Such a combination seems best able to satisfy the requirements of efficient classification and high particle breakage rates in the mill.

3. The detailed analysis of liberation using automated scanning electron microscope technology has indicated a need to review the concept of the current grinding and flotation circuit configuration to try to improve plant performance.

ACKNOWLEDGEMENT

The authors wish to thank Mount Isa Mines Limited for permission to publish this paper.

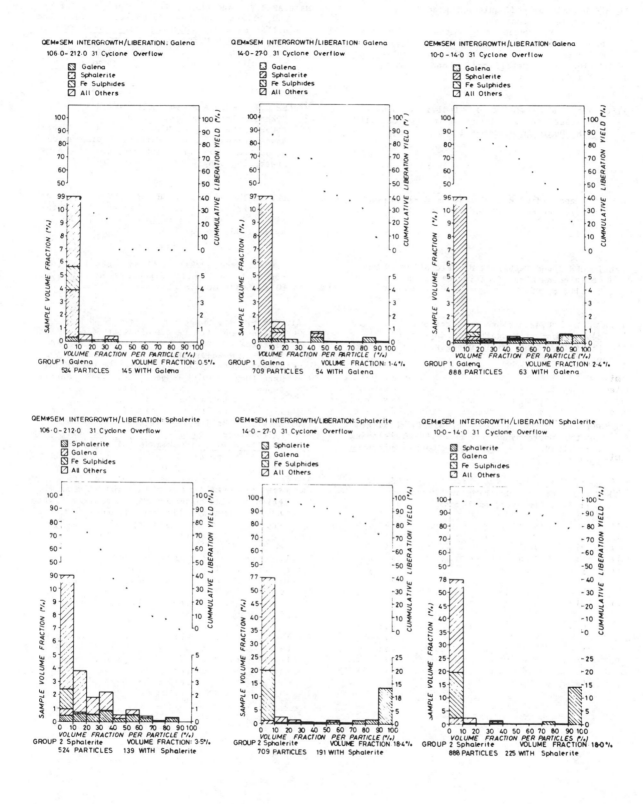

Fig. 15 QEM*SEM Intergrowth and Liberation Curves for the Secondary Grinding Product

REFERENCES

BUSH P.D., "Plant and Equipment practice in
   base-metal concentrators of
   the M.T.M. Group of companies", Mineral
   Processing and Extractive Metallurgy, Jones
   M.J. and Gill P. eds, The Institution of
   Mining and Metallurgy, London 1984, pp
   147-160.

FINCH J.A. and MATWIJENKO O. "Individual Mineral
   Behaviour in a Closed Grinding Circuit"
   C.I.M. Bulletin November 1977.

GOWANS J and SIMKUS R. " Coarse Lead Flotation
   Practice at Polaris", op cit, pp 74-84.

KALLIOINEN J and TARVAINEN M. "Flotation as Part
   of Grinding Classification Circuits",
   op cit, pp 55-73.

KANAU J.L., "Hydrocyclone research and
   development at Bougainville Copper
   Limited", 1982 Mill Operators Conference,
   pp 169-177.

KLIMPEL R.R., "The influence of slurry rheology
   on the performance of mineral/and grinding
   circuirts", 1982 Mill Operators Conference
   pp 1-14.

LYNCH A.J. "Mineral Crushing and Grinding
   Circuit", Elsevier, Amsterdam
   1977, p 119.

MATHIAS B.V., and CLARK, G.J., "Mount Isa Copper
   and Silver-Lead-Zinc Orebodies - Isa and
   Hilton Mines", Economic Geology of Australia
   and Papua New Guinea Vol. 1., Metals,
   Knight C.L., ed., The Australian Institute
   of Mining and Metallurgy, Melbourne, 1975,
   pp 351-372.

MILLER P.R., REID A.F. and ZUIDERWYK M.A.,
   "QEM*SEM Image Analysis in the
   Determination of Modal Assays, Mineral
   Associations and Mineral Liberation",
   presented to XIV International Mineral
   Processing Congress : Toronto 1982.

RAMIREZ-CASTRO J. and FINCH J.A., "Simulation of
   a grinding circuit change to reduce
   sliming", CIM Bulletin, Vol.73, No.816,
   April 1980, pp 132-139.

RAWLING K.R. and GOYMAN J. "Lead flotation from
   rod and ball mill discharge at Nanisivik
   Mines Ltd", op cit, pp 74-84.

WATSFORD R.M.S., SCHACHE I.S., and BUSH R.C.,
   "Recent and future developments in milling
   practice at Mount Isa Mines Limited" Mining
   and Metallurgical Institute of
   Japan/Australasian Institute of Mining and
   Metallurgy Joint Symposium 1983 Sendai,
   Japan.

WHITEN W.J., "Ball Mill Simulation using Small
   Calculators", Proceedings of The Australian
   Institute of Mining and Metallurgy, No. 258,
   June 1976, pp 47-53.

WILLIAMS A.J. "Flotation of base metals from
   grinding mill discharges" Proceedings - 16th
   Annual Meeting of the Canadian Mineral
   Processors, Ottawa 1984, pp 9-32.

Automatic Control of Semi-Autogenous Grinding at Los Bronces

JORGE JEREZ C.

Instrumentation and Control Engineer

HECTOR TORO CH.

Senior Metallurgist

GERHARD VON BORRIES H.

Concentrator Operation Head

Compania Minera Disputada de Las Condes S.A., Los Bronces Mine, Chile, South America

Pedro de Valdivia 291, Santiago, Chile.

Abstract. A brief description of the semi-autogenous grinding (SAG) facilities of Compania Minera Disputada de Las Condes is given. A data-logging and computer system was connected to the plant interfacing the existing analog control system in a supervisory set point control philosophy. The development of a control strategy to maintain maximum power of the SAG mill at steady operation and subjected to process restrictions is shown.

The present control scheme is discussed in detail. A supervisory algorithm was designed based on correlations between variables. Associated with this, a direct level of control was developed based on classical control algorithms, process inter-active tuning has been included. Also, an algorithm to prevent overloading of the mill is discussed. Control scheme tests are presented in comparison to manual operator control under the same conditions. It is shown that the control strategy gives a more stable operation and a higher plant throughput.

KEYWORDS: Computer control, semi autogenous grinding, process control, adaptive control.

INTRODUCTION

The grinding facilities of Compania Minera Disputada de Las Condes (CMD) are located at 3600 m.a.s.l. in the Andean Cordillera 60 km east of Santiago, Chile.

A porphyry copper orebody is mined and milled at a rate of 12000 tons/day. A new semi-autogenous grinding plant was started up in 1981 and in 1983 an automatic computer control project was started to control the grinding plant.

Grinding Circuit

Open pit trucks dump the ore into a primary 1m x 2m jaw crusher. Primary crushed -8" ore is fed to a 20000 ton stockpile. Six variable speed belt feeders reclaim the ore from the stockpile to a main feed belt which feeds the ore into the mill.

The main comminution equipment is a 28´ φ x 15´ ℓ semi-autogenous primary grinding mill with two 3500 HP drives. Mill discharge is classified by a trommel with 3/4" openings. The fine fraction is sent to a 26" diameter cyclone nest. Cyclone underflow goes back to the mill and overflow is final product.

Trommel oversize (pebbles) is sent back to the SAG mill. In September 1984 an on-line 5.5ft. Symons pebble crusher was installed. Crushed pebbles are sent back to the mill. As an alternative, a belt system allows the crushed pebbles to be sent to the secondary mills.

Two secondary 9.5´ φ x 12´ ℓ, 650HP ball mills, are normally fed with part of the SAG mill trommel undersize. These mills operate in closed circuit with 20" diameter cyclones whose overflow is also final product.

At least five circuits can be configured very easily by manipulating some gates and valves.

SA : SAG mill only, in closed circuit (original plant).

SAB : SAG mill with secondaries, without on-line crusher.

SABC : SAG mill and ball mills with on-line crusher sending crushed product back to the mill.

SACB : SAG mill with on-line crusher sending product to the secondary mills.

SAC : SAG mill with on-line crusher, without secondary mills.

The plant was designed for 400tons/hour. At present the plant capacity, including on-line crusher is 500tons/hour. Fig. 1 shows the grinding plant flowsheet.

In November 1984 a new 30" x 48" primary gyratory crusher which delivers a -5.5 inch product was started up.

Automatic Control System

The automatic control system at Los Bronces is configured in a computer supervisory set point control mode as shown in Fig. 2. An existing analog control panel was interfaced with a microcomputer which generates the set points to the final controllers. In that way the analog back up is automatically present in case of computer failure.

Field Instrumentation and Analog Panel

From the start-up on, the plant was almost fully instrumented and at the beginning of the computer control project, the emphasis was placed to complete this instrumentation. Briefly, the plant mass balance including all the streams can be

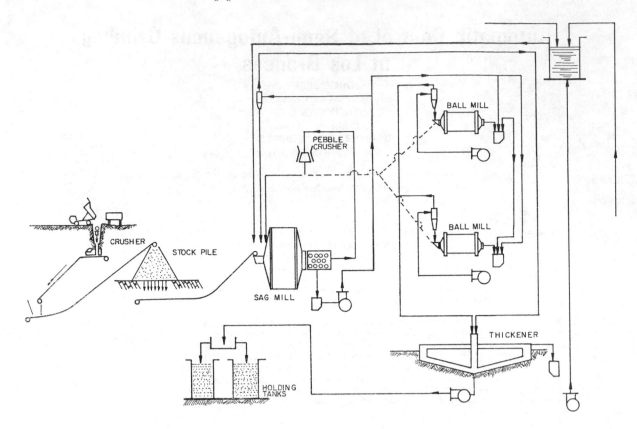

Fig. 1.   Crushing and Grinding Circuit at Los Bronces.

COMPUTER   SUPERVISORY   SET   POINT   CONTROL

MODES OF CONTROL

① MANUAL
② LOCAL SET POINT
③ COMPUTER SUPERVISORY SET POINT
④ COMPUTER SET POINT
⑤ DEVELOPMENT

Fig. 2.   Computer Supervisory Set Point Control
          Structure.

obtained continuously by means of weightometers,
pulp and water magnetic flowmeters and nuclear
pulp density meters. Other important instruments
are a PSM-100 particle size monitor at the
cyclone overflow, a MSD-95 coarse particle size
monitor on the main feed belt to the SAG mill,
bearing back pressure transducers on both SAG
mill main bearings to obtain an indirect measure
of the holdup and power transducers on the
primary and secondary mills.

The manual operator control used from start up on
allows maintenance of a feedrate setpoint. Power
(J) and bearing back pressure (BBP) are monitored
visually by the operator and action on the
feedrate setpoint is taken.

Before the intensive data logging campaign, all
the field instrumentation was rigorously checked
independently from the maintenance routines.
Several (20) sampling campaigns were carried out
using sizing data and direct measurements of pulp
density to check the cyclone feed mass flow
rates. MSD-95 calibration was carried out sizing
1 to 2 ton samples taken from the feed belt. As
an alternative, a pre- determined size
distribution of acrylic sheets were placed on the
running empty belt comparing this distribution
with the output of the instrument. Acceptable
results were obtained.

The actuators are the variable speed feeders,
pneumatic water valves and continuously
adjustable splitter gates in the on-line crushing
system.

All the 4-20mA in and output signals are
centralized in a Foxboro Spec 200 analog panel
which contains PI controllers, recorders,
integrators and indicators.

Digital Configuration

The digital system was a locally configured[1]
microcomputer based on a ISBC 8024 Intel CPU. The
configuration includes A/D converters for 60
inputs, isolated digital input-output channels, a
256Kb bubble memory, 64Kb EPROM memory and a high
speed mathematical logic unit. A DEC VT 100 is
used as operator interface.

A Hewlett Packard 87 personal computer connected
to the front end microcomputer, is used as a
development computer.

----

(1) The computer was made by Fundacion Chile,
     ITT supported Research Centre.

All lower level software such as data
acquisition, date preprocessing, high frequency
filtering, conversion to engineering units or
calculation of some virtual variables,
communication and self- diagnostics, are stored
in EPROM. Minute averages of all the analog and
digital inputs as well as alarms and equipment
status are stored in the bubble memory using a
FIFO structure.

The development computer is used to retrieve the
data stored in the bubble memory, to perform data
analysis or to run control programs during the
test phase. This computer can be used with on-
line data or in stand-alone applications.

Since the main effort was placed in developing a
system which permits data handling and
manipulation by metallurgists and control
engineers, the operator—machine interface is
quite limited. It allows changes in control
strategy parameters, activation of the computer
control and display of two pages of numerical
figures of the main process variables.

Some difficulties arose in the communication
between the computer and the analog controllers,
to provide an effective analog backup and an easy
and bumpless computer-analog control transfer and
vice-versa. A special interface card, designed by
CMD Instrumentation staff, was installed which
transfers the setpoint changes in the form of up
and down pulses to the existing controllers and
maintains the last setpoint value in case of
computer failure (Jerez, 1983).

Main Process Disturbances

Intrinsic ore hardness and feed size distribution
are the main disturbances. Feed size distribution
variations result from the following facts:

- An heterogeneous ore body with variable
  natural clast size and matrix distribution
  which results in different size distribution
  after blasting.

- Segregation in the coarse ore stockpile
  which on one hand has only a 6000 ton live
  charge and on the other is fed from one
  single point causing peripheral segregation
  of coarse particles. Due to the low live
  charge, any lengthy shutdown of the primary
  crusher forces the use of front loaders in
  the stockpile, which present batches of pure
  coarse or fine material to the mill.

- Start up and shutdown of the on-line crusher
  also changes significantly the amount of
  intermediate, say –1", particles fed to the
  mill.

The first fact can be considered as a long term
variation since the open pit operation delivers
batches of 20 to 60Ktons of the same ore to the
plant, usually without blending, meanwhile the
latter are short term, say hour to hour
disturbances.

The effects of changes in size distribution
caused by segregation or circuit configuration
are shown in Figs. 3 and 4 respectively.

Greater than the size distribution is the
influence of ore hardness. Throughput variations
from 300tons/hour to 600tons/hour are usual. Two
clearly defined ore types can be recognized in th
orebody. Both are composed by two main

lithological types (quartz monzonite and
monzodiorite). The hard ore presents a weak
sericitic alteration and a lower matrix

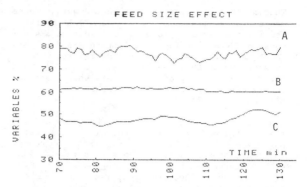

Fig 3.   Size Distribution Change Effect.
          (A = Feed Rate, B = Power Draw,
          C = % <1inch)

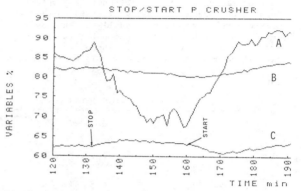

Fig 4.   Start and Stop Pebble Crusher Effect.
          (A = Feed Rate, B = B Pressure,
          C = Power Draw)

percentage. The soft ore presents a moderate
sericitic alteration and a higher matrix
percentage (Walker, 1984).

Attempts to characterize these ore variations
through the in pit drilling rate have been
successful. Good correlations between the
drilling rate and the plant unit grinding power
consumption (kwh/ton) were obtained. Nevertheless it has not been possible to include a
predictive parameter in a feed-forward scheme
into a control strategy.

Finally, environmental factors like winter time,
where the pit operation becomes irregular and
considerable proportions of ice and snow are fed
with the ore to the mill, are important
disturbances. Feeder blockage due to frozen ore
makes the feed to the plant unstable, but any
control scheme, including manual control must be
able to handle these situations.

Relationships Between Operating Variables

During a 6 month data logging campaign the
following relationships between process variables
have been investigated empirically.

Mill Power (J) vs. bearing back pressure (BBP)

At the beginning the typical power—load curve was
found as shown in Fig. 5.

This curve was obtained plotting J vs. BBP during
several overload situations and could be
characterized by a well correlated mathematical
expression.

After the startup of the new primary crusher and
the on-line pebble crusher, both contributing to
a finer feed to the mill, a new behaviour arose.

This is characterized by an almost constant power
and a rising pressure curve at constant feedrate.
Usually the bearing pressure rises during 3-4
hours and when the feedrate is turned off at this
point, an increasing power and decreasing
pressure shows that the mill was overloaded. This
implies that the specific instant, when overload
occurs cannot be detected under these
circumstances. Fig. 6.

Many visual inspections of mill charge showed
that at the same load level the mill takes less
power when the load is finer (or almost no coarse
rocks are present). Hence there is a family of J
vs. BBP curves each one depending on size
distribution of the load. This leads to
interpretation of this overload as shown in Fig.
7.

If feed or load size distribution becomes finer,
then overloading can occur without the typical
power down, BBP up scheme. It must be noted that
ball charge is maintained usually at the lowest
level, that is balls are added until full
available power can be drawn. Three years
experience showed that this ball level is about
9.5% balls by volume (Tarifeno and von Borries,
1984).

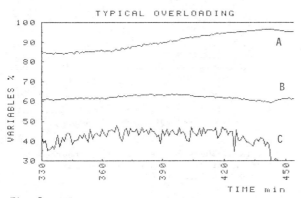

Fig. 5.   Typical Overload Condition.
          (A = B Pressure, B = Power Draw,
          C= Feed Rate)

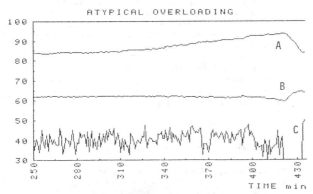

Fig. 6.   Atypical Overload Condition.
          (A = B Pressure, B = Power Draw,
          C= Feed Rate)

Mill Power vs. Fresh Feed Rate (W)

Steady state power was plotted versus throughput,
parameterising the relationship between these two
variables by the average kwh/ton (p) ratio of the
period. This ratio reflects all the effects like
intrinsic ore hardness, feed size distribution,
ball charge, pulp density, etc. Fig. 8 shows a
family of J vs. W curves parameterised by p. It
can be observed that dispersion is low at higher
p values and the average is also higher. This

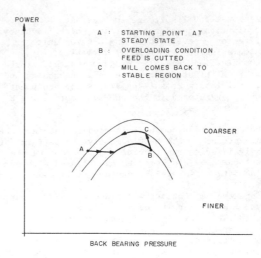

Fig. 7.   Atypical Overload Mechanism.

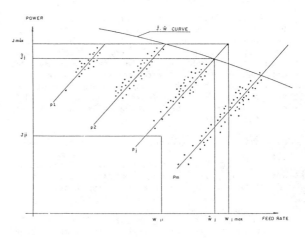

Fig. 8.   Power vs Feed Rate Parameterized by
          p(kwh/ton).

reflects essentially that at manual control, on
one hand operators take less care about power
when ore is "softer" (high throughput), and on
the other hand when ore is "softer" maximum
available power often cannot be reached without
overloading the mill.

This last fact was frequently observed and it is
attributed to the fact that "soft" ore is
characterised by a high breakage rate of coarse
particles, hence the mill load is finer and the
overload described in the previous sections
occurs. The maximum power which can be reached at
each p, is shown in Fig. 8 as an envelope curve.

Fresh Feed Rate vs. Feed Size Distribution

Steady state data of throughput were plotted vs.
one output channel of the MSD-95, again
parameterised by p. Influence was less than
expected but it can be masked since p also
receives a contribution from the feed size
distribution. Anyhow, the relationship shows
that finer feed ore media leads to higher
throughput, provided that there is no lack of
grinding media (see Fig. 9).

Other Variables

Circuit alternatives cause drastic changes in the
steady state and dynamic behaviour of the mill.
At steady state throughput demands changes when
another circuit is used which is quite obvious,

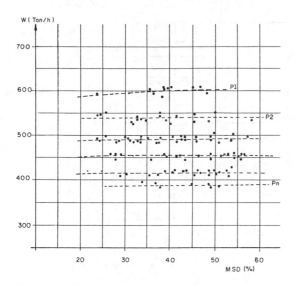

Fig. 9.   Feed Rate vs Feed Size Distribution
           Parameterized by p (kwh/ton).

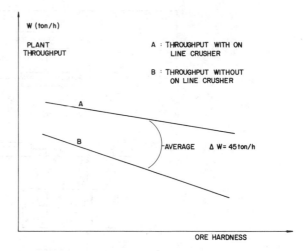

Fig. 11.   Contribution to Throughput by Pebble
            Crusher.

but the mill also shows different response times
when the circuit is perturbed (e.g. changes in
feed characteristics). All the effects due to
circuit changes have still to be studied which
demands a long data logging period.

Figures 10 and 11 show the contribution of the
secondary ball mills and on-line crusher
respectively on the base (SAB) circuit
throughput.

The previously presented relationships between
variables and quantification of effects are the
process support to the control strategy that
will be described in the next section. These
relationships were complemented with a great
number of step response experiments, mainly with
feed rate steps under different operating
conditions.

Control Objectives

In this particular case, control objectives were
stated as follows:

- Maximize mill throughput by using maximum
  available power.

- Maintain the mill operation in the stable
  zone, thus preventing overload.

- Maintain the cyclone overflow particle size
  below a maximum.

Secondary objectives as a specific percent solids
in the mill (or optimum if there is any) can also
be mentioned.

At the development of the control scheme, the
mill and the sump-cyclone system have been
regarded as fairly decoupled systems and
interactions (which surely exist) have been
neglected. All the attention has been focused on
the mill control problem, that is to achieve the
first two objectives.

Control Strategy

The first control strategy which was tested in
August 1984 consisted of a power controller (PID)
cascaded to a feed rate controller. The output
of the power controller was added to an operator
fixed feedrate set point. This strategy resulted
in hunting cycles and could not be adjusted
properly. Mill overload could neither be
prevented nor be successfully carried back to
stable state. From this point on, the
development of a control scheme was divided into
a direct level and a supervisory level.

Figure 12 shows the direct control level. It
consists basically of a PID for the power
control. This controller can receive its set
point from the operator or from the supervisory
routine. Need was detected, to make the
proportional and integral gain factors error

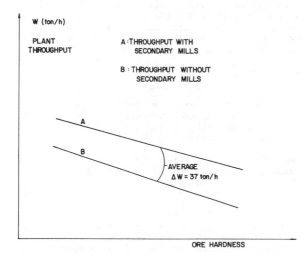

Fig. 10.   Contribution to Throughput by Secondary
            Ball Mills.

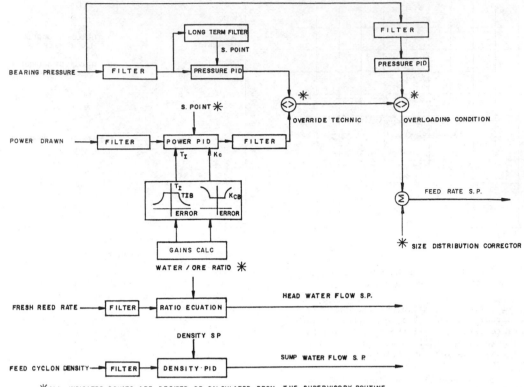

Fig. 12. Direct Control Level Structure.

dependent including a dead band. Linear equations used at the beginning were shown to be insufficient. Thus, logarithmic equations were chosen arbitrarily. The integral base gain showed to be ore dependent, that is with higher p (kwh/ton) higher integral factors were obtained. This has not been definitely included in the control scheme since more step response experiments have to be carried out under different ore hardness conditions.

In parallel with this power controller, there is a BBP controller which takes the control over when it is activated by the supervisory routine. Since the general level of the BBP signal depends on factors like the temperature of the oil, and the settings of some valves in the oil circuit, no absolute BBP value can be established as a set point. Hence a strongly filtered signal is used as set point to this controller (Toro, von Borries and Jerez, 1984).

The direct level generates the following set points:

- Fresh feed rate as explained before.
- Water flow to the SAG mill from water/ore ratio fixed by the supervisory routine.
- Water flow to the sump according to a manual fixed cyclone feed density set point.

In future water addition to the sump will be related to the overflow particle size indicated by the PSM 100.

Supervisory Control Level

This level performs basically the following tasks:

- It avoids overload and checks if the mill is operating in an overload condition (Anderson, 1981), (Mular, 1979), (McManus and Allen, 1977).

- It fixes the water/fresh feed ratio.

- It corrects the feedrate set point if feed size distribution changes.

For the first task the routine checks periodically (each 5 minutes) the sign and value of the J and BBP first derivatives and compares these values with the patterns obtained from the J vs. BBP correlation explained before. In this way, the algorithm finds out if the mill is operating in an overload zone. If overload occurs two actions can be taken, which are chosen previously by the process engineer:

- Override by a BBP PID which brings the mill back to the stable region.

- Decrease the power set point automatically step by step during several 5 minute periods until the mill is again in a stable operating condition.

Once the mill is again in the stable region, control is returned to the power PID taking the last power reading as the set point, then, step by step, the power is again raised to the nominal one.

The atypical overload conditions are avoided by a second routine. BBP is filtered with a long term filter. The output of this filter is compared with the measurement and if the latter exceeds the former by a certain percentage, then the control is transferred to a BBP PID as explained in the direct control level. This PID will generate the ore feed set point until the BBP is again below the long term filter output. If this occurs, control is again transferred to the power control algorithm (Toro, von Borries and Jerez, 1984).

Water/ore ratio is modified by a "hill climbing" routine (Mular, 1979) which computes and compares the p (kwh/ton) value of two consecutive periods of say 30 minutes and increases or decreases the water ratio according to the obtained effect. Large experimental campaigns showed that the mill has a better performance in terms of p with lower dilutions. It seems to be beneficial to evacuate the fine particles from the mill constantly using high water flowrate.

Finally, from the correlation between ore feedrate and the MSD-95 signal a bias is generated when the feed size distribution changes. This bias is added to the feedrate set point.

Control Results

During a trial period manual control was compared with automatic control Three factors were considered:

- Stability: The control strategy maintains the circuit in a more stable condition, that is, feedrate variations are smoother than with manual control. Power deviation from the set point is lower with automatic control. Fig. 1 shows manual and control strategy reaction to an overload condition.

- Throughput: The trial has shown consistent throughput increase with the automatic control strategy. Table 1 shows the comparative results of manual (operator) and automatic control.

- Knowledge: The data logging system and the careful experimental plant work have improved the knowledge of the process. Intuitive control by the operators has become a control based on a real knowledge of the process phenomena.

Table 1 : Manual/Automatic Control Comparison

Period	Feed Rate (MTPH)	Mill Power (HP)	Kwh/Ton
Manual Before Automatic	550.3	6939	9.40
Automatic	585.1	7152	9.12
Manual After Automatic	568.8	7035	9.22

Conclusions and Future Development

A control strategy which produces a marginal throughput increase and more plant stability has been developed successfully at Disputada, solving the power control problem of a semi-autogenous grinding mill.

A combination of conventional control algorithms with override, adjustable tuning and interactive process tuning concepts have proven to be suitable for the SAG control problem. However, an extensive data logging and evaluation program had to be carried out to obtain the necessary process knowledge to support the control strategy. At present a minimum variance control algorithm based on a stochastic plant model for the power control is being tested. As a first step this algorithm has one input and one output. During 1985 the same concept with multivariable control will be applied (Box and Jennings, 1970), (Cuper, 1978) (Astrom and Wittenmark, 1973).

A large sampling campaign has been carried out between May 1984 and January 1985. Complete mass balances were obtained with size distribution of all the mass streams including mill load. The objective is to develop a phenomenological model to be included at the supervisory level into the control scheme. This work is being carried out by the JKMRC at Brisbane (Toro, 1984).

ACKNOWLEDGEMENTS

The authors wish to thank very especially the Operations Vice-President, Mr. Johann G. von Loebenstein and the Los Bronces Area Manager, Mr. Nelson Pizarro C., for having permitted the publication of the present work.

REFERENCES

Jerez, J.C., (1983). Internal Report SAG Process Control Phase I.

Walker, C.A., (1984). Internal Report by CMD Geological Staff.

Tarifeno, E.V. and von Borries, G.H., (1984). Analisis del Desgaste de Bolas en al Molina Semi-Autogeno de CMD. Armco-Chile IV Grinding Symposium.

Anderson, L., (1981). Island Copper Process Control SAG Grinding Circuit.

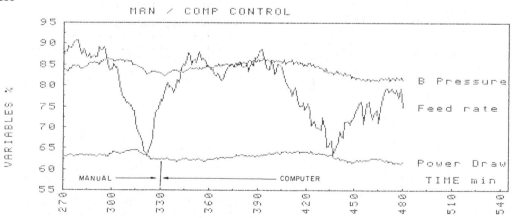

Fig. 13. Manual and Digital Control Comparison.

Mular, A.L., (1979). Automatic Control of
Semi-Autogenous Grinding Circuits in North
America. (Internal Report for Exxon Research
Co.).

McManus, J. and Allan, P., (1977). Process
Control in the Lornex Grinding Circuit. C.I.M.
District 6 Meeting, Victoria, B.C.

Box, G.E.P. and Jennings, G.M., (1970). G.M.
Time Series Analysis, Forecasting and Control
Holden Day.

Cuper, D.D., (1978). Control Suboptimo de
Varianza Minima incluyendo Ajuste de Parametros
para un Sistema de Fase No Minima. Universidad
de Chile.

Astrom, K.J. and Wittenmark, B., (1973). On Self
Tuning Regulators. Automatica, Vol. 9., 185-199.

Toro, H.C., (1984). Internal Report by CMD
Metallurgical Staff.

Toro, H., von Borries, G. and Jerez, J., (1984).
Control Strategy. Internal Report SAG Process
Control Phase II.

Multivariable Control of a Grinding Circuit

P.L. LEE and R.B. NEWELL

Department of Chemical Engineering, University of Queensland, St. Lucia, Qld. 4067

Abstract. Grinding circuits are used in most mineral processing plants. Their widespread application makes the efficient running of the circuits essential to almost all resource projects. This paper will show that, with the application of modern multivariable control techniques, improved plant performance can be obtained when compared with conventional single loop control strategies. Results are presented for a simulated closed-circuit ball mill.

The grinding circuit was controlled using the rod-mill water and primary and secondary sump dilution flowrates as the manipulated variables. Variables controlled were the primary cyclone feed flowrate, the primary cyclone feed density and the particle-size measurement (percentage smaller than 75 μm).

A multivariable dynamic matrix controller was compared with multiloop PI controllers.

Keywords. Grinding; predictive control; multivariable control systems; mining.

SUMMARY

Multivariable control of a grinding circuit has been demonstrated using a simulation study of a combined rod and pebble mill circuit. Dynamic matrix control was shown to produce better regulatory and servo response than conventional single-loop controllers. Responses were only marginally inferior to multivariable controllers designed by the more complex frequency domain techniques.

INTRODUCTION

The control of a grinding circuit is an important operation in most mineral processing plants. Often the grinding circuit can heavily influence the economics of downstream processing facilities, particularly flotation circuits.

Conventional control strategies for grinding circuits involve combinations of single loop controllers each measuring one variable and controlling one other variable. These strategies have been reviewed by Herbst and Rajamani (1982). As stated by these authors, interactions that occur between the single-loop controllers degrade the overall performance of the grinding circuit. These interactions have been overcome by de-tuning one or more of the controllers to maintain stable but "sloppy" control.

Two different approaches can alleviate this problem. The first of these involves the application of multivariable frequency domain methods to design decouplers (Hulbert et al 1980; Hulbert and Woodside 1983). This approach has used the Inverse Nyquist array method due to Rosenbrock (1974) and is characterised by the need for a sophisticated computer design package to efficiently design the decouplers. Once the decouplers have been designed, each loop behaves without affecting other loops and single loop

controllers can be applied to each in turn. Implementation is not usually very complex, especially with today's digital equipment.

The second approach taken to the interaction problem has been to apply state-space multivariable optimal control techniques. This approach has been typified by the work of Herbst and his co-workers (Herbst and Rajamani 1982). This technique relies on a linear model of the process to design the multivariable controller. In addition, as all the 'states' of the system are rarely measurable, implementation usually involves the use of some form of predictor or state-estimator, typically a Kalman filter. Also, the model derived from a mathematical description of the process is usually of high-order and non-linear so that considerable simplification of the model is required to make it amenable for control-system design. A software design package is also required to design these controllers, but again implementation is not difficult with modern instrumentation.

In this paper dynamic matrix control is used to control a simulated grinding circuit. Dynamic matrix control has been used successfully in the chemical process industries and was originally developed by Shell Oil. Descriptions of the technique are given by Cutler and his co-workers (Cutler and Ramaker 1980; Cutler 1981; Prett and Gillette 1980).

PROCESS DESCRIPTION

The process used in this study was that used by Hulbert et al (1980). A diagram of the grinding circuit is shown in Fig. 1. Hulbert et al developed a transfer function model of this process using step testing methods.

$$
\begin{bmatrix} \overline{\Delta PCF} \\ \overline{\Delta PCD} \\ \overline{\Delta PSM} \end{bmatrix}
=
\begin{bmatrix}
\dfrac{6.72}{1+1766s} & \dfrac{1.257}{1+61.3s} & \dfrac{0.1866}{1+573s} \\[2ex]
\dfrac{80.9}{1+1978s} & \dfrac{-3.61}{1+73.7s} & \dfrac{0.854}{1+654s} \\[2ex]
\dfrac{-5.25e^{-942s}}{1+1059s} & \dfrac{(0.255+176.8s)e^{-197s}}{1+302s+20900s^2} & \dfrac{(0.0657+122.4s)e^{-76s}}{1+329s+26300s^2}
\end{bmatrix}
\begin{bmatrix} \overline{\Delta RMF} \\ \overline{\Delta PD} \\ \overline{\Delta SD} \end{bmatrix}
$$

$$
\begin{bmatrix} \overline{\Delta PCF} \\ \overline{\Delta PCD} \\ \overline{\Delta PSM} \end{bmatrix}
=
\begin{bmatrix}
\dfrac{1.006-608s}{1+6850s+7\,800\,000s^2} \\[2ex]
\dfrac{4.57-7990s}{1+3750s+3\,350\,000s^2} \\[2ex]
\dfrac{-0.00406+733s}{1+2320s+1\,330\,000s^2}e^{-390s}
\end{bmatrix}
\overline{\Delta PEB}
$$

This transfer function model represents the regressed best-fit against the true plant data. For the purposes of dynamic matrix control, the actual step test data could be used, but this was not available to these authors.

DYNAMIC MATRIX CONTROL (DMC)

Dynamic matrix control, a multivariable control technique, was developed by the Shell Oil Company and principally expounded in the literature by Cutler and his co-workers (Cutler and Ramaker 1980; Prett and Gillette 1980; Cutler 1981). Marchetti et al (1981) also produced a comparison between DMC and model-algorithmic control which is closely related to DMC. For a detailed explanation of DMC the reader is referred to the above references and only a brief outline is given in this paper.

Consider a model of a process given by

$$\underline{E} = \underline{A} \cdot \underline{I} \tag{1}$$

where $\underline{E}$ is the error projection vector $((p*n*)1)$ of p controlled variables n time steps in the future

$\underline{I}$ is a vector $((m*q)*1)$ of the change in the q manipulated inputs for the m moves of these inputs

and $\underline{A}$ is the dynamic matrix of the summed convolution response of the system ie. the step test response data.

The choice of values for n and m is a design and/or tuning decision discussed by Marchetti et al (1981).

The least squares solution of equation (1) is given by

$$\underline{I} = (\underline{A}^T\underline{A})^{-1}\underline{A}^T\underline{E} \tag{2}$$

This is used as the control algorithm. The solution of the matrix $(\underline{A}^T\underline{A})^{-1}\underline{A}^T$ can be obtained off-line.

Although the method calculates values of the manipulated inputs n time steps into the future, in practice only the first step is calculated and applied to the plant. At the next sampling interval a further one-step calculation is made with an updated input vector.

In addition to using the feedback strategy described above, feedforward control can also be incorporated. This is achieved by augmenting the dynamic matrix $\underline{A}$ with the effect of the measured disturbance input. In addition, the input vector would have to include the change in the disturbance variables.

RESULTS AND DISCUSSION

The model of the grinding circuit was simulated on a PDP 11/44 computer with a program written in FORTRAN. Two different control strategies were employed - a DMC controller and three single-loop Proportional-plus-integral controllers.

A close examination of the model equations in section two reveals that the response of the system outputs to changes in the rod-mill feed rate is slower than that for the other two manipulated inputs. Also the response of the particle size measurement variable to changes in all inputs is slower than that of the other two output variables. These features of the system created difficulties in selecting a suitable time step size to include sufficient detail of the response of the fast output variables, while containing the dimension of the dynamic matrix to a reasonable size. This was resolved in this study by taking a pragmatic view. Since the effects of particle size and rod-mill feed rate were considerably slower, these were paired in a single-loop controller. A DMC was applied to the remaining input and output variables. The DMC controller included feed-forward control by including the effect of changes in the pebble mill feed rate.

The pairings of input and output variables for the three single-loop controllers were determined using the Bristol (1966) relative gain array method. This resulted in the relative gain array shown in Table 1. The pairings were derived by choosing the values that were closest to one. These are shown as the diagonal elements in this array. Ths size of the off-diagonal elements indicate the degree of interaction between the control loops.

TABLE 1   Relative Gain Array

	RMF	PD	SD
PSM	.609	-.158	.549
PCF	.203	.818	-.0208
PCD	.189	.341	.471

The Proportional-plus-integral controllers, both for the DMC study and the three single-loop controller study were tuned empirically. The tuning parameters for the PSM-RMF loop were the same in both studies.

The results of the studies are shown indicatively in Fig. 2 and Fig. 3. Figure 1 shows the response of the three output variables to a step change in the pebble feed rate. Figure 2 shows the response of the same output variables in response to a step change in the setpoint of the particle size leaving the circuit. Both of these studies indicate that the DMC produced better controlled responses than the three single-loop controllers. The DMC responses are less oscillatory and return faster to the chosen setpoint. They also exhibit less interaction amongst the output variables.

The results for the DMC controller were also compared with those obtained by Hulbert et al (1980) using a decoupler designed by the use of inverse Nyquist arrays and 3 PI control loops. The responses with the DMC appear to be only marginally less stable but do possess more interaction amongst the output variables. This slight degradation in performance was obtained at considerable gain in ease of design and implementation.

CONCLUSIONS

There are a number of conclusions to be drawn from this work. These are:

(i)   The DMC produced better control performance than three single-loop controllers.

(ii)   The DMC did not produce as good control as the Inverse Nyquist array decoupler combined with 3 single-loop controllers.

(iii)   The DMC design method is simpler than the Inverse Nyquist Array method.

(iv)   Further work is required to develop guidelines for the tuning parameters in the DMC method. The tuning used in this study was based on the work of Marchetti et al (1981) and on trial and error.

These conclusions must be tempered with the acknowledgement that this work was carried out using a simulation of the system. Full plant verification is required.

REFERENCES

Bristol, E.H. (1966). On a new measure of interaction for multivariable process control. IEEE Trans. Auto. Control, Vol. AC-11, No. 1, pp. 133-134.

Cutler, C.R. (1981). Dynamic matrix control of imbalanced systems. Proc. ISA Conf., St. Louis, USA, pp 51-56.

Cutler, C.R. and Ramaker, B.L. (1980). Dynamic matrix control - a computer algorithm. Joint Automatic Control Conf., San Francisco, USA, p. WP5-B.

Herbst, J.A. and Rajamani, K. (1982). The application of modern control theory to mineral-processing operations. Proc. 12th CMMI Congress, Johannesburg, South Africa, pp. 779-792.

Hulbert, D.G., Koudstaal, J., Braae, M. and Gossman, G.I. (1980). Multivariable control of an industrial grinding circuit. 3rd IFAC Symposium, Montreal, Canada, pp. 311-315.

Hulbert, D.G. and Woodside, E.T. (1983). Multivariable control of a wet-grinding circuit. AIChEJ, Vol. 29, No. 2, pp. 186-191.

Marchetti, J.L., Mellichamp, D.A. and Seborg, D.E. (1981). Predictive control based on convolution models. 74th Annual AIChE Meeting, New Orleans, USA.

Prett, D.M. and Gillette, R.D. (1980). Optimisation and constrained multivariable control of a catalytic cracking unit. Joint Auto. Control Conf., San Francisco, USA, p. WP5-C.

Rosenbrock, H.H. (1974). Computer-aided control system design. Academic Press.

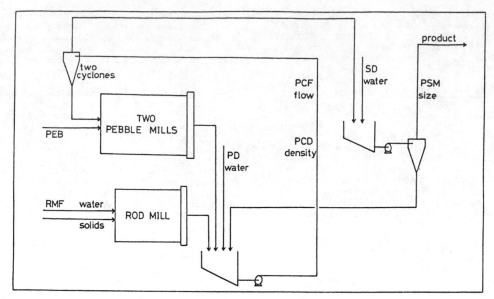

Fig. 1. Diagram of the grinding circuit

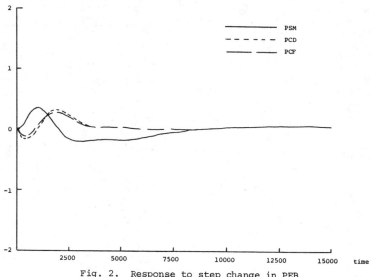

Fig. 2. Response to step change in PEB

(a) DMC control

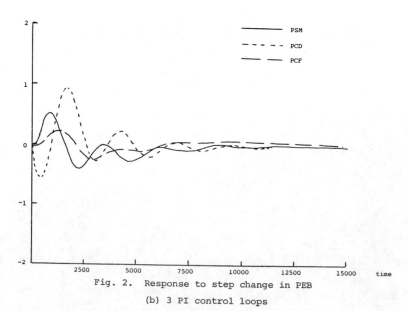

Fig. 2. Response to step change in PEB

(b) 3 PI control loops

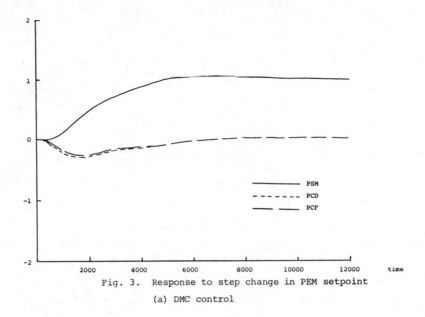

Fig. 3.   Response to step change in PEM setpoint

(a) DMC control

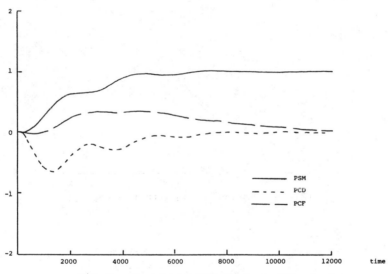

Fig. 3.   Response to step change in PSM setpoint

(b) 3 PI control loops

Parameter Adaptive Modelling of Mineral Beneficiation Processes

T.M. ROMBERG

Division of Mineral Physics, CSIRO, Lucas Heights Research Laboratories, Sutherland, NSW, Australia

Abstract. Data logged on mineral beneficiation processes during normal operation are analysed using time series (correlation and spectral) methods to identify the dynamics of the processes for on-line computer control. Recursive least squares estimation techniques are used to identify parametric models for a cassiterite flotation circuit and a coal washery. Identification problems such as the determination of model order, process time delays and the 'contamination' of the data by process recycle streams (feedback) are discussed. Typical results are presented in which optimised models are identified from data logged during normal operation.

Keywords. Closed loop; correlation; feedback; feedforward; models; multivariate; parameter estimates; spectral estimates.

INTRODUCTION

Data logged on a mineral beneficiation process during normal operation can often provide important information on the dynamics of the process; this information can be analysed to identify models suitable for on-line computer control. The scope for improving the computer control of mineral processes has been considerably enhanced in recent years by the development and installation of radioisotope gauges which enable accurate on-line assays of the minerals to be monitored at key locations in the beneficiation process. However, the natural fluctuations in the process assays, mass flow rates and densities are often 'contaminated' by recycle streams which complicate the identification procedures.

This paper discusses the identifiability of parametric models from data logged on closed loop processes. Particular emphasis is given to recursive parameter estimation procedures, which generally assume a priori knowledge of the process time delays and the order (structure) of the autoregressive moving average (ARMA) feedforward and feedback (recycle) models. A methodology is described for determining this information from operating data, and typical results are presented in which optimised models are identified from data logged on a cassiterite flotation circuit and a coal washery.

DATA ANALYSIS PROCEDURES

In the succeeding sections, the time series data logged on the mineral processes are analysed as follows:

(1) the mean value of each record is computed and subtracted to obtain the fluctuations from the mean operating condition;

(2) each record is then bandpass filtered with a digital filter to minimise nonstationary drift and aliasing;

(3) pairs of records are analysed using correlation and spectral methods to determine the degree and frequency of interaction (Jenkins and Watts, 1969; Priestley, 1981);

(4) this information, coupled with a priori knowledge of the process, is used to formulate multivariate models of the process, which are investigated using multivariate spectral methods (Romberg, 1978, 1979; Romberg and Jacobs, 1980; Romberg and Rees, 1981); and

(5) the dominant interacting mechanisms are then modelled using recursive parameter estimation techniques (Isermann, 1981, 1982).

The relevant equations for computing the covariance and spectral estimates are given in the standard texts and source references.

PARAMETRIC MODELLING

Parametric modelling problems that need to be resolved in practical closed loop systems are the model structure (order) and the magnitude of inherent time delays. The time delays may be estimated by crosscorrelation methods, and the process model structure can be determined from open loop model order indicators when the structure of the feedback process or controller is known. However, the feedback structure is often unknown in practice, and a simulation model or other a priori information of the process is required for the determination of the model structure.

The output of a bivariate process may be represented by an autoregressive moving average (ARMA) model of the form (Isermann, 1981, 1982)

$$
\begin{aligned}
y(k) = \sum_{j=1}^{m} & \left[-a(j)y(k-j) + b(j)u(k-j-d) \right. \\
& \left. + c(j)v(k-j) \right] + v(k) \quad k-j>d \\
= \; & 0, \qquad\qquad\qquad\qquad k-j<d
\end{aligned}
\tag{1}
$$

where $u(k)$ $(k=1,N)$ is the measured input, $v(k)$ is the nonmeasurable residual output noise, and d is the time delay through the process. The second line of equation (1) is the physical realisability condition imposed on the model. Equation (1) can be written in shorthand shift operator notation as

$$y(k) = G(B)u(k) + Z(B)v(k) \qquad (2)$$

where $G(B) = B(B)B^d/A(B)$, $Z(B)=C(B)/A(B)$,

$$A(B) = 1 + a_1B + a_2B^2 + \ldots + a_mB^m$$

$$B(B) = \quad b_1B + b_2B^2 + \ldots + b_mB^m$$

$$C(B) = 1 + c_1B + c_2B^2 + \ldots + c_mB^m \qquad (3)$$

and B is a backward shift operator, $y(k-1) = By(k)$. When $C(B) = 1$, the uncorrelated residual noise is assumed to be 'white', and the parameters are estimated by a recursive least squares (RLS) procedure. If the residual noise is 'coloured', then the RLS parameter estimates are biassed, and they are best calculated by recursive extended least squares (RELS) or recursive maximum likelihood (RML) procedures (Isermann, 1981).

The input-output crosscovariance is given by the convolution equation

$$Cuy(i) = E\left[u*(k)y(k+i)\right] \qquad 0<i<L$$

$$= \sum_{j=1}^{m} \left[-a(j)Cuy(i-j)\right.$$
$$\left. + b(j)Cuu(i-j-d)\right] + Cuv(i) \qquad (4)$$

$$\hspace{4cm} i-j>d$$

$$= Cuu(i)G(B) = Cuv(i)$$

$$= 0, \qquad\qquad\qquad i-j<d$$

where $E[\ ]$ is the expected value operation, and (*) denotes the complex conjugate. In general, the input u(k) is related causally to the residual noise v(k) via the feedback process, and only 'contaminates' the crosscovariance estimates in the negative lag domain of equation (4). This follows from the physical realisability constraint that the present and immediate past values of v(k) are not causally related to past values of the process input u(k), and thus do not contribute to the crosscovariance estimates in the positive lag domain of equation (4). However, there will be a distortion of the crosscovariance estimate at the zero lag axis if the feedback process delay is zero. This constraint requires that v(k) must be 'white' (broadband) or at least 'pink' (bandlimited) with no deterministic components. A further requirement is that the closed loop system be linearly stable.

This rationale differs slightly from that presented by Box and MacGregor (1974), who maintained that past values of the residual noise can affect the positive lag crosscovariance estimates via the feedback process. However, this can only occur in practice if the residual noise is sufficiently 'coloured' to be correlated with the previous residual noise that has passed through the feedback and feedforward processes, that is, the order of the noise model, C(B), is greater than the order of the combined feedback and feedforward processes. This situation can never arise with the ARMA model definition given in equation (1).

CASSITERITE FLOTATION CIRCUIT

The data analysed in this section were logged on the cassiterite circuit at Renison Ltd, Zeehan, Tasmania. A schematic diagram of the circuit and instrumentation is shown in Fig. 1. Cassiterite is an oxide of tin, which does not float as strongly as the more common sulphide minerals. Consequently, the particle size is kept small, entrainment is high, and the circuit operates with high levels of recycle.

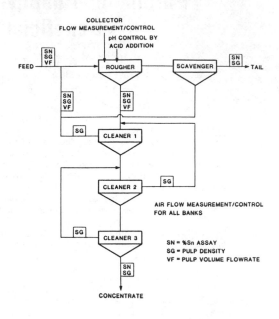

Figure 1. Layout and instrumentation of cassiterite flotation circuit.

Radioisotope on-stream analysis (ROSA) probes were used to give density and tin assays at 5 minute intervals; these were logged together with the air and pulp flow rates. The variables were analysed in a previous study (Sutherland, Romberg and Selby, 1984) using correlation and multivariate spectral analysis techniques (Romberg 1978). The spectral density estimates, computed from the digitised data after removal of their mean values and bandpass digital filtering in the frequency range 0.001-0.6 cycles per hour (cph), indicated that the dominant information was in the 0-0.25 cph bandwidth. There were several possible models for the process variables, but only those that were investigated are listed in Table 1. This

TABLE 1   Cassiterite Flotation Models

INPUTS	OUTPUT
New feed tonnage (tph) New feed concentrate (%Sn) Rougher and scavenger air	Final tails concentrate (%Sn)
New feed tonnage (tph) New feed concentrate (%Sn) Rougher air	Rougher concentrate (%Sn)
Rougher concentrate (%Sn) Cleaner feed (tph) Cleaner air	Final concentrate (%Sn)

study showed that the process dynamics were dominated by the new feed flow variations, which were almost cyclic with a period of 27 hours, and were caused by upstream process changes. This is

demonstrated by Fig. 2, which shows the partial
and multiple coherence estimates computed for the
first model relating feed tonnage, feed assay and
bank air supply (inputs) to the tailings assay
(output). The feed tonnage has a dominant effect
on the tailings assay at low frequencies (<0.1
cph); an increase in tonnage overloads the circuit; and there is a consequent fall in recovery
and a rise in the tailing assay. The bank air
flow influenced the tailings assays at intermediate frequencies (0.1-0.2 cph), but the associated
power spectral density estimates indicated there
was no significant information above 0.15 cph, so
for practical purposes, the effect of the air was
of secondary importance, and may be neglected.
Only parametric aspects of the first model are
discussed in the present context.

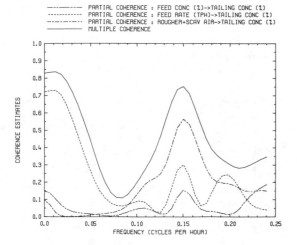

Figure 2.   Partial and multiple coherence
            estimates for MISO model relating feed
            assay, feed tonnage and rougher plus
            scavenger air to tailings assay.

The single input single output (SISO) relationship
between feed tonnage and final tailings assay was
modelled using RLS and RELS parameter estimation
procedures (Isermann, 1981, 1982). The order of
the ARMA model, equation (1), was investigated
using Akaike's information criterion (AIC –
Akaike, 1974), the P-matrix log estimation error
variance norm (EVN), log geometric error variance
norm (GEVN) and the coefficient of determination
(COD – Young, Jakeman and McMurtries, 1980). A
typical case showing the RLS/COD model order
results for a range of time delays is presented in
Fig. 3; the maximum delay was deduced from the
time lag of the maximum input-output crosscovariance estimate. The independence of all models for
a time delay below 20 minutes is caused by the
dominance of the autoregressive terms in the
model, and is characteristic of an unstable system
or, alternatively, one driven by a deterministic
input. For example, an ARMA(2,2) model with a
time delay of 20 minutes gives satisfactory
results, as shown by the one-step ahead model predictions (Fig. 4) based on the asymptotic parameter estimates given in Table 2. The RLS parameter estimates have autoregressive roots outside
the unit circle (0.994 and -1.287), whereas the
corresponding RELS model has one autoregressive
root just inside the unit circle (0.387 and
0.975). The moving average parameters almost
cancel each other, and mainly contribute to the
'dither' on the predicted output.

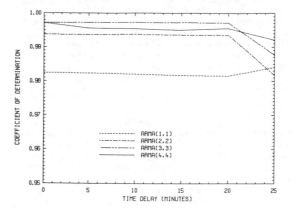

Figure 3.   Cassiterite circuit RLS/COD model order
            results as a function of time delay.

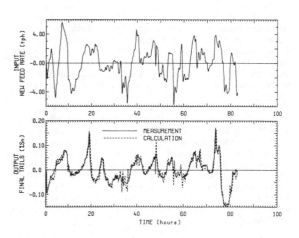

Figure 4.   Comparison of measured and calculated
            tailings assay.

TABLE 2   Feed Tonnage → Final Tails Parameter
          Estimates

Model	Parameter	Stand. Dev.
Recursive Least Squares (RLS):		
Autoregressive	0.2924	0.0034
	-1.2793	0.0046
Moving Average	0.0497	0.0001
	-0.0496	0.0002
Recursive Extended Least Squares:		
Autoregressive	-1.3621	0.0032
	0.3772	0.0019
Moving Average	0.0102	0.0000
	-0.0100	0.0000
Residual Noise	0.9746	0.0607
	0.0044	0.0007
Time Delay	20 minutes	

These results highlight the problems of identify-
ing parametric models from normal operating data.
Ideally the data should be 'persistently
exciting', that is, they should have sufficiently
broadband frequency content to excite all the
dynamic modes of the process. The air flow rates
to the rougher, scavenger and cleaner banks are

the most significant control variables affecting
flotation behaviour, and need to be perturbed
about their set points by independent pseudorandom
binary sequences. This will ensure that the
'control' model has consistent parameter estimates
for the development of a suitable parameter
adaptive control algorithm.

COAL WASHERY

The data analysed here were logged on the coal
washery at Ulan Coal Mines Ltd, Ulan, New South
Wales. The 1200 tonnes per hour washery has two
modules in parallel, each module consisting of
heavy media baths, cyclones and fine coal flotation circuits, and produces steaming coal with a
nominal product ash content of 17 weight per
cent. A COALSCAN on-line ash measurement system
monitored the ash in the product stream of the
washery (Smith and co-workers, 1983), and this was
logged at one minute intervals in conjunction with
the feed rates, heavy media bath and cyclone
densities, and product feed rate.

Similar time series and recursive parameter estimation procedures were adopted for the data analysis. A multivariate model relating feed
rate, heavy media bath density and cyclone density
(inputs) to the ash in product (output) showed
that the feed rate had a minimal influence on the
plant performance, so it was eliminated from the
model. The partial and multiple coherence estimates for the resulting model are presented in
Fig. 5. The variations in bath density (or
specific gravity) and cyclone density both
influence the product ash at low frequencies (< 3
cph). However, inspection of the crosscovariance
estimates between the bath and cyclone densities
showed that they were highly correlated (0.54)
owing to operator action, and are dependent
inputs. These results suggest that the bath
density is the dominant variable affecting the
product ash.

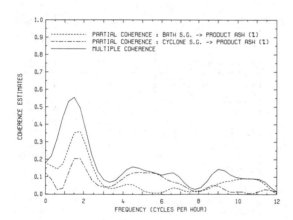

Figure 5.   Partial and multiple coherence
            estimates for MISO model relating heavy
            media bath and cyclone specific
            gravities to product ash.

A survey of RLS and RELS parametric models
relating heavy media bath density to product ash
movements was performed in a similar manner to
that given in the previous section. The RLS/COD
model order results for time delays up to 15
minutes is shown in Fig. 6. The one-step ahead
ARMA(2,2) model predictions with a time delay of
12 minutes is given in Fig. 7; this delay is due

mainly to the COALSCAN sampling system. The RLS
and RELS parameter estimates are given in Table 3.

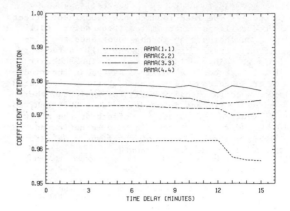

Figure 6.   Coal washery RLS/COD model order
            results as a function of time delay.

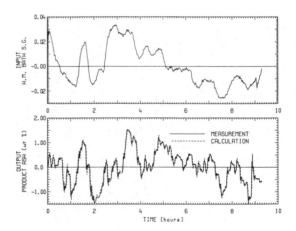

Figure 7.   Comparison of measured and calculated
            product ash.

TABLE 3   Bath Density → Product Ash Parameter
          Estimates

Model	Parameter	Stand. Dev.
Recursive Least Squares (RLS):		
Autoregressive	−1.5219	0.0059
	0.5741	0.0046
Moving Average	6.4106	0.1510
	−4.9023	0.1717
Recursive Extended Least Squares:		
Autoregressive	−1.3593	0.0053
	0.4038	0.0041
Moving Average	7.8517	0.1468
	−6.3400	0.1606
Residual Noise	0.7436	0.0253
	−0.3214	0.0171
Time Delay	12 minutes	

One of the problems encountered in a reported
attempt at real time computer control of a coal
washery (Lyman and co-workers, 1983), has been the
estimation of the local gradient in the heavy
media density-ash relationship at the nominal
steady state operating conditions. If the heavy

media bath density, u(t), is related linearly to
the product ash, y(t), by the equation

$$y(t) = b \, u(t) + v(t) \qquad (5)$$

where v(t) is the uncorrelated noise, it follows
from the above definitions that

$$Cuy(k) = b \, Cuu(k), \qquad 0 < k < L \qquad (6)$$

Thus, the local gradient (or controller gain
factor) can be determined recursively on-line from
a plot of the density-ash crosscovariance as a
function of the density autocovariance, as shown
in Fig. 8. The gradient of 3.4 wt% ash/0.1 change
in specific gravity (S.G.) computed by this method
from the time series data agreed favourably with
that obtained from laboratory assays of 3.2 wt%
ash/0.1 change in S.G..

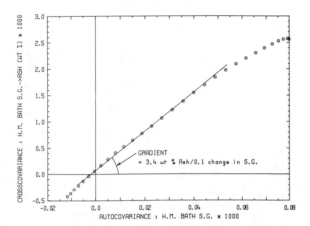

Figure 8. Graph showing the linear relationship
between the specific gravity of the
heavy media baths and the ash in
product coal (weight %) using
correlation techniques.

CONCLUSIONS

The data logged on a cassiterite flotation circuit
and a coal washery during normal operation gave
useful information on the process dynamics when
analysed by time series and recursive parameter
estimation procedures. The cassiterite flotation
circuit was predominantly characterised by its new
feed tonnage to final tailings concentrate
relationship, which was modelled by an ARMA(2,2)
model with a time delay of 20 minutes. However,
the RLS and RELS models were dominated by the
autoregressive parameters, and consistent moving
average parameters could not be estimated from the
deterministic feed tonnage and tailings concentrate data.

The coal washery was predominantly characterised
by its heavy media bath density (specific gravity)
to product ash relationship, which was modelled by
an ARMA(2,2) model with a time delay of 12 minutes
(mainly due to the COALSCAN gauge sampling system). A linear relationship between bath density
and the product ash was also obtained from the

data for on-line computer control applications.

ACKNOWLEDGEMENTS

The author gratefully acknowledges helpful
discussions with Mr. R. Winby and Mr. W. Selby
(Renison Ltd) and Mr. H. Crowden (Ulan Coal Mines
Ltd), and for their co-operation in providing the
data used in this investigation.

REFERENCES

Akaike, H. (1974). A new look at statistical model
    identification. IEEE Trans. Auto. Control
    AC19, 716-722.
Box, G.E.P. and MacGregor, J.F. (1974). Analysis
    of closed-loop dynamic-stochastic systems.
    Technometrics 16(3), 391-398.
Isermann, R. (1981). Digital Control Systems,
    Springer-Verlag (Berlin).
Isermann, R. (1982). Parameter adaptive control
    algorithms - a tutorial. Automatica, 18(5),
    513-528.
Jenkins, G.M. and Watts, D.G. (1969). Spectral
    Analysis and its Applications. Holden-Day,
    San Francisco.
Lyman, G.J., Denney, B., Wood, C.J., Askew, H. and
    Brenchley, R. (1983). Automatic control of
    product ash content using an on-line coal ash
    gauge. Proc. Second Australian Coal
    Preparation Conference, Rockhampton,
    pp. 291-309.
Priestley, M.B. (1981). Spectral Analysis and Time
    Series. Academic Press, London.
Romberg, T.M. (1978). An algorithm for the
    multivariate spectral analysis of linear
    systems. J. Sound & Vibration, 59(3),
    395-404.
Romberg, T.M. (1979). Identification of lead
    flotation plant dynamics from normal
    operating data. Proc. Conf. on Control
    Engineering, Melbourne, pp. 18-21.
Romberg, T.M. and Jacobs, W.S.V. (1980).
    Identification of a mineral processing plant
    from normal operating data. In J. O'Shea and
    M. Polis (Eds.), Automation in Mining
    Mineral and Metal Processing (IFAC). Pergamon
    Press, Oxford. pp. 275-282.
Romberg, T.M. and Rees, N.W. (1981). On the
    modelling of boiling channel dynamics using
    spectral methods. Int. J. Control, 34(2),
    259-284.
Smith, K.G., Crowden, H., Sowerby, B.D., Lyman,
    G.J., Howarth, W.J. (1983). Performance of a
    COALSCAN on-line ash measurement system at
    Ulan Coal Mines Ltd, Australia. EPRI Seminar
    on Principles and Applications of Continuous
    Coal Analysis, EPRI Report CS-989, Vol. 13,
    Project 983.
Sutherland, D.N., Romberg, T.M. and Selby, D.W.
    (1984). Experiences with time series analysis
    techniques used to estimate the dynamics of a
    flotation plant. Proc. 12th Australian
    Chemical Engineering Conference, Melbourne,
    775-781.
Young, P.C., Jakeman, A. and McMurtries, R.
    (1980). An instrumental variable method for
    model order identification. Automatica,
    16(3), 281-294.

Comminution Kinetics for Multi-Mineral Ores

O.N. TIKHONOV

Department of Metallurgy, Leningrad Mining Institute, USSR

Abstract. Comminution Kinetics equations are developed to take into account difference of grindability of mineral components. The equations describing both batch and closed circuit grinding or crushing processes are given. These equations are applied to industrial data to derive comminution kinetic and classification characteristics of the component minerals in a multi-mineral ore. Control strategy based on these concepts is presented.

INTRODUCTION

Comminution Kinetics equations may be made more accurate by means of taking in account difference of grindability (hardness, softness) of mineral components. Instead of usual differential grain size distribution $\gamma_{us}(\ell)$ more detailed two-dimensional grain size (ℓ) and grindability distribution (ξ) is involved. Various measures of grindability (ξ) may be chosen (limit tension, kg/cm^2), consumed energy for grinding from initial to final conventional size, Joule/kg, inverse of Rittinger's constant, $\xi = \frac{1}{a}$ mm.sec., etc.

The usual kinetic comminution equation (A. Zagustin, 1935) is then written in a more general form to explicitly include the grindability of the different minerals making up the ore. Mathematical details of how this is done for batch processes are given in Appendix A.

For closed grinding or crushing circuits the batch equations are rewritten in terms of the fresh feed, circulating load and mill (or crusher) discharges. A complete solution is obtained by including with these equations the classification performance curves. Details of these equations are given in Appendix B.

INDUSTRIAL EXAMPLES

Initial experimental data for an industrial grinding circuit (ball mill $42m^3$, spiral classifier $36m^2$) and a multi-mineral ore are in Table 1 (there are other minerals FeS_2, SiO_2).

Calculated data - through Eq.s (12) and (13) of Appendix B are in Table 2. Corresponding graphs are in Fig. 2a. Two more industrial examples are in Fig. 2b,c. $Q_{rel}(\ell)$ is the tph of particles broken into all sizes less than ℓ, and $\Sigma(\ell)$ is the fraction of particles of size ℓ reporting to classifier final product, i.e., overflow.

Such simplified comminution kinetic (Q_{rel}) and classification (ϵ) characteristics are applicable for prediction grinding behaviour of different minerals of multi-mineral ores.

For $Q_{rel}(\ell)$-curves in Fig. 1a, b it is possible to apply linearization using Rittinger's constants, e.g., in Fig. 1a:

$A_{ore} = 0.007$ mm^{-1}sec^{-1}, $A_{Baso4} = 0.014$ mm^{-1}sec^{-1} etc., $A_{ore} = \sum_{i=1}^{n} C_i A_i$

CONTROL STRATEGY

It follows from the kinetics presented in this paper the ball mill-and-classifier process the rate of grinding i.e., rate of production of the class $\ell < \ell_{50}$ is

$$Q_f C_f = \bar{\epsilon} A_{ore} \ell_{50} M \qquad (14)$$

where ℓ_{50} = cut point of the classifier, $\bar{\epsilon}$ = average classifier efficiency within $0 < \ell < \ell_{50}$.

Control strategy is to maximize

$\bar{\epsilon} A_{ore}/N \rightarrow max$ or to maximize

$\bar{\epsilon} \int_{\ell_{50}}^{\ell_{max}} Q_{rel}(\ell) d\ell/N \rightarrow max$

by means of manipulating such inputs as mill hold up, M, charge of balls M_ℓ, pulp density. This may be performed by stepwise extrema search.

For the crushing closed circuit process (crusher and screen) the computer optimal control additionally to $\bar{\epsilon} A_{ore}/N \rightarrow max$ deals with manipulating the crusher opening ℓ_o and/or the fresh feed rate Q_{fr}. For that purpose predicting kinetic calculations of Q_c, $\gamma_c(\ell)$ $\gamma_f(\ell)$ are performed by computer:

$$Q_c = Q_{f\gamma} \left[\ell_n(\ell_{max}/\ell_o) + \ell_o/\ell_{max}{}^{-1} \right];$$

$$\gamma_c(\ell) = \begin{cases} (1/\ell - 1/\ell_{max})/\left[\ell_n(\ell_{max}/\ell_o) - 1\right] & \text{for } \ell > \ell_o, \\ 0 & \text{for } \ell < \ell_o; \end{cases}$$

$$\gamma_f(\ell) = \begin{cases} 0 & \text{for } \ell > \ell_o. \\ 1/\ell_o - 1/\ell_{max} + (1/\ell_{max})\ell_n(\ell_o/\ell) & \text{for } \ell < \ell_o \end{cases}$$

Then the actual size distribution $\gamma_f(\ell)$ in final product is compared with the desired one $\gamma_{des}(\ell)$ and correction of ℓ_o and Q_{fr} is performed.

TABLE 1 Distribution of Minerals $BaSO_4$, $CaMg(CO_3)_2$, $CaCO_3$ in Grinding Circuit Products

Class ℓ mm	Class content in total product %	Mineral content, %			Class content in mineral, %		
		$CaMg(CO_3)_2$	$CaCO_3$	$BaSO_4$	$CaMg(CO_3)_2$	$CaCO_3$	$BaSO_4$
1	2	3	4	5	6	7	8
Fresh Feed							
-0,074	3,20	50,00	21,60	6,30	2,80	3,40	3,00
0,074-0,315	1,73	56,00	17,70	6,10	1,80	0,70	1,31
0,315-0,80	1,18	54,00	23,80	6,05	1,60	0,80	1,25
0,80-5,0	15,40	55,00	20,00	6,12	14,80	15,60	14,70
+5,0	78,49	58,00	20,10	6,13	80,00	79,90	79,74
Σ	100,0	57,48	20,27	6,16	100,0	100,0	100,0
Mill Discharge							
-0,074	22,40	53,00	17,80	11,30	18,67	21,80	42,70
0,0,74-0,315	20,80	64,50	16,00	4,28	20,58	18,30	14,90
0,315-0,80	21,60	66,50	18,50	4,70	22,58	21,90	16,80
0,80- 5,0	25,20	66,50	20,90	4,48	26,36	28,60	19,00
+5,0	10,50	71,50	17,10	3,90	11,81	9,40	6,60
Σ	100,0	58,00	18,28	5,96	100,0	100,0	100,0
Circulating Product (Classifier Underflow)							
-0,074	8,0	51,00	20,70	11,32	6,40	8,80	21,60
0,074-0,315	21,50	61,50	19,90	6,10	20,80	22,60	26,70
0,315-0,80	22,00	66,50	17,60	5,08	23,00	20,80	22,80
0,80 -5,0	34,20	65,00	19,40	2,34	35,00	35,00	16,00
+5,0	14,30	65,00	18,60	4,32	14,80	12,00	12,70
Σ	100,0	63,40	19,08	4,91	100,0	100,0	100,0
Classifier Overflow (Final Product)							
-0,074	46,00	53,10	17,80	10,82	46,50	36,30	81,40
0,074-0,315	32,00	51,00	28,00	3,04	31,30	37,10	14,80
0,315-0,80	20,00	53,00	26,00	1,08	20,20	23,00	3,40
0,80 -	2,00	51,20	29,00	1,00	2,00	2,60	0,40
Σ	100,0	100,0	22,80	6,15	100,0	100,0	100,0

TABLE 2   Calculations to Fig. 2a

Class ℓ, mm	$Q_m C_m \Gamma_m -$ $-Q_c C_c \Gamma_c -$ $-Q_{fr} C_{fr} \Gamma_{fr}$	$MC_m x$ $x(1-\Gamma_m),$ t	$Q_{rel}(\ell),$ 1/t	$\varepsilon(\ell)$
		BaSo$_4$		
-0,074	5,01	0,45	11,13	0,73
0,074-0,315	6,82	0,82	0,34	0,38
0,315-0,80	7,63	0,20	38,15	0,08
0,80-5,0	5,68	0,05	113,6	0,008
+5,0				
		CaMg(CO$_3$)$_2$		
-0,074	23,60	6,52	3,61	0,92
0,074-0,315	37,03	4,83	7,66	0,53
0,315-0,80	50,95	3,06	16,65	0,32
0,80 -5,0	49,54	0,88	66,29	0,02
+5,0				
		CaCO$_3$		
-0,074	8,71	1,97	4,42	0,95
0,074-0,315	12,14	1,48	8,20	0,78
0,315-0,80	16,21	0,93	17,43	0,40
0,80 -5,0	16,26	0,22	73,90	0,037
+5,0				
		Ore		
-0,074	43,92	10,08	4,35	0,94
0,074-0,315	63,77	7,4	8,6	0.65
0,315-0,80	83,83	4,55	18,42	0,38
0,80 -5,0	89,66	1,74	65,15	0,035
+5,0				

$Q_{fr} = Q_f + 105t/h$, $M = 13t$, $Q_c + 165t$

$\varepsilon(\ell) = Q_f C_f \gamma_f(\ell) \left[Q_m C_m \gamma_m(\ell)\right]^{-1}$

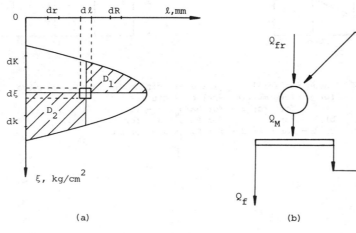

(a)   (b)

Fig. 1.   Integration domains in Eq. (4)

a)

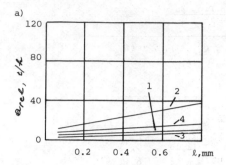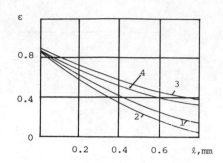

b)

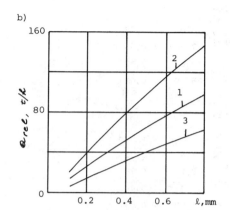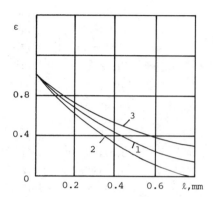

c)

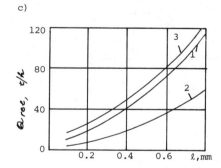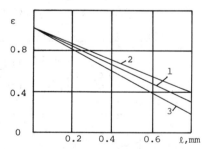

Fig. 2.   Industrial mill (Q_{rel}) and classifier (ε) characteristics
for various multimineral case studies:
a) 1 - for ore; 2 - for $BaSo_4$; 3 - for $CaMg(Co_3)_2$; 4 - for $CaCo_3$;
b) 1 - for ore; 2 - for $BaSo_4$; 3 - for SiO_2;
c) 1 - for ore; 2 - for SiO_2; 3 - for $BaSo_4$.

APPENDIX A: BATCH EQUATIONS

The usual kinetic equation (A. Zagustin, 1935) of periodic grinding

$$\frac{\partial \gamma_{us}(\ell,t)}{\partial t} = \int_{\ell}^{\ell_{max}} \gamma_{us}(R,t) f(R) \gamma_d(R,\ell) dR - \gamma_{us}(\ell,t) f(\ell)$$

(1)

is transferred into more general one

$$\frac{\partial \gamma(\ell,\xi,t)}{\partial t} = \int_{\ell}^{\ell_{max}} \int_{\xi}^{\xi_{max}} \gamma(R,K,t) f(R,K) \gamma_d(R,K,\ell,\xi) x$$

$$xdRdK - \gamma(\ell,\xi,t) f(\ell,\xi)$$

(2)

where R,K - grain size and grindability of initial ore particles; $\gamma_d(R,K,\ell,\xi)$ - distribution for daughter particles after crushing of initial R,K - particles; f(R,K) - rate of crushing of R,K - particles. For ores with ξ= const. Eq. (2) becomes Eq. (1).

For example with $f = AR(\xi_{max} - K)$ and $\gamma_d = 1/R(\xi_{max} - K)$ the solution to Eq. (2) $\gamma(\ell,\xi,t) = \gamma(\ell,\xi,0) \exp$

$$\left[-A\ell(\xi_{max}-\xi)t\right] + AL(\ell,\xi,0)\exp\left[-A\ell(\xi_{max}-\xi)t\right] x$$

$$x\left[\frac{1}{A\xi\ell}(1-\exp(-a\ell\xi t))\right]$$

(3)

where $\gamma(\ell,\xi,0)$ and $L(\ell,0) = \int_{\ell}^{\ell_{max}}\int_{o}^{\xi_{max}}\gamma(\ell,\xi,0)d\ell d\xi$ are initial distributions; A = const. Integrating Eq. (3) by ℓ and by ξ yields cumulative distribution $L(\ell,t) = L(\ell,0)$ x exp $(-A\xi_{max}\ell t)$.

A generalization of Eq. (2)

$$\frac{\partial \gamma(\ell,\xi,t)}{\partial t} = \iint_{D_1} qdRdK - \iint_{D_2} qdrdk$$

(4)

where $q = \gamma(R,K,t)q_{rel}(R,r,K,k)$; q_{rel} is relative mass flow from R,K - particles to r,k - daughter particles, Fig. 1 (compared with Eq. (2) $q_{rel} = f\gamma_d$).

In Eq. (4) term $\partial\gamma/\partial t$ equals the change of relative mass of $d\ell d\xi$ - particles per unit time.

A simplification of Eq. (4) for n - component ore with mineral components having discrete $\xi_1,\xi_2,\ldots,\xi_n$:

$$\frac{\partial \gamma_i(\ell,t)}{\partial t} = \int_{\ell}^{\ell_{max}}\gamma_i(R,t)q_{rel_i}(R,r)dr -$$

$$- \gamma_i(\ell,t)\int_{o}^{\ell}q_{rel_i}(\ell,r)dr; \quad i = 1,2,\ldots,n$$

(5)

where $\gamma_i(\ell,t)$ - partial grain size distribution of i-th mineral.

Through known γ_i from Eq. (5):

$$\gamma(\ell,\xi,t) = \sum_{i=1}^{n}C_i\gamma_i(\ell,t)\delta(\xi-\xi_i)$$

(6)

where $\delta(\xi-\xi_i)$ - Dirack (impulse) function for ξ_i = const.; C_i - mass portion of i-th mineral in ore.

For simplified example $q_{rel_i} = f_i\gamma_{di}, f_i = A_iR$, $\gamma_{di} = 1/R$ the solution to Eq. (5)

$$L_i(\ell,t) = \int_{\ell}^{\ell_{max}}\gamma_i(\ell,t)d\ell = L_i(\ell,0)\exp(-A_i\ell t)$$

(7)

where A_i - Rittinger's parameter, 1/mm sec. Where from the cumulative grain size distribution of the ore as a whole

$$L(\ell,t) = \sum_{i=1}^{n}C_iL_i(\ell,0)\exp(-A_i\ell t)$$

(8)

Finding A_i, q_{rel_i} is possible from experimental data on batch process.

APPENDIX B: CLOSED GRINDING (CRUSHING) CIRCUIT EQUATIONS

For closed circuits main Eq. (4) or (2) is used to obtain:

$$M\left[\int_{\ell}^{\ell_{max}}\int_{\xi_{min}}^{\xi_{max}} q(R,\ell,K,\xi)dRdK - \int_{o}^{\ell}\int_{\xi_{min}}^{\xi_{max}}q(\ell,r,\xi,k) x\right.$$

$$\left. drdk\right] + Q_{fr}\gamma_{fr}(\ell,\xi) + Q_c\gamma_c(\ell,\xi) = Q_m\gamma_m(\ell,\xi)$$

(9)

where Q_{fr}, Q_c and Q_m - fresh feed, circulating and mill (crusher) discharge tonnage; $\gamma_{fr}, \gamma_c, \gamma_m$ - mentioned two-dimensional distributions for these products; M - mill (crusher) hold up, t

Assuming for n - mineral ore

$$q(R,r,K,k) = \sum_{i=1}^{n}q_{rel_i}(R,\xi)C_i\gamma_{mi}(R,\xi)\delta(\xi - \xi_i)$$

(10)

then putting into Eq. (9) and integrating by ξ in vicinity of ξ_i yields for each i-mineral:

$$M\left[\int_{\ell}^{\ell_{max}}C_i\gamma_{mi}(R)q_{rel_i}(\ell)dR - C_i\gamma_{mi}(\ell)\int_{o}^{\ell}q_{rel_i}(r)dr\right] +$$

$$Q_{fr}C_{ifr}\gamma_{ifr}(\ell) + Q_cC_{ic}\gamma_{ic}(\ell) = Q_mC_{im}\gamma_{im}(\ell)$$

(11)

Finally integrating Eq. (11) by ℓ from 0 to ℓ yields

$$Q_{rel_i}(\ell) = \left[Q_mC_{im}\Gamma_{im}(\ell) - Q_{fr}C_{ifr}\Gamma_{ifr}(\ell) -\right.$$

$$\left. Q_cC_{ic}\Gamma_{ic}(\ell)\right]/MC_{im}\left[1-\Gamma_{im}(\ell)\right]$$

(12)

where $Q_{rel_i}(\ell) = \int_{o}^{\ell}q_{rel_i}(\ell)d\ell; \Gamma_i(\ell) = \int_{o}^{\ell}\gamma_i(\ell)d\ell$

Eq. (12) substitute Eq. (9); functions $Q_{rel_i}(\ell)$ may be experimentally found through measurements of Q_i, C_i, Γ_i in Eq. (12).

For complete solution Eq. (11) should be accompanied with classification (screening) performance curves

$$\varepsilon_i(\ell) = Q_fC_{fi}\gamma_{fi}(\ell)/Q_mC_{mi}\gamma_{mi}(\ell)$$

(13)

where Q_f, C_f, γ_f are for final product, i.e. for classifier overflow.

REFERENCES

Tikhonov, O.N. (1981). Experimental methods
    of finding kinetic characteristics for
    multimineral grinding.
    Isvestija Metallurgia, N 1, pp. 3-7.
Tikhonov, O.N. and Toktaganov, S.D. (1982).
    Kinetic characteristics of multi-
    mineral industrial grinding. Gorny
    Journal, N 9, pp. 120-124.

Modern Control Theory Applied to Crushing
Part 1: Development of a Dynamic Model for a
Cone Crusher and Optimal Estimation of Crusher
Operating Variables

J.A. HERBST and A.E. OBLAD

Utah Generic Mineral Technology Centre in Comminution, University of Utah, Salt Lake City, Utah 84112

Abstract. A dynamic model of a cone crusher is proposed for use in automatic control of crushing circuits. The model determines the crusher size distributions for feedrate, feedsize, set and hardness disturbances. Optimal estimation of measured and unmeasured crusher operating variables is made using a Kalman filter.

Keywords. Parameter estimation; Kalman filter; crushing; model; computer control; particle size.

INTRODUCTION

In mineral processing plants which contain size reduction operations much attention has been focused on the improvement of the performance of grinding circuits. This is because these circuits are by far the most energy intensive, often consuming up to one half of the total plant requirement. However, over the last twenty years some effort has also been devoted to the improvement of crushing operations. This has been a result of the fact that, although crushing is not as costly as grinding, in terms of energy and capital investment, the crushing plant product is the feed to the grinding circuits. Thus, the efficiency and consistency of the crushing operation has an impact on all down stream processing, especially on grinding.

The objective of automatic control for crushing is to maintain operating conditions that result in an optimum throughput-product size relationship, this may either be maximum throughput at a constant product size or maximum throughput at the finest product size possible. In addition, constraints such as continuity of flow between circuits must be maintained, and disturbances such as ore feed size and hardness changes be compensated for. The latter type of disturbance is critical because an ore hardness increase can substantially alter the power draw of the crusher, resulting in a possible power overload condition.

Crushing plant control using classical feedback strategies has resulted in reduced crusher power fluctuations, improved product quality, lower labor costs and an increase in production over manual operation.

Classical feedback control strategies attempt to maintain plant outputs at a given level, or set point, by manipulating a number of controllable inputs according to a specified control law. Commonly, the control law is of the proportional-integral-derivative (PID) form used in the majority of control applications. For classical strategies each controlled output is linked to a single manipulated variable with the objective of obtaining the desired output response.

In the case of crushing a possible linking would be that of power draw to feedrate and product size to closed-side setting (set). In some crushing applications it is not easy to change the set of a crusher, and thus the control problem may be reduced to that of a single input-single output system. However, changing either the feedrate or set of a crusher results in changes in both the power draw and product fineness. This condition, termed controller interaction, is a typical problem in the control of complex systems, and conventional methods such as detuning for minimizing controller interactions are less than satisfactory. Besides this difficulty classical control of crushing also suffers from problems arising from the non-linear nature of the process, lack of measurement precision (especially for power draw), instabilities due to delays in the circuit and gradual drift in circuit performance due to wear of equipment.

MODERN CONTROL THEORY

Modern control theory takes advantage of the existence of a mathematical model of the controlled process. The model is used on-line in combination with process measurements to optimally estimate the values of the variables (states) that determine the process behavior. This technique results in improved estimates of process outputs, state variables and model parameters. Based on the information provided by the estimator, a control strategy can be devised which optimizes a performance index.

Experience with classical feedback control of crushing circuits indicates that an improvement in circuit production rate of about 10% can be realized. However, the variation in operating conditions and significant circuit delays would seem to be well suited to control strategies based on modern control techniques.

Indeed, one study (Borisson and Syding, 1976) of a tertiary crushing circuit showed that applying self-tuning control based on an empirical dynamic model for the power draw resulted in a 15% improvement in operating power level and 10% greater throughput over that obtained using classical PI feedback control. In addition, the set point for power draw was raised from 75% to 90% of maximum without causing overload conditions. This was a result of the capability of the self-tuning regulator to recognize and correct a power surge much more rapidly than the classical feedback controller.

In order to apply modern control theory to crushing it is first necessary to develop dynamic models that are sufficiently detailed to reproduce the essential dynamic characteristics of crushing. For a cone crusher this would involve accurately predicting the discharge size distribution, throughput and power consumption as a function of time. Then optimal estimation using a Kalman filter is developed for the dynamic model. Finally, an optimal control algorithm is employed to determine the values of the manipulated variables which optimize the chosen control objectives. The optimal control algorithm will also supervise the set points of the regulatory control loops.

In this study a dynamic model of a cone crusher has been used in conjunction with the extended Kalman filter to optimally estimate the size distributions of the material within the crusher as well as the measured output variables. In addition, estimation of the specific crushing rates has been implemented for the purpose of detecting ore hardness changes.

The dynamic model and estimator are incorporated in a simulator as a first step in developing an optimal control scheme for crushing circuits.

MATHEMATICAL MODEL OF A CONE CRUSHER

Modeling of crushing has undergone development since the 1960's and several crushing plant simulators have appeared in the metallurgical literature (Hatch and Mular, 1982; Karra, 1982). These have followed the basic model of Whiten (Whiten, 1972) for steady-state crushing. The Whiten model views a crusher as a single well-mixed breakage zone with an internal classifier arranged sequentially in closed circuit. Figure 1 represents this concept for postclassification. Alternatives such as preclassification and simultaneous post- and preclassification have been proposed (Austin et al, 1980). In Fig. 1 the classification matrix C gives the fraction of the zone discharge which is recycled, and the breakage matrix B gives the fragment size distribution of broken material. It is assumed that all material which enters the breakage zone is actually broken. A mass balance for the circuit at steady-state yields the following expression for the crusher product size distribution:

$$x_p = (I - C)(I - B C)^{-1} x_f \qquad (1)$$

Equation 1 has been widely used for steady-state simulation of industrial crushing operations. This model has been extended to a dynamic model (Lynch, 1977) by combining the terms of Equation 1 and equating the resulting difference to the time derivative of x_p. This, however, implies that the crusher acts as a single mixer and classifier. This is a dubious premise, considering the interior geometry of the crushing chamber.

A rigorous dynamic model would have to account for particle breakage and transport in both the crusher bowl and crushing chamber. The method applied in ball mill grinding of isolating mill breakage kinetics and particle transport through the use of separate batch and continuous tests is not possible for crushers. The application of a rigorous transport model is therefore considered unadvisable in the initial development of the model.

An approximate model has been developed which avoids some of the difficulties of a more exact model. The breakage kinetics are represented by an equation which gives the net mass rate of production of particles in a specified size interval:

$$\frac{dHx_j}{dt} = - S_j H x_j + H \sum_{k=1}^{j-1} b_{jk} S_k x_k \qquad (2)$$

The first term on the RHS of Equation 2 represents the rate of breakage out of size interval j, and the second term gives the accumulation of mass in size j due to breakage out of all larger sizes.

The continuous flow of material through the crushing chamber can be superimposed onto the breakage kinetics to yield a continuous dynamic model.

The volume of the crushing chamber is divided into several zones as shown in Fig. 2. Each zone is postulated to be well-mixed and size classification is present due to the decrease in chamber volume in the direction of particle flow. The classification curve used is that developed by Lynch and Rao for hydrocyclones (Lynch and Rao, 1975).

$$C_{ij} = \frac{\exp(a_i d_j / D_{50_i}) - 1}{\exp(a_i d_j / D_{50_i}) + \exp(a_i) - 2} \qquad (3)$$

A mass balance for the first crushing zone yields:

$$\frac{dH_1 x_{1j}}{dt} = F x_{oj} - F \frac{(1 - C_{1j}) x_{1j}}{1 - \Sigma C_{1k} x_{1k}}$$
$$- S_{1j} H_1 x_{1j} + H_1 \sum_{k=1}^{j-1} b_{jk} S_{1k} x_{1k}$$

and for the ith (i > 1) zone:

$$\frac{dH_i x_{ij}}{dt} = F \frac{(1 - C_{i-1,j}) x_{i-1,j}}{1 - \Sigma C_{i-1,k} x_{i-1,k}} - F \frac{(1 - C_{ij}) x_{ij}}{1 - \Sigma C_{ik} x_{ik}}$$
$$- S_{ij} H_i x_{ij} + H_i \sum_{k=1}^{j-1} b_{jk} S_{ik} x_{ik} \qquad (4)$$

The size distribution for the crusher discharge is given by:

$$y_j = \frac{(1 - C_{nj}) x_{nj}}{1 - \Sigma C_{nk} x_{nk}} \qquad (5)$$

The flowrate F represents both the internal and discharge flowrate of the crusher. The dynamics of particle transport are, to a great extent, determined by the amount of material contained in the crusher bowl. A linear bowl model has been used which specifies F as a function of the instantaneous weight of material in the bowl. If F_{max} and F_{min} are the flowrates out of the bowl into the crushing chamber at maximum and minimum bowl filling, respectively, then the following differential equation describes, under the given assumptions, the dynamics of the hold-up of material in the bowl at time t :

$$\frac{dV_B}{dt} = F_o - F_{min} - \frac{F_{max} - F_{min}}{V_{B,max}} V_B \qquad (6)$$

The change in bowl filling in a time interval Δt over which the external feedrate to the crusher is maintained at a value of F_o is :

$$\Delta V_B = \left(V_{B,max} \frac{F_o - F_{min}}{F_{max} - F_{min}} - V_B \right) \cdot$$
$$\left(1 - \exp\left[- \frac{F_{max} - F_{min}}{V_{B,max}} \Delta t \right] \right) \qquad (7)$$

The corresponding change in F is :

$$\Delta F = \frac{F_{max} - F_{min}}{V_{B,max}} \Delta V_B \qquad (8)$$

Thus, a step change in feedrate to the crusher will cause a first-order (exponential) change in the discharge flowrate. The operation of the crusher is controlled by the external feedrate and the closed side setting (set). Increasing the feedrate and/or decreasing the set results in an overall finer product size and a larger power draw. These effects are incorporated into the model via the classification parameters. The zone median classification size D_{50} has been related to changes in feedrate and set by the following empirical expression :

$$D_{50} = K \frac{SET^n}{F/(1 - \Sigma C_{ij} x_{ij})} \qquad (9)$$

where the demoninator of Equation 9 represents the flowrate to the classifier of zone i.

The maximum and minimum flowrates are mainly dependent on the set, but also vary with feedsize distribution, particle characteristics, and operating variables. F_{max} and F_{min} were calculated as a function of the set at steady-state for average conditions in the other variables. This was done by varying the feedrate to the steady-state version of the crushing model. The maximum and minimum feedrates were assumed to have been reached when the non-linear equations could no longer be solved because of extreme values in the parameter D_{50} (approximately d_{max} and 0, respectively).

The crusher power consumption is a result of overcoming the breakage resistance of the particles in each zone. For the current model a linear relationship between the power draw and the size consist of the crushing chamber is assumed:

$$P = \sum_{i=1}^{m} p_i = \sum_{i=1}^{m} H_i \sum_{j=1}^{n} K_{ij} x_{ij} \qquad (10)$$

A moderately finer particle size distribution in the crushing chamber is associated with an increase in power consumption, all other factors being constant.

In addition, it is necessary to specify the effect of operating variables on the breakage parameters. The breakage function is assumed to be independent of crusher operating conditions. The selection functions are taken as being proportional to the zone specific power draw, as

is the case for ball mill grinding (Herbst and Fuerstenau, 1973):

$$S_{i1} = S_{i1}^E \left(\frac{p_i}{H_i} \right) \qquad (11)$$

The size dependence of the selection function is given by a single power relationship:

$$S_{ij} = S_{i1} \left(\frac{d_j d_{j+1}}{d_1 d_2} \right)^{a/2} \qquad (12)$$

The steady-state solution of the crushing model requires a tertiary iteration, since the implicit equations for the zonal size distributions depend on the breakage parameters, the power draw, and the classification parameters, all of which are functions of the zonal size distributions. The dynamic model, however, can be directly integrated to give the size distributions in the crushing chamber, thus avoiding the iterations on size distribution necessary to solve the steady-state model. The effects of feedrate, set and feed size distribution on the crusher discharge size distribution and power consumption can be modeled using Equation 4. It was desired to simulate the dynamic response of a pilot-scale cone crusher. The specifications and parameters used in the simulation are approximately for a 0.5 ton per hour unit. Figure 3 shows the transient behavior of the crusher discharge for a feedrate change. Figure 4 represents the dynamics of a feed size distribution change.

Finally, it is very important to be able to predict the effect of an ore hardness change, since this is the type of disturbance that can lead to a power overload and subsequent crusher stoppage. A hardness disturbance is represented as a new ore suddenly being fed to the crusher after a certain time t_o. This new ore is characterized by different selection functions, but is identical in other aspects (with the possible exception of size distribution) to the original material in the crusher.

A set of differential equations analogous to that for a pure material can be written for the dynamics of each different ore in the crusher:

$$\frac{dH_i^m x_{ij}^m}{dt} = F_{i-1}^m \frac{(1 - C_{i-1,j}) x_{ij}^m}{1 - \Sigma C_{i-1,k} x_{i-1,k}^m}$$
$$- F_i^m \frac{(1 - C_{ij}) x_{ij}^m}{1 - \Sigma C_{ik} x_{ik}^m}$$
$$- S_{ij}^m H_i^m x_{ij}^m + \sum_{k=1}^{j-1} b_{jk}^m S_{ik}^m H_i^m x_{ik}^m \qquad (13)$$

where the superscript m refers to the "disturbance" ore (m=1) or the original ore (m=2).

It is required to have models for the zonal hold-up and throughput. From a dynamic mass balance:

$$\frac{dH_i^m}{dt} = F_{i-1}^m - F_i^m \qquad (14)$$

AMR-K

and

$$F_i^m = \frac{1 - \Sigma C_{ik} x_{ik}^m}{1 - \Sigma C_{ik} x_{ik}} \left(\frac{H_i^m}{H_i} \right) F \tag{15}$$

The individual breakage kinetics and power consumption are given by:

$$S_{ij}^m = S_{i1}^{E,m} \frac{P_i}{H_i} \left(\frac{d_j \, d_{j+1}}{d_1 \, d_2} \right)^{a/2} \tag{16}$$

$$P_i^m = \sum K_{ij} \left[\frac{S_{i1}^E}{S_{i1}^{E,m}} \right] H_i^m x_{ij}^m \tag{17}$$

The effect of a 20% increase in ore hardness is shown in Fig. 5. The crusher discharge size distribution is only slightly changed, but the crusher power draw is significantly increased. This is a result of the crusher requiring enough torque to overcome the increased fracture resistance of the harder material. The power draw, and not the product size distribution, is the major indicator of a hardness disturbance.

KALMAN FILTER ESTIMATION

The Kalman filter is an optimal least-squares estimator for linear dynamic systems driven by white noise. Let $\mathbf{x}$ be the system state vector. Then, the system dynamics can be represented by:

$$\frac{d\mathbf{x}}{dt} = \mathbf{A} \, \mathbf{x} + \mathbf{w} \tag{18}$$

In time-discrete form the solution to Equation 18 is:

$$\mathbf{x}_{k+1} = \boldsymbol{\Phi}_k \, \mathbf{x}_k + \mathbf{w}_k \tag{19}$$

where $E(\mathbf{w}_k) = 0$ and $E(\mathbf{w}_k \mathbf{w}_k^T) = Q_k$

The transition matrix $\boldsymbol{\Phi}_k$ estimates the value of the state vector at the next measurement time:

$$\hat{\mathbf{x}}_{k+1}^- = \boldsymbol{\Phi}_k \, \hat{\mathbf{x}}_k \tag{20}$$

The covariance matrix for system noise is Q_k. The system output is given by:

$$\mathbf{z}_k = H_k \, \mathbf{x}_k + \mathbf{v}_k \tag{21}$$

where $E(\mathbf{v}_k) = 0$ and $E(\mathbf{v}_k \mathbf{v}_k^T) = R_k$

It is desired to estimate the state vector at time t_k from the linear relationship:

$$\hat{\mathbf{x}}_k = \hat{\mathbf{x}}_k^- + \mathbf{B}_k \, (\mathbf{z}_k - H_k \, \hat{\mathbf{x}}_k^-) \tag{22}$$

such that the error of estimation is a minimum. The "-" superscript denotes that the estimate has been made before current data were available, while no superscript indicates that the estimate is updated with a current measurement.

The error of measurement is represented by the covariance matrix:

$$P_k = E \left[(\mathbf{x}_k - \hat{\mathbf{x}}_k) (\mathbf{x}_k - \hat{\mathbf{x}}_k)^T \right]$$

The Kalman filter determines $\mathbf{B}_k$ such that the trace of P_k is a minimum:

$$\mathbf{B}_k = P_k^- H_k^T (H_k \, P_k^- H_k^T + R_k)^{-1} \tag{23}$$

$$P_k = (I - \mathbf{B}_k \, H_k) \, P_k^- \tag{24}$$

where P_k^- is the prediction of the estimation error before the measurement at time t_k has occurred and is calculated from information available at time t_{k-1}:

$$P_k^- = \boldsymbol{\Phi}_{k-1} \, P_{k-1} \, \boldsymbol{\Phi}_{k-1}^T + Q_{k-1} \tag{25}$$

For non-linear systems the dynamics would be a specific function of the states (and deterministic inputs, u):

$$\frac{d\mathbf{x}}{dt} = f (\mathbf{x}, u) + \mathbf{w} \tag{26}$$

To apply the Kalman filter the system dynamics are linearized about the current operating point.

$$\boldsymbol{\Phi}_k = \exp (A_k \, \Delta t_k) = I + \left[\frac{\partial f}{\partial \mathbf{x}} \right] \hat{\mathbf{x}}_k \, \Delta t_k \tag{27}$$

for Δt_k small. The system output vector is given by

$$\mathbf{z}_k = h(\mathbf{x}_k) + \mathbf{v}_k \tag{28}$$

$$H_k = \left[\frac{\partial h}{\partial \mathbf{x}} \right] \hat{\mathbf{x}}_k^- \tag{29}$$

ESTIMATION SCHEME FOR CONE CRUSHER

A dynamic simulator, DYNACRUSH, has been developed to evaluate the feasibility of crusher modeling, estimation and control. The dynamic response of the cone crusher model to various kinds of control input and ore hardness disturbances has already been reviewed.

For the cone crusher the states were chosen to be the size distributions of material within the crushing chamber. Since the weight fraction of solids in each size interval in each zone sum to one, only the first n-1 sizes were used. The nth size fraction was calculated by difference. The outputs were taken as being the first n-1 size fractions in the crusher discharge and the power draw. The zonal specific selection functions were considered to be the only unknown parameters.

The plant model was a cone crusher in open circuit and observations were generated at discrete time intervals by adding normally distributed noise to the model predictions. These observed outputs were sampled at chosen time intervals such that the linear approximations in the filter were valid. The on-line model was the same as the plant model except that it had no multicomponent hardness model. The state vector and transition matrix were augmented with the parameters so that all internal size distributions and specific selection functions could be estimated.

The Kalman filter was run with a 2.4 second sampling time, as longer intervals required a more accurate computation of the transition matrix. The simulations initially use reasonable, but incorrect, guesses for the state vector and selection functions. Figures 6 and 7 show typical results for a 15% feedrate change. The filter estimates can be seen to be considerably more accurate than the measurements for the fines production and power draw. Parameter estimation rapidly converges as shown in Fig. 8.

It is apparent that for a crusher model the Kalman filter is advantageous at several levels of use. First, it is possible to estimate the size distributions inside the crusher. This would be infeasible without the filter since these variables are not directly measurable. Secondly, improved estimates are obtained for the output responses of the crusher. This is especially useful for making an estimate of the power draw which, in practice, is quite noisy. The accuracy of power determination obtained with the Kalman filter would be superior to that for analogue filters, since the Kalman filter is optimal. Finally, the improved state and output estimates have strong implications for control. At the very least the filtered output estimates provide better values for conventional output feedback control. With access to state estimates it is also possible to use state feedback control as an alternative to output feedback control. If optimal control is the objective then the Kalman filter is indispensible.

SUMMARY AND CONCLUSIONS

A simulation study has been made for the purpose of developing a dynamic model of a cone crusher and obtaining estimates of operating variables. The study represents the initial phase in developing an optimal control scheme for crushing plants.
The crusher simulator predicted responses that are in agreement with known crusher behavior. Especially significant was the response due to an ore hardness disturbance. Unlike in ball mill grinding there was no large change in discharge size distribution. Rather, a hardness disturbance produced a large variation in power draw.

The use of the Kalman filter in conjunction with the dynamic crushing model was shown to provide estimates of unmeasured variables, improved estimates of measured output responses, and estimates of chosen crushing parameters.

The results of this study indicate that this approach to crushing plant control holds much promise. Tests are currently underway to verify the dynamic model on industrial crushers. A simplified model will be developed for use on-line as a basis for Kalman filter estimation and evaluation of crushing plant control strategies, including optimal control.

ACKNOWLEDGEMENT

The authors would like to acknowledge the support of the USBM (under Grant #G1125149) through the Utah Generic Technology Center in Comminution.

REFERENCES

Austin, L.G., D.R. Van Orden, and J.W. Perez (1980). A Preliminary Analysis of Smooth Roll Crushers. Int. J. Miner. Process. 6, pp. 321-336.

Borisson, U. and R. Syding (1976). Self-Tuning Control of an Ore Crusher. Automatica, 12, pp. 1-7.

Hatch, C.C. and A.L. Mular (1982). Simulation of the Brenda Mines Ltd. Secondary Crushing Plant. Mining Engineering, 34 (9), pp. 1354-1362.

Herbst, J.A. and D.W. Fuerstenau. (1973). Mathematical Simulation of Dry Ball Milling Using Specific Power Information. Transactions SME-AIME. 254, 343-348.

Karra, V.K. (1982). A Process Performance Model for Cone Crushers. XIV Int. Min. Proc. Conf. paper III-6, Toronto.

Lynch, A.J. and T.C. Rao. (1975). Modelling and Scale-Up of Hydrocyclone Classifiers. XI Int. Min. Proc. Cong. Cagliari, paper 9.

Lynch, A.J. (1977). Mineral Crushing and Grinding Circuits. Elsevier Scientific Publishing Co., New York, pp. 45-50.

Whiten, W.J. (1972). The Simulation of Crushing Plants with Models Developed using Multiple Spline Regression. APCOM X, pp. 317-323, Johannesburg.

NOTATION AND SYMBOLS

A	Matrix in linear dynamic equation
b_{jk}	Individual discretized breakage
B	Breakage matrix
B_k	Kalman filter gain matrix at time t_k
C	Classification matrix
C_{ij}	Classification coefficient for material in jth size interval in zone i
d_j	Discretized particle size
D_{50}	Medium classification size
E	Expectation operator
F	Solids flowrate in crusher
F_i^m	Flowrate of m-th component in zone i
H_i	Mass hold-up in zone i
H_k	Measurement matrix at time t_k
I	Identity matrix
P	Total power draw for crusher
P_k	Estimation covariance matrix at time t_k
p_i	Power draw in zone i
Q_k	Process noise covariance matrix at time t_k
R_k	Measurment noise covariance matrix at time t_k
S_{ij}	Selection function for material in jth size interval in zone i
v_k	Measurement noise vector at time t_k

V_B Volume of particles in crusher bowl

$V_{B,max}$ Volume of crusher bowl

w_k Process noise vector at time t_k

x_k State vector at time t_k

x_{ij} Discretized size distribution for zone i

y_j Crusher discharge size distribution

z_k Process output vector at time t_k

ϕ_k State transition matrix at time t_k

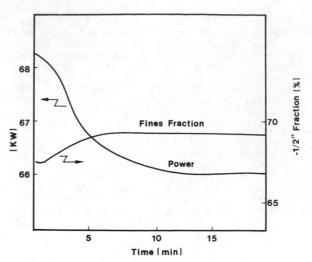

Fig. 3 Dynamic response of crusher to a
        15% decrease in feedrate

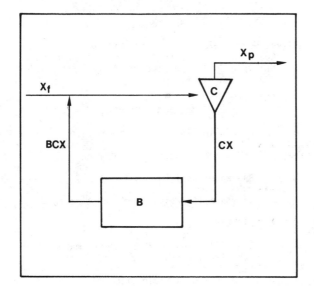

Fig. 1 Conceptual representation of
        breakage in a crusher

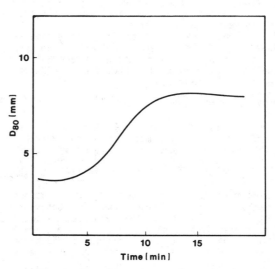

Fig. 4 Dynamic response of crusher to an
        increase in 80% passing size of
        feed

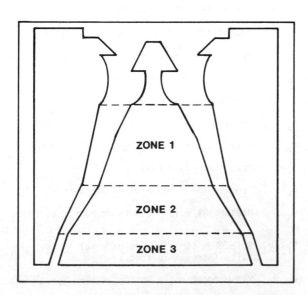

Fig. 2 Schematic representation of the
        zones of a cone crusher

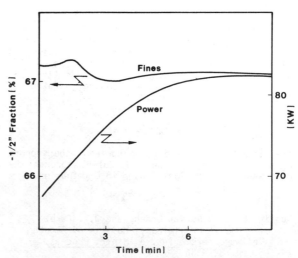

Fig. 5 Effect of a 20% increase in ore
        hardness

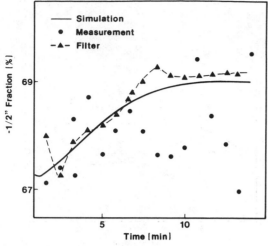

Fig. 6 Kalman filter estimation of fines
in crusher discharge for a 15%
feedrate decrease

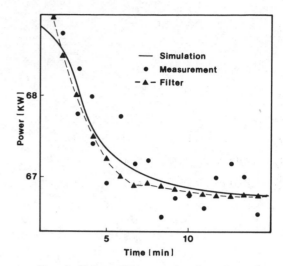

Fig. 7 Kalman filter estimation of crusher
power draw for a 15% feedrate
decrease

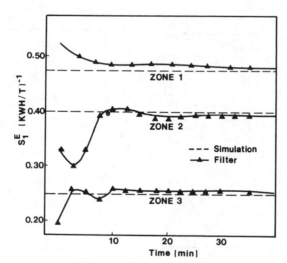

Fig. 8 Kalman filter estimation of zonal
specific selection functions

A Generalised Investigation of Adaptive Optimisation in the Chemical Processing of Minerals

ALAN P. PROSSER

School of Metallurgy, University of New South Wales, Kensington, N.S.W., Australia

Abstract. The strategy of adjusting operating conditions to maintain optimum process performance as the feed quality varies - adaptive optimisation - is being investigated by computer simulation of chemical processes for the treatment of ores. Those ore-process systems that are most likely to yield substantial benefits from adaptive optimisation are being identified as a guide to the selection and development of systems for practical application. The generalised mathematical representation of the cost structure, ore quality variations, process kinetics, reactor characteristics and operating variables for ore-process systems is described. Results are summarised for temperature, reagent concentration or residence time as the manipulated variable. For almost all systems, adaptive optimisation was economically beneficial, although a few exceptions have been identified where there is a significant synergistic effect between poor-performance and good-performance materials in a blended feed. The sensitivity of the benefits of adaptive optimisation to a range of economic and technical factors is summarised.

Keywords. Metallurgy; mineral processing; chemical variables control; modelling; optimisation; adaptive control; computer evaluation.

INTRODUCTION

Variable raw materials are an unavoidable characteristic of mineral processing operations. Blending the raw materials in a stockpile, prior to physical or chemical treatment processes, is the common way of minimising the disturbances caused by the variations. However, the benefits of steady operation of the treatment processes are achieved at a cost. The optimum performance of a blend of two or more materials must be less than the sum of the optimum performance of the separately treated materials, unless there is a synergistic effect between poor-performance and good-performance materials. Alternatively, adaptive optimisation, in which an operating variable such as temperature, reagent concentration or flowrate is adjusted to maintain optimum conditions for each raw material in turn, requires a more sophisticated control procedure than that normally used with blended feeds. At the present time it is not clear that the cost of developing and installing the control system can be justified by the improved overall performance of a particular ore-process system.

The objectives of this project are to estimate, by inexpensive means, what improvements may be expected from processes run with adaptive optimisation, and which ore-process systems are likely to yield the greatest benefits. The results are not expected to be accurate for any particular system; they are expected to be useful guidelines for people considering adaptive optimisation. Results reported earlier (Prosser, 1982; Prosser and Jefferson, 1984) indicate that the objectives are being achieved.

The methods used to generalise the investigation for a range of chemical treatment processes are described here, together with a summary of recent results. The project is not presently concerned with practical methods whereby adaptive optimisation can be implemented.

The outcomes of the investigation are relevant to processes such as the various versions of gold ore leaching, the leaching of uranium ore or low-grade concentrates, the processing of bauxite and laterites, and the several chemical processes for the treatment of oxidised copper ores.

BASIC RELATIONS

The 'inexpensive means' of investigating a range of ore-process systems is computer simulation using mathematical representations of the characteristics of processes which affect the potential benefits of adaptive optimisation. In effect, this means the technical characteristics that influence recovery of the valuable component, quality of the product, consumption of consumables, and throughput. Also, because the final criterion is economic benefit, the cost characteristics are required as well. The cost structure necessary for the simulation is not that normally used in an economic evaluation; the form which has been adopted is as follows.

The capital cost of instrumentation, etc. for adaptive optimisation has been assumed to be the same as that for blending. Thus, the only characteristic investigated has been the operating cost. The basic relation used is:-

(value of output) = (cost of input) + (operating cost related to the manipulated variable) + (other operating costs) + (fixed charges apportioned to the process) + (profit)

or

$$Z_O = Z_I + A(X) + D(X) \qquad (1)$$

where $A(X)$ is the operating cost related to the manipulated variable, $D(X)$ is the sum of the last three terms above, and X is the manipulated variable, viz. energy consumption, reagent consumption or residence time.

The value of the output is obtained from

$$Z_o = \bar{R} I V \qquad (2)$$

The recovery of the valuable component in the ore, $\bar{R}$, from each block of ore is different and determined by the manipulated variable. Thus both $\bar{R}$ and Z_o are functions of X.

The cost of the input, Z_I, may be calculated either as a fixed value per tonne or as the value of the required substance in each block of ore. For a large quantity of ore of variable composition, the overall benefits of adaptive optimisation were identical using each method of accounting for the cost of the input.

The objective function is

$$S = Z_o - Z_I - A(X) = D(X) \qquad (3)$$

and the maximum also corresponds to maximum profit since all the costs contained in D(X) are the same for each block of ore and independent of X.

ORE PROCESS SYSTEM CHARACTERISTICS

Cost Structure

The cost structures of different ore-process systems are represented by different parameters in equations (1) to (3). The ranges of values were set using the following guidelines. The cost of mining ranges from $3 to $80 per tonne of ore. The cost of transportation ranges from $1 to $2 per tonne, and the cost of comminution prior to chemical processing ranges from $1 to $5 per tonne. Thus, the extreme range of the cost of input is $5 to $87 per tonne of ore. The cost of a single chemical process for the treatment of ore ranges from $4 to $12 per tonne of ore. Thus, the extreme range for the value of the output, V, is $9 to $99. In the simulation the values used were $5 to $95 for the cost of the input and $10 to $100 for the value of the output.

The cost related to the manipulated variable, A(X), is particularly significant to the benefits of adaptive optimisation. Constant mean values between $0.17 and $1.33 per tonne of ore were investigated.

The optimum values of Z_o, S and profit are obtained via the recovery; see equations (2) and (3). Many technical characteristics of the ore and the process affect the relation between the recovery and the manipulated variable. Different ores and processes are represented in the simulation by different mathematical relations for each technical characteristic, as follows.

Ore Qualities

A common characteristic of chemical processes for the treatment of ores (as distinct from high grade concentrates) is the substantial decrease in the rate of reaction as the reaction approaches completion, beyond that due to the depletion of the reactants. The exact shape of the recovery versus time curve varies from one sample of the ore to another. Two strategies are used in the simulation to represent these variations. The first is to assume that the variable product is obtained from one reactive component and one refractory component in the ore. The variations from one block of ore to another are partly represented by changes in the proportion of reactive component. Different ores are represented by different *distributions* of this proportion and by different *distributions* of the assay for the valuable substance. Another variable

characteristic of the ores is the relative rates of reaction of the two components. The basic equation for this strategy is

$$\frac{d\bar{R}}{dt} = k_1 Q (1-R_1)^m c^n + k_2 (1-Q)(1-R_2)^m c^n \qquad (4)$$

The second strategy is the Brittan model (Brittan, 1975) in which *continuous* variations in the reactivity are represented by an activation energy which increases as the recovery increases. The basic equation is

$$\frac{d\bar{R}}{dt} = (1-\bar{R})^m c^n \exp\left(j - \frac{k+\ell\bar{R}}{RT}\right) \qquad (5)$$

Different ores are represented by different distributions of sets of values of j, k and ℓ.

The distributions of the assay and the proportion as reactive component have been illustrated previously (Prosser, 1982). Table 1 summarises the distribution data which have been used to date. The absolute values of the assays are unimportant. The polarised distribution has peaks at the extreme ends of the range, all others have a single peak. The distributions of assay and the proportion as reactive component are treated as independent characteristics of ores.

TABLE I Distribution Statistics

variable	mean	range	standard deviation
assay	100	36 – 180	±25
	100	36 – 180	±32
	100	36 – 180	±39
reactive proportion	0.6	0.06 – 0.96	±0.15
	0.6	0.06 – 0.96	±0.16
	0.6	0.06 – 0.96	±0.25
	0.6	0.06 – 0.96	±0.33*

*polarised distribution

Process Kinetics

Only chemical processes have been simulated so far. Although the equations have originated from ore-aqueous solution reactions they are equally relevant to ore-gas reactions. Different processes are represented by different temperature coefficients (activation energies), the apparent orders of reaction for the solid reactant and the reagent, and the relative rates of reaction of reactive and refractory components (or j, k and ℓ in the Brittan model). The effect of particle size distribution has been incorporated in the other kinetic parameters. The concentration of the reagent in the reactor is related to the extent of reaction and not assumed constant. When the two-component model is used, the different reactivities are modelled by different activation energies.

Operating Variables

The operating variables are also the manipulated variables in the search for the optimum; they are temperature (or energy consumption), reagent concentration (or consumption) and mean residence time. As each is related to the recovery of the valuable product by a different form of equation it was anticipated that the potential benefits of adaptive optimisation may be significantly different for these operating variables.

Reactor Characteristics

The precise movements of the reactants within the reactor are a significant characteristic of a process. The simplest method of representing these movements is through a distribution of residence times. So far the extreme characteristics of a batch or plug-flow reactor and a single stirred tank reactor have been investigated.

Cost Factors

The method of representing the cost structure of
the ore-process system was summarised above. Different systems are represented by different sets
of parameters in equations (1) to (3). The
variations are best represented in terms of the
value of the output, the *difference* between the
value of the output and the cost of the input (i.e.
the total cost of the process plus profit), and the
cost related to the manipulated variable, A(X).

The form of A(X) depends on the manipulated
variable:-

$$A(X) = \alpha X \qquad (6)$$

for the reagent concentration and mean residence
time; and

$$A(X) = \beta(X-T_0) \qquad (7)$$

for temperature. In the early stages of the
investigation, results were compared for constant
α, but more recently results have been compared
for constant A(X); a clearer pattern has emerged
as a consequence.

The Complete Ore-Process System

An ore-process system is represented in the simulation by specifying the following information:-

  (i) which operating variable is to be manipu-
      lated;

 (ii) the cost of the input; the cost of the
      process which is related to the manipulated
      variable; the value of the output;

(iii) the distribution of the assays; the dist-
      ribution of the proportion as the reactive
      form;

 (iv) if the Brittan model is used, the distrib-
      ution of the parameters, j, k and ℓ;

  (v) the kinetic parameters, i.e. activation
      energies, rate constants and orders of
      reaction;

 (vi) the stoichiometry factors which relate the
      reagent consumption to the recovery of the
      product; and

(vii) the values of the operating variables which
      are not being manipulated.

COMPUTATIONAL METHODS

A Fibonacci search was used to find the optimum
operating conditions for 208 types in a total of
1000 blocks of ore. An accuracy of <0.1% was set
for locating the optimum value of X and the
objective function. This ensured an accuracy no
worse than approximately 3% for the benefits of
adaptive optimisation, and normally much better.

Two problems peculiar to this simulation were
overcome as follows. When the Brittan model was
used, the performance of the blended feed could
not be calculated from weighted mean values of j,
k and ℓ in equation (5). The extent of reaction
for each ore type in the blended feed had to be
calculated separately and the weighted mean of
these values determined. The procedure was checked
using the two-component model where a simpler
calculation was possible.

The recovery, $\bar{R}$, could not be accurately calculated

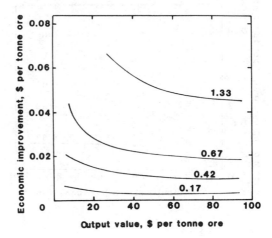

Fig. 1. Economic improvement achieved with temperature as the manipulated variable in a
single stirred tank reactor. Curves
correspond to constant total cost for
energy consumption, in $ per tonne ore as
indicated.

from the mean residence time when stirred tank
reactor models were used for systems which were
not first order with respect to the solid reactant,
i.e. m $\neq$ 1. Satisfactory accuracy was obtained by
dividing the material emerging from a tank into
many packages, each corresponding to a small
increment in recovery and residence time. A geometric sequence of recovery values gave better
results for a specific number of packages. A
Runge-Kutta procedure was used to calculate the
increment in residence time from the specified
increment in recovery. The weighted mean value of
recovery was calculated from the proportions of the
feed emerging with the various residence times.
Independently, Li and Chen (1983) presented the
theoretical basis for this method of computation.

RESULTS

Hundreds of ore-process systems have been simulated
through different combinations of the information
listed in (i) to (vii) above. Some of the results
obtained for batch or plug-flow reactors have been
reported earlier (Prosser, 1982; Prosser and
Jefferson, 1984). Here, the results obtained with
a single stirred tank reactor system are summarised.
In almost all systems, adaptive optimisation is
potentially advantageous. (The exceptions are
discussed below.) Typically, the potential economic improvement was less than 2% of the total
operating cost of the process, or about $0.10 per
tonne of ore. However, by deliberate selection of
the variables almost double this benefit was
achieved. (The improvement obtained with batch or
plug-flow reactors was slightly greater.) The
recovery of the valuable product was virtually the
same from the blended feed and adaptive optimisation strategies. The economic improvement was
achieved by a reduction in reagent consumption,
etc., but there was not a simple relation between
the two. The economic benefit could be dispro-
portionally greater or less depending on how a
given amount of reagent, etc. was utilised in

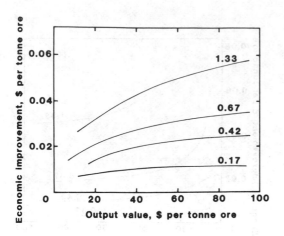

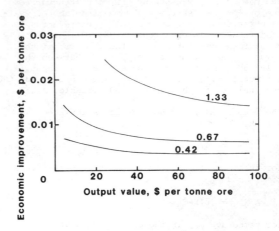

Fig. 2. Economic improvement achieved with
         initial reagent concentration as the
         manipulated variable. Curves correspond
         to constant total cost for reagent con-
         sumption, in $ per tonne ore as indicated.

Fig. 3. Economic improvement achieved with resid-
         ence time as the manipulated variable.
         Curves correspond to constant total cost
         for time-dependent consumables and
         services, in $ per tonne ore as indicated.

achieving the optimum for each block of ore in
adaptive optimisation. All of the potential
improvements were sensitive to a number of the
variables included in the simulation.

Figs. 1 to 3 illustrate some of the trends in the
results. Using initial reagent concentration as
the manipulated variable (Fig. 2), the economic
benefit of adaptive optimisation *increased* with
the value of the output, for a constant operating
cost associated with the reagent. On the other
hand, using temperature or residence time as the
manipulated variable (Figs. 1 and 3), the economic
benefit *decreased* with the value of the output.
None of the curves was affected by the cost of the
input or the total operating cost of the process.
As was expected, the greater the cost associated
with the manipulated variable the greater was the
potential economic benefit.

The distribution of the proportion of the value as
the reactive component had a greater effect than
the distribution of the assay of the value. (See
below for the special effect with residence time
as the manipulated variable.)

The trends illustrated in Figs. 1 to 3 persisted
for combinations of the other ore-process system
variables, but there was some interaction between
the variables. Table 2 illustrates the sensitivity
of the economic benefit to each variable. As an
example, the first row of Table 2 indicates that
for an output value of 3.11x$31=$96.4, the benefit
of adaptive optimisation decreased to 0.75x$0.025=
$0.018 when temperature or residence time was the
manipulated variable, but increased to 1.41x$0.025=
$0.035 when reagent concentration was the manipula-
ted variable. The directions of the changes are
different for the different manipulated variables.
The variables to which the economic benefit is most
sensitive are the operating cost associated with
the manipulated variable, A(X), the distribution
of the proportion of value in the reactive component,
the initial reagent concentration and the order of
reaction with respect to the reagent.

Table 2 shows some negative results when residence
time was used as the manipulated variable. A
negative result means that adaptive optimisation is
economically *disadvantageous*. The reason for these
results is a beneficial interaction between poor-
performance and good-performance ore blocks in the
blended feed. The synergistic effect occurs because
reagent that is not required for low-assay ore is
available for use with high-assay ore. When the
various ore types were treated separately by
adaptive optimisation this transfer of reagent
was not possible. The magnitude of the synergistic
effect is indicated by the result for a zero order
of reaction with respect to reagent, i.e. when the
reagent concentration has no effect on the rate of
reaction and performance of the system. The
economic benefit was 6.3 times larger than the
'normal' benefit. A similar improvement was
expected and found when a large excess of reagent
was utilised, i.e. when the initial concentration
was 2.6 times normal and the benefit increased 5.5
times.

The results in the Figs. and Table 2 are those
obtained without any limit placed on the manipulated variables. In reality it would be difficult
to avoid restricting the range of values of the
manipulated variable for optimum processing of the
different types of ore block. For instance, the
temperature for optimum performance ranged from
30C to 110C, the initial reagent concentration
from 0.6M to 3.2M, and the residence time from
20,000 to 180,000 units, as a direct consequence
of the ranges listed in Table 1. When restrictions
were placed on the range of the manipulated
variable, the economic benefit was reduced. For
example, restricting the temperature range to 30C
to 85C reduced the benefit to 0.8 times normal.

SIGNIFICANCE OF THE RESULTS

The objective function, S in equation (3), changed
only slightly with the manipulated variable in the
vicinity of the optimum. Thus, finding the
optimum in a real ore-process system by an empirical search of the operating costs for the plant

TABLE 2 Changes in the Economic Benefit of Adaptive Optimisation[*]

Variable (normal value)	change	Manipulated variable		
		temperature	reagent concn.	residence time
output value ($31)	x 3.11	0.75	1.41	0.75
	x 0.311	1.67	0.63	1.72
total process cost ($3)	x 5	1.00	1.00	1.00
manipd. variable cost ($0.67)	x 2	2.7	1.55	2.7
	x 0.625	0.56	0.73	0.54
	x 0.25	0.15	0.37	0.18
assay distribution (std. devn. ±32)	x 1.22	1.07	1.16	-0.47
	x 0.78	0.96	0.88	2.6
reactivity distribution (std. devn. ±0.33)	x 0.76	0.55	0.69	-1.00
	x 0.48	0.25	0.46	-2.56
	x 0.46	0.25	0.44	-2.61
residence time (20,000)	x 1.5	1.08	0.87	-
	x 0.6	0.88	1.18	-
temperature (88C)	+ 14	-	0.76	0.78
	- 13	-	1.29	1.22
reagent concn. (2.5 M)	x 2.6	1.83	-	5.5
	x 1.4	1.27	-	3.9
	x 0.88	0.82	-	-1.57
rate constant (300,000)	x 2	1.19	0.77	0.98
	x 0.5	0.87	1.24	1.00
activation energies[**] (55 and 49.2 kJ)	+ 18	0.78	-	-
	- 18	1.46	-	-
relative rates (~8 times)	x 2	1.39	1.27	3.2
	x 0.5	0.57	0.70	-1.49
order w.r.t. reagent (1)	x 1.5	0.88	0.91	-2.4
	x 0.5	1.08	0.73	3.9
	x 0.2	-	0.41	-
	x 0	1.12	-	6.3
order w.r.t. solid (1)	x 1.33	0.90	1.13	1.00
	x 0.67	1.07	0.90	1.00

[*]   The changes are reported as multiples of the benefit calculated when all normal values were used, viz. $0.025 per tonne of ore.

[**] The rate constants were adjusted to give the same rates at 88C.

would be subject to appreciable error. This characteristic is significant for two reasons. First, it suggests that existing processes with blended feeds are not being operated at the optimum. Second, it shows that the optimum would be best sought through a simulation based on empirically accurate models of the cost structure, kinetic and reactor engineering characteristics of the ore-process system rather than on measurement of recovery, reagent and energy consumption, etc., or on idealised mechanistic models.

Thus, although the calculated economic benefit of adaptive optimisation over the optimum blended feed strategy may not be great, the benefit of adaptive optimisation over *current practice* with blended feed and less than optimum conditions may be significantly greater.

SYMBOLS

$A(X)$	operating cost related to the manipulated variable
C	concentration of reagent
$D(X)$	sum of constant or fixed costs and profit
I	mass of ore processed
j, k, ℓ	parameters in Brittan model
k_1	rate constant of reactive component
k_2	rate constant of refractory component
m	apparent order of reaction for the ore

314 A. P. Prosser

	components
n	apparent order of reaction for the reagent
Q	proportion of the value of the reactive component
R	gas constant
$\bar{R}$	overall recovery of the valuable product
R_1	recovery from the reactive component
R_2	recovery from the refractory component
S	objective function, equal to profit plus constant costs
t	time
T	temperature
T_o	temperature attained without energy consumption
V	value of output per unit mass of ore processed
X	manipulated variable
Z_I	cost of input
Z_o	value of output
α	constant
β	constant

ACKNOWLEDGEMENTS

The financial support of Western Mining Corporation Ltd., Alcoa (Australia) Ltd. and C.R.A. Research for this project, and the work of G. Check and M. Stefulj in the development of the stirred tank simulation are gratefully acknowledged.

REFERENCES

Brittan, M.I. (1975). Variable activation energy model for leaching kinetics. *International J. Mineral Process.*, 2, 321-331.

Li Zushu and Chen Jiayong (1983). A method for calculating the extent of reaction of solid particles with fluid in completely mixed reactors connected in series. *Selected Papers. J. Chem. Indy and Eng. (China)*, June 1983, 88-98.

Prosser, A.P. (1982). Identification of the significant parameters for adaptive optimisation in the chemical processing of ores. *Preprints Second Conference on Control Engineering:* Institution of Engineers, Australia, Canberra, 219-223.

Prosser, A.P., and C.B. Jefferson. (1984). Sensitivity of adaptive optimisation to reaction kinetics in leaching processes. *Preprints Extractive Metallurgy Symposium;* Australasian Inst. Mining Metallurgy, Melbourne, 133-138.

Dynamic Simulators for Training Personnel in the Control of Grinding/Flotation Systems

OSVALDO A. BASCUR and JOHN A. HERBST

Department of Metallurgy and Metallurgical Engineering, University of Utah, Salt Lake City, Utah 84112-1183

Abstract. During the last decade, it has been established that conventional mineral-processing control strategies based on classical control theory result in significant increases in plant throughput and operating-cost saving. The proper education and training of personnel using dynamic simulators of the principal unit operations is vital for proper design and maintenance of control systems. Severe multivariable interactions and numerous constraints on both manipulated and state variables in grinding/classification and flotation systems can be studied using time domain analysis with proper dynamic simulators. Two interactive programs are described, and the basis of each simulator and its structure are discussed. Two examples concerned with choosing, tuning, and decoupling of manipulated and controlled variables for a grinding/classification circuit (DYNA-MILL II) and for a rougher flotation circuit (DYNAFLOAT II) are presented. The way in which these dynamic simulators can be used for the design of model-based control strategies is also outlined. A brief description of the use of these dynamic simulators for screening alternative strategies and circuit optimization is given.

Keywords. Grinding; flotation; process control; computer-aided instruction; computer-aided analysis; dynamic modeling; simulation; model-based control.

INTRODUCTION

Increasing energy costs and low-grade ore deposits have forced the modern mineral engineer to make better use of the available capacity of existing plants without much additional investment.

Aside from methods that involve modifications of the process itself, as for example new classification systems, other ways are being sought to exploit the available degrees of freedom and improve plant performance. For instance, by continuously maintaining the plant at its steady-state optimum despite changing ore characteristics and other process disturbances, it is possible to achieve significant performance improvement.

Ore hardness, feed moisture, wear of classification and grinding equipment, and changes in feed composition offer a few examples of the disturbances that can have a lasting economic impact on the operation. The continuous tracking and driving of the process to its best operating conditions when such changes occur is termed optimizing control.

Advances in process control hardware, computer systems, and measurement instrumentation have occurred much faster than the introduction of these systems to the mineral-processing engineer. It takes time, effort, and much education to produce high-enough numbers of scientists and engineers with the expertise to take advantage of these developments. It is well known that mineral-processing systems constitute highly non-linear systems and are essentially different from those of other process industry systems, primarily because of the lack of precise-enough data on the dynamics of mineral-processing systems.

The use of distributed digital control (DDC) currently dominates industrial process control technology. DDC's importance stems from the abrupt enlargement of scope it affords control technology through integration of functions beyond the primary control loop. Process control is no longer restricted to physical control of processing units. It now includes the various levels of supervisory control in a plant or group of plants, upper management policy control, production scheduling to match present or anticipated sales, and optimization of operations at all levels.

As the level of control becomes more advanced, greater process efficiency can be achieved, but greater process knowledge and additional sensing and control instruments are required to obtain the benefits.

The major limitations involved in mineral processing, to truly optimize its systems, are sensors and manpower. One reason that mineral-process technology has not advanced as rapidly as other process technologies is that minerals in processing streams are contained in solid particles of widely varying characteristics. In order to use the current technology, the mineral-processing engineer needs to have an overall understanding of computers, instrumentation and optimization techniques, as well as an intimate understanding of the processes themselves.

These needs can be partially met by educating the engineer using dynamic simulators which permit the trainee to "drive" the plant to different operating conditions under manual or controlled modes. Two levels of simulators have been developed at the University of Utah. The first level includes a sophisticated dynamic simulation involving many features that do not allow an easy interaction with the engineer (Rajamani and Herbst, 1980; Bascur and Herbst, 1982). A second level of simulation has been made in which fixed plants' flowsheets are available for interactive self-training. By subjecting the dynamic simulator to disturbances, the trainee gets a "feeling" for the most important interactions between the magnitude and rate of change of the variables. After producing a list of disturbances, manipulated, possible measurements,

controlled variables, the trainee produces a process matrix of the system, which gives him/her a
complete understanding of the possible interactions. The analysis of the process matrix will
allow him/her to link manipulated and controlled
variables to produce possible control alternatives.
The selection of the proper objective function will
finally help him/her to decide which strategy seems
to be more advantageous.

This paper describes the use of two interactive
programs developed for simplified dynamic simulation and control for a grinding/classification circuit and a flotation cell. Both dynamic simulators
have been developed to be used in a personal computer for easy personal access.

A brief description of both dynamic simulators will
be given, and two application examples will be presented. The first application involves the use of
simulation tuning as a means of reducing the experimental effort required for multiloop tuning. The
second application involves the use of the simulator for selecting alternative control strategies
prior to their implementation in the plant.

GENERAL STRUCTURE AND USE OF THE DYNAMIC SIMULATORS

The dynamics of a grinding circuit and a flotation
cell are both highly nonlinear. One of the most
appropriate methods of analyzing their behavior
with a view toward control is by simulation in the
time domain. Simulations can be used for screening
of alternative control strategies and, in some instances, for control-loop tuning. Manual control
of selected variables can be exercised using
"joysticks." Thus the comparison of computer-
controlled responses with manual operation can be
made. The comparison can be made based on the evaluation of performance indices.

The fact that the models used here are relatively
simple does not limit in any significant way their
usefulness for control system development. These
models have been shown to provide a good approximation to dynamic circuit performance and are particularly useful for screening of strategies and for
operator/engineer training.

DYNAMILL II Description

DYNAMILL II is a program written in BASIC for simulation of grinding circuit dynamics and control
using simplified models with a personal computer.
The single-stage grinding circuit being modeled

here is shown in Fig. 1. After identification of
the important subsystems in the circuit (ball mill,
sump, hydrocyclone, pump, and piping), it is necessary to specify an appropriate model for each of
them.

Only subsystems with long response times need to be
modeled as dynamic elements. In this regard, the
hydrocyclone, which has a very rapid dynamic response (retention time of a few seconds), requires
only a steady-state model for its description.

This simplified dynamic simulator for a grinding-
classification circuit is obtained from a series of
mass balances for solids, water, and the amount of
material that is coarser than a certain size, written around the mill, the sump, and cyclones. Instead of considering a large number of size fractions, as is done with more detailed simulators,
one variable characterizes the entire size distribution, i.e., cumulative fraction coarser than a
specified size. The mathematical model of the ball
mill is based on the assumption that the mill acts
as a single perfect mixer. The solids are assumed
to have the same residence time distribution as the
liquid. The ball mill type is an overflow mill;
thus constant volume is assumed. The kinetics of
breakage is assumed to follow the simplified first-
order rate law described elsewhere (Herbst and
Fuerstenau, 1968; Herbst and Bascur, 1979). The
major equations used and a detailed description are
found in the user manual of DYNAMILL II. With
these assumptions, a two-size-fraction model with a
single grinding-rate parameter can be used to accurately describe the dynamics of the ball mill.

For steady-state simulation of grinding performance, a model for the sump and pump is not required, since at steady state the solid/liquid
ratio in the inlet and outlet of the sump are the
same, and, since no grinding or classification occurs in the sump, the steady-state size distribution in the inlet and outlet are identical. The
sump is, however, a very important dynamic element
of a grinding circuit and therefore must be properly modeled. Its dynamic importance arises from
the fact that a disturbance in the feed to a typical ball-mill sump will persist for several minutes. A persistent disturbance in the sump has a
significant dynamic impact on the entire circuit
through the cyclone and resulting recycle. The
sump was modeled as a single perfect mixer, although the actual sump has more complicated behavior. The default version of the simulator assumes
constant volume of slurry in the sump. This assumption is premised on the availability of a variable-speed pump plus a fast level-control loop, so
that increases or decreases in slurry flow to the
sump are compensated for immediately by an equivalent change in pumping rate.

In the modeling of the hydrocyclone, the empirical
models of Rao and Lynch (Lynch, 1977) are used
along with the Plitt equation for the corrected
efficiency (Plitt, 1976). This latter relationship
is used to obtain an appropriate approximation for
the classification of a feed consisting of two size
fractions (Herbst and Bascur, 1979).

Automatic control of the circuit is simulated by
introducing the digital equivalent of two PI (proportional and integral action) controllers. The
manipulated variables in this case are fresh feedrate and the water addition to the sump. The controlled variables are the circulating load and the
cumulative fraction finer than a specific size.

The output of the program is obtained via a CRT.
Here the dynamics are updated with time on a flow
diagram. Selected variables are stored in a file
in a tabular form for later analysis and plotting.

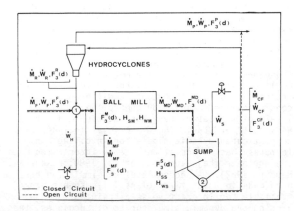

Fig. 1. Circuit simulation and notation
for DYNAMILL II.

The program is exercised in an interactive manner and can be stopped at any time in order to obtain more information and can be restarted at the same point.

Use of DYNAMILL II

A flowchart of DYNAMILL II is given in Fig. 2. The program consists of five modules: (1) Data Input; (2) Circuit Calculations; 3) Disturbance Initiation; (4) Control Initialization; and (5) Output. A run is initiated in the Data Input module with a series of "reads" to establish circuit conditions prior to the disturbances. This is taken from a data file, or it asks for the necessary input. The program takes these initial conditions and establishes a steady state with them in the Circuit Calculations module. This module involves a node, mixing calculation, ball mill product, a sump product calculation, and a cyclone subroutine calculation. The system of nonlinear differential equations is solved by a fourth-order Runge-Kutta integration subroutine.

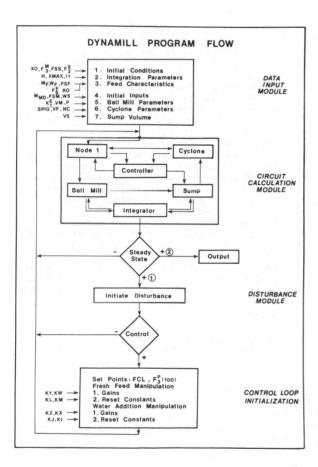

Fig. 2. DYNAMILL II flowchart

After the initial steady state has been achieved, the Disturbance Initiation module is used to identify the type and magnitude of the disturbance whose response is of interest. Step changes in water, feedrate, and feed size distributions are all options.

Once the disturbance is defined, a decision must be made as to whether an uncontrolled response, a manually controlled response, or an automatically controlled response is desired. If the uncontrolled response is desired, the program branches

back to the Circuit Calculation module and proceeds to find a new steady state. Intermediate results are printed out at the end of each minute of elapsed circuit time after the disturbance.

If a manually controlled response is desired, the use of a special joystick is necessary to manipulate the feed rate of solids to the ball mill or to close or open the sump water addition valve. To exercise this option, two trainees establish their objectives and then try to drive the plant as efficiently as possible to the new setpoints using their practical experience. Two performance indexes are stored for final comparison of their manipulations.

If automatic control options are to be exercised once the disturbance is defined, the program transfers to the control initialization module. Here a total of four gains and four reset times are entered to establish various linkages between the controlled variables, product particle size, and recycle mass flow (circulating load), and manipulated variables, sump water, and fresh feedrate. Once the tuning constants have been selected, program control is transferred to the Circuit Calculation module, which computes the controlled approach to steady state. Once again, intermediate results are printed out at the end of each minute of elapsed time after the disturbance. An objective function for comparison of the results of different selected tuning constants and strategies is calculated. The objective function evaluates the performance, trying to maximize the grinding production rate, $\dot{M}_F$, subject to penalty for producing a product that does not meet the size specification for flotation, y_{size}^{SP}, or for a circulating load that deviates from setpoint, y_{CL}^{SP}. The mathematical form of the performance index described is:

$$\phi = \int_0^T [\dot{M}_F(t) - \alpha_1(y_{size}(t) - y_{size}^{SP})^2$$
$$- \alpha_2(y_{CL}(t) - y_{CL}^{SP})^2]dt \qquad (1)$$

where the values of the penalty coefficients α_1 and α_2 are chosen according to desired conditions.

At the completion of either controlled, manual, or uncontrolled options, program control is transferred to the output for final display. A table of the time evolution of the manipulated and controlled variables is obtained for further analysis and/or for plotting.

Examples Using DYNAMILL II

In the examples which follow, the use of DYNAMILL II is described for (1) evaluating the uncontrolled dynamic response of a grinding circuit to a sump water disturbance and the development of a process matrix for control and manipulated-variable linking, and (2) tuning a control loop to maintain constant particle size during a setpoint change.

1. Problem Statement (Uncontrolled Dynamics):
   Consider a 10x15-ft ball mill which is operated in closed-circuit with five 20" cyclones at a fresh feed rate of 80 tph in a slurry containing 55% solids with F_3^F (100 μm) = .25 fraction. The sump water addition rate is 150 tph, and a variable-speed drive pump with good sump-level control is available.

   Determine the dynamic response of the circuit to a step change in sump water from 150 to 200 tph. Obtain a simplified process matrix for this circuit (Shinskey, 1967).

DYNAMILL II Solution: The first part of the solution requires the analysis of the circuit in terms of the disturbances, the measured variables, the manipulated variables, and the controlled variables. This analysis provides a table of grinding variables (Table 1).

Table 1. Classification of Variables in Grinding/Classification Control

Disturbance	Measured
Hardness	Densities
Feed size	Volume flowrates
Uncontrolled water	Levels
Uncontrolled feedrate	Power
Uncontrolled ball charge	Particle size
	Hold-up

Manipulated	Controlled
Sump water rate	Product size
Feedrate	Sump level
Pumping rate	Circulating load
Mill feed water rate	Hold-up of solids
Mill speed	

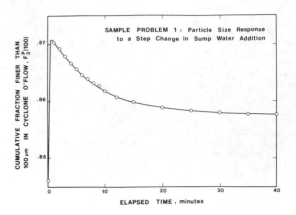

Fig. 3. Dynamic response of product size distribution to a step change in sump water addition.

Load DYNAMILL II into the memory of the computer, read the desired data file, and check steady state. Since the values entered are from a known steady state, the overall balance will appear on the CRT (Fig. 5), and the program will ask immediately for a step change in the manipulated variables. Only a step change in sump water from 150 to 200 tph is entered. The program will evolve to the next steady state. The evolution of $F_3^P(100)$ versus time is then obtained. Fig. 3 shows the approach to the new steady state of the particle size distribution obtained from the tabulation of $F_3^P(100)$ values displayed at the termination of the run. The process matrix for this circuit is shown in Table 2. This table gives a qualitative indication of how these four manipulated variables influence the four principal controlled variables of this circuit. The process matrix contains considerable useful information for the design of the control system. For feedback control strategies, the process matrix can be inverted to identify the most appropriate actions (outputs) for a given set of devia-

tions between measured variables of the controlled variables and their setpoint values. According to Table 2 there are several manipulated variables that can be made to yield large and rapid corrections in the controlled variables. The proper way to link the variables will depend on the interactions and objectives for this particular system. A more elaborate way of evaluating the process matrix has been reported by Tung and Edgar (1981). Table 3 enumerates the most appropriate options for linking manipulated variables with the controlled variables of interest. The various strategies involving the feedback control loops that are in use in industry today have been critically reviewed elsewhere (Herbst and Rajamani, 1979; Gault and others, 1978). New control strategies involving mill speed as a manipulated variable for ball mill grinding control have been analyzed by Herbst, Robertson, and Rajamani (1983).

2. Problem Statement (Controlled Response): From the steady state given in Problem No. 1, con-

TABLE 2 Response of Controlled Variables to Changes in Manipulated Variables as Indicated by the Process Matrix.

	Controlled Variable			
Manipulated Variable	y_1 = product fineness	y_2 = circulating load	y_3 = sump level	y_4 = mill % solids
u_1 = sump water addition rate	+ fast	+ fast	+ fast	- + slow
u_2 = fresh feed solids rate	- slow	+ slow	+ slow	+ fast
u_3 = cyclone feed pumping rate	+ fast	+ fast	- fast	+ fast
u_4 = feed water addition rate	+ - slow	+ slow	+ slow	- fast

TABLE 3   Linking of Controlled and Manipulated
          Variables for Classical Control Strate-
          gies for Grinding/Classification Control

CONTROLLED	MANIPULATED
Sump level	(1) Pumping rate
	(2) Sump water addition rate
	(3) Fresh feed solids rate
Circulating load	(1) Fresh feed solids rate
	(2) Sump water addition rate
Particle size	(1) Fresh feed solids rate
	(2) Sump water addition rate
Mill percent of solids	(1) Fresh feed water rate

figure a product particle size to sump water
addition controller with proportional gain KP =
450 and reset constant KI = 180. Examine the
response of the system to a setpoint change in
$F_3^P(100)$ to .86 (from .84). Use the sump-level
controller with the variable-speed drive pump
already existing in the plant.

DYNAMILL II Solution: The values from the previous steady state are retained for both the
manipulated variable and perturbation variable.
Select the control scheme desired by entering
the setpoint and necessary gains. Fig. 4 shows
the controlled response of the particle size
distribution as obtained from the tabulation of
$F_3^P(100)$ values displayed at the termination of
the run. The effect of other tuning constants
is shown. Note that, by doubling the proportional gain, a very fast rise time but a longer
settling time are obtained. An over-damped
response is obtained when half of the proportional gain is used.

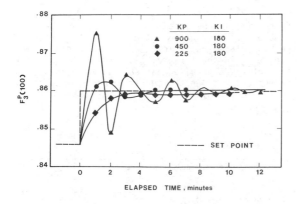

	KP	KI
▲	900	180
●	450	180
◆	225	180

Fig. 4.   Particle size distribution
          response to a setpoint change
          for various tuning constants.

DYNAFLOAT II Description

DYNAFLOAT II is a program written in BASIC for the
dynamic simulation of a flotation cell using a phenomenological model developed by Bascur and Herbst
(1982). In the development of this model, an
attempt was made to include all the geometrical,
manipulated, and controlled variables. The flota-

tion process involves the interaction of three
phases: solid, liquid, and gas. For the purpose
of model development, an abstraction of the process
was made. The flotation cell is divided into two
volumes: the pulp volume, in which intimate parti-
cle/bubble contact is induced by the turbulent
action of the impeller; and the froth volume, which
acts as a separating medium to segregate and to
remove the valuable minerals.

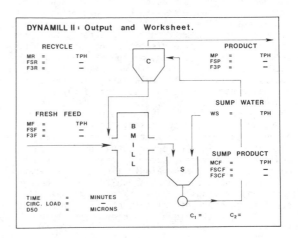

Fig. 5.   DYNAMILL II CRT output.

In summary, the general flotation model is based on
a population balance model and the hydraulic characteristics of the three-phase contacting device.
The model represents the behavior of each mineralogical species selected and any number of particle
sizes. Each of the particle types can exist in one
of four states in the cell (free in the pulp, attached in the pulp, free in the froth, attached in
the froth), and water can exist in one of two
states (in the pulp or in the froth). Mechanisms
of interphase transfer represented in the kinetic
equations include attachment/detachment and en-
trainment/drainage. In each case the influence of
important manipulated variables of the flotation
process, such as aeration rate, frother addition,
agitation, pulp level, froth level, or interphase
transfer are included in the model equations.

Early workers concerned with the modeling of the
flotation process generally concentrated on the
behavior of valuable particles and tended to ne-
glect the behavior of gangue particles and water.
Recent investigators have begun to recognize the
importance of including water behavior in flotation
models (Lynch and others, 1981). Empirical entrainment models for gangue recovery have been used
by Lynch and others (1981), while Moys (1978) has
given a first approximation model for the distribution of water in a flotation cell. One of the most
important factors that controls the water distribution in a flotation cell is the froth. Unfortunately there has been very little consideration of
water and particle transport in the froth, even
though the froth and associated drainage phenomena
are considered to be the principal reasons for the
cleaning action in the cell. The approach used
here to describe the hydraulic transport through
the froth involves the assumption that bubbles
leaving the pulp carry a sheath of water, which
moves into the froth. The size of the bubbles and
the thickness of the water sheath are determined by
frother concentration and agitation conditions in
the pulp volume. Drainage of water back into the

pulp is modeled by analogy with drainage through bubble plateau borders in foams. The height of the froth is determined by the froth removal mechanism. For natural overflow, results of studies of flow over a weir were used. A detailed description of the flotation model can be found in the references (Bascur and Herbst, 1982; Bascur, 1982).

The dynamic simulator program was written incorporating the particle model, the hydraulic model, the hydrodynamic relationships of the flotation cell, and geometric considerations. The models were parameterized with steady-state and dynamic experimental data. The simulator is capable of handling disturbances such as feed mass flowrate, feed percent solid, feed size distribution, and mineralogical species distribution, manipulated variables such as aeration rate, froth addition, impeller speed, tailings pumping rate, and pulp level, and various controlled variables. The simulator is represented in a general state representation, so nothing is said specifically about output variables, which can be any variable that is observed, being a state or a combination of states. In the simulator, typical observed output or technical indices are provided, such as grade of major element and valuable mineralogical species, flowrates of solids and water, percent solids in the concentrate and tailings streams, recovery of valuable species, recovery of valuable element, concentrate mass flowrate of valuable species, froth height, and water recovery.

The direct digital control (DDC) functions of the minicomputer are also incorporated in the simulator. The flexibility desired for simulation using this algorithm is demonstrated below.

Use of DYNAFLOAT II

A flowchart of the dynamic simulator, DYNAFLOAT, is shown in Fig. 6. The simulator consists of four main modules; each of these is subdivided into appropriate subroutines. The data input module handles all information required to exercise a particular option of the program. The calculation module consists of three major subroutines: the flotation mode; the controller; and the integrator. The flotation model routine combines the particle flotation model with the model for cell hydraulics. Each size and mineralogical species is described by a set of four differential equations plus the water model. The states calculated in this routine are used in an auxiliary routine to perform all the output variable calculations each time they are required by the main program. The symmetry of the model equations was exploited to optimize the computer code making it as efficient as possible. In spite of this optimization the complexity of the equations produces code which runs rather slowly on standard personal computers. A computationally simpler version of the particle model is now being incorporated as an option for DYNAFLOAT II to speed up calculation. This simplified flotation model has been described by Bascur and Herbst (1984).

The controller subroutine consists of a general proportional/integral controller that allows the coupling of manipulated variables including aeration rate, impeller speed, frother addition, pumping rate of tailings, together with controlled variables, such as recovery of valuable species, grade, mass flowrate in the concentrate of valuable species, pulp level in the flotation cell, froth height, etc. There are also options to cascade the grade or recovery of the valuable species with the pulp level to obtain the froth height that will produce the desired grade or recovery for the operation conditions set by manipulation of any of the variables described earlier.

The PI control algorithm is given by:

$$u_i(t) = u_{i,S} + K_{p,i}\,\varepsilon_i + \frac{K_{p,i}}{\tau_{I,i}} \int_0^\tau \varepsilon_i\, dt \qquad (2)$$

where $u_i(t)$ is the instantaneous value of the manipulated variable, $u_{i,S}$ is the steady-state value of the manipulated variable, $k_{p,i}$ is the proportional gain, $\tau_{I,i}$ is the integral time, and ε is the deviation of the controlled variable from setpoint. The subscript i denotes the ith control loop. Under direct digital control (DDC), control action

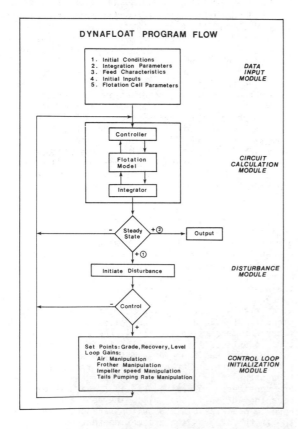

Fig. 6. DYNAFLOAT II flowchart.

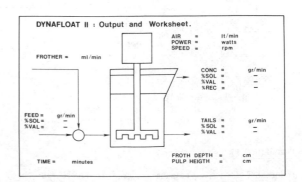

Fig. 7. DYNAFLOAT II CRT output.

is taken at discrete intervals of time as opposed to continuous control action in the case of an analog device. The corresponding DDC control algorithm is given by:

$$u_i(j\Delta T) = u_{i,S} + K_{P,i}\ \epsilon_i(j\Delta T)$$
$$+ \frac{K_{P,i}}{\tau_{I,i}} \sum_{\ell=0}^{j} \epsilon_i(\ell\Delta T) \qquad (3)$$

For large values of process error over prolonged periods of time, it is possible that the value of a manipulated variable might become unrealistically high or low. Therefore, an upper and lower bound for each of the manipulated variables was programmed in the simulator.

The integrator subroutine uses a fourth-order Runge-Kutta method for solving the overall system of nonlinear differential equations.

The disturbance module allows for introduction of random and/or deterministic disturbances to analyze dynamic responses or for testing of control strategies.

The control loop initialization module is called by the main program when different possible controlled strategies are to be analyzed.

DYNAFLOAT II is written for use in an interactive mode. The computer asks for options to be exercised, i.e., dynamic simulation, control, etc., and parameters input. A CRT display shows the evolution of the process in a schematic representation of a flotation cell (Fig. 6). The pulp and froth volumes have different colors and pulp level and froth depth change depending upon the operating conditions.

The model framework is very general and can in principle be used to represent the behavior of any flotation separation by simply adapting the subprocesses appropiately for a given ore. The parameterization and verification of the model for Illinois No. 6 coal and copper sulfide ore has been reported in the references (Bascur and Herbst, 1982, 1984; Bascur, 1982).

Examples Using DYNAFLOAT II

In the examples which follow the use of DYNAFLOAT II is described for 1) evaluating the dynamic response of a flotation cell to an aeration rate change and 2) comparing control strategies in which product grade is controlled by manipulating pulp level and by manipulating aeration rate.

1. Problem Statement (Process Matrix Development): Consider a one-cubic-foot pilot-plant flotation cell for coal cleaning. The coal contains 23% ash and is all -30 mesh material. The reagent schedule is 0.3 kg/ton (0.6 lb/ton) of frother MIBC and 0.76 kg/ton (1.5 lb/ton) of kerosene at pH of 6.5. The coal slurry feed is at 7.5 percent solids by weight. The flotation cell operates at 900 rpm, the aeration rate is 40 lt/min, and the frother addition rate is 50 ml/min.

    Determine the dynamic response of the circuit to step change to each of the manipulated variables with constant pulp level using a variable-speed pump.

    DYNAFLOAT II Solution: As suggested in the grinding/classification analysis, the first step is to classify the variables of the system under study. The listing of the available manipulated and controlled variables is shown in Table 4.

    To obtain the process matrix, start by tuning the pulp-level controller. Make individual step changes to aeration rate, impeller speed, and frother addition rate using the disturbance module and the pulp-level controller. Each time the program will evolve to the next steady state. The evolution of the recovery of clean coal, clean coal grade, froth depth, and pulp level are recorded. Fig. 8 shows the approach to the new steady state of the recovery, grade, and froth height. The process matrix obtained for this system is shown in Table 5. This table reveals the strong interactions between the controlled and manipulated variables for this flotation system.

2. Problem Statement (Comparison of control strategies): Configure and compare the performance of a grade/level controller and a grade/ air controller with constant pulp levels. Examine the response of the system to a setpoint change in the grade of valuable in the froth product from 55.5% clean coal to 53% clean coal.

    DYNAFLOAT II Solution: In the grade/level, the setpoint for the pulp level in the flotation cell is cascaded from the grade-controller loop, that defines the pulp level necessary to obtain a desired grade.

TABLE 4 Variables of Importance for Coal Flotation Control

Disturbance		Measured
Coal rank and type		Coal composition
Coal oxidation		Volume flowrate
Feed size distribution		Pulp densities
Presence of clay slimes		Level
Water composition		Power input
Feed rate		
Manipulated		**Controlled**
Physical	Chemical	Recovery (yield)
		Grade (BTU value)
Aeration	Reagent addition	Circulating loads
Impeller speed	(frother, fuel oil)	Froth level
Pulp level	Frother addition ratio	Percent solids
Froth sprinkling rate	Reagent addition points	
Gross/Fine split		
Conditioning time		

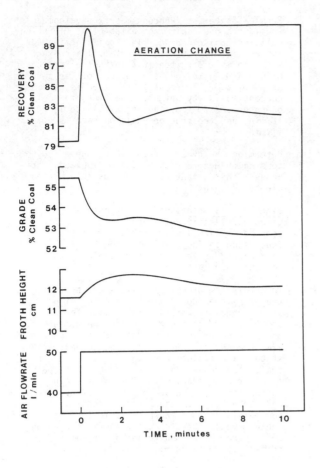

Fig. 8. Simulated response in grade,
         recovery, and froth depth for
         a step change in aeration rate.

TABLE 5   Linking of Controlled and Manipulated
           Variables for Classifical Control
           Strategies for Coal Flotation

CONTROLLED		MANIPULATED
Grade	(1)	Aeration rate*
	(2)	Pulp level
Recovery	(1)	Pulp level
	(2)	Aeration rate*

*Impeller speed coupling to keep specific power
constant has been suggested by Herbst and Bascur
(1983).

For evaluation and tuning of feedback control
strategies, a systematic search minimizing a
performance index is recommended. The integral
of the square error between the actual and desired setpoint can be evaluated, and the controller indices such as rise time, settling
time, and offset error can be used to decide
which controller performs better (Dorf, 1980).

Figure 9 shows the responses of the coal system
to a grade setpoint change from 55.5% to 53%

clean coal, with air and level as the manipulated variables. The grade/air controller response gives a very desirable response with a
short settling time. The grade/level controller, due to is cascaded nature, is somewhat
more oscillatory and displays a longer settling
time. Notice that both controllers produce a
very fast response in recovery with the grade/
air controller producing a stable response more
rapidly. From the analysis of these two alternatives, it can be concluded that grade/air
with constant level produces a better control
response. The description of several more elaborate control alternatives for coal flotation
have been reported by Herbst and Bascur (1983).

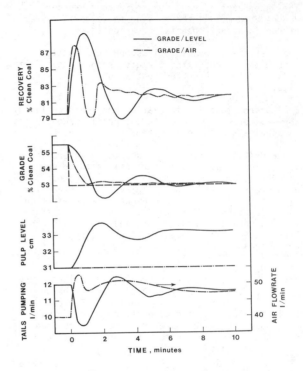

Fig. 9. Comparison between grade/air
         and grade/level control loop
         responses of recovery and grade.

SUMMARY AND CONCLUSIONS

Two simplified dynamic simulators, DYNAMILL II and
DYNAFLOAT II, have been presented to aid in the introduction of process control concepts for mineral-
processing systems. The use of simplified phenomenological models allows the principal characteristic of the systems under analysis to be represented. The interactions of the most important
variables for process optimization are included.

The combination of color graphics, simple models,
interactive computer access, and personal computer
simplicity associated with this type of dynamic
simulator will allow a non-computer-skilled operating engineer to "get his feet wet" before moving on
to more sophisticated dynamic simulators.

Systematic analysis for process control synthesis
is clearly enhanced by the use of semi-empirical
dynamic models that include the principal subprocesses of the system under study. A useful technique, using simplified phenomenological models for

control system strategy development, has been reported by Godvind and Power (1982). A more sophisticated analysis for synthesis of control structures, based on aspects of structural controllability and observability to yield a controllable and observable system, has been presented (Morari et al., 1980). The educational power that is offered to students by a console dynamic simulator system is far greater than anything devised in a hands-on unit operation. The amount of knowledge stored in the equations describing the fundamental processes occurring in each unit allows the trainee to develop a real feel for process dynamics and control. Sensitivity analyses can be conducted and a variety of alternative strategies can be tried in a short time. The adaptive nature of modeling should always be considered for updating the dynamic simulator as experience accumulates. The communication of information using the type of simulators described in this article is believed to be enhanced through "user friendliness." The more knowledge about the process acquired by the trainee, the better will be the controls he/she will develop for it.

The development of dynamic models for other unit operations, such as thickening and leaching, is currently underway at the University of Utah. Similar dynamic simulators are being planned for dynamic analysis using personal computers.

The interested reader can either develop his own simulators using simplified model equations described here or can acquire the code and a user's manual at a nominal cost from the University of Utah.

The proper interconnection of the processes to synthesize regulatory control structures for a large-scale mineral-processing plant is clearly required in the future. A hierarchical structure with different level of optimization has been proposed by Umeda, Kuriyama, and Ichikawa (1978). The first level of design activity starts by generating alternative control structures for each single unit in the plant. The second level of design activities is concerned with coordinating the unit operations' control structures to arrive at a plant control strategy by minimizing the conflicts among the control strategies in the various unit operations. A more sophisticated approach developed by Morari (1977) and Morari, Arkun, and Stephanopoulos (1980) incorporates, within the hierarchical control concepts, useful criteria to classify the disturbances, translate economic objectives into controlled variables, and decompose the control tasks and the process.

An acquaintance with such dynamic simulators is a must for the development and implementation of model-based control for hierarchical plant optimization. The availability of distributed digital control systems together with adequate on-line models open a new approach to the control of mineral-processing operations (Herbst and Bascur, 1984). Making use of the process knowledge/information imbedded in the models promises to approach optimal use of our resources and will prove economically beneficial for the operation of mineral-processing plants in the future.

REFERENCES

Bascur, O. A. (1980). General analysis of separation efficiency relationships: Application to hydrocyclones. Tech. Notes in Metallurgy, Dept. of Metallurgy, University of Utah, Salt Lake City, Utah, 4, No. 2 (Winter).

Bascur, O. A. (1982). Modeling and computer control of a flotation cell. Ph.D. Dissertation, University of Utah, Salt Lake City, Utah (University Microfilms International, Ann Arbor, Michigan, U.S.A.), December.

Bascur, O. A., and J. A. Herbst (1982). Dynamic modeling of a flotation cell with a view toward automatic control. Preprints of the XIV International Mineral Processing Congress, Canadian I.M.M., Toronto, Canada.

Bascur, O. A., and J. A. Herbst (1984). On the development of a model-based control strategy for copper-ore flotation. In E. Forssberg (Ed.), Flotation of Sulphide Minerals Workshop preprints, Stockholm, Sweden, pp. 295-316.

Dorf, R. C. (1980). Modern Control Systems. Addison Wesley Publishing Co., 3rd Edition, Massachusetts.

Gault, G. A., N. J. Howarth, A. J. Lynch, and W. J. Whiten (1978). Automatic control of grinding circuits. AMDEL, Brisbane, Australia.

Godvind, R., and G. J. Powers (1982). Control system synthesis strategies. AIChE Journal, 28, 60-73.

Herbst, J. A., and O. A. Bascur (1979). STEMILL: A program for the simulation of steady-state grinding behavior using a small computer. Program description and user manual, University of Utah, Salt Lake City, Utah.

Herbst, J. A., and O. A. Bascur (1983). Alternative control strategies for coal flotation. 112th SME-AIME Annual Meeting, preprint #83-614, SME, Colorado.

Herbst, J. A., and O. A. Bascur (1984). Mineral processing control in the 1980s -- Realities and dreams. In J. A. Herbst (Ed.), Control 84, SME/TMS of AIME, Colorado. pp. 197-215.

Herbst, J. A., O. A. Bascur, and K. Rajamani (1979). DYNAMILL II: A program for the simulation of grinding circuit dynamics and control using a small computer. Program description and user manual, University of Utah, Salt Lake City, Utah.

Herbst, J. A., and K. Rajamani (1979). Control of grinding circuits. Computer Methods in the 80's. A. Weiss (Ed.), SME Press, New York, pp. 770-786.

Herbst, J. A., K. Robertson, and K. Rajamani (1983). Mill speed as a manipulated variable for ball mill grinding control. 4th IFAC Automation in Mining, Mineral and Metal Processing, Helsinki, Finland, Pergamon Press, New York, 153-160.

Lynch, A. J. (1977). Developments in Mineral Processing. 1. Mineral Crushing and Grinding Circuits. 2. Their Simulation, Optimization, Design and Control. Elsevier Scientific Publishing Company, New Jersey.

Lynch, A. J. (1984). Computers in mineral processing -- The first twenty-five years. 18th APCOM International Symposium, IMM, London, pp. 1-10.

Lynch, A. J., N. W. Johnson, E. V. Manlapig, and C. G. Thorne (1981). Mineral and Coal Flotation Circuits, Their Simulation and Control. Elsevier Scientific Publishing Company, New York.

Morari, M., Y. Arkun, and G. Stephanopoulos (1980). Studies in the synthesis of control structures for chemical processes. AIChE Journal, 26, 220-232.

Plitt, L. R. (1976). A mathematical model of the
hydrocyclone classifier. CIM Bulletin, December, 114-123.

Rajamani, K., and J. A. Herbst (1980). A dynamic
simulator for the evaluation of grinding circuit
control alternatives. 5th European Symposium on
Comminution, Amsterdam, pp. 64-81.

Shinskey, F. G. (1967). Process Control Systems.
McGraw-Hill, New York.

Tung, L. S., and T.F. Edgar (1981). Analysis of
control-output interactions in dynamic systems.
AIChE Journal, 27, 690-693.

Umeda, T., T. Kuriyama, and A. Ichikawa (1978).
A logical structure for process control system
syntheses. Proc. IFAC Congress, Helsinki.

Computer Aided Mineral Identification

DR. GEORGE L. KOVACS

Computer and Automation Institute, Hungarian Academy of Sciences, Budapest, Hungary

Abstract. A microcomputer based expert system is presented for fast and cheap mineral identification. Because of the increasing pipularity of and interest for minerals and precious stones (selling, hobby polishing, collecting, teaching etc.) we pointedly deal with gemstone identification. In most cases only some fundamental data are needed to identify a mineral sample (or precious stone) if we know how to use these data. The important, characterizing data of the well-known minerals are stored in the data-base of our program-system. The visible and measured optical and physical properties of the sample are compared adequately with the stored data to get the result, the name of the examined mineral. The system works in a simple, friendly way — asking questions, understanding answers — so that no special computer knowledge is necessary to use it, only the answers should be typed in. The main advantage of the system is that only relatively simple and fast measurements are required. Based on these measurements a fast, reliable identification is processed on a cheap computer running a sophisticated program-system. The MINTEST system handles even those cases when the tested mineral cannot be determined based on the given input data.

Keywords. Microcomputer; mineral identification; expert system.

1. INTRODUCTION

The revolutionary fast decrease of microcomputer prices made it possible to use computers not only in big factories, universities and research institutes, but even in household. Small capacity microcomputers (e.g. Commodore 64) are now available for 100—400 $ or high capacity ones (e.g. IBM PC) can be purchased for 1000—10000 $. The prices depend mostly on disc, other peripherals and software support.

The mentioned computer prices are comparable with that of one or two pieces of gemstone or with the price of an X-ray equipment or any other special mineral identification equipment of big laboratories.

If an appropriate program-system is available — microcomputers can be used for mineral identification. Such a program-system is now under development in the Computer and Automation Institute of the Hungarian Academy of Sciences. Special knowledge in minerology is supplied by the Mineralogical Dept. of the Lorand Eotvos University of Sciences, Budapest.

The MINTEST system has some learning facilities to handle new, unexpected data, thus its capacity should always be increasing. Based on its knowledge base and learning facility the system is a really expert one.

As most small mineralogical laboratories, jewellers and even mineral collectors, furthermore hobby polishers may afford purchasing a microcomputer, it seems to be reasonable to supply the above mentioned program-system.

2. IDENTIFICATION DATA — VALUE LIMITS OF MINERAL PROPERTIES

The following section enumerates the main properties which should be determined in order to identify a polished precious stone or any mineral sample. Some of these properties may be determined by visual observations, others with simple, fast, not too expensive measurements. Some of the data are used mostly in case of gemstones (g), others give information on non-polished minerals and grown-up crystals (m) only, but most of them are commonly used.

- The knowledge of all the possible values (for transparency, colour, streak, fracture, cleavage, crystal form, pleochroism and UV sensibility) and the lower and upper limits (for hardness, specific gravity, light refraction, birefrigence and dispersion) is necessary for input data checking when the data base is generated. (See section 3.1.1.)

- For every mineral which is put into the data-base the appropriate enumerations and limit values should be available. (See section 3.1.2.)

- When a sample is tested there is only one colour, streak or hardness value which has to correspond to one of the valid values found in the data base during the identification process. (See sections 3.2 and 4.)

The main properties are as follows:

- Transparency:
  transparent
  semi-transparent (translucent)
  non-transparent (opaque)

- Colour, streak:
  there are about 15—20 different, possible colours

- Hardness:
  full or half numbers between 1 and 10

- Fracture (m):
  conchoidal (shell-like), subconchoidal, splintery, hackly, smooth, irregular, uneven, earthy

- Cleavage (m):
  very good (perfect), good, poor, none; along twin boundaries

- Cleavage direction:
  basal, octahedral, prismatic, pinacoidal, cubic

- Crystal system:
  cubic (isometric), trigonal, tetragonal, hexagonal, orthorombic, monoclinic, triclinic, amorphous, (with about 32 subclasses)

- Crystal habit:
  columnar, prismatic, tabular, bladed, foliated, botryoidal, reniform, granular, massive

- Specific gravity (SG): between 1 and 22

- Luster: metallic, adamantine, greasy, resinous, silky, pearly, non-metallic

- Light refraction: between 1 and 4

- Birefrigence: between - 0,3 and + 0.3; or none

- Dispersion: between 0 and 0.3

- Pleochroism (g): none, weak, strong in some colours

- UV sensibility: none, weak, strong in some colours

- Absorption spectrum: between 400 Å and 7000 Å

- Rare light phenomena: asterism, adularescence, fluorescence, thermoluminescence etc.

- Other properties: radioactivity, magnetic and electrical properties, toxicity solubility, sublimation, hygroscopic properties etc.

The exact values (or their lower and upper limits) are well-known for all minerals, thus having the appropriate program system and on the other hand the measured data of an examined sample are available, the identification can be done.

It can be seen that the number of all the possible data and value ranges are rather limited. That fact gives a great assistance in organizing our program-system. It stands for the Data Input part and for the Interrogation System, as well.

It has to be mentioned that the MINTEST system recently doesn't deal with absorption spectrum measurements, and with any "other properties". For gemstone identification an additional microscope test is suggested.

3. STRUCTURE OF THE MINTEST PROGRAM–SYSTEM

The system consists of the main parts:

- Data Input of Minerals
- Interrogation System for Identification.

3.1 Data Input

This program part makes it possible to load the computer with all the necessary data which characterize the minerals/gemstones that we want to identify. The data should be read in easily with validity tests.

The input program is organized in form of dialogue systems and it consists of two parts:

3.1.1 Input of possible values and value limits. The operator should answer the questions of the computer. The questions ask for data as given in the previous section ("Identification data...") of this paper. For translucency, colour, streak, fracture, cleavage, crystal form, fluorescence and pleochroism the operator has to enumerate all the valid answers (e.g. all the colours) and for hardness, specific gravity, refraction, birefrigence, dispersion the possible maximum and minimum values should be typed in.

The system checks whether the number of values typed in corresponds to the question. E.g. if three hardness limits are given or if the lower limit is higher than the higher one etc., an error message appears and the question is repeated.

The enumerations of possible values and value limits are used in the following program parts not only for input

data checking but as menus as well.

3.1.2 Input of mineral data. This part of the input system gives questions to the operator starting with the name and synonyms of the mineral, and then the same questions are asked as for the first input part. The given answers should be based on a good handbook. (In our case the "BLV Bestimmungsbuch 17: Edelsteine und Schmucksteine", W.Schumann, 1981 was used.) Typing in all the specific data of all important minerals is a rather tedious and time consuming activity. However, it's worth-while doing it.

To make the input process faster and easier the operator doesn't have to type in all possible values of enumerative properties (e.g. colour, streak) but the program gives assistance offering the previously (3.1.1) learnt data in form of menus.

All the possible answers to all questions appear on the monitor of the computer and the operator has only to choose the valid ones by using the cursor.

When exact values should be typed in (e.g. refraction, SG), the operator has to do it carefully, as these data will be compared only with the previously defined limit values (e.g. a hardness larger than 10 or smaller than 1 is not accepted).

Recently the program has been built up for 200 minerals (mostly gemstones) but it still can be enlarged. All the data of all minerals can easily be checked and corrected or completed if it is necessary. Further minerals with all their data can be added to the system and even new properties can be built in.

The Data Input parts of the system should be used only once. All the data of all minerals are stored in an appropriate data base which is copied to and stored on disc, when data input is completed.

3.2 Interrogation System for Identification

This program part contains search and compare procedures – looking for the measured data of the sample in the data base.

The interrogation program is organized in such a way that after each question and answer the range of possible resulting minerals is getting smaller, and finally the sample is identified. The program structure can be enlarged by asking further properties, too, which are not active at the moment (e.g. absorption spectrum).

4. MINERAL IDENTIFICATION WITH THE MINTEST SYSTEM

The work of the Identification System starts with loading in the data base. (See section 3.1.2.) It is followed by the question-answer part (3.2).

The computer puts questions to the user of the system and waits for exact numbers as answer in case of S.G., light refraction etc., or the system provides a menu to choose one possibility in the case of colour, crystal form etc. The questions are the same as they were at Data Input. If a measurement result is missing or uncertain, only a Return should be typed in.

It is supposed that in most cases the mineral tests were previously done, however the operator may process the test just when the appropriate question appears.

The validity of all answers is checked and if error appears the question is repeated with the message: "invalid answer".

Having a valid answer the next question appears. The sequence of questions is determined to assist logical testing procedure, starting with easily defined properties, however the operator may choose any other order he wants.

As it was previously mentioned the range of possible resulting minerals is getting smaller after each answer. If some questions are not yet answered and the range is so narrow that the mineral is already identified the process could be finished.

However it has to be underlined that the skipped questions might contain contradictionary answers, thus no questions are suggested to be omitted if the answers are available. This way some errors may be avoided.

Some measurements are rarely done (e.g. absorption spectrum), some properties are impossible or hard to define in certain cases (e.g. crystal form for massive minerals), or some values are missing, even if they could be determined (e.g. the hardness of gemstones, as examination may damage the stone).

If there are too many missing data for a given sample, it may happen that the system doesn't give the exact solution, but 2−3 or more minerals are chosen as possible results. In most of these cases further tests are suggested automatically to arrive to a correct, unambiguous conclusion − as the system knows which missing data will be decisive. Sometimes the human intelligence may help to select the only solution. For precious stones an examination by using a microscope almost always helps as the characteristics of microscopic inclusions are assisting to lead to correct diagnosis. This test is the most important in distinguishing between natural and synthetic stones.

It is possible that even if all questions are answered the system results in no solution. This might have the following reasons:

- failure in measurement
- failure in data input
- failure in the data base
- appearence of an unknown mineral.

After having checked whether none of the first three reasons exists the learning feasibility of the system steps into action. The new data will be incorporated into the system − when their validity is checked in a mineralogy handbook. We suppose that in most cases there is a rare mineral, the data of which weren't coded and now it is added to the system. However, it may occur that really a new mineral is identified this way.

If a mineral is identified the system asks whether the next identification should start or not.

5. CONCLUSION

A system was proposed for (mineral and) precious stone identification. The system consists of a good mineralogy handbook, of measurement instruments, and of a microcomputer equipped with a sophisticated program-system, called MINTEST:

The basic measurement instruments are as follows:

- hammer (m)
- magnifying glass
- white ceramics plate (m)
- polariscope (g)
- Mohr-Westphal beam balance or Jolly spring balance (for S.G.)
- refractometer
- Mohs-hardness scale needles
- microscope
- spectroscope is not yet used.

Using all these instruments (or some of them) most optical and physical properties of a sample can be determined. Based on these data is takes a long time to decide which kind of mineral is present if only a handbook is used as assistance.

Having a computer and our appropriate, friendly useable program-system, the identification may be carried out easily in a couple of minutes.

The cheapest computer on which the MINTEST program runs is a Sinclair Spectrum or a Commodore 64 but on these computers only a restricted number (approx. 150) of minerals can be identified. On an IBM PC category computer there is no restriction in the number of minerals to be identified.

ACKNOWLEDGEMENT

I should like to express my thanks to Dr I.Gatter of the Mineralogy Dept. for his assistance by discussing some problems and by correcting the manuscript.

Educational Plans and Programmes for Automatics and Metrology of Mining Technology at the Polish Universities

L. KRUSZECKI

Institute of Mining and Dressing Machines and Automation, University of Mining and Metallurgy, Cracow, Poland

Abstract. About 60 persons are admitted to the graduate course every year after passing the entrance examinations. The entire course of studies is free of charge and the universities provide accommodation for the students who need it. The five years of studies are divided into 10 semesters. One semester lasts 15 weeks. Automatics can be studied at three separate faculties. At the University of Mining and Metallurgy, it is offered by the Faculty of Electrical Engineering, Automatics and Electronics as well as the Faculty of Mining and Metallurgical Machines (metrology can also be studied there); at the Silesian Technical University in Gliwice, the Faculty of Mining offers the course. The educational plans and programmes comprise 3 groups of subjects: common for a given course of studies (foreign languages, political science and the basic technical knowledge) common for the speciality and subjects of diploma courses. The whole time of studies covers about 4400 hours excluding 3 periods of professional training and the physical training. The classes, laboratory classes, seminars and design classes are obligatory and the attendance is strictly checked. Students may extend their knowledge by participating and working in the Science Clubs. The students´ work is checked not only during the classes but also by the examinations held at the end of respective semesters. Those who fail the exams can repeat them only if very important reasons are presented. At the very end of the studies there is a diploma examination together with a discussion on the diploma thesis, written by the student. The successful result entitles the student to recieve the title of "magister inzynier" M.Sc. of his speciality. The graduates, after working for 2 years, can attend another course of studies lasting 2 semesters and called "Postgraduate Course", or start "Doctoral Studies" finished by the defence of a doctor´s thesis. In these cases the expenses are usually covered by the graduates´ employers.

INTRODUCTION

Every year about 60 students are admitted to the graduate course on automatics and metrology of mining in Poland. The course of study lasts five years and consists of ten semesters. At the end there is a diploma examination with a diploma master´s thesis. Up till now about 1260 students have graduated in this specialization. Their participation as engineers in the medium and larger enterprises of the Ministry of Mining and Power Engineering (mines, factories, design offices, research laboratories and others) amounts from 4 to 12%. Beside the graduates having degrees directly in the mining branches, the mining enterprises also employ graduates of other disciplines, such as automatics, metrology, information science and electronics graduating at other universities and technical colleges. For these graduates the time for achieving a necessary professional skill is longer because of their lack of knowledge in mining engineering.

ENTRANCE REQUIREMENTS AND HELP PROVIDED

Candidates who have completed 8-year course of Primary Education and 4-year course of Secondary Education or 5-year vocational course at Technical schools are eligible for admission to the course. The number of places is limited, so only those students are admitted who obtain the highest score during the qualifying entrance examination for the following subjects: mathematics, physics and a foreign language (one of the four languages – English, French, German or Russian). The education is free of charge. The universities provide accommodation for the students. The range and amount of economic help depends on the progress in studies of a student and the student's parents' income. The universities have at their disposal quite a large number of rooms at the student hostels and sufficient scholarship funds. In addition, a student may obtain a scholarship from a state enterprise which plans to appoint him after graduating in a particular field of specialization.

COURSES OF STUDIES

Organisation of Academic Year

The organisation of the academic year can be presented as follows:

. 3rd October – 23 January winter semester (the winter vacation is from 24th December to 2nd January)

. 24th January – 4 February winter examination session

. 5th–14th February a break between the semesters

. 15th February – 5th June summer semester

. the spring vacation is from 20th – 26th April

. 6th – 23rd June the summer examination session

. 24th June – 9th September summer vacation and practical training

Three Kinds of Courses

The education of automatics in mining is provided in three courses of studies:

. Electrical Engineering
. Mining and Geology
. Mechanical Engineering

In Poland there are two universities where the automatic control engineering in mining is held. They are the University of Mining and Metallurgy in Cracow and the Silesian Technical University in Gliwice. At the University of Mining and Metallurgy courses are provided according to the plans of studies applicable for "electrical engineering" in the Faculty of Electrical Engineering, Automatics and Electronics (from 1965) and for "mechanical engineering" in the Faculty of Mining and Metallurgical Machines (from 1969). At the Silesian Technical University instructions are provided for "mining and geology" in the Faculty of Mining (from 1969). Educating of automatic control engineers for mining on the basis of different course of studies leads to the fact that their knowledge would differ for different courses.

Course Organisation

The basic courses are made compulsory by the Ministry of Science, Higher Education and Technology. Each plan of studies comprises three groups of subjects:

. common subjects for the given course of studies,
. common subjects of the speciality,
. subject of diploma courses (final course of studies).

The common subjects of the given course of study, foreign languages, political science and the basic technical knowledge, are taught during the first five semesters and partly in the following semesters, and are compulsory in all the universities in Poland providing a given course. The group of common subjects for a given speciality is taught during the period from the sixth to ninth semester. The subjects of this group are specific for the given faculty and university, and in the considered cases these are "automation and electrification of mines" at the University of Mining and Metallurgy and the Silesian Technical University, and "automatics and metrology" at the University of Mining and Metallurgy. The subjects of diploma courses are introduced from the seventh or eighth semester and extended till the tenth semester. The speciaility "automation and electrification in mines" has two diploma courses: "underground mines" and "open cast mines".

Table 1 represents the number of hours of classes provided for each of the three groups of subjects for all the three courses of studies. There may by some deviations in the field of specialization and diploma course in matter as well as in duration of study period for the more talented students who decide to study according to an individual plan of studies. The students who undertake such a method of studies draw special attention of professors and assistant professors.

Updating and Improving Courses

Such a method of education for automatics and metrology with the five year course duration is regarded to be suitable at the present stage of mechanization and automation in the Polish mining industry. In the course of development of a

TABLE 1   Number of hours of classes for the three groups of subjects

Group of subjects	Course of study		
	electrical engineering	mining and geology	mechanical engineering
Common for the course (hours)	2895	2850	3131
Common for the speciality (hours)	810	1470	735
Diploma course (hours)	585 570*	–	555
Total	4290 4275*	4320	4425

*for open cast mines

given discipline and according to the needs of the national economy and methods of teaching, education plans are amended and improved every 5 to 10 years. In contrast to this, programme materials of particular subjects can always be changed to improve a given plan of studies. The educational plans and programmes are printed in one volume and are meant for the general use of the teachers and students. Also, the students are supplied at the beginning of every semester with the detailed programme material for every subject together with the list of the compulsory literature.

METHODS OF TEACHING

Teaching of each subject consists of several forms, such as :lectures, classes, laboratory classes or seminars and design classes. The contribution of the different forms of teachings to the whole course of studies is shown in Table 2.

TABLE 2   Contribution of the different forms of teaching

Course of study	Teaching forms				Total
	lecture	classes	laboratory classes or seminars	design classes	
Electrical engineering	1905	1500	765	120	4290
Mining and geology	1800	660	1590	270	4320
Mechanical engineering	1695	915	1425	390	4425

Lectures

The lectures are held in halls equipped with audiovisual aids. It is not compulsory for a student to attend lectures, but the lecturer, with the permission of the Faculty Council, may demand compulsory attendance for all students. The lectures can be read by professors, assistant professors, senior lecturers, lecturers and tutors (teachers with a Ph.D. degree), with the agreement of the Faculty Council. Other classes are compulsory.

Classes

The classes are held for a group of 16-30 students. The students are required to take part in these exercises and are subjected to evaluation tests.

Laboratory Classes

The laboratory classes are held for a group of 7-15 students. The classes are generally conducted by a teaching staff member and a laboratory assistant. The basic requirement for a student to use a laboratory stand is that the student must know the working instructions of a given equipment and substantially the experiment itself. The students have also access to the laboratory stands used by the members of the staff for industrial, or individual Ph.D. or D.Sc. research works. Every class is completed by a written and evaluated report. The students who were absent because of serious reasons, (illness, etc.), must have the class during their spare time.

Seminars

Seminars are held for selected subjects of speciality and diploma courses. In these classes scientific discussions are held on problems of a particular topic prepared by a student or a group of students. The seminars take place with a group of 7-15 students.

Design Classes

The design classes are provided for several subjects, such as engineering drawing, basic machine construction and mechanical engineering. The students prepare their first and second paper. These classes are conducted in rooms equipped with drawing facilities models and original objects and other aids. The topics of the first and second papers generally serve as the introduction to the problem of the future diploma thesis work.

Professional Training

Three periods of professional training are also included in the teaching process. The first term of training takes place in mines, the second in factories producing automation equipment and the third one is the diploma practical training which is arranged in an enterprise having facilities for the training in the field close to the diploma thesis. It is compulsory for a student to write a report of the training and to get credits for every practical training period.

Science Clubs

During the studies, the students can take part in different scientific activities in the Science Clubs. The science club of "Mining Automatic Engineers" at the University of Mining and Metallurgy organises every year on the occasion of the "Mining Festival Day" a students' scientific session, where the students deliver lecures on the subjects related to their seminar papers or diploma theses. The lectures and the way of their presentation are evaluated by the board consisting of the teaching staff. The best lectures are rewarded.

There is also a possibility for a student to earn some money by taking part in consultant works which are commissioned to the university by the industrial enterprises.

Another activity of the Science Clubs is to organise some scientific meetings during vacations in mining enterprises. The participating students at the scientific camp perform specified duties in the field of their speciality. For that work, the enterprise pays them money, provides accommodation and so on.

ASSESSEMENT

Examinations

Every semester has an examination session at its end. To be allowed to take part in the examination session a student must have credits for all subjects of a given semester.

Essentially, to get the credit a student must collect good marks during the semester of the classes, laboratory classes, seminars and design classes. After every semester, except the tenth, a student is obliged to pass the examinations in several subjects, most often three. In such an organised study system, a student is required to work more systematically throughout each semester. In a case of not obtaining credits of a certain semester or year by a student due to some personal or other important reasons, he may apply to repeat the semester or year. In a case of a lack of sufficient justifiable reasons for failure, the student may be struck off from studies and would be bound to return the money already received as a scholarship.

At the end of the study there is a diploma examination and a discussion of the diploma thesis. For the diploma examination it is necessary to know the problems of the common subjects of the given specialization and the diploma course. The subject of the diploma thesis is fixed by Institute authorities and connected with the research subjects carried out at the Institute or also with subjects forwarded by industrial enterprises.

A diploma thesis must contain: a discussion about the state of knowledge of a given problem, arguments and the main idea of the thesis, and the justification of the arguments. The thesis is evaluated by the supervising member of the teaching staff and a reviewer who may be a teaching staff member of the university or from outside. The diploma examination and discussion of the diploma thesis takes place before the Examining Board of Master Degrees accepted by Faculty Council. The members of the Board may be the thesis supervising members and reviewers. The graduate receives the degree of "Master of Science (in Polish "magister inzynier") of his speciality and the diploma course.

POSTGRADUATE TRAINING

The teaching activities of the universities are not only limited to graduate courses. The graduates, after a minimum of two years of employment in their profession, may contest for the titles of "engineer professional specialization". One of the conditions for obtaining a given professional specialization position is to complete "Postgraduate Courses" at the university. Such courses are provided according to the plans set by the University Senate. These courses are conducted periodically depending on the number of candidates.

The candidates are sponsored by their employers who are obliged to cover the costs of studies. The duration of the course is two semesters, having about 250 teaching hours. The course material contains the subjects closely connected with the given specialization and some basic subjects such as selected problems of higher mathematics, selected problems of theory of systems, etc. The classes are conducted by the teaching staff of the university and specialists from the industry and are held once or twice a week, usually on working days but numbering no more than 28.

The course is completed by the examination conducted by the Board appointed by the Faculty Council. The university also offers 3-year doctorate courses. The candidates for such courses are sponsored by their employers and they are entitled to the all employee´s services in their enterprises.

CONCLUSION

The need for the specialists-graduates from the University increases simultaneously with the constant development of the automation of mining technologies and management in mining. Young people willingly start such studies not only because of the importance and fashion of this branch but also because of higher salaries and better living conditions offered by the mining industry.

Considering the state and needs of automation in the perspective of the coming years, it can be said that the education of electrical engineering, mechanics, mining and geology is proper and very useful. The educational plans and programmes of studies as well as the requirements for the teachers and students, together with well equipped laboratories, contribute to the good training and preparation of future graduates.

Considerable attention is paid to students´ activities in Scientific Clubs or students´ scientific sessions.

The postgraduate studies are quite popular and common among the graduates of the University of Mining and Metallurgy as well as of univiersities who want on one hand to increase their professional knowledge and, on the other, to fulfil the requirements necessary to apply for their "professional specialization".

Problem of Limitation of Oscillation of Winder Ropes

LUDGER M. SZKLARSKI

University of Mining and Metallurgy, Al. Mickiewicza 30, Cracow, Poland

KAZIMIERZ JARACZ

University of Mining and Metallurgy, Al. Mickiewicza 30, Cracow, Poland

Abstract. At the operation of mine winders in deep shafts considerable oscillations of winding ropes may occur due to elasticity of both winding and balance ropes. The amplitudes of longitudinal oscillations may reach ± lm at relatively low frequencies, ranging from 0.5 to 2 c/s. These oscillations may be dangereus for the rope and produce several hazards to the winding installation as a whole. They may cause considerable instant accelerations on the pulley and dangerous tensions in rope wires. The excessive acceleration may cause the uncontrolled rope slip on the pulley. Several methods of limitation of these accelerations were developed, some of them have been already in control system of electric drive of the winder. The most simple ones are: a/ limitation of the rate of change of acceleration and retardation of the winder /so called "jerk"/ and b/ the proper selection of regulators of the winder control system to get the optimal control. The trouble in solving the problem of rope oscillations is that the dynamic system of the winder is a nonstationaryone owing to changing distance of conveyance from the pulley. Besides, the system is described by the equations with distributed parameters. The brief discussion of the theory and its practical aspects are enclosed as well.

Keywords. Electricdrives; electric variable control; optimal control.

1. INTRODUCTION

At the operation of winders in deep shaft considerable oscillations of winding ropes may occur due to the elasticity of both winding and balance ropes. The amplitudes of oscillations of the conveyances may reach ± 1 m at relatively low frequencies, ranging from 0.5 to 2 c/s. These oscillations may cause considerable instant accelerations on the pulley and dangerous tensions in the rope wires. The excessive instant accelerations in turn cause the rope slip on the pulley. Several solutions of the limitation of rope oscillations problem were suggested. The simplest one is the limitation of the rate of change of winder acceleration /"shock" or "jerk"/.

The osiillations may occur not only at starting or braking, but also at balance speed of the winder due to all kinds of disturbances such as change of friction in shaft guides, variations of winder acceleration, excentricity of the pulley due to the wear of the grooves, during the loading and discharging operations, et c.

In 1951 T.H. Petch published a very detailed study of the problem of rope oscillations in South African mines [3].

One of the plots of conveyance oscillations as well as stresses in winding ropes recorded by T.H. Petch are shown in Figs. 1 and 2. In his paper the author suggested the limitation of acceleration rate as the most effective means of suppresion of rope oscillations. Unfortunately, the W.-L. drives with the open-loop control systems were not quite suitable

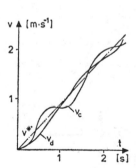

Fig. 1. Plot of the rope oscillations at the acceleration period. v - desired speed of the winder, v_d - velocity of the winder drum, v_c - velocity of the conveyance

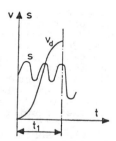

Fig. 2. Plot of the rope stresses at acceleration period. t_1 - acceleration period, S - rope stresses, v_d - velocity of the winder drum

333

for this purpose owing to considerable time constants and low accuracy of the contro control of these drives.
In the later dates several solutions of the probmem were proposed owing to the application of modern control systems of the winders /e.g. [19]/.
The authors of this paper intend to present the discussinn of some methods and general conclusions concerning the control of mine winders, enabling the limitation of rope oscillations.

2. THE METHODS OF LIMITATION OF ROPE OSCILLATIONS

2.1. Application of Kessler's Modular Optimum.
The modern electric drives of the winders are mainly the d.c. ones with thyristor control, or with a.c. motors /both synchronous or induction/ with trequency regulation based on the application of the thyristor cycloconverter frequency changers. In this latter case the control system is based on the so-called "Transvector" principle. In both cases however the driving motor static and dynamic characteristics are much alike. In all these drives the control system containes at least two closed-loops: a/ the tachometric, and b/ current one.
To choose the proper solution of the problem it is necessary to describe the dynamical system by means of some differential equations, at least with some simplifying assumptions.
The real winder is a nonstationary system due to the variation of rope length and the position of the load /i.e. the conveyance / during the winding cycle. This makes the exact solution of the differential equations of the system very difficult. Moreover the determining of the parameters of the regulators is even more difficult.
The dynamical system of the winder may be represented by the diagram shown in Fig.3.

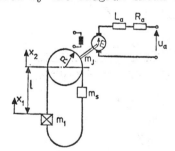

Fig. 3. Simplified diagram of winder

This model may be replaced by a simplified one by division of both winding and balance ropes into some definite mumber of discrete elements, consisting of elementary mass m_i, elastic member c_i, and damping member μ_i each /Fig. 4/. The elementary displacements are denoted x_i.
The following assumptions were made:
1. Elastic and damping elements were imponderable, their characteristics being linear,
2. The pulley and driving shafts of the rotor were stiff,
3. The linear displacement x_i of the centre of the elementary masses was assumed to be the co-ordinate system, with its direction in accordance with the direction of action of longitudinal forces,

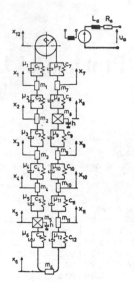

Fig. 4. Diagram of mine winder with discrete simulation of ropes /model with distributed parameters/

with zero point in the equilibrium centre of the elements.

At the beginning the system may be regarded as a stationary one at least during the acceleration and retardation periods of the winding cycle. During these periods the relative change of rope length is small and may be neglected.
The winder with two conventional regulators in the feedback loops with the d.c. driving motor will be discussed, mainly for simplicity reasons. The regulators are: 1. the proportional type /P/ in the tachometric loop, and 2. the proportional-integratortype /PI/ in the current feedback loop of control system. The system of electric winders with such control and optimization based on the above mentioned criteria were described in [12 & 13].

List of Symbols.
m_s - mass of conveyance,
m_u - net mass /net load/,
m_{wl} - mass of the left part of balance rope
m_j - effective total mass of all rotating parts of the winder, reduced to the pulley diameter,
E - elasticity modulus of the rope,
F - balance rope cross-section,
c - per unit length mass of the rope,
h_1, h_2 - factors of viscous friction of conveyances,
x_1 - linear displacement of loaded conveyance at the shaft bottom,
x_2 - linear displacement of the point on the pulley,
u_a - winder motor armature voltage,
i_a - armature current,
k_{PT}, τ_p - average gain and time delay of the thyristor converter, resp.
u_{st} - converter control signal /control voltage/,
k_{TG} - tachometer gain,
k_e, k_m - winding motor constants /e.m.f. and torque constants, resp./,
R_a, L_a - resistance and inductance of winding motor armature circuit, resp.
$T_a=L_a/R_a$ - electromagnetic time constant of winding motor armature circuit,
$C_I=EF/l$ - rope stiffness cactor,
$D=2R$ - pulley diameter,
M - winding motor torque,

$m_L = \delta l$ - mass of winding rope,
l_L - rope length,
$m_1 = m_s + m_u + m_{w1}$
$m_2 = m_j + m_3$
$m_3 = m_{L2} + m_s + m_{w2}$ - the mass of the right hand branch of the winding rope, lalance rope and the conveyance,
$x = x_2 - x_1$ - dynamical elongation of the rope.

Applying the Hamiltonian variational principle, the simulation model of the winder may be obtained corresponding to the given system of equations [13]:

$$\left[(m_1 + \frac{m_L}{2})s^2 + h_1 s\right]X_2(s) - \left[(m_1 + \frac{m_L}{3})s^2 + h_1 s + c_L\right]X(s) = 0 \qquad (1)$$

$$\left[(m_2 + \frac{m_L}{2})s^2 + h_2 s\right]X(s) + \left[-\frac{m_L}{6}s^2 + c_L\right]X(s) = \frac{M s}{R} - m_u g \qquad (2)$$

$$U_a(s) = \frac{k_{PT}}{1+s\tau_0}U_{st}(s) \qquad (3)$$

$$U_a(s) - \frac{k_e}{R}X_2(s)\cdot s = R_a(1+sT_a)I_a(s) \qquad (4)$$

$$M(s) = k_m I_a(s) \qquad (5)$$

where "s" is Laplace's operator.
The block diagram corresponding to this system is presented in Fig. 5.

$$= \frac{K_1 R(m_1 + \frac{m_L}{2})s}{k_{TG}(m_1 + \frac{m}{3})(s^3 + K_1 s^2 + \omega_e^2 s + K_1 \omega_F^2)} \qquad (10)$$

Where

$$K_1 = \frac{kk_m k_{TG}\omega_F^2}{k_i R^2(m_1 + m_2 + m_L)\omega_e^2}$$

$$\omega_e = \sqrt{\frac{c_L(m_1 + m_2 + m_L)}{m_1 m_2 + \frac{m_L}{3}(m_1 + m_2 + \frac{m_L}{4})}} \qquad (11)$$

$$h_1 = h_2 = 0 \;, \quad \omega_F = \sqrt{\frac{c_L}{m_1 + \frac{m_L}{3}}}$$

ω_F and ω_e are angular frequencies of rope oscillations at pulley standstill /brakes applied/ and at start of the winder, respectively.
The root-locus plot of the transfer function poles of the denominator of the closed-loop control system given by the equations (10) and (11) is shown in Fig.6.
PI type regulator will have the gain:

$$R_\omega(s) = k_{\omega i}\left(1 + \frac{1}{s\tau_\omega}\right) \qquad (12)$$

The closed-loop control system gains are:

$$G_{PI}(s) = \frac{X_2(s)}{U_{x_2}^*(s)} =$$

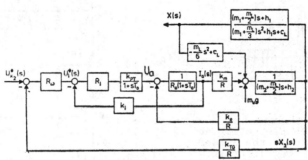

Fig. 5. Block diagram of the winder

The structure of the current-loop regulator may be determined from Kessler's modular optimum. Application of this optimum results in the equivalent transfer function $G_{zi}(s)$ of the current feed-back loop [18]:

$$G_{zi}(s) = \frac{I_a(s)}{U_1^*(s)} = \frac{1}{k_i(2\tau_0^2 s^2 + 2\tau_0 s + 1)} \qquad (6)$$

Regarding τ_0 negligible, the expression for $G_{zi}(s)$ will be simplified to:

$$G_{zi}(s) \approx \frac{1}{k_i} \qquad (7)$$

Transfer functions of the system with the speed regulators of P and PI types will be, respectively:
For P type regulator:

$$R_\omega(s) = k_\omega \qquad (8)$$

For closed-loop control system:

$$G_p(s) = \frac{X_2(s)}{U_{x_2}^*(s)} = \frac{K_1 R(s^2 + \omega_F^2)}{k_{TG}(s^3 + K_1 s^2 + \omega_e^2 s + K_1 \omega_F^2)} \qquad (9)$$

$$G_{PX}(s) = \frac{X(s)}{U_{x_2}^*(s)} =$$

$$= \frac{K_{1i}R(1 + 1/\tau_\omega)(s^2 + \omega_F^2)}{k_{TG}(s^4 + a_3 s^3 + a_2 s^2 + a_1 s + a_0)} \qquad (13)$$

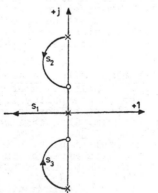

Fig. 6. Root-locus curves for the winder with P-type regulator of velocity loop

$$G_{PIX}(s) = \frac{X(s)}{U_{x_2}^*(s)} =$$

$$= \frac{K_{1i}R\left(m_1 + \frac{m_L}{2}\right)\left(s + 1/\tau_\omega\right)s}{k_{TG}\left(m_1 + \frac{m_L}{3}\right)\left(s^4 + a_3 s^3 + a_2 s^2 + a_1 s + a_0\right)} \quad (14)$$

$$K_{1i} = \frac{k_m k_{TG} k_{\omega i} \omega_e^2}{k_i R^2 \left(m_1 + m_2 + m_L\right)\omega_F^2} \quad (15)$$

$$a_3 = K_{1i}$$

$$a_2 = \omega_e^2 + \frac{K_{1i}}{\tau_\omega}$$

$$a_1 = K_{1i}\omega_F^2$$

$$a_0 = \frac{K_{1i}\omega_F^2}{\tau_\omega}$$

$$h_1 = h_2 = 0$$

From the root-locus plots the values of k_ω, $k_{\omega i}$, τ_ω may be determined which will produce the maximum damping effect on the rope oscillations. Fig. 7 shows the elongation x of the rope of the bottom conveyance versus the time diagram at the acceleration period $t_a = 14{,}5s$ and part of

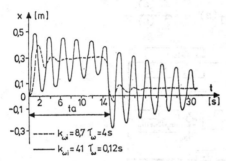

Fig. 7. The diagram of rope elongation at acceleration period of the winder with PI-type speed regulator for the parameters calculated from modular optimum.

the balance speed period. The solid line corresponds to the wrong selection of time constants of the regulators, while the dotted line corresponds to the optimal control. One may notice that with a proper choice of parameters, the winder operates practically without oscillations, whereas the oscillations occur due to the wrong choice of parameters $/k_{\omega i} = 41$, $\tau_\omega = 0{,}12s/$.

2.2. The Limitation of the Winder "Jerk". Let us assume the application of kinematic driving function to the pulley i.e. the realization of its rotational speed versus the time diagram. This case was discussed in [5] and [17].
Applying the digital simulation technique the following results were obtained.
The winder under consideration had the data given in [17].
The following cases were investigated:
a/ the acceleration applied was a jump function at the beginning of the winding cycle; after the end of acceleration period the acceleration jump function declined /i.e. the "jerk" was not limited/, b/, c/ and d/ – the jerk was limited to 7, 2 and 1 m/s^3, respectively. The results obtained are shown in the graph of Figs. 8, 9 and 10. The solid lines in these

graphs concern the skip, the dotted line – the pulley.

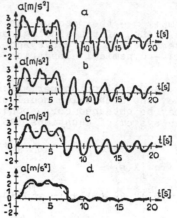

Fig. 8. Acceleration curves of the conveyance with "jerk" limitation:
a/ no limitation, b/ limit= 7 m/s^3, c/ limit= 2 m/s^3, d/ limit =1 m/s^3.

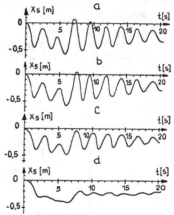

Fig. 9. Amplitude diagrams of the conveyance with "jerk" limitation, according to Fig. 8.

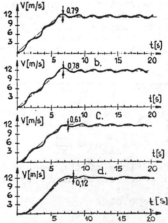

Fig. 10. Velocity graphs of the conveyance with "jerk" limitation, according to Fig. 8.

As one may notice the best results may be obtained with the jerk limitation to da/dt 2 m/s^3.
The effect of rope elasticity modulus E was investigated by several authors /e.g. [9]/. The elasticity modulus of the rope E depends on the material elasticity mo-

dulus E_m:

$$E = \propto E_m \qquad \text{where} \qquad \propto \, < 1$$

It may be noticed that the oscillation amplitudes increase with the decrease of $\propto$. The oscillation frequency decreases with the decline of $\propto$ /Fig. 11./.

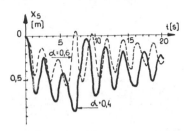

Fig. 11. Amplitude diagrams of the conveyance at different values of rope elasticity modulus

The above results relate to the acceleration period with the loaded conveyance starting from the shaft bottom. The above investigation proved that the jerk limitation is one of the most effective and simplest methods of diminishing the oscillation of the conveyance.
At the present most of modern winders apply the jerk limitation to 1 m/s³ or 0.6 m/s³.
Both methods described in paragraphs 2.1 and 2.2 may be applied with the closed-loop control of the winder electric drive. Some more sophisticated methods were investigated, they were described in [4], [5], [17] and [10].

3. FINAL REMARKS

1. The investigations of the winder performance by means of the simulation model to select the most effective method of limitation of rope oscillations is a very convenient way of solving the problem, provided that the identification of the winder is accurate enough.
2. The division of the rope into elementary sections, even not numerous ones /practically about 20/, provides just a very small error when compared with the measurments carried out on the real winder.
3. The simulation model provides the opportunity of investigation of a nonstationary model of the winder, which in turn enables examining the behaviour of the winder in all the winding cycle, not only in the beginning and final parts of the cycle, as with the stationary model. This also enables the calculation of the parameters of the regulators to achieve the optimal or suboptimal control of the winder.
Both the methods described provide the simple and effective methods of the rope oscillation limitation.
4. Further investigations in the field of the limitation of the rope oscillations are very important and should be carried out.

REFERENCES

[1]. Gierlotka, K. 1977 . Dynamics of electric drives of mine winders of deep shafts /in Polish/. Doctor's Thesis, Silesian Polytechnic, Glivice.
[2]. Kiszka, J. 1975 . Mathematical model and control of mine winders as a system with distributed parameters /in Polish/. Doctor's Thesis, Min. and Metallurgical Academy, Cracow.
[3]. Petch, F.H. 1951 . Proceedings IEE, Paper 1100-U, 98, part II, 65.
[4]. Szklarski, L., Zając M., A. Dziadecki. 1978 . Scientific Papers of the Min. and Metall. Academy in Cracow, Section Electrification and Mechanization in Minning and Metallurgy /in Polish.
[5]. Szklarski, L., M. Zając, A. Dziadecki 1977 . Papers of ICAMC Congress, Ostrava
[6]. Szklarski, L., R.Wojnicki, A. Stankiewicz 1969 . Papers of ICAMC Congress in Cracow.
[7]. Szklarski, L., M. Zając, A. Dziadecki 1977 . Scientific Papers of Silesian Polytechnic "Mining", 80/81, Glivice.
[8]. Stypuła, R. 1979 . Problem of discretization at simulation of mine winder electric drives in deep shafts /in Polish/. Diploma Thesis, Min. and Metall. Academy, Cracow.
[9]. Bukhstein, 1961 . Steel ropes /in Russian/. Metallurgizdat, Moscow.
[10]. Zając, M., T. Stankiewicz 1979 . Reverse problem of mechanics applied to the problem of control of driving system with elastic elements /in Polish/. The 5th symposium at the Min. and Metall. Academy in Cracow.
[11]. Szklarski, L., A. Skalny 1975 1977 Theoretical problems in mine winders /in Polish/, publication of the Polish Academy of Sciences, p.I, p.II.
[12]. Gierlotka, K. 1980 . The influence of system and parameters of speed regulator on the transients of thyristor fed mine winder /in Polish/, Scientific Papers of Silesian Polytechnic "Mining" 80.
[13]. Gierlotka, K. 1980 . Synthesis of control system of mine winders for deep shafts /in Polish/. Papers of Congress ICAMC, p.II, Katowice.
[14]. Liberus, Z., K. Kalinowski 1977 . Problem of optimization of dynamic torque of mine winder of deep shafts /in Polish/. Scient. Papers of Silesian Polytechnic "Mining" 81, Glivice.
[15]. Wolski, A., Z. Mrozek 1980 . Dynamics of mine winder with induction winding motor at starting with elasticity of ropes taken into account. Papers of Congress ICAMC, Katowice.
[16]. Szklarski, L., R. Wojnicki, A. Stankiewicz 1970 . Analysis of driving systems with elastic elements /in Polish/, Zesz. Probl. Górnictwa PAN, 2 /Publication of the Polish Academy of Sciences/.
[17]. Szklarski, L., M. Zając, A. Dziadecki 1980 . Problem of minimalization of rope oscillations at winder control /in Polish/, Archiwum Górnictwa PAN, 25, 3 /Publication of the Polish Academy of Scs/.
[18]. Szklarski, L., K. Jaracz 1982 . Selected problems of dynamics drives /in Polish/, Publication of the Polish Acad. of Scs.

Choice of Electric Driving Systems for Mining Equipment

A. HORODECKI

Department of Fundamental Researches in Electrotechnics of the Ministry of Metallurgy and Machine Industries and
Polish Academy of Sciences at Electrical Institute. Warszawa, Poland

Abstract. The author presents various procedures for the choice of driving systems for mining, including the method of indeterminate Lagrange factors and linear programming. The possible application of the points-awarding method is demonstrated. It allows for taking into account both surd and commensurable characteristics of a driving system for mining. Also, the method of economic efficiency factors is outlined. All these methods make possible the best choice of a driving system, from the aspect of electric energy saving. Two examples are given.

Keywords. Optimal control; electric drives; mining; economics.

INTRODUCTION

Electric energy consumption of electro-mechanical energy conversion systems, in particular, by electric driving systems, is evaluated by some authors at 60-70 %, of the total (Horodecki, 1983). This high percentage of the energy consumed by driving systems emphasizes their importance both technical and economic. This applies particularly to coal mines, in which power demand reaches up to 100 MW in large mines.

In mines, electric driving systems are used in the following groups of machines: coal getting; transportation by belt conveyors or vehicles ; hoisting; fans; and pumps.

Among these appliances, the electric drives of fans and pumps are important for energy management. The driving motors of pumps are sometimes rated up to 2000 kW or more. These devices are continuous duty ones and their driving motors have a substantially constant load. This differentiates fans and pumps from the other groups of operating machines. However, as far as fans are concerned, the changing ventilation resistances in a mine causes a shift of the working point on the load curve of these devices. These changes are due to the opening or closing of ventilation doors, as well as the gradual shift of coal working sites. The installed power of fan driving motors for the main ventilation as well as pumps in wet mines may reach 30-50 % and more of the total installed power of a mine.

These numbers indicate that the choice of an optimum driving system requires consideration of both technical and economic features.

This paper surveys several methods of choosing driving systems.

THE METHOD OF LAGRANGE INDETERMINATE FACTORS

This method is high accuracy and can take into account all limitations expressed by equations. The optimum solution is found by determining the coordinates of the point $D(x_{10}, x_{20}, \ldots, x_{n0})$ at which the extreme of the objective function $z = cX$ is obtained with limitations $AX = b$. In the latter relation : $c = (c_1, c_2, \ldots, c_n)$ is the line vector and $X = (x_1, x_2, \ldots, x_n)$ the column vector, A is the matrix of factors $[a_{ij}]$ and $b = (b_1, b_2, \ldots, b_n)$ is also the column vector. For problem formulated in this way, Lagrange's L will take the form:

$$L(x_1, x_2, \ldots, x_n) = z(x_1, x_2, \ldots, x_n) +$$
$$+ \sum_{i=1}^{n} \lambda_i \varphi_i (x_1, x_2, \ldots, x_n)$$

where : λ_i - is the Lagrange indeterminate factor and φ_i - represents the particular limiting equations AX = b. The coordinates of point D are found by solving the partial differential equations

$$\frac{\partial L}{\partial x_n} = 0$$

and the relations describing the numbers of particular variants of the driving systems constitute the decisive variables (Foulds, 1981; Pank, 1979).

An example of the application of this method may be a mine in which fans of output Q_1 and Q_2 are used. The number x_1 of fans of output Q_1 and the number x_2 of fans of output Q_2 is to be determined to assure that the daily cost C_e of the electric energy consumed will be the lowest. Because of the supply network, this is limited by the number d of fans allowed to work simultaneously. Thus, the daily cost of electric energy used, according to statistical studies carried out in the mine, may be determined by the function having the following form :

$$C_e = \sum_{i=1}^{2} \sum_{j=1}^{2} c_{ij}\, x_i\, x_j \quad (1)$$

or

$$C_e = c_{11}x_1^2 + (c_{12} + c_{21})\, x_1 x_2 + c_{22}x_2^2$$

Denoting $c_{11} = a$, $\frac{1}{2}(c_{12} + c_{21}) = b$ and $c_{22} = c$, the function z takes the form

$$C_e = \begin{bmatrix} x_1 & x_2 \end{bmatrix} \begin{bmatrix} a & b \\ b & c \end{bmatrix} \begin{bmatrix} x_1 \\ x_2 \end{bmatrix} = X^T A_x$$

where : X^T - matrix transposed of elements x_1 and x_2, A - symmetric matrix of square form (1). Admitting certain data from the actual case where : a = 2, b = 1, c = 3 and d = 12, Lagrange's function

$$L(x_1 x_2) = 2x_1^2 + 2x_1 x_2 + 3x_2^2 + \lambda(x_1 + x_2 - 12)$$

is obtained. After solving the partial differential equations

$$\frac{\partial L}{\partial x_1} = 0, \quad \frac{\partial L}{\partial x_2} = 0$$

and after determining the limiting condition as $\varphi(x) = x_1 + x_2 - 12 = 0$, the numbers of fans $x_1 = 8$ and $x_2 = 4$ are obtained. Lagrange's multiplication factor $\lambda = -40$. It is easy to prove that, for the determined values x_1 and x_2, the function z reaches its minimum.

SIMPLEX METHOD OF LINEAR PROGRAMMING

In most practical cases, the decisive variables of the objective function appear in the first power and the simplex procedure based on linear programming is useful. However, to obtain practical results it is necessary to apply the so-called Land-Doig interpretation. This gives optimum solutions in the form of integers representing the number of particular variants. Application of the simlex method contributes also to large energy savings (Panik, 1979).

POINTS-AWARDING METHOD DERIVED FROM THE STATISTICAL APPROACH TO THE PROBLEM OF ASSESSMENT

This method is very useful for solving innovation problems, in particular, concerned with the choice of driving systems. It is simple, and its application requires neither the use of costly technological means nor large labour expenses. The points-awarding method covers features that can be determined quantitatively and economical effects that cannot be determined quantitatively. These features may be the name-plate data and catalogue data as well as statistical operational data. The idea of the points-awarding method is to quantify the assessment of either quantitative or qualitative features by awarding points from a preestablished total number range. To determine the resulting assessment by points, the formulae proven best are :

- the quotient expression

$$\Theta = \frac{\sum_{1}^{n} k_{ij}\, G_j}{n} \quad (2)$$

- the product expression

$$\Theta = \prod_{1}^{n} \frac{k_{ij}}{\left(k_{wzj}\right)^{G_j}} \quad (3)$$

where : Θ_1 - denotes the results of assessment according to the points-wawarding method for i = 1, 2,...,m , k_{ij} - is the

j-feature of the innovation variant j=1,2,
...,n where n is the number of features of
novation variant, G_j - is the by-weight
factor of the j-features and k_{wzj} - is the
j-features of the master variant. The by-
weight factor G_j takes into account the effect of particular features on the value
of the driving systems. In both expres -
sions (2) and (3), the optimum variant is
chosen when $\Theta = \Theta_{i\ max}$. The expression(2)
gives an absolute assessment of each var-
iant analysed, while the expression (3)
furnishes on assessment of the particular
variant in reference to a master variant.
Commeasurable features may be, for instance : the power of driving motor, the electric consumption per tonne of winned coal,
the power factor, outlays for buying the
driving system, and its operation costs.
Surd features may be: the reliability of
the system . easy repair and transport .
novelty of design, etc.

METHOD OF ECONOMIC EFFICIENCY
FACTORS

This is a simple method for choosing a
driving system (Horodecki, 1983) .From a-
mong different forms of these factors E ,
it is convenient to use the quotient :

$$E = \frac{F_e (r + s) + C_e}{Q} \qquad (4)$$

where : F_e - are the outlays for the purchase of equipment, r - is the discount
rate, s - is the amortisation rate, C_e -
expenses for the consumed electric energy.
The optimum variant is chosen, taking into account a suitable criterion used for
minimizing the factor E determined for
equipment. It should be kept in mind that,
for each variant, the operational effect
Q might be either the same or different.
Studies and calculation were carried out
with the use of this procedure to determine the factor E for various driving systems working with machines having different mechanical characteristics. The values
of the exponent k in the equation expres-
sing the load torque m_w of the working machines varies from k = 0 to k = 4, according to the relation :

$$m_w = m_o + (1 - m_o) \left(\frac{1 - s}{1 - s_n} \right)^k$$

where : m_o - load torque at slip s = 1 of
the driving motor, s_n - rated slip of driving systems covered were a system provided with induction slipping clutch and a
thyristor cascade system with constant
torque. The difference between operatio -
nal costs of the above two systems, from
the aspect of the time of operation$\mathcal{T}$, is
displayed by Fig. 1, while Fig. 2 displays
the difference of the costs C_e as depending on the exponent k. Both fans and
pumps are characterised by the exponent
k = 2. Figure 3 displays the relation of
the difference in outlays F_e for both the
above driving systems versus the power P_w
of driving motors.
Transforming the relation(4)for both the
driving systems, we obtain for the same
outputs Q of both types of fans and
pumps :

$$\frac{\Delta C_e}{\Delta F_e (s + r)} \geqslant 1 \qquad (5)$$

where : $\Delta C_e = C_{e1} - C_{e2}$. and $\Delta F_e = F_{e2} - F_{e1}$
If the calculations according to(4)lead
to a result exceeding unity, the choice
of the system should be for higher out -
lays F_e .

CRITICAL COMPARISON OF
METHODS

The methods presented above are applied
depending on the numerical data being at
disposal. Thus the method of Lagrange indeterminate factors is a high precision
method. It has the advantage over other
methods that it allows to take into ac -
count the existing limitations concerning
the coefficients appearing in the objecti
ve function. This necessitates a formu -
lation of a set of equations describing
these limitations.
From among the methods of linear programming used for the choice of an electro -
mechanical system the simplex method is
applied. It allows, in particular, for
the determination of the division of the
possessed financial means for the realisation of respective variants of these systems or for the determination of the number of particular variants at a minimum
involvement of these means.
The points-awarding method being a sta -

tistical method, requires the assessment
of the weight of the particular features
of the electromachine system with the use
of a suitable scale of point awarding, giving rise to possible errors. In comparison with the former methods this method
is simple but does not take into account
the restrictions appearing in the assessment and choice of driving systems.
The method widest spread in engineering
practice is the method of economics efficiency factors. It is the simplest one
from the methods considered in the paper.
It renders the choice of the driving system stritly dependent on its operation
costs, its price and its particular technical features. It accuracy is sufficient
for engineering purposes.

CONCLUSION

To assess driving systems for mines, as
well as to choose, according to preset
assumptions, the appropriate variant, the

following steps should be taken : - list
features that are to be taken into ac -
count in the calculations, - determine
the necessary relations between the features, - formulate criteria and methods ,
which should them give data for the assessment and the choice of driving systems.

REFERENCES

Foulds, L.R. (1981). Optimization tech -
    niques. An introduction. Springer
    Verlag, Berlin - New York .
Horodecki, A. (1983). Mode possible de
    l'évaluation technico-économique et
    du choix des systèms de conversion
    électromécanique de l'énergie. Archi-
    wum Górnictwa, Academy Polonaise des
    Sciences, 1, 45 - 61 .
Panik, M.J. (1979). Classical optimiza-
    tion. North - Holland Publ.

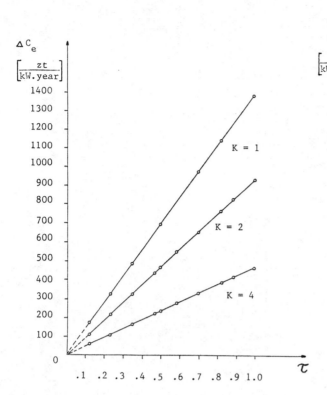

Figure 1. The difference ΔC_e of the costs of the above two driving systems, from the aspect of the time of operations τ.

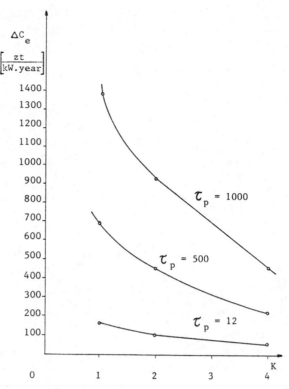

Figure 2. The difference ΔC_e of the costs of the two driving systems, from the aspect of the exponent k.

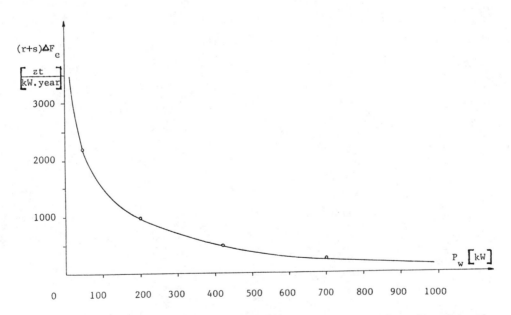

Figure 3. The relation of the difference in outlays F_e from the aspect of the power P_w of the driving motors.

INDEX OF AUTHORS, TITLES AND KEYWORDS

345